Introduction to finite fields and their applications

RUDOLF LIDL

University of Tasmania, Hobart, Australia

HARALD NIEDERREITER

Austrian Academy of Sciences, Vienna, Austria

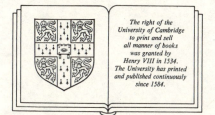

The right of the
University of Cambridge
to print and sell
all manner of books
was granted by
Henry VIII in 1534.
The University has printed
and published continuously
since 1584.

CAMBRIDGE UNIVERSITY PRESS

Cambridge

London   New York   New Rochelle

Melbourne   Sydney

Published by the Press Syndicate of the University of Cambridge
The Pitt Building, Trumpington Street, Cambridge CB2 1RP
32 East 57th Street, New York, NY 10022, USA
10 Stamford Road, Oakleigh, Melbourne 3166, Australia

First published 1986

Printed in Great Britain at the University Press, Cambridge

British Library Cataloguing in Publication Data

Lidl, Rudolf
Introduction to finite fields and their
applications.
1. Finite Fields (Algebra)
I. Title   II. Niederreiter, Harald
512′.3   QA247.3

Library of Congress Cataloging in Publication Data

Lidl, Rudolf.
Introduction to finite fields and their applications.

Bibliography: p.
Includes index.
1. Finite fields (Algebra)   I. Niederreiter, Harald,
1944–   II. Title.
QA247.3.L54   1985   512′.3   85-9704
ISBN 0-521-30706-6

T.M.

Contents

Preface   vii

Chapter 1 **Algebraic Foundations**   1
   1 Groups   2
   2 Rings and Fields   11
   3 Polynomials   18
   4 Field Extensions   30
     Exercises   37

Chapter 2 **Structure of Finite Fields**   43
   1 Characterization of Finite Fields   44
   2 Roots of Irreducible Polynomials   47
   3 Traces, Norms, and Bases   50
   4 Roots of Unity and Cyclotomic Polynomials   59
   5 Representation of Elements of Finite Fields   62
   6 Wedderburn's Theorem   65
     Exercises   69

Chapter 3 **Polynomials over Finite Fields**   74
   1 Order of Polynomials and Primitive Polynomials   75
   2 Irreducible Polynomials   82

3 Construction of Irreducible Polynomials   87
4 Linearized Polynomials   98
5 Binomials and Trinomials   115
   Exercises   122

Chapter 4 Factorization of Polynomials   **129**
1 Factorization over Small Finite Fields   130
2 Factorization over Large Finite Fields   139
3 Calculation of Roots of Polynomials   150
   Exercises   159

Chapter 5 Exponential Sums   **162**
1 Characters   163
2 Gaussian Sums   168
   Exercises   181

Chapter 6 Linear Recurring Sequences   **185**
1 Feedback Shift Registers, Periodicity Properties   186
2 Impulse Response Sequences, Characteristic Polynomial   193
3 Generating Functions   202
4 The Minimal Polynomial   210
5 Families of Linear Recurring Sequences   215
6 Characterization of Linear Recurring Sequences   228
7 Distribution Properties of Linear Recurring Sequences   235
   Exercises   245

Chapter 7 Theoretical Applications of Finite Fields   **251**
1 Finite Geometries   252
2 Combinatorics   262
3 Linear Modular Systems   271
4 Pseudorandom Sequences   281
   Exercises   294

Chapter 8 Algebraic Coding Theory   **299**
1 Linear Codes   300
2 Cyclic Codes   311
3 Goppa Codes   325
   Exercises   332

Chapter 9 Cryptology   **338**
1 Background   339

2 Stream Ciphers   342
3 Discrete Logarithms   346
4 Further Cryptosystems   360
   Exercises   363

Chapter 10 Tables   **367**
1 Computation in Finite Fields   367
2 Tables of Irreducible Polynomials   377

Bibliography   **392**

List of Symbols   **397**

Index   **401**

To Pamela *and* Gerlinde

Preface

This book is designed as a textbook edition of our monograph *Finite Fields* which appeared in 1983 as Volume 20 of the *Encyclopedia of Mathematics and Its Applications.* Several changes have been made in order to tailor the book to the needs of the student. The historical and bibliographical notes at the end of each chapter and the long bibliography have been omitted as they are mainly of interest to researchers. The reader who desires this type of information may consult the original edition. There are also changes in the text proper, with the present book having an even stronger emphasis on applications. The increasingly important role of finite fields in cryptology is reflected by a new chapter on this topic. There is now a separate chapter on algebraic coding theory containing material from the original edition together with a new section on Goppa codes. New material on pseudorandom sequences has also been added. On the other hand, topics in the original edition that are mainly of theoretical interest have been omitted. Thus, a large part of the material on exponential sums and the chapters on equations over finite fields and on permutation polynomials cannot be found in the present volume.

  The theory of finite fields is a branch of modern algebra that has come to the fore in the last 50 years because of its diverse applications in combinatorics, coding theory, cryptology, and the mathematical study of switching circuits, among others. The origins of the subject reach back into the 17th and 18th centuries, with such eminent mathematicians as Pierre de Fermat (1601–1665), Leonhard Euler (1707–1783), Joseph-Louis Lagrange (1736–1813), and Adrien-Marie Legendre (1752–1833) contributing to the structure theory of special finite fields—namely, the so-called finite prime fields. The general theory of finite fields may be said to begin with the work of

Carl Friedrich Gauss (1777–1855) and Evariste Galois (1811–1832), but it only became of interest for applied mathematicians in recent decades with the emergence of discrete mathematics as a serious discipline.

In this book we have aimed at presenting both the classical and the applications-oriented aspects of the subject. Thus, in addition to what has to be considered the essential core of the theory, the reader will find results and techniques that are of importance mainly because of their use in applications. Because of the vastness of the subject, limitations had to be imposed on the choice of material. In trying to make the book as self-contained as possible, we have refrained from discussing results or methods that belong properly to algebraic geometry or to the theory of algebraic function fields. Applications are described to the extent to which this can be done without too much digression. The only noteworthy prerequisite for the book is a background in linear algebra, on the level of a first course on this topic. A rudimentary knowledge of analysis is needed in a few passages. Prior exposure to abstract algebra is certainly helpful, although all the necessary information is summarized in Chapter 1.

Chapter 2 is basic for the rest of the book as it contains the general structure theory of finite fields as well as the discussion of concepts that are used throughout the book. Chapter 3 on the theory of polynomials and Chapter 4 on factorization algorithms for polynomials are closely linked and should best be studied together. Chapter 5 on exponential sums uses only the elementary structure theory of finite fields. Chapter 6 on linear recurring sequences depends mostly on Chapters 2 and 3. Chapters 7, 8, and 9 are devoted to applications and draw on various material in the previous chapters. Chapter 10 supplements parts of Chapters 2, 3, and 9. Each chapter starts with a brief description of its contents, hence it should not be necessary to give a synopsis of the book here.

In order to enhance the attractiveness of this book as a textbook, we have inserted worked-out examples at appropriate points in the text and included lists of exercises for Chapters 1–9. These exercises range from routine problems to alternative proofs of key theorems, but contain also material going beyond what is covered in the text.

With regard to cross-references, we have numbered all items in the main text consecutively by chapters, regardless of whether they are definitions, theorems, examples, and so on. Thus, "Definition 2.41" refers to item 41 in Chapter 2 (which happens to be a definition) and "Remark 6.23" refers to item 23 in Chapter 6 (which happens to be a remark). In the same vein, "Exercise 5.21" refers to the list of exercises in Chapter 5.

We gratefully acknowledge the help of Mrs. Melanie Barton and Mrs. Betty Golding who typed the manuscript with great care and efficiency.

R. LIDL

H. NIEDERREITER

Chapter 1

Algebraic Foundations

This introductory chapter contains a survey of some basic algebraic concepts that will be employed throughout the book. Elementary algebra uses the operations of arithmetic such as addition and multiplication, but replaces particular numbers by symbols and thereby obtains formulas that, by substitution, provide solutions to specific numerical problems. In modern algebra the level of abstraction is raised further: instead of dealing with the familiar operations on real numbers, one treats general operations —processes of combining two or more elements to yield another element—in general sets. The aim is to study the common properties of all systems consisting of sets on which are defined a fixed number of operations interrelated in some definite way—for instance, sets with two binary operations behaving like $+$ and $\cdot$ for the real numbers.

Only the most fundamental definitions and properties of algebraic systems—that is, of sets together with one or more operations on the set—will be introduced, and the theory will be discussed only to the extent needed for our special purposes in the study of finite fields later on. We state some standard results without proof. With regard to sets we adopt the naive standpoint. We use the following sets of numbers: the set $\mathbb{N}$ of natural numbers, the set $\mathbb{Z}$ of integers, the set $\mathbb{Q}$ of rational numbers, the set $\mathbb{R}$ of real numbers, and the set $\mathbb{C}$ of complex numbers.

1. GROUPS

In the set of all integers the two operations addition and multiplication are well known. We can generalize the concept of operation to arbitrary sets. Let S be a set and let $S \times S$ denote the set of all ordered pairs (s, t) with $s \in S$, $t \in S$. Then a mapping from $S \times S$ into S will be called a (*binary*) *operation* on S. Under this definition we require that the image of $(s, t) \in S \times S$ must be in S; this is the *closure property* of an operation. By an *algebraic structure* or *algebraic system* we mean a set S together with one or more operations on S.

In elementary arithmetic we are provided with two operations, addition and multiplication, that have associativity as one of their most important properties. Of the various possible algebraic systems having a single associative operation, the type known as a group has been by far the most extensively studied and developed. The theory of groups is one of the oldest parts of abstract algebra as well as one particularly rich in applications.

1.1.   Definition.   A *group* is a set G together with a binary operation $*$ on G such that the following three properties hold:

1. $*$ is *associative*; that is, for any $a, b, c \in G$,

$$a * (b * c) = (a * b) * c.$$

2.   There is an *identity* (or *unity*) *element* e in G such that for all $a \in G$,

$$a * e = e * a = a.$$

3.   For each $a \in G$, there exists an *inverse element* $a^{-1} \in G$ such that

$$a * a^{-1} = a^{-1} * a = e.$$

If the group also satisfies
4.   For all $a, b \in G$,

$$a * b = b * a,$$

then the group is called *abelian* (or *commutative*).

It is easily shown that the identity element e and the inverse element a^{-1} of a given element $a \in G$ are uniquely determined by the properties above. Furthermore, $(a * b)^{-1} = b^{-1} * a^{-1}$ for all $a, b \in G$. For simplicity, we shall frequently use the notation of ordinary multiplication to designate the operation in the group, writing simply ab instead of $a * b$. But it must be emphasized that by doing so we do not assume that the operation actually is ordinary multiplication. Sometimes it is also convenient to write $a + b$ instead of $a * b$ and $-a$ instead of a^{-1}, but this additive notation is usually reserved for abelian groups.

The associative law guarantees that expressions such as $a_1 a_2 \cdots a_n$ with $a_j \in G$, $1 \leqslant j \leqslant n$, are unambiguous, since no matter how we insert parentheses, the expression will always represent the same element of G. To indicate the n-fold composite of an element $a \in G$ with itself, where $n \in \mathbb{N}$, we shall write

$$a^n = aa \cdots a \qquad (n \text{ factors } a)$$

if using multiplicative notation, and we call a^n the nth power of a. If using additive notation for the operation $*$ on G, we write

$$na = a + a + \cdots + a \qquad (n \text{ summands } a).$$

Following customary notation, we have the following rules:

Multiplicative Notation	Additive Notation
$a^{-n} = (a^{-1})^n$	$(-n)a = n(-a)$
$a^n a^m = a^{n+m}$	$na + ma = (n+m)a$
$(a^n)^m = a^{nm}$	$m(na) = (mn)a$

For $n = 0 \in \mathbb{Z}$, one adopts the convention $a^0 = e$ in the multiplicative notation and $0a = 0$ in the additive notation, where the last "zero" represents the identity element of G.

1.2. Examples

  (i)   Let G be the set of integers with the operation of addition. The ordinary sum of two integers is a unique integer and the associativity is a familiar fact. The identity element is 0 (zero), and the inverse of an integer a is the integer $-a$. We denote this group by $\mathbb{Z}$.

  (ii) The set consisting of a single element e, with the operation $*$ defined by $e * e = e$, forms a group.

  (iii) Let G be the set of remainders of all the integers on division by 6—that is, $G = \{0, 1, 2, 3, 4, 5\}$—and let $a * b$ be the remainder on division by 6 of the ordinary sum of a and b. The existence of an identity element and of inverses is again obvious. In this case, it requires some computation to establish the associativity of $*$. This group can be readily generalized by replacing the integer 6 by any positive integer n. $\square$

These examples lead to an interesting class of groups in which every element is a power of some fixed element of the group. If the group operation is written as addition, we refer to "multiple" instead of "power" of an element.

1.3. Definition. A multiplicative group G is said to be *cyclic* if there is an element $a \in G$ such that for any $b \in G$ there is some integer j with $b = a^j$.

Such an element a is called a *generator* of the cyclic group, and we write $G = \langle a \rangle$.

It follows at once from the definition that every cyclic group is commutative. We also note that a cyclic group may very well have more than one element that is a generator of the group. For instance, in the additive group $\mathbb{Z}$ both 1 and -1 are generators.

With regard to the "additive" group of remainders of the integers on division by n, the generalization of Example 1.2(iii), we find that the type of operation used there leads to an equivalence relation on the set of integers. In general, a subset R of $S \times S$ is called an *equivalence relation* on a set S if it has the following three properties:

(a) $(s, s) \in R$ for all $s \in S$ (*reflexivity*).
(b) If $(s, t) \in R$, then $(t, s) \in R$ (*symmetry*).
(c) If $(s, t), (t, u) \in R$, then $(s, u) \in R$ (*transitivity*).

The most obvious example of an equivalence relation is that of equality. It is an important fact that an equivalence relation R on a set S induces a partition of S—that is, a representation of S as the union of nonempty, mutually disjoint subsets of S. If we collect all elements of S equivalent to a fixed $s \in S$, we obtain the *equivalence class* of s, denoted by

$$[s] = \{t \in S : (s, t) \in R\}.$$

The collection of all distinct equivalence classes forms then the desired partition of S. We note that $[s] = [t]$ precisely if $(s, t) \in R$. Example 1.2(iii) suggests the following concept.

1.4. Definition. For arbitrary integers a, b and a positive integer n, we say that a is *congruent* to b modulo n, and write $a \equiv b \bmod n$, if the difference $a - b$ is a multiple of n—that is, if $a = b + kn$ for some integer k.

It is easily verified that "congruence modulo n" is an equivalence relation on the set $\mathbb{Z}$ of integers. The relation is obviously reflexive and symmetric. The transitivity also follows easily: if $a = b + kn$ and $b = c + ln$ for some integers k and l, then $a = c + (k + l)n$, so that $a \equiv b \bmod n$ and $b \equiv c \bmod n$ together imply $a \equiv c \bmod n$.

Consider now the equivalence classes into which the relation of congruence modulo n partitions the set $\mathbb{Z}$. These will be the sets

$$[0] = \{\ldots, -2n, -n, 0, n, 2n, \ldots\},$$

$$[1] = \{\ldots, -2n + 1, -n + 1, 1, n + 1, 2n + 1, \ldots\},$$

$$\vdots$$

$$[n - 1] = \{\ldots, -n - 1, -1, n - 1, 2n - 1, 3n - 1, \ldots\}.$$

We may define on the set $\{[0], [1], \ldots, [n - 1]\}$ of equivalence classes a binary operation (which we shall again write as +, although it is certainly not ordinary addition) by

$$[a]+[b]=[a+b], \tag{1.1}$$

where a and b are any elements of the respective sets $[a]$ and $[b]$ and the sum $a + b$ on the right is the ordinary sum of a and b. In order to show that we have actually defined an operation—that is, that this operation is well defined—we must verify that the image element of the pair $([a],[b])$ is uniquely determined by $[a]$ and $[b]$ alone and does not depend in any way on the representatives a and b. We leave this proof as an exercise. Associativity of the operation in (1.1) follows from the associativity of ordinary addition. The identity element is $[0]$ and the inverse of $[a]$ is $[-a]$. Thus the elements of the set $\{[0],[1],\ldots,[n-1]\}$ form a group.

1.5. Definition. The group formed by the set $\{[0],[1],\ldots,[n-1]\}$ of equivalence classes modulo n with the operation (1.1) is called the *group of integers modulo n* and denoted by $\mathbb{Z}_n$.

$\mathbb{Z}_n$ is actually a cyclic group with the equivalence class $[1]$ as a generator, and it is a group of order n according to the following definition.

1.6. Definition. A group is called *finite* (resp. *infinite*) if it contains finitely (resp. infinitely) many elements. The number of elements in a finite group is called its *order*. We shall write $|G|$ for the order of the finite group G.

There is a convenient way of presenting a finite group. A table displaying the group operation, nowadays referred to as a *Cayley table*, is constructed by indexing the rows and the columns of the table by the group elements. The element appearing in the row indexed by a and the column indexed by b is then taken to be ab.

1.7. Example. The Cayley table for the group $\mathbb{Z}_6$ is:

+	[0]	[1]	[2]	[3]	[4]	[5]
[0]	[0]	[1]	[2]	[3]	[4]	[5]
[1]	[1]	[2]	[3]	[4]	[5]	[0]
[2]	[2]	[3]	[4]	[5]	[0]	[1]
[3]	[3]	[4]	[5]	[0]	[1]	[2]
[4]	[4]	[5]	[0]	[1]	[2]	[3]
[5]	[5]	[0]	[1]	[2]	[3]	[4]

□

A group G contains certain subsets that form groups in their own right under the operation of G. For instance, the subset $\{[0],[2],[4]\}$ of $\mathbb{Z}_6$ is easily seen to have this property.

1.8.   Definition.   A subset H of the group G is a *subgroup* of G if H is itself a group with respect to the operation of G. Subgroups of G other than the *trivial subgroups* $\{e\}$ and G itself are called *nontrivial subgroups* of G.

One verifies at once that for any fixed a in a group G, the set of all powers of a is a subgroup of G.

1.9.   Definition.   The subgroup of G consisting of all powers of the element a of G is called the subgroup *generated by* a and is denoted by $\langle a \rangle$. This subgroup is necessarily cyclic. If $\langle a \rangle$ is finite, then its order is called the *order* of the element a. Otherwise, a is called an element of *infinite order*.

Thus, a is of finite order k if k is the least positive integer such that $a^k = e$. Any other integer m with $a^m = e$ is then a multiple of k. If S is a nonempty subset of a group G, then the subgroup H of G consisting of all finite products of powers of elements of S is called the subgroup *generated by* S, denoted by $H = \langle S \rangle$. If $\langle S \rangle = G$, we say that S *generates* G, or that G is *generated by* S.

For a positive element n of the additive group $\mathbb{Z}$ of integers, the subgroup $\langle n \rangle$ is closely associated with the notion of congruence modulo n, since $a \equiv b \bmod n$ if and only if $a - b \in \langle n \rangle$. Thus the subgroup $\langle n \rangle$ defines an equivalence relation on $\mathbb{Z}$. This situation can be generalized as follows.

1.10.   Theorem.   *If H is a subgroup of G, then the relation R_H on G defined by $(a, b) \in R_H$ if and only if $a = bh$ for some $h \in H$, is an equivalence relation.*

The proof is immediate. The equivalence relation R_H is called *left congruence* modulo H. Like any equivalence relation, it induces a partition of G into nonempty, mutually disjoint subsets. These subsets (= equivalence classes) are called the *left cosets* of G modulo H and they are denoted by

$$aH = \{ah : h \in H\}$$

(or $a + H = \{a + h : h \in H\}$ if G is written additively), where a is a fixed element of G. Similarly, there is a decomposition of G into *right cosets* modulo H, which have the form $Ha = \{ha : h \in H\}$. If G is abelian, then the distinction between left and right cosets modulo H is unnecessary.

1.11.   Example.   Let $G = \mathbb{Z}_{12}$ and let H be the subgroup $\{[0], [3], [6], [9]\}$. Then the distinct (left) cosets of G modulo H are given by:

$$[0] + H = \{[0], [3], [6], [9]\},$$
$$[1] + H = \{[1], [4], [7], [10]\},$$
$$[2] + H = \{[2], [5], [8], [11]\}. \qquad \square$$

1.12.   Theorem.   *If H is a finite subgroup of G, then every (left or right) coset of G modulo H has the same number of elements as H.*

1.13. **Definition.** If the subgroup H of G only yields finitely many distinct left cosets of G modulo H, then the number of such cosets is called the *index* of H in G.

Since the left cosets of G modulo H form a partition of G, Theorem 1.12 implies the following important result.

1.14. **Theorem.** *The order of a finite group G is equal to the product of the order of any subgroup H and the index of H in G. In particular, the order of H divides the order of G and the order of any element $a \in G$ divides the order of G.*

The subgroups and the orders of elements are easy to describe for cyclic groups. We summarize the relevant facts in the subsequent theorem.

1.15. **Theorem**

(i) *Every subgroup of a cyclic group is cyclic.*

(ii) *In a finite cyclic group $\langle a \rangle$ of order m, the element a^k generates a subgroup of order $m/\gcd(k,m)$, where $\gcd(k,m)$ denotes the greatest common divisor of k and m.*

(iii) *If d is a positive divisor of the order m of a finite cyclic group $\langle a \rangle$, then $\langle a \rangle$ contains one and only one subgroup of index d. For any positive divisor f of m, $\langle a \rangle$ contains precisely one subgroup of order f.*

(iv) *Let f be a positive divisor of the order of a finite cyclic group $\langle a \rangle$. Then $\langle a \rangle$ contains $\phi(f)$ elements of order f. Here $\phi(f)$ is Euler's function and indicates the number of integers n with $1 \leq n \leq f$ that are relatively prime to f.*

(v) *A finite cyclic group $\langle a \rangle$ of order m contains $\phi(m)$ generators — that is, elements a^r such that $\langle a^r \rangle = \langle a \rangle$. The generators are the powers a^r with $\gcd(r,m)=1$.*

Proof. (i) Let H be a subgroup of the cyclic group $\langle a \rangle$ with $H \neq \{e\}$. If $a^n \in H$, then $a^{-n} \in H$; hence H contains at least one power of a with a positive exponent. Let d be the least positive exponent such that $a^d \in H$, and let $a^s \in H$. Dividing s by d gives $s = qd + r$, $0 \leq r < d$, and $q, r \in \mathbb{Z}$. Thus $a^s(a^{-d})^q = a^r \in H$, which contradicts the minimality of d, unless $r = 0$. Therefore the exponents of all powers of a that belong to H are divisible by d, and so $H = \langle a^d \rangle$.

(ii) Put $d = \gcd(k,m)$. The order of $\langle a^k \rangle$ is the least positive integer n such that $a^{kn} = e$. The latter identity holds if and only if m divides kn, or equivalently, if and only if m/d divides n. The least positive n with this property is $n = m/d$.

(iii) If d is given, then $\langle a^d \rangle$ is a subgroup of order m/d, and so of index d, because of (ii). If $\langle a^k \rangle$ is another subgroup of index d, then its order is m/d, and so $d = \gcd(k, m)$ by (ii). In particular, d divides k, so that $a^k \in \langle a^d \rangle$ and $\langle a^k \rangle$ is a subgroup of $\langle a^d \rangle$. But since both groups have the same order, they are identical. The second part follows immediately because the subgroups of order f are precisely the subgroups of index m/f.

(iv) Let $|\langle a \rangle| = m$ and $m = df$. By (ii), an element a^k is of order f if and only if $\gcd(k, m) = d$. Hence, the number of elements of order f is equal to the number of integers k with $1 \leqslant k \leqslant m$ and $\gcd(k, m) = d$. We may write $k = dh$ with $1 \leqslant h \leqslant f$, the condition $\gcd(k, m) = d$ being now equivalent to $\gcd(h, f) = 1$. The number of these h is equal to $\phi(f)$.

(v) The generators of $\langle a \rangle$ are precisely the elements of order m, so that the first part is implied by (iv). The second part follows from (ii). $\square$

When comparing the structures of two groups, mappings between the groups that preserve the operations play an important role.

1.16. Definition. A mapping $f: G \rightarrow H$ of the group G into the group H is called a *homomorphism* of G into H if f preserves the operation of G. That is, if $*$ and $\cdot$ are the operations of G and H, respectively, then f preserves the operation of G if for all $a, b \in G$ we have $f(a*b) = f(a) \cdot f(b)$. If, in addition, f is onto H, then f is called an *epimorphism* (or *homomorphism* "*onto*") and H is a *homomorphic image* of G. A homomorphism of G into G is called an *endomorphism*. If f is a one-to-one homomorphism of G onto H, then f is called an *isomorphism* and we say that G and H are *isomorphic*. An isomorphism of G onto G is called an *automorphism*.

Consider, for instance, the mapping f of the additive group $\mathbb{Z}$ of the integers onto the group $\mathbb{Z}_n$ of the integers modulo n, defined by $f(a) = [a]$. Then

$$f(a + b) = [a + b] = [a] + [b] = f(a) + f(b) \quad \text{for } a, b \in \mathbb{Z},$$

and f is a homomorphism.

If $f: G \rightarrow H$ is a homomorphism and e is the identity element in G, then $ee = e$ implies $f(e)f(e) = f(e)$, so that $f(e) = e'$, the identity element in H. From $aa^{-1} = e$ we get $f(a^{-1}) = (f(a))^{-1}$ for all $a \in G$.

The automorphisms of a group G are often of particular interest, partly because they themselves form a group with respect to the usual composition of mappings, as can be easily verified. Important examples of automorphisms are the *inner automorphisms*. For fixed $a \in G$, define f_a by $f_a(b) = aba^{-1}$ for $b \in G$. Then f_a is an automorphism of G of the indicated type, and we get all inner automorphisms of G by letting a run through all elements of G. The elements b and aba^{-1} are said to be *conjugate*, and for a nonempty subset S of G the set $aSa^{-1} = \{asa^{-1} : s \in S\}$ is called a *conjugate* of S. Thus, the conjugates of S are just the images of S under the various inner automorphisms of G.

1.17.   Definition.   The *kernel* of the homomorphism $f: G \to H$ of the group G into the group H is the set

$$\ker f = \{a \in G : f(a) = e'\},$$

where e' is the identity element in H.

1.18.   Example.   For the homomorphism $f: \mathbb{Z} \to \mathbb{Z}_n$ given by $f(a) = [a]$, $\ker f$ consists of all $a \in \mathbb{Z}$ with $[a] = [0]$. Since this condition holds exactly for all multiples a of n, we have $\ker f = \langle n \rangle$, the subgroup of $\mathbb{Z}$ generated by n.   □

It is easily checked that $\ker f$ is always a subgroup of G. Moreover, $\ker f$ has a special property: whenever $a \in G$ and $b \in \ker f$, then $aba^{-1} \in \ker f$. This leads to the following concept.

1.19.   Definition.   The subgroup H of the group G is called a *normal subgroup* of G if $aha^{-1} \in H$ for all $a \in G$ and all $h \in H$.

Every subgroup of an abelian group is normal since we then have $aha^{-1} = aa^{-1}h = eh = h$. We shall state some alternative characterizations of the property of normality of a subgroup.

1.20.   Theorem

  (i)   *The subgroup H of G is normal if and only if H is equal to its conjugates, or equivalently, if and only if H is invariant under all the inner automorphisms of G.*
  (ii)   *The subgroup H of G is normal if and only if the left coset aH is equal to the right coset Ha for every $a \in G$.*

One important feature of a normal subgroup is the fact that the set of its (left) cosets can be endowed with a group structure.

1.21.   Theorem.   *If H is a normal subgroup of G, then the set of (left) cosets of G modulo H forms a group with respect to the operation $(aH)(bH) = (ab)H$.*

1.22.   Definition.   For a normal subgroup H of G, the group formed by the (left) cosets of G modulo H under the operation in Theorem 1.21 is called the *factor group* (or *quotient group*) of G modulo H and denoted by G/H.

If G/H is finite, then its order is equal to the index of H in G. Thus, by Theorem 1.14, we get for a finite group G,

$$|G/H| = \frac{|G|}{|H|}.$$

Each normal subgroup of a group G determines in a natural way a homomorphism of G and vice versa.

1.23.   Theorem (Homomorphism Theorem).   *Let $f: G \to f(G) = G_1$ be a homomorphism of a group G onto a group G_1. Then $\ker f$ is a normal subgroup of G, and the group G_1 is isomorphic to the factor group $G/\ker f$. Conversely, if H is any normal subgroup of G, then the mapping $\psi: G \to G/H$ defined by $\psi(a) = aH$ for $a \in G$ is a homomorphism of G onto G/H with $\ker \psi = H$.*

We shall now derive a relation known as the *class equation* for a finite group, which will be needed in Chapter 2, Section 6.

1.24.   Definition.   Let S be a nonempty subset of a group G. The *normalizer* of S in G is the set $N(S) = \{a \in G: aSa^{-1} = S\}$.

1.25.   Theorem.   *For any nonempty subset S of the group G, $N(S)$ is a subgroup of G and there is a one-to-one correspondence between the left cosets of G modulo $N(S)$ and the distinct conjugates aSa^{-1} of S.*

Proof.   We have $e \in N(S)$, and if $a, b \in N(S)$, then a^{-1} and ab are also in $N(S)$, so that $N(S)$ is a subgroup of G. Now

$$aSa^{-1} = bSb^{-1} \Leftrightarrow S = a^{-1}bSb^{-1}a = (a^{-1}b)S(a^{-1}b)^{-1}$$

$$\Leftrightarrow a^{-1}b \in N(S) \Leftrightarrow b \in aN(S).$$

Thus, conjugates of S are equal if and only if they are defined by elements in the same left coset of G modulo $N(S)$, and so the second part of the theorem is shown. $\square$

If we collect all elements conjugate to a fixed element a, we obtain a set called the *conjugacy class* of a. For certain elements the corresponding conjugacy class has only one member, and this will happen precisely for the elements of the center of the group.

1.26.   Definition.   For any group G, the *center* of G is defined as the set $C = \{c \in G: ac = ca$ for all $a \in G\}$.

It is straightforward to check that the center C is a normal subgroup of G. Clearly, G is abelian if and only if $C = G$. A counting argument leads to the following result.

1.27.   Theorem (Class Equation).   *Let G be a finite group with center C. Then*

$$|G| = |C| + \sum_{i=1}^{k} n_i,$$

where each n_i is ≥ 2 and a divisor of $|G|$. In fact, $n_1, n_2, \ldots, n_k$ are the numbers of elements of the distinct conjugacy classes in G containing more than one member.

Proof. Since the relation "a is conjugate to b" is an equivalence relation on G, the distinct conjugacy classes in G form a partition of G. Thus, $|G|$ is equal to the sum of the numbers of elements of the distinct conjugacy classes. There are $|C|$ conjugacy classes (corresponding to the elements of C) containing only one member, whereas $n_1, n_2, \ldots, n_k$ are the numbers of elements of the remaining conjugacy classes. This yields the class equation. To show that each n_i divides $|G|$, it suffices to note that n_i is the number of conjugates of some $a \in G$ and so equal to the number of left cosets of G modulo $N(\langle a \rangle)$ by Theorem 1.25. $\square$

2. RINGS AND FIELDS

In most of the number systems used in elementary arithmetic there are two distinct binary operations: addition and multiplication. Examples are provided by the integers, the rational numbers, and the real numbers. We now define a type of algebraic structure known as a ring that shares some of the basic properties of these number systems.

1.28. Definition. A *ring* $(R, +, \cdot)$ is a set R, together with two binary operations, denoted by $+$ and $\cdot$, such that:

1. R is an abelian group with respect to $+$.
2. $\cdot$ is associative—that is, $(a \cdot b) \cdot c = a \cdot (b \cdot c)$ for all $a, b, c \in R$.
3. The *distributive laws* hold; that is, for all $a, b, c \in R$ we have $a \cdot (b + c) = a \cdot b + a \cdot c$ and $(b + c) \cdot a = b \cdot a + c \cdot a$.

We shall use R as a designation for the ring $(R, +, \cdot)$ and stress that the operations $+$ and $\cdot$ are not necessarily the ordinary operations with numbers. In following convention, we use 0 (called the *zero element*) to denote the identity element of the abelian group R with respect to addition, and the additive inverse of a is denoted by $- a$; also, $a + (- b)$ is abbreviated by $a - b$. Instead of $a \cdot b$ we will usually write ab. As a consequence of the definition of a ring one obtains the general property $a0 = 0a = 0$ for all $a \in R$. This, in turn, implies $(- a)b = a(- b) = - ab$ for all $a, b \in R$.

The most natural example of a ring is perhaps the ring of ordinary integers. If we examine the properties of this ring, we realize that it has properties not enjoyed by rings in general. Thus, rings can be further classified according to the following definitions.

1.29. Definition

(i) A ring is called a *ring with identity* if the ring has a multiplicative identity—that is, if there is an element e such that $ae = ea = a$ for all $a \in R$.

(ii) A ring is called *commutative* if $\cdot$ is commutative.

(iii)   A ring is called an *integral domain* if it is a commutative ring with identity $e \neq 0$ in which $ab = 0$ implies $a = 0$ or $b = 0$.

(iv)   A ring is called a *division ring* (or *skew field*) if the nonzero elements of R form a group under $\cdot$.

(v)   A commutative division ring is called a *field*.

Since our study is devoted to fields, we emphasize again the definition of this concept. In the first place, a *field* is a set F on which two binary operations, called addition and multiplication, are defined and which contains two distinguished elements 0 and e with $0 \neq e$. Furthermore, F is an abelian group with respect to addition having 0 as the identity element, and the elements of F that are $\neq 0$ form an abelian group with respect to multiplication having e as the identity element. The two operations of addition and multiplication are linked by the distributive law $a(b + c) = ab + ac$. The second distributive law $(b + c)a = ba + ca$ follows automatically from the commutativity of multiplication. The element 0 is called the *zero element* and e is called the *multiplicative identity element* or simply the *identity*. Later on, the identity will usually be denoted by 1.

The property appearing in Definition 1.29(iii)—namely, that $ab = 0$ implies $a = 0$ or $b = 0$—is expressed by saying that there are *no zero divisors*. In particular, a field has no zero divisors, for if $ab = 0$ and $a \neq 0$, then multiplication by a^{-1} yields $b = a^{-1}0 = 0$.

In order to give an indication of the generality of the concept of ring, we present some examples.

1.30.   Examples

(i)   Let R be any abelian group with group operation $+$. Define $ab = 0$ for all $a, b \in R$; then R is a ring.

(ii)   The integers form an integral domain, but not a field.

(iii)   The even integers form a commutative ring without identity.

(iv)   The functions from the real numbers into the real numbers form a commutative ring with identity under the definitions for $f + g$ and fg given by $(f + g)(x) = f(x) + g(x)$ and $(fg)(x) = f(x)g(x)$ for $x \in \mathbb{R}$.

(v)   The set of all 2×2 matrices with real numbers as entries forms a noncommutative ring with identity with respect to matrix addition and multiplication. $\square$

We have seen above that a field is, in particular, an integral domain. The converse is not true in general (see Example 1.30(ii)), but it will hold if the structures contain only finitely many elements.

1.31.   Theorem.   *Every finite integral domain is a field.*

Proof.   Let the elements of the finite integral domain R be $a_1, a_2, \ldots, a_n$. For a fixed nonzero element $a \in R$, consider the products $aa_1, aa_2, \ldots, aa_n$. These are distinct, for if $aa_i = aa_j$, then $a(a_i - a_j) = 0$, and since $a \neq 0$ we must have $a_i - a_j = 0$, or $a_i = a_j$. Thus each element of R is of the form aa_i, in particular, $e = aa_i$ for some i with $1 \leq i \leq n$, where e is the identity of R. Since R is commutative, we have also $a_i a = e$, and so a_i is the multiplicative inverse of a. Thus the nonzero elements of R form a commutative group, and R is a field.   □

1.32. Definition. A subset S of a ring R is called a *subring* of R provided S is closed under $+$ and $\cdot$ and forms a ring under these operations.

1.33. Definition. A subset J of a ring R is called an *ideal* provided J is a subring of R and for all $a \in J$ and $r \in R$ we have $ar \in J$ and $ra \in J$.

1.34. Examples

(i) Let R be the field $\mathbb{Q}$ of rational numbers. Then the set $\mathbb{Z}$ of integers is a subring of $\mathbb{Q}$, but not an ideal since, for example, $1 \in \mathbb{Z}$, $\frac{1}{2} \in \mathbb{Q}$, but $\frac{1}{2} \cdot 1 = \frac{1}{2} \notin \mathbb{Z}$.

(ii) Let R be a commutative ring, $a \in R$, and let $J = \{ra : r \in R\}$, then J is an ideal.

(iii) Let R be a commutative ring. Then the smallest ideal containing a given element $a \in R$ is the ideal $(a) = \{ra + na : r \in R, n \in \mathbb{Z}\}$. If R contains an identity, then $(a) = \{ra : r \in R\}$.   □

1.35. Definition. Let R be a commutative ring. An ideal J of R is said to be *principal* if there is an $a \in R$ such that $J = (a)$. In this case, J is also called the principal ideal *generated by* a.

Since ideals are normal subgroups of the additive group of a ring, it follows immediately that an ideal J of the ring R defines a partition of R into disjoint cosets, called *residue classes* modulo J. The residue class of the element a of R modulo J will be denoted by $[a] = a + J$, since it consists of all elements of R that are of the form $a + c$ for some $c \in J$. Elements $a, b \in R$ are called *congruent* modulo J, written $a \equiv b \bmod J$, if they are in the same residue class modulo J, or equivalently, if $a - b \in J$ (compare with Definition 1.4). One can verify that $a \equiv b \bmod J$ implies $a + r \equiv b + r \bmod J$, $ar \equiv br \bmod J$, and $ra \equiv rb \bmod J$ for any $r \in R$ and $na \equiv nb \bmod J$ for any $n \in \mathbb{Z}$. If, in addition, $r \equiv s \bmod J$, then $a + r \equiv b + s \bmod J$ and $ar \equiv bs \bmod J$.

It is shown by a straightforward argument that the set of residue classes of a ring R modulo an ideal J forms a ring with respect to the operations

$$(a + J) + (b + J) = (a + b) + J, \tag{1.2}$$

$$(a + J)(b + J) = ab + J. \tag{1.3}$$

1.36. Definition. The ring of residue classes of the ring R modulo the ideal J under the operations (1.2) and (1.3) is called the *residue class ring* (or *factor ring*) of R modulo J and is denoted by R/J.

1.37.   Example (The residue class ring $\mathbb{Z}/(n)$).   As in the case of groups (compare with Definition 1.5), we denote the coset or residue class of the integer a modulo the positive integer n by $[a]$, as well as by $a+(n)$, where (n) is the principal ideal generated by n. The elements of $\mathbb{Z}/(n)$ are

$$[0] = 0+(n),\, [1] = 1+(n),\dots,[n-1] = n-1+(n).\qquad\square$$

1.38.   Theorem. $\mathbb{Z}/(p)$, *the ring of residue classes of the integers modulo the principal ideal generated by a prime p, is a field.*

Proof.   By Theorem 1.31 it suffices to show that $\mathbb{Z}/(p)$ is an integral domain. Now $[1]$ is an identity of $\mathbb{Z}/(p)$, and $[a][b] = [ab] = [0]$ if and only if $ab = kp$ for some integer k. But since p is prime, p divides ab if and only if p divides at least one of the factors. Therefore, either $[a] = [0]$ or $[b] = [0]$, so that $\mathbb{Z}/(p)$ contains no zero divisors. $\square$

1.39.   Example.   Let $p = 3$. Then $\mathbb{Z}/(p)$ consists of the elements $[0]$, $[1]$, and $[2]$. The operations in this field can be described by operation tables that are similar to Cayley tables for finite groups (see Example 1.7):

$+$	$[0]$	$[1]$	$[2]$		$\cdot$	$[0]$	$[1]$	$[2]$
$[0]$	$[0]$	$[1]$	$[2]$		$[0]$	$[0]$	$[0]$	$[0]$
$[1]$	$[1]$	$[2]$	$[0]$		$[1]$	$[0]$	$[1]$	$[2]$
$[2]$	$[2]$	$[0]$	$[1]$		$[2]$	$[0]$	$[2]$	$[1]$

$\square$

The residue class fields $\mathbb{Z}/(p)$ are our first examples of *finite fields* — that is, of fields that contain only finitely many elements. The general theory of such fields will be developed later on.

The reader is cautioned not to assume that in the formation of residue class rings all the properties of the original ring will be preserved in all cases. For example, the lack of zero divisors is not always preserved, as may be seen by considering the ring $\mathbb{Z}/(n)$, where n is a composite integer.

There is an obvious extension from groups to rings of the definition of a homomorphism. A mapping $\varphi: R \to S$ from a ring R into a ring S is called a *homomorphism* if for any $a, b \in R$ we have

$$\varphi(a+b) = \varphi(a)+\varphi(b) \quad \text{and} \quad \varphi(ab) = \varphi(a)\varphi(b).$$

Thus a homomorphism $\varphi: R \to S$ preserves both operations $+$ and $\cdot$ of R and induces a homomorphism of the additive group of R into the additive group of S. The set

$$\ker \varphi = \{a \in R: \varphi(a) = 0 \in S\}$$

is called the *kernel* of φ. Other concepts, such as that of an *isomorphism*, are analogous to those in Definition 1.16. The homomorphism theorem for rings, similar to Theorem 1.23 for groups, runs as follows.

1.40.   Theorem (Homomorphism Theorem for Rings).   *If φ is a homomorphism of a ring R onto a ring S, then $\ker \varphi$ is an ideal of R and S is* *isomorphic to the factor ring $R/\ker \varphi$. Conversely, if J is an ideal of the ring R, then the mapping $\psi: R \to R/J$ defined by $\psi(a) = a + J$ for $a \in R$ is a homomorphism of R onto R/J with kernel J.*

Mappings can be used to transfer a structure from an algebraic system to a set without structure. For instance, let R be a ring and let φ be a one-to-one and onto mapping from R to a set S; then by means of φ one can define a ring structure on S that converts φ into an isomorphism. In detail, let s_1 and s_2 be two elements of S and let r_1 and r_2 be the elements of R uniquely determined by $\varphi(r_1) = s_1$ and $\varphi(r_2) = s_2$. Then one defines $s_1 + s_2$ to be $\varphi(r_1 + r_2)$ and $s_1 s_2$ to be $\varphi(r_1 r_2)$, and all the desired properties are satisfied. This structure on S may be called the ring structure *induced by* φ. In case R has additional properties, such as being an integral domain or a field, then these properties are inherited by S. We use this principle in order to arrive at a more convenient representation for the finite fields $\mathbb{Z}/(p)$.

1.41. Definition. For a prime p, let $\mathbb{F}_p$ be the set $\{0, 1, \ldots, p-1\}$ of integers and let $\varphi: \mathbb{Z}/(p) \to \mathbb{F}_p$ be the mapping defined by $\varphi([a]) = a$ for $a = 0, 1, \ldots, p-1$. Then $\mathbb{F}_p$, endowed with the field structure induced by φ, is a finite field, called the *Galois field of order p*.

By what we have said before, the mapping $\varphi: \mathbb{Z}/(p) \to \mathbb{F}_p$ is then an isomorphism, so that $\varphi([a] + [b]) = \varphi([a]) + \varphi([b])$ and $\varphi([a][b]) = \varphi([a])\varphi([b])$. The finite field $\mathbb{F}_p$ has zero element 0, identity 1, and its structure is exactly the structure of $\mathbb{Z}/(p)$. Computing with elements of $\mathbb{F}_p$ therefore means ordinary arithmetic of integers with reduction modulo p.

1.42. Examples

(i) Consider $\mathbb{Z}/(5)$, isomorphic to $\mathbb{F}_5 = \{0, 1, 2, 3, 4\}$, with the isomorphism given by: $[0] \to 0$, $[1] \to 1$, $[2] \to 2$, $[3] \to 3$, $[4] \to 4$. The tables for the two operations $+$ and $\cdot$ for elements in $\mathbb{F}_5$ are as follows:

+	0	1	2	3	4
0	0	1	2	3	4
1	1	2	3	4	0
2	2	3	4	0	1
3	3	4	0	1	2
4	4	0	1	2	3

$\cdot$	0	1	2	3	4
0	0	0	0	0	0
1	0	1	2	3	4
2	0	2	4	1	3
3	0	3	1	4	2
4	0	4	3	2	1

(ii) An even simpler and more important example is the finite field $\mathbb{F}_2$. The elements of this field of order two are 0 and 1, and the operation tables have the following form:

+	0	1
0	0	1
1	1	0

$\cdot$	0	1
0	0	0
1	0	1

In this context, the elements 0 and 1 are called *binary elements*. □

If b is any nonzero element of the ring $\mathbb{Z}$ of integers, then the additive order of b is infinite; that is, $nb = 0$ implies $n = 0$. However, in the ring $\mathbb{Z}/(p)$, p prime, the additive order of every nonzero element b is p; that is, $pb = 0$, and p is the least positive integer for which this holds. It is of interest to formalize this property.

1.43.   Definition.   If R is an arbitrary ring and there exists a positive integer n such that $nr = 0$ for every $r \in R$, then the least such positive integer n is called the *characteristic* of R and R is said to have (positive) characteristic n. If no such positive integer n exists, R is said to have characteristic 0.

1.44.   Theorem.   *A ring $R \neq \{0\}$ of positive characteristic having an identity and no zero divisors must have prime characteristic.*

Proof.   Since R contains nonzero elements, R has characteristic $n \geq 2$. If n were not prime, we could write $n = km$ with $k, m \in \mathbb{Z}$, $1 < k, m < n$. Then $0 = ne = (km)e = (ke)(me)$, and this implies that either $ke = 0$ or $me = 0$ since R has no zero divisors. It follows that either $kr = (ke)r = 0$ for all $r \in R$ or $mr = (me)r = 0$ for all $r \in R$, in contradiction to the definition of the characteristic n.   □

1.45.   Corollary.   *A finite field has prime characteristic.*

Proof.   By Theorem 1.44 it suffices to show that a finite field F has a positive characteristic. Consider the multiples $e, 2e, 3e, \ldots$ of the identity. Since F contains only finitely many distinct elements, there exist integers k and m with $1 \leq k < m$ such that $ke = me$, or $(m - k)e = 0$, and so F has a positive characteristic.   □

The finite field $\mathbb{Z}/(p)$ (or, equivalently, $\mathbb{F}_p$) obviously has characteristic p, whereas the ring $\mathbb{Z}$ of integers and the field $\mathbb{Q}$ of rational numbers have characteristic 0. We note that in a ring R of characteristic 2 we have $2a = a + a = 0$, hence $a = -a$ for all $a \in R$. A useful property of commutative rings of prime characteristic is the following.

1.46.   Theorem.   *Let R be a commutative ring of prime characteristic p. Then*

$$(a + b)^{p^n} = a^{p^n} + b^{p^n} \quad and \quad (a - b)^{p^n} = a^{p^n} - b^{p^n}$$

for $a, b \in R$ and $n \in \mathbb{N}$.

Proof.   We use the fact that

$$\binom{p}{i} = \frac{p(p-1)\cdots(p-i+1)}{1\cdot 2\cdot\,\cdots\,\cdot i} \equiv 0 \bmod p$$

for all $i \in \mathbb{Z}$ with $0 < i < p$, which follows from $\binom{p}{i}$ being an integer and the observation that the factor p in the numerator cannot be cancelled. Then by the binomial theorem (see Exercise 1.8),

$$(a+b)^p = a^p + \binom{p}{1}a^{p-1}b + \cdots + \binom{p}{p-1}ab^{p-1} + b^p = a^p + b^p,$$

and induction on n completes the proof of the first identity. By what we have shown, we get

$$a^{p^n} = ((a-b)+b)^{p^n} = (a-b)^{p^n} + b^{p^n},$$

and the second identity follows. $\qquad\Box$

Next we will show for the case of commutative rings with identity which ideals give rise to factor rings that are integral domains or fields. For this we need some definitions from ring theory.

Let R be a commutative ring with identity. An element $a \in R$ is called a *divisor* of $b \in R$ if there exists $c \in R$ such that $ac = b$. A *unit* of R is a divisor of the identity; two elements $a, b \in R$ are said to be *associates* if there is a unit ε of R such that $a = b\varepsilon$. An element $c \in R$ is called a *prime element* if it is no unit and if it has only the units of R and the associates of c as divisors. An ideal $P \neq R$ of the ring R is called a *prime ideal* if for $a, b \in R$ we have $ab \in P$ only if either $a \in P$ or $b \in P$. An ideal $M \neq R$ of R is called a *maximal ideal* of R if for any ideal J of R the property $M \subseteq J$ implies $J = R$ or $J = M$. Furthermore, R is said to be a *principal ideal domain* if R is an integral domain and if every ideal J of R is principal—that is, if there is a generating element a for J such that $J = (a) = \{ra : r \in R\}$.

1.47. Theorem. *Let R be a commutative ring with identity. Then:*

(i) *An ideal M of R is a maximal ideal if and only if R/M is a field.*

(ii) *An ideal P of R is a prime ideal if and only if R/P is an integral domain.*

(iii) *Every maximal ideal of R is a prime ideal.*

(iv) *If R is a principal ideal domain, then $R/(c)$ is a field if and only if c is a prime element of R.*

Proof.

(i) Let M be a maximal ideal of R. Then for $a \notin M$, $a \in R$, the set $J = \{ar + m : r \in R, m \in M\}$ is an ideal of R properly containing M, and therefore $J = R$. In particular, $ar + m = 1$ for some suitable $r \in R$, $m \in M$, where 1 denotes the multiplicative identity element of R. In other words, if $a + M \neq 0 + M$ is an element of R/M different from the zero element in R/M, then it possesses a multiplicative inverse, because $(a + M)(r + M) = ar + M = (1 - m) + M = 1 + M$. Therefore, R/M is a field. Conversely, let R/M be a field and let $J \supseteq M$, $J \neq M$, be an ideal of R. Then for $a \in J$, $a \notin M$, the residue class $a + M$ has a multiplicative inverse, so that $(a + M)(r + M) = 1 + M$ for some $r \in R$. This implies $ar + m = 1$ for some $m \in M$. Since J is an ideal, we have $1 \in J$ and therefore $(1) = R \subseteq J$, hence $J = R$. Thus M is a maximal ideal of R.

(ii) Let P be a prime ideal of R; then R/P is a commutative ring with identity $1 + P \neq 0 + P$. Let $(a + P)(b + P) = 0 + P$, hence $ab \in P$. Since P is a prime ideal, either $a \in P$ or $b \in P$; that is, either $a + P = 0 + P$ or $b + P = 0 + P$. Thus, R/P has no zero divisors and is therefore an integral domain. The converse follows immediately by reversing the steps of this proof.

(iii) This follows from (i) and (ii) since every field is an integral domain.

(iv) Let $c \in R$. If c is a unit, then $(c) = R$ and the ring $R/(c)$ consists only of one element and is no field. If c is neither a unit nor a prime element, then c has a divisor $a \in R$ that is neither a unit nor an associate of c. We note that $a \neq 0$, for if $a = 0$, then $c = 0$ and a would be an associate of c. We can write $c = ab$ with $b \in R$. Next we claim that $a \notin (c)$. For otherwise $a = cd = abd$ for some $d \in R$, or $a(1 - bd) = 0$. Since $a \neq 0$, this would imply $bd = 1$, so that d would be a unit, which contradicts the fact that a is not an associate of c. It follows that $(c) \subseteq (a) \subseteq R$, where all containments are proper, and so $R/(c)$ cannot be a field because of (i). Finally, we are left with the case where c is a prime element. Then $(c) \neq R$ since c is no unit. Furthermore, if $J \supseteq (c)$ is an ideal of R, then $J = (a)$ for some $a \in R$ since R is a principal ideal domain. It follows that $c \in (a)$, and so a is a divisor of c. Consequently, a is either a unit or an associate of c, so that either $J = R$ or $J = (c)$. This shows that (c) is a maximal ideal of R. Hence $R/(c)$ is a field by (i).   □

As an application of this theorem, let us consider the case $R = \mathbb{Z}$. We note that $\mathbb{Z}$ is a principal ideal domain since the additive subgroups of $\mathbb{Z}$ are already generated by a single element because of Theorem 1.15(i). A prime number p fits the definition of a prime element, and so Theorem 1.47(iv) yields another proof of the known result that $\mathbb{Z}/(p)$ is a field. Consequently, (p) is a maximal ideal and a prime ideal of $\mathbb{Z}$. For a composite integer n, the ideal (n) is not a prime ideal of $\mathbb{Z}$, and so $\mathbb{Z}/(n)$ is not even an integral domain. Other applications will follow in the next section when we consider residue class rings of polynomial rings over fields.

3. POLYNOMIALS

In elementary algebra one regards a polynomial as an expression of the form $a_0 + a_1 x + \cdots + a_n x^n$. The a_i's are called coefficients and are usually real or complex numbers; x is viewed as a variable: that is, substituting an arbitrary number α for x, a well-defined number $a_0 + a_1\alpha + \cdots + a_n\alpha^n$ is obtained. The arithmetic of polynomials is governed by familiar rules. The concept of polynomial and the associated operations can be generalized to a formal algebraic setting in a straightforward manner.

Let R be an arbitrary ring. A *polynomial* over R is an expression of the form

$$f(x) = \sum_{i=0}^{n} a_i x^i = a_0 + a_1 x + \cdots + a_n x^n,$$

where n is a nonnegative integer, the *coefficients* a_i, $0 \leqslant i \leqslant n$, are elements of R, and x is a symbol not belonging to R, called an *indeterminate* over R. Whenever it is clear which indeterminate is meant, we can use f as a designation for the polynomial $f(x)$. We adopt the convention that a term $a_i x^i$ with $a_i = 0$ need not be written down. In particular, the polynomial $f(x)$ above may then also be given in the equivalent form $f(x) = a_0 + a_1 x + \cdots + a_n x^n + 0x^{n+1} + \cdots + 0x^{n+h}$, where h is any positive integer. When comparing two polynomials $f(x)$ and $g(x)$ over R, it is therefore possible to assume that they both involve the same powers of x. The polynomials

$$f(x) = \sum_{i=0}^{n} a_i x^i \quad \text{and} \quad g(x) = \sum_{i=0}^{n} b_i x^i$$

over R are considered equal if and only if $a_i = b_i$ for $0 \leqslant i \leqslant n$. We define the *sum* of $f(x)$ and $g(x)$ by

$$f(x) + g(x) = \sum_{i=0}^{n} (a_i + b_i) x^i.$$

To define the *product* of two polynomials over R, let

$$f(x) = \sum_{i=0}^{n} a_i x^i \quad \text{and} \quad g(\dot{x}) = \sum_{j=0}^{m} b_j x^j$$

and set

$$f(x)g(x) = \sum_{k=0}^{n+m} c_k x^k, \quad \text{where } c_k = \sum_{\substack{i+j=k \\ 0 \leqslant i \leqslant n, 0 \leqslant j \leqslant m}} a_i b_j.$$

It is easily seen that with these operations the set of polynomials over R forms a ring.

1.48. Definition. The ring formed by the polynomials over R with the above operations is called the *polynomial ring* over R and denoted by $R[x]$.

The zero element of $R[x]$ is the polynomial all of whose coefficients are 0. This polynomial is called the *zero polynomial* and denoted by 0. It should always be clear from the context whether 0 stands for the zero element of R or the zero polynomial.

1.49. Definition.   Let $f(x) = \sum_{i=0}^{n} a_i x^i$ be a polynomial over R that is not the zero polynomial, so that we can suppose $a_n \neq 0$. Then a_n is called the *leading coefficient* of $f(x)$ and a_0 the *constant term*, while n is called the *degree* of $f(x)$, in symbols $n = \deg(f(x)) = \deg(f)$. By convention, we set $\deg(0) = -\infty$. Polynomials of degree $\leqslant 0$ are called *constant polynomials*. If R has the identity 1 and if the leading coefficient of $f(x)$ is 1, then $f(x)$ is called a *monic polynomial*.

By computing the leading coefficient of the sum and the product of two polynomials, one finds the following result.

1.50. Theorem.   *Let $f, g \in R[x]$. Then*

$$\deg(f + g) \leqslant \max(\deg(f), \deg(g)),$$

$$\deg(fg) \leqslant \deg(f) + \deg(g).$$

If R is an integral domain, we have

$$\deg(fg) = \deg(f) + \deg(g). \tag{1.4}$$

If one identifies constant polynomials with elements of R, then R can be viewed as a subring of $R[x]$. Certain properties of R are inherited by $R[x]$. The essential step in the proof of part (iii) of the subsequent theorem depends on (1.4).

1.51. Theorem.   *Let R be a ring. Then:*

   (i)   *$R[x]$ is commutative if and only if R is commutative.*
   (ii) *$R[x]$ is a ring with identity if and only if R has an identity.*
   (iii) *$R[x]$ is an integral domain if and only if R is an integral domain.*

In the following chapters we will deal almost exclusively with polynomials over fields. Let F denote a field (not necessarily finite). The concept of divisibility, when specialized to the ring $F[x]$, leads to the following. The polynomial $g \in F[x]$ *divides* the polynomial $f \in F[x]$ if there exists a polynomial $h \in F[x]$ such that $f = gh$. We also say that g is a *divisor* of f, or that f is a *multiple* of g, or that f is *divisible* by g. The units of $F[x]$ are the divisors of the constant polynomial 1, which are precisely all nonzero constant polynomials.

As for the ring of integers, there is a division with remainder in polynomial rings over fields.

1.52. Theorem (Division Algorithm).   *Let $g \neq 0$ be a polynomial in $F[x]$. Then for any $f \in F[x]$ there exist polynomials $q, r \in F[x]$ such that*

$$f = qg + r, \quad \text{where } \deg(r) < \deg(g).$$

1.53. Example.   Consider $f(x) = 2x^5 + x^4 + 4x + 3 \in \mathbb{F}_5[x]$, $g(x) = 3x^2 + 1 \in \mathbb{F}_5[x]$. We compute the polynomials $q, r \in \mathbb{F}_5[x]$ with $f = qg + r$ by using long division:

$$
\begin{array}{r}
4x^3 + 2x^2 + 2x + 1 \\
\hline
3x^2 + 1 \overline{\smash{\big)}\ 2x^5 + x^4 \qquad\qquad\qquad + 4x + 3} \\
-2x^5 \qquad -4\ x^3 \\
\hline
x^4 + x^3 \\
-x^4 \qquad\qquad -2x^2 \\
\hline
x^3 + 3x^2 + 4x \\
-x^3 \qquad\qquad -2x \\
\hline
3x^2 + 2x + 3 \\
-3x^2 \qquad -1 \\
\hline
2x + 2
\end{array}
$$

Thus $q(x) = 4x^3 + 2x^2 + 2x + 1$, $r(x) = 2x + 2$, and obviously $\deg(r) < \deg(g)$. $\quad\square$

The fact that $F[x]$ permits a division algorithm implies by a standard argument that every ideal of $F[x]$ is principal.

1.54. Theorem. *$F[x]$ is a principal ideal domain. In fact, for every ideal $J \neq (0)$ of $F[x]$ there exists a uniquely determined monic polynomial $g \in F[x]$ with $J = (g)$.*

Proof. $F[x]$ is an integral domain by Theorem 1.51(iii). Suppose $J \neq (0)$ is an ideal of $F[x]$. Let $h(x)$ be a nonzero polynomial of least degree contained in J, let b be the leading coefficient of $h(x)$, and set $g(x) = b^{-1}h(x)$. Then $g \in J$ and g is monic. If $f \in J$ is arbitrary, the division algorithm yields $q, r \in F[x]$ with $f = qg + r$ and $\deg(r) < \deg(g) = \deg(h)$. Since J is an ideal, we get $f - qg = r \in J$, and by the definition of h we must have $r = 0$. Therefore, f is a multiple of g, and so $J = (g)$. If $g_1 \in F[x]$ is another monic polynomial with $J = (g_1)$, then $g = c_1 g_1$ and $g_1 = c_2 g$ with $c_1, c_2 \in F[x]$. This implies $g = c_1 c_2 g$, hence $c_1 c_2 = 1$, and c_1 and c_2 are constant polynomials. Since both g and g_1 are monic, it follows that $g = g_1$, and the uniqueness of g is established. $\quad\square$

1.55. Theorem. *Let $f_1, \ldots, f_n$ be polynomials in $F[x]$ not all of which are 0. Then there exists a uniquely determined monic polynomial $d \in F[x]$ with the following properties: (i) d divides each f_j, $1 \leq j \leq n$; (ii) any polynomial $c \in F[x]$ dividing each f_j, $1 \leq j \leq n$, divides d. Moreover, d can be expressed in the form*

$$d = b_1 f_1 + \cdots + b_n f_n \quad \text{with } b_1, \ldots, b_n \in F[x]. \tag{1.5}$$

Proof. The set J consisting of all polynomials of the form $c_1 f_1 + \cdots + c_n f_n$ with $c_1, \ldots, c_n \in F[x]$ is easily seen to be an ideal of $F[x]$. Since not all f_j are 0, we have $J \neq (0)$, and Theorem 1.54 implies that $J = (d)$

for some monic polynomial $d \in F[x]$. Property (i) and the representation (1.5) follow immediately from the construction of d. Property (ii) follows from (1.5). If d_1 is another monic polynomial in $F[x]$ satisfying (i) and (ii), then these properties imply that d and d_1 are divisible by each other, and so $(d) = (d_1)$. An application of the uniqueness part of Theorem 1.54 yields $d = d_1$. $\square$

The monic polynomial d appearing in the theorem above is called the *greatest common divisor* of $f_1,\ldots,f_n$, in symbols $d = \gcd(f_1,\ldots,f_n)$. If $\gcd(f_1,\ldots,f_n) = 1$, then the polynomials $f_1,\ldots,f_n$ are said to be *relatively prime*. They are called *pairwise relatively prime* if $\gcd(f_i, f_j) = 1$ for $1 \leqslant i < j \leqslant n$.

The greatest common divisor of two polynomials $f, g \in F[x]$ can be computed by the *Euclidean algorithm*. Suppose, without loss of generality, that $g \neq 0$ and that g does not divide f. Then we repeatedly use the division algorithm in the following manner:

$$f = q_1 g + r_1 \qquad\qquad 0 \leqslant \deg(r_1) < \deg(g)$$

$$g = q_2 r_1 + r_2 \qquad\qquad 0 \leqslant \deg(r_2) < \deg(r_1)$$

$$r_1 = q_3 r_2 + r_3 \qquad\qquad 0 \leqslant \deg(r_3) < \deg(r_2)$$

$$\vdots \qquad\qquad\qquad\qquad \vdots$$

$$r_{s-2} = q_s r_{s-1} + r_s \qquad\qquad 0 \leqslant \deg(r_s) < \deg(r_{s-1})$$

$$r_{s-1} = q_{s+1} r_s.$$

Here $q_1,\ldots,q_{s+1}$ and $r_1,\ldots,r_s$ are polynomials in $F[x]$. Since $\deg(g)$ is finite, the procedure must stop after finitely many steps. If the last nonzero remainder r_s has leading coefficient b, then $\gcd(f, g) = b^{-1} r_s$. In order to find $\gcd(f_1,\ldots,f_n)$ for $n > 2$ and nonzero polynomials f_i, one first computes $\gcd(f_1, f_2)$, then $\gcd(\gcd(f_1, f_2), f_3)$, and so on, by the Euclidean algorithm.

1.56.   Example.   The Euclidean algorithm applied to

$$f(x) = 2x^6 + x^3 + x^2 + 2 \in \mathbb{F}_3[x], \qquad g(x) = x^4 + x^2 + 2x \in \mathbb{F}_3[x]$$

yields:

$$2x^6 + x^3 + x^2 + 2 = (2x^2 + 1)(x^4 + x^2 + 2x) + x + 2$$

$$x^4 + x^2 + 2x = (x^3 + x^2 + 2x + 1)(x + 2) + 1$$

$$x + 2 = (x + 2)1.$$

Therefore $\gcd(f, g) = 1$ and f and g are relatively prime. $\square$

A counterpart to the notion of greatest common divisor is that of least common multiple. Let $f_1,\ldots,f_n$ be nonzero polynomials in $F[x]$. Then one shows (see Exercise 1.25) that there exists a uniquely determined monic polynomial $m \in F[x]$ with the following properties: (i) m is a multiple of each f_j, $1 \leqslant j \leqslant n$; (ii) any polynomial $b \in F[x]$ that is a multiple of each f_j, $1 \leqslant j \leqslant n$, is a multiple of m. The polynomial m is called the *least common multiple* of $f_1,\ldots,f_n$ and denoted by $m = \operatorname{lcm}(f_1,\ldots,f_n)$. For two nonzero polynomials $f, g \in F[x]$ we have

$$a^{-1}fg = \operatorname{lcm}(f, g)\operatorname{gcd}(f, g), \tag{1.6}$$

where a is the leading coefficient of fg. This relation conveniently reduces the calculation of $\operatorname{lcm}(f, g)$ to that of $\operatorname{gcd}(f, g)$. There is no direct analog of (1.6) for three or more polynomials. In this case, one uses the identity $\operatorname{lcm}(f_1,\ldots,f_n) = \operatorname{lcm}(\operatorname{lcm}(f_1,\ldots,f_{n-1}), f_n)$ to compute the least common multiple.

The prime elements of the ring $F[x]$ are usually called irreducible polynomials. To emphasize this important concept, we give the definition again for the present context.

1.57. Definition.   A polynomial $p \in F[x]$ is said to be *irreducible over F* (or *irreducible in $F[x]$*, or *prime in $F[x]$*) if p has positive degree and $p = bc$ with $b, c \in F[x]$ implies that either b or c is a constant polynomial.

Briefly stated, a polynomial of positive degree is irreducible over F if it allows only trivial factorizations. A polynomial in $F[x]$ of positive degree that is not irreducible over F is called *reducible over F*. The reducibility or irreducibility of a given polynomial depends heavily on the field under consideration. For instance, the polynomial $x^2 - 2 \in \mathbb{Q}[x]$ is irreducible over the field $\mathbb{Q}$ of rational numbers, but $x^2 - 2 = (x + \sqrt{2})(x - \sqrt{2})$ is reducible over the field of real numbers.

Irreducible polynomials are of fundamental importance for the structure of the ring $F[x]$ since the polynomials in $F[x]$ can be written as products of irreducible polynomials in an essentially unique manner. For the proof we need the following result.

1.58. Lemma.   *If an irreducible polynomial p in $F[x]$ divides a product $f_1 \cdots f_m$ of polynomials in $F[x]$, then at least one of the factors f_j is divisible by p.*

Proof.   Since p divides $f_1 \cdots f_m$, we get the identity $(f_1 + (p)) \cdots (f_m + (p)) = 0 + (p)$ in the factor ring $F[x]/(p)$. Now $F[x]/(p)$ is a field by Theorem 1.47(iv), and so $f_j + (p) = 0 + (p)$ for some j; that is, p divides f_j.   □

1.59. Theorem (Unique Factorization in $F[x]$).   *Any polynomial $f \in F[x]$ of positive degree can be written in the form*

$$f = ap_1^{e_1} \cdots p_k^{e_k}, \tag{1.7}$$

where $a \in F$, $p_1,\ldots,p_k$ are distinct monic irreducible polynomials in $F[x]$, and $e_1,\ldots,e_k$ are positive integers. Moreover, this factorization is unique apart from the order in which the factors occur.

Proof. The fact that any nonconstant $f \in F[x]$ can be represented in the form (1.7) is shown by induction on the degree of f. The case $\deg(f) = 1$ is trivial since any polynomial in $F[x]$ of degree 1 is irreducible over F. Now suppose the desired factorization is established for all nonconstant polynomials in $F[x]$ of degree $< n$. If $\deg(f) = n$ and f is irreducible over F, then we are done since we can write $f = a(a^{-1}f)$, where a is the leading coefficient of f and $a^{-1}f$ is a monic irreducible polynomial in $F[x]$. Otherwise, f allows a factorization $f = gh$ with $1 \leqslant \deg(g) < n$, $1 \leqslant \deg(h) < n$, and $g, h \in F[x]$. By the induction hypothesis, g and h can be factored in the form (1.7), and so f can be factored in this form.

To prove uniqueness, suppose f has two factorizations of the form (1.7), say

$$f = ap_1^{e_1} \cdots p_k^{e_k} = bq_1^{d_1} \cdots q_r^{d_r}. \qquad (1.8)$$

By comparing leading coefficients, we get $a = b$. Furthermore, the irreducible polynomial p_1 in $F[x]$ divides the right-hand side of (1.8), and so Lemma 1.58 shows that p_1 divides q_j for some $j, 1 \leqslant j \leqslant r$. But q_j is also irreducible in $F[x]$, so that we must have $q_j = cp_1$ with a constant polynomial c. Since q_j and p_1 are both monic, it follows that $q_j = p_1$. Thus we can cancel p_1 against q_j in (1.8) and continue in the same manner with the remaining identity. After finitely many steps of this type, we obtain that the two factorizations are identical apart from the order of the factors. $\qquad \square$

We shall refer to (1.7) as the *canonical factorization* of the polynomial f in $F[x]$. If $F = \mathbb{Q}$, there is a method due to Kronecker for finding the canonical factorization of a polynomial in finitely many steps. This method is briefly described in Exercise 1.30. For polynomials over finite fields, factorization algorithms will be discussed in Chapter 4.

A central question about polynomials in $F[x]$ is to decide whether a given polynomial is irreducible or reducible over F. For our purposes, irreducible polynomials over $\mathbb{F}_p$ are of particular interest. To determine all monic irreducible polynomials over $\mathbb{F}_p$ of fixed degree n, one may first compute all monic reducible polynomials over $\mathbb{F}_p$ of degree n and then eliminate them from the set of monic polynomials in $\mathbb{F}_p[x]$ of degree n. If p or n is large, this method is not feasible, and we will develop more powerful methods in Chapter 3, Sections 2 and 3.

1.60. Example. Find all irreducible polynomials over $\mathbb{F}_2$ of degree 4 (note that a nonzero polynomial in $\mathbb{F}_2[x]$ is automatically monic). There are $2^4 = 16$ polynomials in $\mathbb{F}_2[x]$ of degree 4. Such a polynomial is reducible over $\mathbb{F}_2$ if and only if it has a divisor of degree 1 or 2. Therefore, we compute all products $(a_0 + a_1 x + a_2 x^2 + x^3)(b_0 + x)$ and $(a_0 + a_1 x + x^2)(b_0 + b_1 x + x^2)$ with $a_i, b_j \in \mathbb{F}_2$ and obtain all reducible polynomials over $\mathbb{F}_2$ of degree 4. Comparison with the 16 polynomials of degree 4 leaves us with the irreducible polynomials $f_1(x) = x^4 + x + 1$, $f_2(x) = x^4 + x^3 + 1$, $f_3(x) = x^4 + x^3 + x^2 + x + 1$ in $\mathbb{F}_2[x]$.   □

Since the irreducible polynomials over a field F are exactly the prime elements of $F[x]$, the following result, one part of which was already used in Lemma 1.58, is an immediate consequence of Theorems 1.47(iv) and 1.54.

1.61. Theorem. *For $f \in F[x]$, the residue class ring $F[x]/(f)$ is a field if and only if f is irreducible over F.*

As a preparation for the next section, we shall take a closer look at the structure of the residue class ring $F[x]/(f)$, where f is an arbitrary nonzero polynomial in $F[x]$. We recall that as a residue class ring $F[x]/(f)$ consists of residue classes $g + (f)$ (also denoted by $[g]$) with $g \in F[x]$, where the operations are defined as in (1.2) and (1.3). Two residue classes $g + (f)$ and $h + (f)$ are identical precisely if $g \equiv h \bmod f$ — that is, precisely if $g - h$ is divisible by f. This is equivalent to the requirement that g and h leave the same remainder after division by f. Each residue class $g + (f)$ contains a unique representative $r \in F[x]$ with $\deg(r) < \deg(f)$, which is simply the remainder in the division of g by f. The process of passing from g to r is called *reduction* $\bmod f$. The uniqueness of r follows from the observation that if $r_1 \in g + (f)$ with $\deg(r_1) < \deg(f)$, then $r - r_1$ is divisible by f and $\deg(r - r_1) < \deg(f)$, which is only possible if $r = r_1$. The distinct residue classes comprising $F[x]/(f)$ can now be described explicitly; namely, they are exactly the residue classes $r + (f)$, where r runs through all polynomials in $F[x]$ with $\deg(r) < \deg(f)$. Thus, if $F = \mathbb{F}_p$ and $\deg(f) = n \geqslant 0$, then the number of elements of $\mathbb{F}_p[x]/(f)$ is equal to the number of polynomials in $\mathbb{F}_p[x]$ of degree $< n$, which is p^n.

1.62. Examples

(i)   Let $f(x) = x \in \mathbb{F}_2[x]$. The $p^n = 2^1$ polynomials in $\mathbb{F}_2[x]$ of degree <1 determine all residue classes comprising $\mathbb{F}_2[x]/(x)$. Thus, $\mathbb{F}_2[x]/(x)$ consists of the residue classes $[0]$ and $[1]$ and is isomorphic to $\mathbb{F}_2$.

(ii) Let $f(x) = x^2 + x + 1 \in \mathbb{F}_2[x]$. Then $\mathbb{F}_2[x]/(f)$ has the $p^n = 2^2$ elements $[0]$, $[1]$, $[x]$, $[x + 1]$. The operation tables for this residue class ring are obtained by performing the required operations with the polynomials determining the residue classes and by carrying out reduction $\bmod f$ if necessary:

+	[0]	[1]	[x]	[x + 1]
[0]	[0]	[1]	[x]	[x + 1]
[1]	[1]	[0]	[x + 1]	[x]
[x]	[x]	[x + 1]	[0]	[1]
[x + 1]	[x + 1]	[x]	[1]	[0]

	[0]	[1]	[x]	[x+1]
[0]	[0]	[0]	[0]	[0]
[1]	[0]	[1]	[x]	[x+1]
[x]	[0]	[x]	[x+1]	[1]
[x+1]	[0]	[x+1]	[1]	[x]

By inspecting these tables, or from the irreducibility of f over $\mathbb{F}_2$ and Theorem 1.61, it follows that $\mathbb{F}_2[x]/(f)$ is a field. This is our first example of a finite field for which the number of elements is not a prime.

(iii)   Let $f(x) = x^2 + 2 \in \mathbb{F}_3[x]$. Then $\mathbb{F}_3[x]/(f)$ consists of the $p^n = 3^2$ residue classes [0], [1], [2], [x], [x + 1], [x + 2], [2x], [2x + 1], [2x + 2]. The operation tables for $\mathbb{F}_3[x]/(f)$ are again produced by performing polynomial operations and using reduction mod f whenever necessary. Since $\mathbb{F}_3[x]/(f)$ is a commutative ring, we only have to compute the entries on and above the main diagonal.

+	[0]	[1]	[2]	[x]	[x+1]	[x+2]	[2x]	[2x+1]	[2x+2]
[0]	[0]	[1]	[2]	[x]	[x+1]	[x+2]	[2x]	[2x+1]	[2x+2]
[1]		[2]	[0]	[x+1]	[x+2]	[x]	[2x+1]	[2x+2]	[2x]
[2]			[1]	[x+2]	[x]	[x+1]	[2x+2]	[2x]	[2x+1]
[x]				[2x]	[2x+1]	[2x+2]	[0]	[1]	[2]
[x+1]					[2x+2]	[2x]	[1]	[2]	[0]
[x+2]						[2x+1]	[2]	[0]	[1]
[2x]							[x]	[x+1]	[x+2]
[2x+1]								[x+2]	[x]
[2x+2]									[x+1]

	[0]	[1]	[2]	[x]	[x+1]	[x+2]	[2x]	[2x+1]	[2x+2]
[0]	[0]	[0]	[0]	[0]	[0]	[0]	[0]	[0]	[0]
[1]		[1]	[2]	[x]	[x+1]	[x+2]	[2x]	[2x+1]	[2x+2]
[2]			[1]	[2x]	[2x+2]	[2x+1]	[x]	[x+2]	[x+1]
[x]				[1]	[x+1]	[2x+1]	[2]	[x+2]	[2x+2]
[x+1]					[2x+2]	[0]	[2x+2]	[0]	[x+1]
[x+2]						[x+2]	[x+2]	[2x+1]	[0]
[2x]							[1]	[2x+1]	[x+1]
[2x+1]								[x+2]	[0]
[2x+2]									[2x+2]

Note that $\mathbb{F}_3[x]/(f)$ is not a field (and not even an integral domain). This is in accordance with Theorem 1.61 since $x^2 + 2 = (x + 1)(x + 2)$ is reducible over $\mathbb{F}_3$.   □

If F is again an arbitrary field and $f(x) \in F[x]$, then replacement of the indeterminate x in $f(x)$ by a fixed element of F yields a well-defined element of F. In detail, if $f(x) = a_0 + a_1 x + \cdots + a_n x^n \in F[x]$ and $b \in F$, then replacing x by b we get $f(b) = a_0 + a_1 b + \cdots + a_n b^n \in F$. In any polynomial identity in $F[x]$ we can substitute a fixed $b \in F$ for x and obtain a valid identity in F (*principle of substitution*).

1.63. Definition. An element $b \in F$ is called a *root* (or a *zero*) of the polynomial $f \in F[x]$ if $f(b) = 0$.

An important connection between roots and divisibility is given by the following theorem.

1.64. Theorem. *An element $b \in F$ is a root of the polynomial $f \in F[x]$ if and only if $x - b$ divides $f(x)$.*

Proof. We use the division algorithm (see Theorem 1.52) to write $f(x) = q(x)(x - b) + c$ with $q \in F[x]$ and $c \in F$. Substituting b for x, we get $f(b) = c$, hence $f(x) = q(x)(x - b) + f(b)$. The theorem follows now from this identity. $\square$

1.65. Definition. Let $b \in F$ be a root of the polynomial $f \in F[x]$. If k is a positive integer such that $f(x)$ is divisible by $(x - b)^k$, but not by $(x - b)^{k+1}$, then k is called the *multiplicity* of b. If $k = 1$, then b is called a *simple root* (or a *simple zero*) of f, and if $k \geqslant 2$, then b is called a *multiple root* (or a *multiple zero*) of f.

1.66. Theorem. *Let $f \in F[x]$ with $\deg f = n \geqslant 0$. If $b_1, \ldots, b_m \in F$ are distinct roots of f with multiplicities $k_1, \ldots, k_m$, respectively, then $(x - b_1)^{k_1} \cdots (x - b_m)^{k_m}$ divides $f(x)$. Consequently, $k_1 + \cdots + k_m \leqslant n$, and f can have at most n distinct roots in F.*

Proof. We note that each polynomial $x - b_j$, $1 \leqslant j \leqslant m$, is irreducible over F, and so $(x - b_j)^{k_j}$ occurs as a factor in the canonical factorization of f. Altogether, the factor $(x - b_1)^{k_1} \cdots (x - b_m)^{k_m}$ appears in the canonical factorization of f and is thus a divisor of f. By comparing degrees, we get $k_1 + \cdots + k_m \leqslant n$, and $m \leqslant k_1 + \cdots + k_m \leqslant n$ shows the last statement. $\square$

1.67. Definition. If $f(x) = a_0 + a_1 x + a_2 x^2 + \cdots + a_n x^n \in F[x]$, then the *derivative* f' of f is defined by $f' = f'(x) = a_1 + 2a_2 x + \cdots + na_n x^{n-1} \in F[x]$.

1.68. Theorem. *The element $b \in F$ is a multiple root of $f \in F[x]$ if and only if it is a root of both f and f'.*

There is a relation between the nonexistence of roots and irreducibility. If f is an irreducible polynomial in $F[x]$ of degree $\geqslant 2$, then Theorem 1.64 shows that f has no root in F. The converse holds for polynomials of degree 2 or 3, but not necessarily for polynomials of higher degree.

1.69. Theorem.   *The polynomial $f \in F[x]$ of degree 2 or 3 is irreducible in $F[x]$ if and only if f has no root in F.*

Proof. The necessity of the condition was already noted. Conversely, if f has no root in F and were reducible in $F[x]$, we could write $f = gh$ with $g, h \in F[x]$ and $1 \leqslant \deg(g) \leqslant \deg(h)$. But $\deg(g) + \deg(h) = \deg(f) \leqslant 3$, hence $\deg(g) = 1$; that is, $g(x) = ax + b$ with $a, b \in F$, $a \neq 0$. Then $-ba^{-1}$ is a root of g, and so a root of f in F, a contradiction.   □

1.70. Example.   Because of Theorem 1.69, the irreducible polynomials in $\mathbb{F}_2[x]$ of degree 2 or 3 can be obtained by eliminating the polynomials with roots in $\mathbb{F}_2$ from the set of all polynomials in $\mathbb{F}_2[x]$ of degree 2 or 3. The only irreducible polynomial in $\mathbb{F}_2[x]$ of degree 2 is $f(x) = x^2 + x + 1$, and the irreducible polynomials in $\mathbb{F}_2[x]$ of degree 3 are $f_1(x) = x^3 + x + 1$ and $f_2(x) = x^3 + x^2 + 1$.   □

In elementary analysis there is a well-known method for constructing a polynomial with real coefficients which assumes certain assigned values for given values of the indeterminate. The same method carries over to any field.

1.71. Theorem (Lagrange Interpolation Formula). *For $n \geqslant 0$, let $a_0, \ldots, a_n$ be $n + 1$ distinct elements of F, and let $b_0, \ldots, b_n$ be $n + 1$ arbitrary elements of F. Then there exists exactly one polynomial $f \in F[x]$ of degree $\leqslant n$ such that $f(a_i) = b_i$ for $i = 0, \ldots, n$. This polynomial is given by*

$$f(x) = \sum_{i=0}^{n} b_i \prod_{\substack{k=0 \\ k \neq i}}^{n} (a_i - a_k)^{-1}(x - a_k).$$

One can also consider polynomials in several indeterminates. Let R denote a commutative ring with identity and let $x_1, \ldots, x_n$ be symbols that will serve as indeterminates. We form the polynomial ring $R[x_1]$, then the polynomial ring $R[x_1, x_2] = R[x_1][x_2]$, and so on, until we arrive at $R[x_1, \ldots, x_n] = R[x_1, \ldots, x_{n-1}][x_n]$. The elements of $R[x_1, \ldots, x_n]$ are then expressions of the form

$$f = f(x_1, \ldots, x_n) = \sum a_{i_1 \cdots i_n} x_1^{i_1} \cdots x_n^{i_n}$$

with coefficients $a_{i_1 \cdots i_n} \in R$, where the summation is extended over finitely many n-tuples $(i_1, \ldots, i_n)$ of nonnegative integers and the convention $x_j^0 = 1$ $(1 \leqslant j \leqslant n)$ is observed. Such an expression is called a *polynomial in $x_1, \ldots, x_n$ over R*. Two polynomials $f, g \in R[x_1, \ldots, x_n]$ are equal if and only if all corresponding coefficients are equal. It is tacitly assumed that the indeterminates $x_1, \ldots, x_n$ commute with each other, so that, for instance, the expressions $x_1 x_2 x_3 x_4$ and $x_4 x_1 x_3 x_2$ are identified.

1.72. Definition.   Let $f \in R[x_1, \ldots, x_n]$ be given by

$$f(x_1, \ldots, x_n) = \sum a_{i_1 \cdots i_n} x_1^{i_1} \cdots x_n^{i_n}.$$

If $a_{i_1 \cdots i_n} \neq 0$, then $a_{i_1 \cdots i_n} x_1^{i_1} \cdots x_n^{i_n}$ is called a *term* of f and $i_1 + \cdots + i_n$ is the degree of the term. For $f \neq 0$ one defines the *degree* of f, denoted by $\deg(f)$, to be the maximum of the degrees of the terms of f. For $f = 0$ one sets $\deg(f) = -\infty$. If $f = 0$ or if all terms of f have the same degree, then f is called *homogeneous*.

Any $f \in R[x_1, \ldots, x_n]$ can be written as a finite sum of homogeneous polynomials. The degrees of polynomials in $R[x_1, \ldots, x_n]$ satisfy again the inequalities in Theorem 1.50, and if R is an integral domain, then (1.4) is valid and $R[x_1, \ldots, x_n]$ is an integral domain. If F is a field, then the polynomials in $F[x_1, \ldots, x_n]$ of positive degree can again be factored uniquely into a constant factor and a product of "monic" prime elements (using a suitable definition of "monic"), but for $n \geqslant 2$ there is no analog of the division algorithm (in the case of commuting indeterminates) and $F[x_1, \ldots, x_n]$ is not a principal ideal domain.

An important special class of polynomials in n indeterminates is that of symmetric polynomials.

1.73. Definition. A polynomial $f \in R[x_1, \ldots, x_n]$ is called *symmetric* if $f(x_{i_1}, \ldots, x_{i_n}) = f(x_1, \ldots, x_n)$ for any permutation $i_1, \ldots, i_n$ of the integers $1, \ldots, n$.

1.74. Example. Let z be an indeterminate over $R[x_1, \ldots, x_n]$, and let $g(z) = (z - x_1)(z - x_2) \cdots (z - x_n)$. Then

$$g(z) = z^n - \sigma_1 z^{n-1} + \sigma_2 z^{n-2} + \cdots + (-1)^n \sigma_n$$

with

$$\sigma_k = \sigma_k(x_1, \ldots, x_n) = \sum_{1 \leqslant i_1 < \cdots < i_k \leqslant n} x_{i_1} \cdots x_{i_k} \quad (k = 1, 2, \ldots, n).$$

Thus:

$$\sigma_1 = x_1 + x_2 + \cdots + x_n,$$
$$\sigma_2 = x_1 x_2 + x_1 x_3 + \cdots + x_1 x_n + x_2 x_3 + \cdots + x_2 x_n + \cdots + x_{n-1} x_n,$$
$$\vdots$$
$$\sigma_n = x_1 x_2 \cdots x_n.$$

As g remains unaltered under any permutation of the x_i, all the σ_k are symmetric polynomials; they are also homogeneous. The polynomial $\sigma_k = \sigma_k(x_1, \ldots, x_n) \in R[x_1, \ldots, x_n]$ is called the kth *elementary symmetric polynomial* in the indeterminates $x_1, \ldots, x_n$ over R. The adjective "elementary" is used because of the so-called "fundamental theorem on symmetric polynomials," which states that for any symmetric polynomial $f \in R[x_1, \ldots, x_n]$ there exists a uniquely determined polynomial $h \in R[x_1, \ldots, x_n]$ such that $f(x_1, \ldots, x_n) = h(\sigma_1, \ldots, \sigma_n)$. □

1.75. Theorem (Newton's Formula). *Let $\sigma_1, \ldots, \sigma_n$ be the elementary symmetric polynomials in $x_1, \ldots, x_n$ over R, and let $s_0 = n \in \mathbb{Z}$ and*

$s_k = s_k(x_1, \ldots, x_n) = x_1^k + \cdots + x_n^k \in R[x_1, \ldots, x_n]$ for $k \geqslant 1$. Then the formula

$$s_k - s_{k-1}\sigma_1 + s_{k-2}\sigma_2 + \cdots + (-1)^{m-1}s_{k-m+1}\sigma_{m-1} + (-1)^m \frac{m}{n}s_{k-m}\sigma_m = 0$$

holds for $k \geqslant 1$, where $m = \min(k, n)$.

1.76. Theorem (Waring's Formula). *With the same notation as in Theorem 1.75, we have*

$$s_k = \sum (-1)^{i_2 + i_4 + i_6 + \cdots} \frac{(i_1 + i_2 + \cdots + i_n - 1)!k}{i_1!i_2! \cdots i_n!} \sigma_1^{i_1}\sigma_2^{i_2} \cdots \sigma_n^{i_n}$$

for $k \geqslant 1$, where the summation is extended over all n-tuples $(i_1, \ldots, i_n)$ of nonnegative integers with $i_1 + 2i_2 + \cdots + ni_n = k$. The coefficient of $\sigma_1^{i_1}\sigma_2^{i_2} \cdots \sigma_n^{i_n}$ is always an integer.

4. FIELD EXTENSIONS

Let F be a field. A subset K of F that is itself a field under the operations of F will be called a *subfield* of F. In this context, F is called an *extension* (*field*) of K. If $K \neq F$, we say that K is a *proper subfield* of F.

If K is a subfield of the finite field $\mathbb{F}_p$, p prime, then K must contain the elements 0 and 1, and so all other elements of $\mathbb{F}_p$ by the closure of K under addition. It follows that $\mathbb{F}_p$ contains no proper subfields. We are thus led to the following concept.

1.77. Definition. A field containing no proper subfields is called a *prime field*.

By the above argument, any finite field of order p, p prime, is a prime field. Another example of a prime field is the field $\mathbb{Q}$ of rational numbers.

The intersection of any nonempty collection of subfields of a given field F is again a subfield of F. If we form the intersection of *all* subfields of F, we obtain the *prime subfield* of F. It is obviously a prime field.

1.78. Theorem. *The prime subfield of a field F is isomorphic to either $\mathbb{F}_p$ or $\mathbb{Q}$, according as the characteristic of F is a prime p or 0.*

1.79. Definition. Let K be a subfield of the field F and M any subset of F. Then the field $K(M)$ is defined as the intersection of all subfields of F containing both K and M and is called the extension (field) of K obtained by *adjoining* the elements in M. For finite $M = \{\theta_1, \ldots, \theta_n\}$ we write $K(M) = K(\theta_1, \ldots, \theta_n)$. If M consists of a single element $\theta \in F$, then $L = K(\theta)$ is said to be a *simple extension* of K and θ is called a *defining element* of L over K.

Obviously, $K(M)$ is the smallest subfield of F containing both K and M. We define now an important type of extension.

1.80. Definition. Let K be a subfield of F and $\theta \in F$. If θ satisfies a nontrivial polynomial equation with coefficients in K, that is, if $a_n\theta^n + \cdots + a_1\theta + a_0 = 0$ with $a_i \in K$ not all being 0, then θ is said to be *algebraic* over K. An extension L of K is called *algebraic* over K (or an *algebraic extension* of K) if every element of L is algebraic over K.

Suppose $\theta \in F$ is algebraic over K, and consider the set $J = \langle f \in K[x]: f(\theta) = 0 \rangle$. It is easily checked that J is an ideal of $K[x]$, and we have $J \neq (0)$ since θ is algebraic over K. It follows then from Theorem 1.54 that there exists a uniquely determined monic polynomial $g \in K[x]$ such that J is equal to the principal ideal (g). It is important to note that g is irreducible in $K[x]$. For, in the first place, g is of positive degree since it has the root θ; and if $g = h_1 h_2$ in $K[x]$ with $1 \leqslant \deg(h_i) < \deg(g)$ $(i = 1, 2)$, then $0 = g(\theta) = h_1(\theta)h_2(\theta)$ implies that either h_1 or h_2 is in J and so divisible by g, which is impossible.

1.81. Definition. If $\theta \in F$ is algebraic over K, then the uniquely determined monic polynomial $g \in K[x]$ generating the ideal $J = \{f \in K[x]: f(\theta) = 0\}$ of $K[x]$ is called the *minimal polynomial* (or *defining polynomial*, or *irreducible polynomial*) of θ over K. By the *degree* of θ over K we mean the degree of g.

1.82. Theorem. *If $\theta \in F$ is algebraic over K, then its minimal polynomial g over K has the following properties:*

   (i)   *g is irreducible in $K[x]$.*
   (ii) *For $f \in K[x]$ we have $f(\theta) = 0$ if and only if g divides f.*
   (iii) *g is the monic polynomial in $K[x]$ of least degree having θ as a root.*

Proof. Property (i) was already noted and (ii) follows from the definition of g. As to (iii), it suffices to note that any monic polynomial in $K[x]$ having θ as a root must be a multiple of g, and so it is either equal to g or its degree is larger than that of g. $\square$

We note that both the minimal polynomial and the degree of an algebraic element θ depend on the field K over which it is considered, so that one must be careful not to speak of the minimal polynomial or the degree of θ without specifying K, unless the latter is amply clear from the context.

If L is an extension field of K, then L may be viewed as a vector space over K. For the elements of L (= "vectors") form, first of all, an abelian group under addition. Moreover, each "vector" $\alpha \in L$ can be multiplied by a "scalar" $r \in K$ so that $r\alpha$ is again in L (here $r\alpha$ is simply the product of the field elements r and α of L) and the laws for multiplication by scalars are satisfied: $r(\alpha+\beta)=r\alpha+r\beta$, $(r+s)\alpha=r\alpha+s\alpha$, $(rs)\alpha=r(s\alpha)$, and $1\alpha=\alpha$, where $r,s\in K$ and $\alpha,\beta\in L$.

1.83. Definition. Let L be an extension field of K. If L, considered as a vector space over K, is finite-dimensional, then L is called a *finite extension* of K. The dimension of the vector space L over K is then called the *degree* of L over K, in symbols $[L:K]$.

1.84. Theorem. *If L is a finite extension of K and M is a finite extension of L, then M is a finite extension of K with*

$$[M:K]=[M:L][L:K].$$

Proof. Put $[M:L]=m$, $[L:K]=n$, and let $\{\alpha_1,\ldots,\alpha_m\}$ be a basis of M over L and $\{\beta_1,\ldots,\beta_n\}$ a basis of L over K. Then every $\alpha\in M$ is a linear combination $\alpha=\gamma_1\alpha_1+\cdots+\gamma_m\alpha_m$ with $\gamma_i\in L$ for $1\leqslant i\leqslant m$, and writing each γ_i in terms of the basis elements β_j we get

$$\alpha=\sum_{i=1}^{m}\gamma_i\alpha_i=\sum_{i=1}^{m}\left(\sum_{j=1}^{n}r_{ij}\beta_j\right)\alpha_i=\sum_{i=1}^{m}\sum_{j=1}^{n}r_{ij}\beta_j\alpha_i$$

with coefficients $r_{ij}\in K$. If we can show that the mn elements $\beta_j\alpha_i$, $1\leqslant i\leqslant m$, $1\leqslant j\leqslant n$, are linearly independent over K, then we are done. So suppose we have

$$\sum_{i=1}^{m}\sum_{j=1}^{n}s_{ij}\beta_j\alpha_i=0$$

with coefficients $s_{ij}\in K$. Then

$$\sum_{i=1}^{m}\left(\sum_{j=1}^{n}s_{ij}\beta_j\right)\alpha_i=0,$$

and from the linear independence of the α_i over L we infer

$$\sum_{j=1}^{n}s_{ij}\beta_j=0\quad\text{for }1\leqslant i\leqslant m.$$

But since the β_j are linearly independent over K, we conclude that all s_{ij} are 0. $\square$

1.85. Theorem. *Every finite extension of K is algebraic over K.*

Proof. Let L be a finite extension of K and put $[L:K]=m$. For $\theta\in L$, the $m+1$ elements $1,\theta,\ldots,\theta^m$ must then be linearly dependent over K, and so we get a relation $a_0+a_1\theta+\cdots+a_m\theta^m=0$ with $a_i\in K$ not all being 0. This just says that θ is algebraic over K. $\square$

For the study of the structure of a simple extension $K(\theta)$ of K obtained by adjoining an algebraic element, let F be an extension of K and let $\theta \in F$ be algebraic over K. It turns out that $K(\theta)$ is a finite (and therefore an algebraic) extension of K.

1.86. Theorem. *Let $\theta \in F$ be algebraic of degree n over K and let g be the minimal polynomial of θ over K. Then:*

(i) *$K(\theta)$ is isomorphic to $K[x]/(g)$.*
(ii) *$[K(\theta): K] = n$ and $\{1, \theta, \ldots, \theta^{n-1}\}$ is a basis of $K(\theta)$ over K.*
(iii) *Every $\alpha \in K(\theta)$ is algebraic over K and its degree over K is a divisor of n.*

Proof. (i) Consider the mapping $\tau: K[x] \to K(\theta)$, defined by $\tau(f) = f(\theta)$ for $f \in K[x]$, which is easily seen to be a ring homomorphism. We have $\ker \tau = \{f \in K[x]: f(\theta) = 0\} = (g)$ by the definition of the minimal polynomial. Let S be the image of τ; that is, S is the set of polynomial expressions in θ with coefficients in K. Then the homomorphism theorem for rings (see Theorem 1.40) yields that S is isomorphic to $K[x]/(g)$. But $K[x]/(g)$ is a field by Theorems 1.61 and 1.82(i), and so S is a field. Since $K \subseteq S \subseteq K(\theta)$ and $\theta \in S$, it follows from the definition of $K(\theta)$ that $S = K(\theta)$, and (i) is thus shown.

(ii) Since $S = K(\theta)$, any given $\alpha \in K(\theta)$ can be written in the form $\alpha = f(\theta)$ for some $f \in K[x]$. By the division algorithm, $f = qg + r$ with $q, r \in K[x]$ and $\deg(r) < \deg(g) = n$. Then $\alpha = f(\theta) = q(\theta)g(\theta) + r(\theta) = r(\theta)$, and so α is a linear combination of $1, \theta, \ldots, \theta^{n-1}$ with coefficients in K. On the other hand, if $a_0 + a_1\theta + \cdots + a_{n-1}\theta^{n-1} = 0$ for certain $a_i \in K$, then the polynomial $h(x) = a_0 + a_1 x + \cdots + a_{n-1}x^{n-1} \in K[x]$ has θ as a root and is thus a multiple of g by Theorem 1.82(ii). Since $\deg(h) < n = \deg(g)$, this is only possible if $h = 0$—that is, if all $a_i = 0$. Therefore, the elements $1, \theta, \ldots, \theta^{n-1}$ are linearly independent over K and (ii) follows.

(iii) $K(\theta)$ is a finite extension of K by (ii), and so $\alpha \in K(\theta)$ is algebraic over K by Theorem 1.85. Furthermore, $K(\alpha)$ is a subfield of $K(\theta)$. If d is the degree of α over K, then (ii) and Theorem 1.84 imply that $n = [K(\theta): K] = [K(\theta): K(\alpha)][K(\alpha): K] = [K(\theta): K(\alpha)]d$, hence d divides n. $\square$

The elements of the simple algebraic extension $K(\theta)$ of K are therefore polynomial expressions in θ. Any element of $K(\theta)$ can be uniquely represented in the form $a_0 + a_1\theta + \cdots + a_{n-1}\theta^{n-1}$ with $a_i \in K$ for $0 \leqslant i \leqslant n-1$.

It should be pointed out that Theorem 1.86 operates under the assumption that both K and θ are embedded in a larger field F. This is necessary in order that algebraic expressions involving θ make sense. We now want to construct a simple algebraic extension *ab ovo* —that is, without reference to a previously given larger field. The clue to this is contained in part (i) of Theorem 1.86.

1.87.   Theorem.   *Let $f \in K[x]$ be irreducible over the field K. Then there exists a simple algebraic extension of K with a root of f as a defining element.*

Proof.   Consider the residue class ring $L = K[x]/(f)$, which is a field by Theorem 1.61. The elements of L are the residue classes $[h] = h + (f)$ with $h \in K[x]$. For any $a \in K$ we can form the residue class $[a]$ determined by the constant polynomial a, and if $a, b \in K$ are distinct, then $[a] \neq [b]$ since f has positive degree. The mapping $a \mapsto [a]$ gives an isomorphism from K onto a subfield K' of L, so that K' may be identified with K. In other words, we can view L as an extension of K. For every $h(x) = a_0 + a_1 x + \cdots + a_m x^m \in K[x]$ we have $[h] = [a_0 + a_1 x + \cdots + a_m x^m] = [a_0] + [a_1][x] + \cdots + [a_m][x]^m = a_0 + a_1[x] + \cdots + a_m[x]^m$ by the rules for operating with residue classes and the identification $[a_i] = a_i$. Thus, every element of L can be written as a polynomial expression in $[x]$ with coefficients in K. Since any field containing both K and $[x]$ must contain these polynomial expressions, L is a simple extension of K obtained by adjoining $[x]$. If $f(x) = b_0 + b_1 x + \cdots + b_n x^n$, then $f([x]) = b_0 + b_1[x] + \cdots + b_n[x]^n = [b_0 + b_1 x + \cdots + b_n x^n] = [f] = [0]$, so that $[x]$ is a root of f and L is a simple algebraic extension of K.   □

1.88.   Example.   As an example of the formal process of root adjunction in Theorem 1.87, consider the prime field $\mathbb{F}_3$ and the polynomial $f(x) = x^2 + x + 2 \in \mathbb{F}_3[x]$, which is irreducible over $\mathbb{F}_3$. Let θ be a "root" of f; that is, θ is the residue class $x + (f)$ in $L = \mathbb{F}_3[x]/(f)$. The other root of f in L is then $2\theta + 2$, since $f(2\theta + 2) = (2\theta + 2)^2 + (2\theta + 2) + 2 = \theta^2 + \theta + 2 = 0$. By Theorem 1.86(ii), or by the known structure of a residue class field, the simple algebraic extension $L = \mathbb{F}_3(\theta)$ consists of the nine elements $0, 1, 2, \theta, \theta + 1, \theta + 2, 2\theta, 2\theta + 1, 2\theta + 2$. The operation tables for L can be constructed as in Example 1.62.   □

We observe that in the above example we may adjoin either the root θ or the root $2\theta + 2$ of f and we would still obtain the same field. This situation is covered by the following result, which is easily established.

1.89.   Theorem.   *Let α and β be two roots of the polynomial $f \in K[x]$ that is irreducible over K. Then $K(\alpha)$ and $K(\beta)$ are isomorphic under an isomorphism mapping α to β and keeping the elements of K fixed.*

We are now asking for an extension field to which all roots of a given polynomial belong.

1.90.   Definition.   Let $f \in K[x]$ be of positive degree and F an extension field of K. Then f is said to *split in F* if f can be written as a product of linear factors in $F[x]$—that is, if there exist elements $\alpha_1, \alpha_2, \ldots, \alpha_n \in F$ such that

$$f(x) = a(x - \alpha_1)(x - \alpha_2) \cdots (x - \alpha_n),$$

where a is the leading coefficient of f. The field F is a *splitting field* of f over K if f splits in F and if, moreover, $F = K(\alpha_1, \alpha_2, \ldots, \alpha_n)$.

It is clear that a splitting field F of f over K is in the following sense the smallest field containing all the roots of f: no proper subfield of F that is an extension of K contains all the roots of f. By repeatedly applying the process used in Theorem 1.87, one obtains the first part of the subsequent result. The second part is an extension of Theorem 1.89.

1.91.   Theorem (Existence and Uniqueness of Splitting Field).   *If K is a field and f any polynomial of positive degree in $K[x]$, then there exists a splitting field of f over K. Any two splitting fields of f over K are isomorphic under an isomorphism which keeps the elements of K fixed and maps roots of f into each other.*

Since isomorphic fields may be identified, we can speak of *the* splitting field of f over K. It is obtained from K by adjoining finitely many algebraic elements over K, and therefore one can show on the basis of Theorems 1.84 and 1.86(ii) that the splitting field of f over K is a finite extension of K.

As an illustration of the usefulness of splitting fields, we consider the question of deciding whether a given polynomial has a multiple root (compare with Definition 1.65).

1.92.   Definition.   Let $f \in K[x]$ be a polynomial of degree $n \geq 2$ and suppose that $f(x) = a_0(x - \alpha_1) \cdots (x - \alpha_n)$ with $\alpha_1, \ldots, \alpha_n$ in the splitting field of f over K. Then the *discriminant* $D(f)$ of f is defined by

$$D(f) = a_0^{2n-2} \prod_{1 \leq i < j \leq n} (\alpha_i - \alpha_j)^2.$$

It is obvious from the definition of $D(f)$ that f has a multiple root if and only if $D(f) = 0$. Although $D(f)$ is defined in terms of elements of an extension of K, it is actually an element of K itself. For small n this can be seen by direct calculation. For instance, if $n = 2$ and $f(x) = ax^2 + bx + c = a(x - \alpha_1)(x - \alpha_2)$, then $D(f) = a^2(\alpha_1 - \alpha_2)^2 = a^2((\alpha_1 + \alpha_2)^2 - 4\alpha_1\alpha_2) = a^2(b^2 a^{-2} - 4ca^{-1})$, hence

$$D(ax^2 + bx + c) = b^2 - 4ac,$$

a well-known expression from the theory of quadratic equations. If $n = 3$ and $f(x) = ax^3 + bx^2 + cx + d = a(x - \alpha_1)(x - \alpha_2)(x - \alpha_3)$, then $D(f) = a^4(\alpha_1 - \alpha_2)^2(\alpha_1 - \alpha_3)^2(\alpha_2 - \alpha_3)^2$, and a more involved computation yields

$$D(ax^3 + bx^2 + cx + d) = b^2 c^2 - 4b^3 d - 4ac^3 - 27a^2 d^2 + 18abcd. \quad (1.9)$$

In the general case, consider first the polynomial $s \in K[x_1,\ldots,x_n]$ given by

$$s(x_1,\ldots,x_n) = a_0^{2n-2} \prod_{1 \leqslant i < j \leqslant n} (x_i - x_j)^2.$$

Then s is a symmetric polynomial, and by a result in Example 1.74 it can be written as a polynomial expression in the elementary symmetric polynomials $\sigma_1,\ldots,\sigma_n$—that is, $s = h(\sigma_1,\ldots,\sigma_n)$ for some $h \in K[x_1,\ldots,x_n]$. If $f(x) = a_0 x^n + a_1 x^{n-1} + \cdots + a_n = a_0(x - \alpha_1)\cdots(x - \alpha_n)$, then the definition of the elementary symmetric polynomials (see again Example 1.74) implies that $\sigma_k(\alpha_1,\ldots,\alpha_n) = (-1)^k a_k a_0^{-1} \in K$ for $1 \leqslant k \leqslant n$. Thus,

$$D(f) = s(\alpha_1,\ldots,\alpha_n) = h(\sigma_1(\alpha_1,\ldots,\alpha_n),\ldots,\sigma_n(\alpha_1,\ldots,\alpha_n))$$
$$= h(-a_1 a_0^{-1},\ldots,(-1)^n a_n a_0^{-1}) \in K.$$

Since $D(f) \in K$, it should be possible to calculate $D(f)$ without having to pass to an extension field of K. This can be done via the notion of resultant. We note first that if a polynomial $f \in K[x]$ is given in the form $f(x) = a_0 x^n + a_1 x^{n-1} + \cdots + a_n$ and we accept the possibility that $a_0 = 0$, then n need not be the degree of f. We speak of n as the *formal degree* of f; it is always greater than or equal to $\deg(f)$.

1.93. Definition.   Let $f(x) = a_0 x^n + a_1 x^{n-1} + \cdots + a_n \in K[x]$ and $g(x) = b_0 x^m + b_1 x^{m-1} + \cdots + b_m \in K[x]$ be two polynomials of formal degree n resp. m with $n, m \in \mathbb{N}$. Then the *resultant* $R(f, g)$ of the two polynomials is defined by the determinant

$$R(f,g) = \left. \begin{vmatrix} a_0 & a_1 & \cdots & a_n & 0 & \cdots & 0 \\ 0 & a_0 & a_1 & \cdots & a_n & 0 & \cdots & 0 \\ \vdots & & & & & & \vdots \\ 0 & \cdots & 0 & a_0 & a_1 & \cdots & a_n \\ b_0 & b_1 & \cdots & & b_m & 0 & \cdots & 0 \\ 0 & b_0 & b_1 & \cdots & & b_m & \cdots & 0 \\ \vdots & & & & & & \vdots \\ 0 & \cdots & 0 & b_0 & b_1 & & \cdots & b_m \end{vmatrix} \right\} \begin{matrix} m \text{ rows} \\ \\ n \text{ rows} \end{matrix}$$

of order $m + n$.

If $\deg(f) = n$ (i.e., if $a_0 \neq 0$) and $f(x) = a_0(x - \alpha_1)\cdots(x - \alpha_n)$ in the splitting field of f over K, then $R(f, g)$ is also given by the formula

$$R(f, g) = a_0^m \prod_{i=1}^{n} g(\alpha_i). \tag{1.10}$$

In this case, we obviously have $R(f, g) = 0$ if and only if f and g have a common root, which is the same as saying that f and g have a common divisor in $K[x]$ of positive degree.

Theorem 1.68 suggests a connection between the discriminant $D(f)$ and the resultant $R(f, f')$. Let $f \in K[x]$ with $\deg(f) = n \geqslant 2$ and leading coefficient a_0. Then we have, in fact, the identity

$$D(f) = (-1)^{n(n-1)/2} a_0^{-1} R(f, f'), \qquad (1.11)$$

where f' is viewed as a polynomial of formal degree $n - 1$. The last remark is needed since we may have $\deg(f') < n - 1$ and even $f' = 0$ in case K has prime characteristic. At any rate, the identity (1.11) shows that we can obtain $D(f)$ by calculating a determinant of order $2n - 1$ with entries in K.

EXERCISES

1.1. Prove that the identity element of a group is uniquely determined.
1.2. For a multiplicative group G, prove that a nonempty subset H of G is a subgroup of G if and only if $a, b \in H$ implies $ab^{-1} \in H$. If H is finite, then the condition can be replaced by: $a, b \in H$ implies $ab \in H$.
1.3. Let a be an element of finite order k in the multiplicative group G. Show that for $m \in \mathbb{Z}$ we have $a^m = e$ if and only if k divides m.
1.4. For $m \in \mathbb{N}$, Euler's function $\phi(m)$ is defined to be the number of integers k with $1 \leqslant k \leqslant m$ and $\gcd(k, m) = 1$. Show the following properties for $m, n, s \in \mathbb{N}$ and a prime p:

(a) $\phi(p^s) = p^s\left(1 - \dfrac{1}{p}\right)$;

(b) $\phi(mn) = \phi(m)\phi(n)$ if $\gcd(m, n) = 1$;

(c) $\phi(m) = m\left(1 - \dfrac{1}{p_1}\right) \cdots \left(1 - \dfrac{1}{p_r}\right)$, where $m = p_1^{e_1} \cdots p_r^{e_r}$ is the prime factor decomposition of m.

1.5. Calculate $\phi(490)$ and $\phi(768)$.
1.6. Use the class equation to show the following: if the order of a finite group is a prime power p^s, p prime, $s \geqslant 1$, then the order of its center is divisible by p.
1.7. Prove that in a ring R we have $(-a)(-b) = ab$ for all $a, b \in R$.
1.8. Prove that in a commutative ring R the formula

$$(a + b)^n = a^n + \binom{n}{1} a^{n-1} b + \cdots + \binom{n}{n-1} ab^{n-1} + b^n$$

holds for all $a, b \in R$ and $n \in \mathbb{N}$. (Binomial Theorem)

1.9. Let p be a prime number in $\mathbb{Z}$. For all integers a not divisible by p, show that p divides $a^{p-1} - 1$. (Fermat's Little Theorem)
1.10. Prove that for any prime p we have $(p - 1)! \equiv -1 \bmod p$. (Wilson's Theorem)

1.11.   Prove: if p is a prime, we have $\binom{p-1}{j} \equiv (-1)^j \bmod p$ for $0 \leqslant j \leqslant$ $p-1$, $j \in \mathbb{Z}$.

1.12.   A conjecture of Fermat stated that for all $n \geqslant 0$ the integer $2^{2^n}+1$ is a prime. Euler found to the contrary that 641 divides $2^{32}+1$. Confirm this by using congruences.

1.13.   Prove: if $m_1, \ldots, m_k$ are positive integers that are pairwise relatively prime—that is, $\gcd(m_i, m_j) = 1$ for $1 \leqslant i < j \leqslant k$—then for any integers $a_1, \ldots, a_k$ the system of congruences $y \equiv a_i \bmod m_i$, $i = 1, 2, \ldots, k$, has a simultaneous solution y that is uniquely determined modulo $m = m_1 \cdots m_k$. (Chinese Remainder Theorem)

1.14.   Solve the system of congruences $5x \equiv 20 \bmod 6$, $6x \equiv 6 \bmod 5$, $4x \equiv 5 \bmod 77$.

1.15.   For a commutative ring R of prime characteristic p, show that

$$(a_1 + \cdots + a_s)^{p^n} = a_1^{p^n} + \cdots + a_s^{p^n}$$

for all $a_1, \ldots, a_s \in R$ and $n \in \mathbb{N}$.

1.16.   Deduce from Exercise 1.11 that in a commutative ring R of prime characteristic p we have

$$(a-b)^{p-1} = \sum_{j=0}^{p-1} a^j b^{p-1-j} \quad \text{for all } a, b \in R.$$

1.17.   Let F be a field and $f \in F[x]$. Prove that $\langle g(f(x)) : g \in F[x] \rangle$ is equal to $F[x]$ if and only if $\deg(f) = 1$.

1.18.   Show that $p^2(x) - xq^2(x) = xr^2(x)$ for $p, q, r \in \mathbb{R}[x]$ implies $p = q = r = 0$.

1.19.   Show that if $f, g \in F[x]$, then the principal ideal (f) is contained in the principal ideal (g) if and only if g divides f.

1.20.   Prove: if $f, g \in F[x]$ are relatively prime and not both constant, then there exist $a, b \in F[x]$ such that $af + bg = 1$ and $\deg(a) < \deg(g)$, $\deg(b) < \deg(f)$.

1.21.   Let $f_1, \ldots, f_n \in F[x]$ with $\gcd(f_1, \ldots, f_n) = d$, so that $f_i = dg_i$ with $g_i \in F[x]$ for $1 \leqslant i \leqslant n$. Prove that $g_1, \ldots, g_n$ are relatively prime.

1.22.   Prove that $\gcd(f_1, \ldots, f_n) = \gcd(\gcd(f_1, \ldots, f_{n-1}), f_n)$ for $n \geqslant 3$.

1.23.   Prove: if $f, g, h \in F[x]$, f divides gh, and $\gcd(f, g) = 1$, then f divides h.

1.24.   Use the Euclidean algorithm to compute $\gcd(f, g)$ for the polynomials f and g with coefficients in the indicated field F:
   (a) $F = \mathbb{Q}$, $f(x) = x^7 + 2x^5 + 2x^2 - x + 2$, $g(x) = x^6 - 2x^5 - x^4 + x^2 + 2x + 3$
   (b) $F = \mathbb{F}_2$, $f(x) = x^7 + 1$, $g(x) = x^5 + x^3 + x + 1$
   (c) $F = \mathbb{F}_2$, $f(x) = x^5 + x + 1$, $g(x) = x^6 + x^5 + x^4 + 1$
   (d) $F = \mathbb{F}_3$, $f(x) = x^8 + 2x^5 + x^3 + x^2 + 1$, $g(x) = 2x^6 + x^5 + 2x^3 + 2x^2 + 2$

1.25. Let $f_1,\ldots,f_n$ be nonzero polynomials in $F[x]$. By considering the intersection $(f_1)\cap \cdots \cap(f_n)$ of principal ideals, prove the existence and uniqueness of the monic polynomial $m \in F[x]$ with the properties attributed to the least common multiple of $f_1,\ldots,f_n$.

1.26. Prove (1.6).

1.27. If $f_1,\ldots,f_n \in F[x]$ are nonzero polynomials that are pairwise relatively prime, show that $\mathrm{lcm}(f_1,\ldots,f_n) = a^{-1}f_1 \cdots f_n$, where a is the leading coefficient of $f_1 \cdots f_n$.

1.28. Prove that $\mathrm{lcm}(f_1,\ldots,f_n) = \mathrm{lcm}(\mathrm{lcm}(f_1,\ldots,f_{n-1}), f_n)$ for $n \geqslant 3$.

1.29. Let $f_1,\ldots,f_n \in F[x]$ be nonzero polynomials. Write the canonical factorization of each f_i, $1 \leqslant i \leqslant n$, in the form

$$f_i = a_i \prod_p p^{e_i(p)},$$

where $a_i \in F$, the product is extended over all monic irreducible polynomials p in $F[x]$, the $e_i(p)$ are nonnegative integers, and for each i we have $e_i(p) > 0$ for only finitely many p. For each p set $m(p) = \min(e_1(p),\ldots,e_n(p))$ and $M(p) = \max(e_1(p),\ldots,e_n(p))$. Prove that

$$\gcd(f_1,\ldots,f_n) = \prod_p p^{m(p)},$$

$$\mathrm{lcm}(f_1,\ldots,f_n) = \prod_p p^{M(p)}.$$

1.30. Kronecker's method for finding divisors of degree $\leqslant s$ of a nonconstant polynomial $f \in \mathbb{Q}[x]$ proceeds as follows:
   (1) By multiplying f by a constant, we can assume $f \in \mathbb{Z}[x]$.
   (2) Choose distinct elements $a_0,\ldots,a_s \in \mathbb{Z}$ that are not roots of f and determine all divisors of $f(a_i)$ for each $i, 0 \leqslant i \leqslant s$.
   (3) For each $(s+1)$-tuple $(b_0,\ldots,b_s)$ with b_i dividing $f(a_i)$ for $0 \leqslant i \leqslant s$, determine the polynomial $g \in \mathbb{Q}[x]$ with $\deg(g) \leqslant s$ and $g(a_i) = b_i$ for $0 \leqslant i \leqslant s$ (for instance, by the Lagrange interpolation formula).
   (4) Decide which of these polynomials g in (3) are divisors of f. If $\deg(f) = n \geqslant 1$ and s is taken to be the greatest integer $\leqslant n/2$, then f is irreducible in $\mathbb{Q}[x]$ in case the method only yields constant polynomials as divisors. Otherwise, Kronecker's method yields a nontrivial factorization. By applying the method again to the factors and repeating the process, one eventually gets the canonical factorization of f. Use this procedure to find the canonical factorization of

$$f(x) = \tfrac{1}{3}x^6 - \tfrac{5}{3}x^5 + 2x^4 - x^3 + 5x^2 - \tfrac{17}{3}x - 1 \in \mathbb{Q}[x].$$

1.31. Construct the addition and multiplication table for $\mathbb{F}_2[x]/(x^3 + x^2 + x)$. Determine whether or not this ring is a field.

1.32. Let $[x + 1]$ be the residue class of $x + 1$ in $\mathbb{F}_2[x]/(x^4 + 1)$. Find the residue classes comprising the principal ideal $([x + 1])$ in $\mathbb{F}_2[x]/(x^4 + 1)$.

1.33. Let F be a field and $a, b, g \in F[x]$ with $g \neq 0$. Prove that the congruence $af \equiv b \bmod g$ has a solution $f \in F[x]$ if and only if $\gcd(a, g)$ divides b.

1.34. Solve the congruence $(x^2 + 1)f(x) \equiv 1 \bmod (x^3 + 1)$ in $\mathbb{F}_3[x]$, if possible.

1.35. Solve $(x^4 + x^3 + x^2 + 1)f(x) \equiv (x^2 + 1)\bmod (x^3 + 1)$ in $\mathbb{F}_2[x]$, if possible.

1.36. Prove that $R[x]/(x^4 + x^3 + x + 1)$ cannot be a field, no matter what the commutative ring R with identity is.

1.37. Prove: given a field F, nonzero polynomials $f_1, \ldots, f_k \in F[x]$ that are pairwise relatively prime, and arbitrary polynomials $g_1, \ldots, g_k \in F[x]$, then the simultaneous congruences $h \equiv g_i \bmod f_i, i = 1, 2, \ldots, k$, have a unique solution $h \in F[x]$ modulo $f = f_1 \cdots f_k$. (Chinese Remainder Theorem for $F[x]$)

1.38. Evaluate $f(3)$ for $f(x) = x^{214} + 3x^{152} + 2x^{47} + 2 \in \mathbb{F}_5[x]$.

1.39. Let p be a prime and $a_0, \ldots, a_n$ integers with p not dividing a_n. Show that $a_0 + a_1 y + \cdots + a_n y^n \equiv 0 \bmod p$ has at most n different solutions y modulo p.

1.40. If $p > 2$ is a prime, show that there are exactly two elements $a \in \mathbb{F}_p$ such that $a^2 = 1$.

1.41. Show: if $f \in \mathbb{Z}[x]$ and $f(0) \equiv f(1) \equiv 1 \bmod 2$, then f has no roots in $\mathbb{Z}$.

1.42. Let p be a prime and $f \in \mathbb{Z}[x]$. Show: $f(a) \equiv 0 \bmod p$ holds for all $a \in \mathbb{Z}$ if and only if $f(x) = (x^p - x)g(x) + ph(x)$ with $g, h \in \mathbb{Z}[x]$.

1.43. Let p be a prime integer and c an element of the field F. Show that $x^p - c$ is irreducible over F if and only if $x^p - c$ has no root in F.

1.44. Show that for a polynomial $f \in F[x]$ of positive degree the following conditions are equivalent:
(a) f is irreducible over F;
(b) the principal ideal (f) of $F[x]$ is a maximal ideal;
(c) the principal ideal (f) of $F[x]$ is a prime ideal.

1.45. Show the following properties of the derivative for polynomials in $F[x]$:
(a) $(f_1 + \cdots + f_m)' = f_1' + \cdots + f_m'$;
(b) $(fg)' = f'g + fg'$;
(c) $(f_1 \cdots f_m)' = \sum_{i=1}^{m} f_1 \cdots f_{i-1} f_i' f_{i+1} \cdots f_m$.

1.46. For $f \in F[x]$ and F of characteristic 0, prove that $f' = 0$ if and only if f is a constant polynomial. If F has prime characteristic p, prove that $f' = 0$ if and only if $f(x) = g(x^p)$ for some $g \in F[x]$.

1.47. Prove Theorem 1.68.

1.48. Prove that the nonzero polynomial $f \in F[x]$ has a multiple root (in some extension field of F) if and only if f and f' are not relatively prime.

1.49. Use the criterion in the previous exercise to determine whether the

following polynomials have a multiple root:
(a) $f(x) = x^4 - 5x^3 + 6x^2 + 4x - 8 \in \mathbb{Q}[x]$
(b) $f(x) = x^6 + x^5 + x^4 + x^3 + 1 \in \mathbb{F}_2[x]$

1.50. The nth derivative $f^{(n)}$ of $f \in F[x]$ is defined recursively as follows: $f^{(0)} = f, f^{(n)} = (f^{(n-1)})'$ for $n \geqslant 1$. Prove that for $f, g \in F[x]$ we have

$$(fg)^{(n)} = \sum_{i=0}^{n} \binom{n}{i} f^{(n-i)} g^{(i)}.$$

1.51. Let F be a field and k a positive integer such that $k < p$ in case F has prime characteristic p. Prove: $b \in F$ is a root of $f \in F[x]$ of multiplicity k if and only if $f^{(i)}(b) = 0$ for $0 \leqslant i \leqslant k-1$ and $f^{(k)}(b) \neq 0$.

1.52. Show that the Lagrange interpolation formula can also be written in the form

$$f(x) = \sum_{i=0}^{n} b_i (g'(a_i))^{-1} \frac{g(x)}{x - a_i} \quad \text{with } g(x) = \prod_{k=0}^{n} (x - a_k).$$

1.53. Determine a polynomial $f \in \mathbb{F}_5[x]$ with $f(0) = f(1) = f(4) = 1$ and $f(2) = f(3) = 3$.

1.54. Determine a polynomial $f \in \mathbb{Q}[x]$ of degree $\leqslant 3$ such that $f(-1) = -1, f(0) = 3, f(1) = 3$, and $f(2) = 5$.

1.55. Express $s_5(x_1, x_2, x_3, x_4) = x_1^5 + x_2^5 + x_3^5 + x_4^5 \in \mathbb{F}_3[x_1, x_2, x_3, x_4]$ in terms of the elementary symmetric polynomials $\sigma_1, \sigma_2, \sigma_3, \sigma_4$.

1.56. Prove that a subset K of a field F is a subfield if and only if the following conditions are satisfied:
(a) K contains at least two elements;
(b) if $a, b \in K$, then $a - b \in K$;
(c) if $a, b \in K$ and $b \neq 0$, then $ab^{-1} \in K$.

1.57. Prove that an extension L of the field K is a finite extension if and only if L can be obtained from K by adjoining finitely many algebraic elements over K.

1.58. Prove: if θ is algebraic over L and L is an algebraic extension of K, then θ is algebraic over K. Thus show that if F is an algebraic extension of L, then F is an algebraic extension of K.

1.59. Prove: if the degree $[L : K]$ is a prime, then the only fields F with $K \subseteq F \subseteq L$ are $F = K$ and $F = L$.

1.60. Construct the operation tables for the field $L = \mathbb{F}_3(\theta)$ in Example 1.88.

1.61. Show that $f(x) = x^4 + x + 1 \in \mathbb{F}_2[x]$ is irreducible over $\mathbb{F}_2$. Then construct the operation tables for the simple extension $\mathbb{F}_2(\theta)$, where θ is a root of f.

1.62. Calculate the discriminant $D(f)$ and decide whether or not f has a multiple root:
(a) $f(x) = 2x^3 - 3x^2 + x + 1 \in \mathbb{Q}[x]$

(b) $f(x) = 2x^4 + x^3 + x^2 + 2x + 2 \in \mathbb{F}_3[x]$

1.63. Deduce (1.9) from (1.11).

1.64. Prove that $f, g \in K[x]$ have a common root (in some extension field of K) if and only if f and g have a common divisor in $K[x]$ of positive degree.

1.65. Determine the common roots of the polynomials $x^7 - 2x^4 - x^3 + 2$ and $x^5 - 3x^4 - x + 3$ in $\mathbb{Q}[x]$.

1.66. Prove: if f and g are as in Definition 1.93, then $R(f, g) = (-1)^{mn} R(g, f)$.

1.67. Let $f, g \in K[x]$ be of positive degree and suppose that $f(x) = a_0(x - \alpha_1) \cdots (x - \alpha_n)$, $a_0 \neq 0$, and $g(x) = b_0(x - \beta_1) \cdots (x - \beta_m)$, $b_0 \neq 0$, in the splitting field of fg over K. Prove that

$$R(f, g) = (-1)^{mn} b_0^n \prod_{j=1}^{m} f(\beta_j) = a_0^m b_0^n \prod_{i=1}^{n} \prod_{j=1}^{m} (\alpha_i - \beta_j),$$

where n and m are also taken as the formal degrees of f and g, respectively.

1.68. Calculate the resultant $R(f, g)$ of the two given polynomials f and g (with the formal degree equal to the degree) and decide whether or not f and g have a common root:

(a) $f(x) = x^3 + x + 1$, $g(x) = 2x^5 + x^2 + 2 \in \mathbb{F}_3[x]$

(b) $f(x) = x^4 + x^3 + 1$, $g(x) = x^4 + x^2 + x + 1 \in \mathbb{F}_2[x]$

1.69. For $f \in K[x_1, \ldots, x_n]$, $n \geq 2$, an n-tuple $(\alpha_1, \ldots, \alpha_n)$ of elements α_i belonging to some extension L of K may be called a *zero* of f if $f(\alpha_1, \ldots, \alpha_n) = 0$. Now let $f, g \in K[x_1, \ldots, x_n]$ with x_n actually appearing in f and g. Then f and g can be regarded as polynomials $\bar{f}(x_n)$ and $\bar{g}(x_n)$ in $K[x_1, \ldots, x_{n-1}][x_n]$ of positive degree. Their resultant with respect to x_n (with formal degree = degree) is $R(\bar{f}, \bar{g}) = R_{x_n}(f, g)$, which is a polynomial in $x_1, \ldots, x_{n-1}$. Show that f and g have a common zero $(\alpha_1, \ldots, \alpha_{n-1}, \alpha_n)$ if and only if $(\alpha_1, \ldots, \alpha_{n-1})$ is a zero of $R(\bar{f}, \bar{g})$.

1.70. Using the result of the previous exercise, determine the common zeros of the polynomials $f(x, y) = x(y^2 - x)^2 + y^5$ and $g(x, y) = y^4 + y^3 - x^2$ in $\mathbb{Q}[x, y]$.

Chapter 2

Structure of Finite Fields

This chapter is of central importance since it contains various fundamental properties of finite fields and a description of methods for constructing finite fields.

The field of integers modulo a prime number is, of course, the most familiar example of a finite field, but many of its properties extend to arbitrary finite fields. The characterization of finite fields (see Section 1) shows that every finite field is of prime-power order and that, conversely, for every prime power there exists a finite field whose number of elements is exactly that prime power. Furthermore, finite fields with the same number of elements are isomorphic and may therefore be identified. The next two sections provide information on roots of irreducible polynomials, leading to an interpretation of finite fields as splitting fields of irreducible polynomials, and on traces, norms, and bases relative to field extensions.

Section 4 treats roots of unity from the viewpoint of general field theory, which will be needed occasionally in Section 6 as well as in Chapter 5. Section 5 presents different ways of representing the elements of a finite field. In Section 6 we give two proofs of the famous theorem of Wedderburn according to which every finite division ring is a field.

Many discussions in this chapter will be followed up, continued, and partly generalized in later chapters.

1. CHARACTERIZATION OF FINITE FIELDS

In the previous chapter we have already encountered a basic class of finite fields—that is, of fields with finitely many elements. For every prime p the residue class ring $\mathbb{Z}/(p)$ forms a finite field with p elements (see Theorem 1.38), which may be identified with the Galois field $\mathbb{F}_p$ of order p (see Definition 1.41). The fields $\mathbb{F}_p$ play an important role in general field theory since every field of characteristic p must contain an isomorphic copy of $\mathbb{F}_p$ by Theorem 1.78 and can thus be thought of as an extension of $\mathbb{F}_p$. This observation, together with the fact that every finite field has prime characteristic (see Corollary 1.45), is fundamental for the classification of finite fields. We first establish a simple necessary condition on the number of elements of a finite field.

2.1. Lemma. *Let F be a finite field containing a subfield K with q elements. Then F has q^m elements, where $m = [F:K]$.*

Proof. F is a vector space over K, and since F is finite, it is finite-dimensional as a vector space over K. If $[F:K] = m$, then F has a basis over K consisting of m elements, say $b_1, b_2, \ldots, b_m$. Thus every element of F can be uniquely represented in the form $a_1b_1 + a_2b_2 + \cdots + a_mb_m$, where $a_1, a_2, \ldots, a_m \in K$. Since each a_i can have q values, F has exactly q^m elements.   □

2.2. Theorem. *Let F be a finite field. Then F has p^n elements, where the prime p is the characteristic of F and n is the degree of F over its prime subfield.*

Proof. Since F is finite, its characteristic is a prime p according to Corollary 1.45. Therefore the prime subfield K of F is isomorphic to $\mathbb{F}_p$ by Theorem 1.78 and thus contains p elements. The rest follows from Lemma 2.1.   □

Starting from the prime fields $\mathbb{F}_p$, we can construct other finite fields by the process of root adjunction described in Chapter 1, Section 4. If $f \in \mathbb{F}_p[x]$ is an irreducible polynomial over $\mathbb{F}_p$ of degree n, then by adjoining a root of f to $\mathbb{F}_p$ we get a finite field with p^n elements. However, at this stage it is not clear whether for every positive integer n there exists an irreducible polynomial in $\mathbb{F}_p[x]$ of degree n. In order to establish that for every prime p and every $n \in \mathbb{N}$ there is a finite field with p^n elements, we use an approach suggested by the following results.

2.3. Lemma. *If F is a finite field with q elements, then every $a \in F$ satisfies $a^q = a$.*

Proof. The identity $a^q = a$ is trivial for $a = 0$. On the other hand, the nonzero elements of F form a group of order $q - 1$ under multiplication.

Thus $a^{q-1} = 1$ for all $a \in F$ with $a \neq 0$, and multiplication by a yields the desired result.　　　　　　　　　　　　　　　　　　　　　　　　　　□

2.4.　Lemma. *If F is a finite field with q elements and K is a subfield of F, then the polynomial $x^q - x$ in $K[x]$ factors in $F[x]$ as*

$$x^q - x = \prod_{a \in F} (x - a)$$

and F is a splitting field of $x^q - x$ over K.

Proof. The polynomial $x^q - x$ of degree q has at most q roots in F. By Lemma 2.3 we know q such roots—namely, all the elements of F. Thus the given polynomial splits in F in the indicated manner, and it cannot split in any smaller field.　　　　　　　　　　　　　　　　　　　　　　　□

We are now able to prove the main characterization theorem for finite fields, the leading idea being contained in Lemma 2.4.

2.5.　Theorem (Existence and Uniqueness of Finite Fields).　*For every prime p and every positive integer n there exists a finite field with p^n elements. Any finite field with $q = p^n$ elements is isomorphic to the splitting field of $x^q - x$ over $\mathbb{F}_p$.*

Proof. (*Existence*) For $q = p^n$ consider $x^q - x$ in $\mathbb{F}_p[x]$, and let F be its splitting field over $\mathbb{F}_p$. This polynomial has q distinct roots in F since its derivative is $qx^{q-1} - 1 = -1$ in $\mathbb{F}_p[x]$ and so can have no common root with $x^q - x$ (compare with Theorem 1.68). Let $S = \{a \in F : a^q - a = 0\}$. Then S is a subfield of F since: (i) S contains 0 and 1; (ii) $a, b \in S$ implies by Theorem 1.46 that $(a - b)^q = a^q - b^q = a - b$, and so $a - b \in S$; (iii) for $a, b \in S$ and $b \neq 0$ we have $(ab^{-1})^q = a^q b^{-q} = ab^{-1}$, and so $ab^{-1} \in S$. But, on the other hand, $x^q - x$ must split in S since S contains all its roots. Thus $F = S$, and since S has q elements, F is a finite field with q elements.

(*Uniqueness*) Let F be a finite field with $q = p^n$ elements. Then F has characteristic p by Theorem 2.2 and so contains $\mathbb{F}_p$ as a subfield. It follows from Lemma 2.4 that F is a splitting field of $x^q - x$ over $\mathbb{F}_p$. Thus the desired result is a consequence of the uniqueness (up to isomorphisms) of splitting fields, which was noted in Theorem 1.91.　　　　　□

The uniqueness part of Theorem 2.5 provides the justification for speaking of *the* finite field (or *the* Galois field) with q elements, or of *the* finite field (or *the* Galois field) of order q. We shall denote this field by $\mathbb{F}_q$, where it is of course understood that q is a power of the prime characteristic p of $\mathbb{F}_q$. The notation $GF(q)$ is also used by many authors.

2.6.　Theorem (Subfield Criterion).　*Let $\mathbb{F}_q$ be the finite field with $q = p^n$ elements. Then every subfield of $\mathbb{F}_q$ has order p^m, where m is a positive divisor of n. Conversely, if m is a positive divisor of n, then there is exactly one subfield of $\mathbb{F}_q$ with p^m elements.*

Proof. It is clear that a subfield K of $\mathbb{F}_q$ has order p^m for some positive integer $m \leqslant n$. Lemma 2.1 shows that $q = p^n$ must be a power of p^m, and so m is necessarily a divisor of n.

Conversely, if m is a positive divisor of n, then $p^m - 1$ divides $p^n - 1$, and so $x^{p^m-1} - 1$ divides $x^{p^n-1} - 1$ in $\mathbb{F}_p[x]$. Consequently, $x^{p^m} - x$ divides $x^{p^n} - x = x^q - x$ in $\mathbb{F}_p[x]$. Thus, every root of $x^{p^m} - x$ is a root of $x^q - x$ and so belongs to $\mathbb{F}_q$. It follows that $\mathbb{F}_q$ must contain as a subfield a splitting field of $x^{p^m} - x$ over $\mathbb{F}_p$, and as we have seen in the proof of Theorem 2.5, such a splitting field has order p^m. If there were two distinct subfields of order p^m in $\mathbb{F}_q$, they would together contain more than p^m roots of $x^{p^m} - x$ in $\mathbb{F}_q$, an obvious contradiction. $\square$

The proof of Theorem 2.6 shows that the unique subfield of $\mathbb{F}_{p^n}$ of order p^m, where m is a positive divisor of n, consists precisely of the roots of the polynomial $x^{p^m} - x \in \mathbb{F}_p[x]$ in $\mathbb{F}_{p^n}$.

2.7.   Example.   The subfields of the finite field $\mathbb{F}_{2^{30}}$ can be determined by listing all positive divisors of 30. The containment relations between these various subfields are displayed in the following diagram.

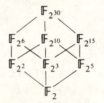

By Theorem 2.6, the containment relations are equivalent to divisibility relations among the positive divisors of 30. $\square$

For a finite field $\mathbb{F}_q$ we denote by $\mathbb{F}_q^*$ the multiplicative group of nonzero elements of $\mathbb{F}_q$. The following result enunciates a useful property of this group.

2.8.   Theorem.   *For every finite field $\mathbb{F}_q$ the multiplicative group $\mathbb{F}_q^*$ of nonzero elements of $\mathbb{F}_q$ is cyclic.*

Proof. We may assume $q \geqslant 3$. Let $h = p_1^{r_1} p_2^{r_2} \cdots p_m^{r_m}$ be the prime factor decomposition of the order $h = q - 1$ of the group $\mathbb{F}_q^*$. For every i, $1 \leqslant i \leqslant m$, the polynomial $x^{h/p_i} - 1$ has at most h/p_i roots in $\mathbb{F}_q$. Since $h/p_i < h$, it follows that there are nonzero elements in $\mathbb{F}_q$ that are not roots of this polynomial. Let a_i be such an element and set $b_i = a_i^{h/p_i^{r_i}}$. We have $b_i^{p_i^{r_i}} = 1$, hence the order of b_i is a divisor of $p_i^{r_i}$ and is therefore of the form $p_i^{s_i}$ with $0 \leqslant s_i \leqslant r_i$. On the other hand,

$$b_i^{p_i^{r_i-1}} = a_i^{h/p_i} \neq 1,$$

and so the order of b_i is $p_i^{r_i}$. We claim that the element $b = b_1 b_2 \cdots b_m$ has order h. Suppose, on the contrary, that the order of b is a proper divisor of h

and is therefore a divisor of at least one of the m integers h/p_i, $1 \le i \le m$, say of h/p_1. Then we have

$$1 = b^{h/p_1} = b_1^{h/p_1} b_2^{h/p_1} \cdots b_m^{h/p_1}.$$

Now if $2 \le i \le m$, then $p_i^{r_i}$ divides h/p_1, and hence $b_i^{h/p_1} = 1$. Therefore $b_1^{h/p_1} = 1$. This implies that the order of b_1 must divide h/p_1, which is impossible since the order of b_1 is $p_1^{r_1}$. Thus, $\mathbb{F}_q^*$ is a cyclic group with generator b. $\square$

2.9.   Definition.   A generator of the cyclic group $\mathbb{F}_q^*$ is called a *primitive element* of $\mathbb{F}_q$.

It follows from Theorem 1.15(v) that $\mathbb{F}_q$ contains $\phi(q-1)$ primitive elements, where ϕ is Euler's function. The existence of primitive elements can be used to show a result that implies, in particular, that every finite field can be thought of as a simple algebraic extension of its prime subfield.

2.10.   Theorem.   *Let $\mathbb{F}_q$ be a finite field and $\mathbb{F}_r$ a finite extension field. Then $\mathbb{F}_r$ is a simple algebraic extension of $\mathbb{F}_q$ and every primitive element of $\mathbb{F}_r$ can serve as a defining element of $\mathbb{F}_r$ over $\mathbb{F}_q$.*

Proof.   Let ζ be a primitive element of $\mathbb{F}_r$. We clearly have $\mathbb{F}_q(\zeta) \subseteq \mathbb{F}_r$. On the other hand, $\mathbb{F}_q(\zeta)$ contains 0 and all powers of ζ, and so all elements of $\mathbb{F}_r$. Therefore $\mathbb{F}_r = \mathbb{F}_q(\zeta)$. $\square$

2.11.   Corollary.   *For every finite field $\mathbb{F}_q$ and every positive integer n there exists an irreducible polynomial in $\mathbb{F}_q[x]$ of degree n.*

Proof.   Let $\mathbb{F}_r$ be the extension field of $\mathbb{F}_q$ of order q^n, so that $[\mathbb{F}_r : \mathbb{F}_q] = n$. By Theorem 2.10 we have $\mathbb{F}_r = \mathbb{F}_q(\zeta)$ for some $\zeta \in \mathbb{F}_r$. Then the minimal polynomial of ζ over $\mathbb{F}_q$ is an irreducible polynomial in $\mathbb{F}_q[x]$ of degree n, according to Theorems 1.82(i) and 1.86(ii). $\square$

2.   ROOTS OF IRREDUCIBLE POLYNOMIALS

In this section we collect some information about the set of roots of an irreducible polynomial over a finite field.

2.12.   Lemma.   *Let $f \in \mathbb{F}_q[x]$ be an irreducible polynomial over a finite field $\mathbb{F}_q$ and let α be a root of f in an extension field of $\mathbb{F}_q$. Then for a polynomial $h \in \mathbb{F}_q[x]$ we have $h(\alpha) = 0$ if and only if f divides h.*

Proof.   Let a be the leading coefficient of f and set $g(x) = a^{-1}f(x)$. Then g is a monic irreducible polynomial in $\mathbb{F}_q[x]$ with $g(\alpha) = 0$ and so it is the minimal polynomial of α over $\mathbb{F}_q$ in the sense of Definition 1.81. The rest follows from Theorem 1.82(ii). $\square$

2.13.   Lemma.   Let $f \in \mathbb{F}_q[x]$ be an irreducible polynomial over $\mathbb{F}_q$ of degree m. Then $f(x)$ divides $x^{q^n} - x$ if and only if m divides n.

Proof.   Suppose $f(x)$ divides $x^{q^n} - x$. Let α be a root of f in the splitting field of f over $\mathbb{F}_q$. Then $\alpha^{q^n} = \alpha$, so that $\alpha \in \mathbb{F}_{q^n}$. It follows that $\mathbb{F}_q(\alpha)$ is a subfield of $\mathbb{F}_{q^n}$. But since $[\mathbb{F}_q(\alpha):\mathbb{F}_q] = m$ and $[\mathbb{F}_{q^n}:\mathbb{F}_q] = n$, Theorem 1.84 shows that m divides n.

Conversely, if m divides n, then Theorem 2.6 implies that $\mathbb{F}_{q^n}$ contains $\mathbb{F}_{q^m}$ as a subfield. If α is a root of f in the splitting field of f over $\mathbb{F}_q$, then $[\mathbb{F}_q(\alpha):\mathbb{F}_q] = m$, and so $\mathbb{F}_q(\alpha) = \mathbb{F}_{q^m}$. Consequently, we have $\alpha \in \mathbb{F}_{q^n}$, hence $\alpha^{q^n} = \alpha$, and thus α is a root of $x^{q^n} - x \in \mathbb{F}_q[x]$. We infer then from Lemma 2.12 that $f(x)$ divides $x^{q^n} - x$. $\square$

2.14.   Theorem.   If f is an irreducible polynomial in $\mathbb{F}_q[x]$ of degree m, then f has a root α in $\mathbb{F}_{q^m}$. Furthermore, all the roots of f are simple and are given by the m distinct elements $\alpha, \alpha^q, \alpha^{q^2}, \ldots, \alpha^{q^{m-1}}$ of $\mathbb{F}_{q^m}$.

Proof.   Let α be a root of f in the splitting field of f over $\mathbb{F}_q$. Then $[\mathbb{F}_q(\alpha):\mathbb{F}_q] = m$, hence $\mathbb{F}_q(\alpha) = \mathbb{F}_{q^m}$, and in particular $\alpha \in \mathbb{F}_{q^m}$. Next we show that if $\beta \in \mathbb{F}_{q^m}$ is a root of f, then β^q is also a root of f. Write $f(x) = a_m x^m + \cdots + a_1 x + a_0$ with $a_i \in \mathbb{F}_q$ for $0 \leqslant i \leqslant m$. Then, using Lemma 2.3 and Theorem 1.46, we get

$$f(\beta^q) = a_m \beta^{qm} + \cdots + a_1 \beta^q + a_0 = a_m^q \beta^{qm} + \cdots + a_1^q \beta^q + a_0^q$$

$$= (a_m \beta^m + \cdots + a_1 \beta + a_0)^q = f(\beta)^q = 0.$$

Therefore, the elements $\alpha, \alpha^q, \alpha^{q^2}, \ldots, \alpha^{q^{m-1}}$ are roots of f. It remains to prove that these elements are distinct. Suppose, on the contrary, that $\alpha^{q^j} = \alpha^{q^k}$ for some integers j and k with $0 \leqslant j < k \leqslant m - 1$. By raising this identity to the power q^{m-k}, we get

$$\alpha^{q^{m-k+j}} = \alpha^{q^m} = \alpha.$$

It follows then from Lemma 2.12 that $f(x)$ divides $x^{q^{m-k+j}} - x$. By Lemma 2.13, this is only possible if m divides $m - k + j$. But we have $0 < m - k + j < m$, and so we arrive at a contradiction. $\square$

2.15.   Corollary.   Let f be an irreducible polynomial in $\mathbb{F}_q[x]$ of degree m. Then the splitting field of f over $\mathbb{F}_q$ is given by $\mathbb{F}_{q^m}$.

Proof.   Theorem 2.14 shows that f splits in $\mathbb{F}_{q^m}$. Furthermore, $\mathbb{F}_q(\alpha, \alpha^q, \alpha^{q^2}, \ldots, \alpha^{q^{m-1}}) = \mathbb{F}_q(\alpha) = \mathbb{F}_{q^m}$ for a root α of f in $\mathbb{F}_{q^m}$, where the second identity is taken from the proof of Theorem 2.14. $\square$

2.16.   Corollary.   Any two irreducible polynomials in $\mathbb{F}_q[x]$ of the same degree have isomorphic splitting fields.

We introduce a convenient terminology for the elements appearing in Theorem 2.14, regardless of whether $\alpha \in \mathbb{F}_{q^m}$ is a root of an irreducible polynomial in $\mathbb{F}_q[x]$ of degree m or not.

2.17. Definition. Let $\mathbb{F}_{q^m}$ be an extension of $\mathbb{F}_q$ and let $\alpha \in \mathbb{F}_{q^m}$. Then the elements $\alpha, \alpha^q, \alpha^{q^2}, \ldots, \alpha^{q^{m-1}}$ are called the *conjugates* of α with respect to $\mathbb{F}_q$.

The conjugates of $\alpha \in \mathbb{F}_{q^m}$ with respect to $\mathbb{F}_q$ are distinct if and only if the minimal polynomial of α over $\mathbb{F}_q$ has degree m. Otherwise, the degree d of this minimal polynomial is a proper divisor of m, and then the conjugates of α with respect to $\mathbb{F}_q$ are the distinct elements $\alpha, \alpha^q, \ldots, \alpha^{q^{d-1}}$, each repeated m/d times.

2.18. Theorem. *The conjugates of $\alpha \in \mathbb{F}_q^*$ with respect to any subfield of $\mathbb{F}_q$ have the same order in the group $\mathbb{F}_q^*$.*

Proof. Since $\mathbb{F}_q^*$ is a cyclic group by Theorem 2.8, the result follows from Theorem 1.15(ii) and the fact that every power of the characteristic of $\mathbb{F}_q$ is relatively prime to the order $q-1$ of $\mathbb{F}_q^*$.   □

2.19. Corollary. *If α is a primitive element of $\mathbb{F}_q$, then so are all its conjugates with respect to any subfield of $\mathbb{F}_q$.*

2.20. Example. Let $\alpha \in \mathbb{F}_{16}$ be a root of $f(x) = x^4 + x + 1 \in \mathbb{F}_2[x]$. Then the conjugates of α with respect to $\mathbb{F}_2$ are $\alpha, \alpha^2, \alpha^4 = \alpha + 1$, and $\alpha^8 = \alpha^2 + 1$, each of them being a primitive element of $\mathbb{F}_{16}$. The conjugates of α with respect to $\mathbb{F}_4$ are α and $\alpha^4 = \alpha + 1$.   □

There is an intimate relationship between conjugate elements and certain automorphisms of a finite field. Let $\mathbb{F}_{q^m}$ be an extension of $\mathbb{F}_q$. By an *automorphism* σ of $\mathbb{F}_{q^m}$ over $\mathbb{F}_q$ we mean an automorphism of $\mathbb{F}_{q^m}$ that fixes the elements of $\mathbb{F}_q$. Thus, in detail, we require that σ be a one-to-one mapping from $\mathbb{F}_{q^m}$ onto itself with $\sigma(\alpha + \beta) = \sigma(\alpha) + \sigma(\beta)$ and $\sigma(\alpha\beta) = \sigma(\alpha)\sigma(\beta)$ for all $\alpha, \beta \in \mathbb{F}_{q^m}$ and $\sigma(a) = a$ for all $a \in \mathbb{F}_q$.

2.21. Theorem. *The distinct automorphisms of $\mathbb{F}_{q^m}$ over $\mathbb{F}_q$ are exactly the mappings $\sigma_0, \sigma_1, \ldots, \sigma_{m-1}$, defined by $\sigma_j(\alpha) = \alpha^{q^j}$ for $\alpha \in \mathbb{F}_{q^m}$ and $0 \leqslant j \leqslant m-1$.*

Proof. For each σ_j and all $\alpha, \beta \in \mathbb{F}_{q^m}$ we obviously have $\sigma_j(\alpha\beta) = \sigma_j(\alpha)\sigma_j(\beta)$, and also $\sigma_j(\alpha + \beta) = \sigma_j(\alpha) + \sigma_j(\beta)$ because of Theorem 1.46, so that σ_j is an endomorphism of $\mathbb{F}_{q^m}$. Furthermore, $\sigma_j(\alpha) = 0$ if and only if $\alpha = 0$, and so σ_j is one-to-one. Since $\mathbb{F}_{q^m}$ is a finite set, σ_j is an epimorphism and therefore an automorphism of $\mathbb{F}_{q^m}$. Moreover, we have $\sigma_j(a) = a$ for all $a \in \mathbb{F}_q$ by Lemma 2.3, and so each σ_j is an automorphism of $\mathbb{F}_{q^m}$ over $\mathbb{F}_q$.

The mappings $\sigma_0, \sigma_1, \ldots, \sigma_{m-1}$ are distinct since they attain distinct values for a primitive element of $\mathbb{F}_{q^m}$.

Now suppose that σ is an arbitrary automorphism of $\mathbb{F}_{q^m}$ over $\mathbb{F}_q$. Let β be a primitive element of $\mathbb{F}_{q^m}$ and let $f(x) = x^m + a_{m-1}x^{m-1} + \cdots + a_0 \in \mathbb{F}_q[x]$ be its minimal polynomial over $\mathbb{F}_q$. Then

$$0 = \sigma\left(\beta^m + a_{m-1}\beta^{m-1} + \cdots + a_0\right)$$
$$= \sigma(\beta)^m + a_{m-1}\sigma(\beta)^{m-1} + \cdots + a_0,$$

so that $\sigma(\beta)$ is a root of f in $\mathbb{F}_{q^m}$. It follows from Theorem 2.14 that $\sigma(\beta) = \beta^{q^j}$ for some j, $0 \leqslant j \leqslant m-1$. Since σ is a homomorphism, we get then $\sigma(\alpha) = \alpha^{q^j}$ for all $\alpha \in \mathbb{F}_{q^m}$. $\qquad\square$

On the basis of Theorem 2.21 it is evident that the conjugates of $\alpha \in \mathbb{F}_{q^m}$ with respect to $\mathbb{F}_q$ are obtained by applying all automorphisms of $\mathbb{F}_{q^m}$ over $\mathbb{F}_q$ to the element α. The automorphisms of $\mathbb{F}_{q^m}$ over $\mathbb{F}_q$ form a group with the operation being the usual composition of mappings. The information provided in Theorem 2.21 shows that this group of automorphisms of $\mathbb{F}_{q^m}$ over $\mathbb{F}_q$ is a cyclic group of order m generated by σ_1.

3. TRACES, NORMS, AND BASES

In this section we adopt again the viewpoint of regarding a finite extension $F = \mathbb{F}_{q^m}$ of the finite field $K = \mathbb{F}_q$ as a vector space over K (compare with Chapter 1, Section 4). Then F has dimension m over K, and if $\{\alpha_1, \ldots, \alpha_m\}$ is a basis of F over K, each element $\alpha \in F$ can be uniquely represented in the form

$$\alpha = c_1\alpha_1 + \cdots + c_m\alpha_m \quad \text{with } c_j \in K \quad \text{for } 1 \leqslant j \leqslant m.$$

We introduce an important mapping from F to K which will turn out to be linear.

2.22. Definition. For $\alpha \in F = \mathbb{F}_{q^m}$ and $K = \mathbb{F}_q$, the *trace* $\mathrm{Tr}_{F/K}(\alpha)$ of α over K is defined by

$$\mathrm{Tr}_{F/K}(\alpha) = \alpha + \alpha^q + \cdots + \alpha^{q^{m-1}}.$$

If K is the prime subfield of F, then $\mathrm{Tr}_{F/K}(\alpha)$ is called the *absolute trace* of α and simply denoted by $\mathrm{Tr}_F(\alpha)$.

In other words, the trace of α over K is the sum of the conjugates of α with respect to K. Still another description of the trace may be obtained as follows. Let $f \in K[x]$ be the minimal polynomial of α over K; its degree d is a divisor of m. Then $g(x) = f(x)^{m/d} \in K[x]$ is called the *characteristic polynomial* of α over K. By Theorem 2.14, the roots of f in F are given by

$\alpha, \alpha^q, \ldots, \alpha^{q^{d-1}}$, and then a remark following Definition 2.17 implies that the roots of g in F are precisely the conjugates of α with respect to K. Hence

$$g(x) = x^m + a_{m-1}x^{m-1} + \cdots + a_0$$
$$= (x - \alpha)(x - \alpha^q)\ldots(x - \alpha^{q^{m-1}}), \qquad (2.1)$$

and a comparison of coefficients shows that

$$\mathrm{Tr}_{F/K}(\alpha) = -a_{m-1}. \qquad (2.2)$$

In particular, $\mathrm{Tr}_{F/K}(\alpha)$ is always an element of K.

2.23.   Theorem.   *Let $K = \mathbb{F}_q$ and $F = \mathbb{F}_{q^m}$. Then the trace function $\mathrm{Tr}_{F/K}$ satisfies the following properties:*

   (i) $\mathrm{Tr}_{F/K}(\alpha + \beta) = \mathrm{Tr}_{F/K}(\alpha) + \mathrm{Tr}_{F/K}(\beta)$ *for all $\alpha, \beta \in F$;*
   (ii) $\mathrm{Tr}_{F/K}(c\alpha) = c\,\mathrm{Tr}_{F/K}(\alpha)$ *for all $c \in K$, $\alpha \in F$;*
   (iii) $\mathrm{Tr}_{F/K}$ *is a linear transformation from F onto K, where both F and K are viewed as vector spaces over K;*
   (iv) $\mathrm{Tr}_{F/K}(a) = ma$ *for all $a \in K$;*
   (v) $\mathrm{Tr}_{F/K}(\alpha^q) = \mathrm{Tr}_{F/K}(\alpha)$ *for all $\alpha \in F$.*

Proof.

   (i)   For $\alpha, \beta \in F$ we use Theorem 1.46 to get

$$\mathrm{Tr}_{F/K}(\alpha + \beta) = \alpha + \beta + (\alpha + \beta)^q + \cdots + (\alpha + \beta)^{q^{m-1}}$$
$$= \alpha + \beta + \alpha^q + \beta^q + \cdots + \alpha^{q^{m-1}} + \beta^{q^{m-1}}$$
$$= \mathrm{Tr}_{F/K}(\alpha) + \mathrm{Tr}_{F/K}(\beta).$$

   (ii)   For $c \in K$ we have $c^{q^j} = c$ for all $j \geq 0$ by Lemma 2.3. Therefore we obtain for $\alpha \in F$,

$$\mathrm{Tr}_{F/K}(c\alpha) = c\alpha + c^q\alpha^q + \cdots + c^{q^{m-1}}\alpha^{q^{m-1}}$$
$$= c\alpha + c\alpha^q + \cdots + c\alpha^{q^{m-1}}$$
$$= c\,\mathrm{Tr}_{F/K}(\alpha).$$

   (iii)   The properties (i) and (ii), together with the fact that $\mathrm{Tr}_{F/K}(\alpha) \in K$ for all $\alpha \in F$, show that $\mathrm{Tr}_{F/K}$ is a linear transformation from F into K. To prove that this mapping is onto, it suffices then to show the existence of an $\alpha \in F$ with $\mathrm{Tr}_{F/K}(\alpha) \neq 0$. Now $\mathrm{Tr}_{F/K}(\alpha) = 0$ if and only if α is a root of the polynomial $x^{q^{m-1}} + \cdots + x^q + x \in K[x]$ in F. But since this polynomial can have at most q^{m-1} roots in F and F has q^m elements, we are done.

   (iv)   This follows immediately from the definition of the trace function and Lemma 2.3.

   (v)   For $\alpha \in F$ we have $\alpha^{q^m} = \alpha$ by Lemma 2.3, and so $\mathrm{Tr}_{F/K}(\alpha^q) = \alpha^q + \alpha^{q^2} + \cdots + \alpha^{q^m} = \mathrm{Tr}_{F/K}(\alpha)$. $\square$

The trace function $\mathrm{Tr}_{F/K}$ is not only in itself a linear transformation from F onto K, but serves for a description of all linear transformations from F into K (or, in an equivalent terminology, of all linear functionals on F) that has the advantage of being independent of a chosen basis.

2.24.   Theorem.   *Let F be a finite extension of the finite field K, both considered as vector spaces over K. Then the linear transformations from F into K are exactly the mappings $L_\beta, \beta \in F$, where $L_\beta(\alpha) = \mathrm{Tr}_{F/K}(\beta\alpha)$ for all $\alpha \in F$. Furthermore, we have $L_\beta \neq L_\gamma$ whenever β and γ are distinct elements of F.*

Proof.   Each mapping L_β is a linear transformation from F into K by Theorem 2.23(iii). For $\beta, \gamma \in F$ with $\beta \neq \gamma$, we have $L_\beta(\alpha) - L_\gamma(\alpha) = \mathrm{Tr}_{F/K}(\beta\alpha) - \mathrm{Tr}_{F/K}(\gamma\alpha) = \mathrm{Tr}_{F/K}((\beta - \gamma)\alpha) \neq 0$ for suitable $\alpha \in F$ since $\mathrm{Tr}_{F/K}$ maps F onto K, and so the mappings L_β and L_γ are different. If $K = \mathbb{F}_q$ and $F = \mathbb{F}_{q^m}$, then the mappings L_β yield q^m different linear transformations from F into K. On the other hand, every linear transformation from F into K can be obtained by assigning arbitrary elements of K to the m elements of a given basis of F over K. Since this can be done in q^m different ways, the mappings L_β already exhaust all possible linear transformations from F into K.   □

2.25.   Theorem.   *Let F be a finite extension of $K = \mathbb{F}_q$. Then for $\alpha \in F$ we have $\mathrm{Tr}_{F/K}(\alpha) = 0$ if and only if $\alpha = \beta^q - \beta$ for some $\beta \in F$.*

Proof.   The sufficiency of the condition is obvious by Theorem 2.23(v). To prove the necessity, suppose $\alpha \in F = \mathbb{F}_{q^m}$ with $\mathrm{Tr}_{F/K}(\alpha) = 0$ and let β be a root of $x^q - x - \alpha$ in some extension field of F. Then $\beta^q - \beta = \alpha$ and

$$0 = \mathrm{Tr}_{F/K}(\alpha) = \alpha + \alpha^q + \cdots + \alpha^{q^{m-1}}$$

$$= (\beta^q - \beta) + (\beta^q - \beta)^q + \cdots + (\beta^q - \beta)^{q^{m-1}}$$

$$= (\beta^q - \beta) + (\beta^{q^2} - \beta^q) + \cdots + (\beta^{q^m} - \beta^{q^{m-1}})$$

$$= \beta^{q^m} - \beta,$$

so that $\beta \in F$.   □

In case a chain of extension fields is considered, the composition of trace functions proceeds according to a very simple rule.

2.26.   Theorem (Transitivity of Trace).   *Let K be a finite field, let F be a finite extension of K and E a finite extension of F. Then*

$$\mathrm{Tr}_{E/K}(\alpha) = \mathrm{Tr}_{F/K}\left(\mathrm{Tr}_{E/F}(\alpha)\right) \quad \textit{for all } \alpha \in E.$$

Proof. Let $K = \mathbb{F}_q$, let $[F:K] = m$ and $[E:F] = n$, so that $[E:K] = mn$ by Theorem 1.84. Then for $\alpha \in E$ we have

$$\text{Tr}_{F/K}\left(\text{Tr}_{E/F}(\alpha)\right) = \sum_{i=0}^{m-1} \text{Tr}_{E/F}(\alpha)^{q^i} = \sum_{i=0}^{m-1} \left(\sum_{j=0}^{n-1} \alpha^{q^{jm}}\right)^{q^i}$$

$$= \sum_{i=0}^{m-1}\sum_{j=0}^{n-1} \alpha^{q^{jm+i}} = \sum_{k=0}^{mn-1} \alpha^{q^k} = \text{Tr}_{E/K}(\alpha). \qquad \square$$

Another interesting function from a finite field to a subfield is obtained by forming the product of the conjugates of an element of the field with respect to the subfield.

2.27.   Definition.   For $\alpha \in F = \mathbb{F}_{q^m}$ and $K = \mathbb{F}_q$, the *norm* $N_{F/K}(\alpha)$ of α over K is defined by

$$N_{F/K}(\alpha) = \alpha \cdot \alpha^q \cdot \cdots \cdot \alpha^{q^{m-1}} = \alpha^{(q^m-1)/(q-1)}.$$

By comparing the constant terms in (2.1), we see that $N_{F/K}(\alpha)$ can be read off from the characteristic polynomial g of α over K—namely,

$$N_{F/K}(\alpha) = (-1)^m a_0. \qquad (2.3)$$

It follows, in particular, that $N_{F/K}(\alpha)$ is always an element of K.

2.28.   Theorem.   *Let* $K = \mathbb{F}_q$ *and* $F = \mathbb{F}_{q^m}$. *Then the norm function* $N_{F/K}$ *satisfies the following properties*:

    (i) $N_{F/K}(\alpha\beta) = N_{F/K}(\alpha)N_{F/K}(\beta)$ *for all* $\alpha, \beta \in F$;
   (ii) $N_{F/K}$ *maps* F *onto* K *and* F^* *onto* K^*;
  (iii) $N_{F/K}(a) = a^m$ *for all* $a \in K$;
  (iv) $N_{F/K}(\alpha^q) = N_{F/K}(\alpha)$ *for all* $\alpha \in F$.

Proof.   (i) follows immediately from the definition of the norm. We have already noted that $N_{F/K}$ maps F into K. Since $N_{F/K}(\alpha) = 0$ if and only if $\alpha = 0$, $N_{F/K}$ maps F^* into K^*. Property (i) shows that $N_{F/K}$ is a group homomorphism between these multiplicative groups. Since the elements of the kernel of $N_{F/K}$ are exactly the roots of the polynomial $x^{(q^m-1)/(q-1)} - 1$ $\in K[x]$ in F, the order d of the kernel satisfies $d \leqslant (q^m-1)/(q-1)$. By Theorem 1.23, the image of $N_{F/K}$ has order $(q^m-1)/d$, which is $\geqslant q-1$. Therefore, $N_{F/K}$ maps F^* onto K^* and so F onto K. Property (iii) follows from the definition of the norm and the fact that for $a \in K$ the conjugates of a with respect to K are all equal to a. Finally, we have $N_{F/K}(\alpha^q) = N_{F/K}(\alpha)^q = N_{F/K}(\alpha)$ because of (i) and $N_{F/K}(\alpha) \in K$, and so (iv) is shown. $\qquad \square$

2.29.   Theorem (Transitivity of Norm).   *Let K be a finite field, let F be a finite extension of K and E a finite extension of F. Then*

$$N_{E/K}(\alpha) = N_{F/K}\left(N_{E/F}(\alpha)\right) \quad \text{for all } \alpha \in E.$$

Proof.   With the same notation as in the proof of Theorem 2.26, we have for $\alpha \in E$,

$$N_{F/K}\left(N_{E/F}(\alpha)\right) = N_{F/K}\left(\alpha^{(q^{mn}-1)/(q^m-1)}\right)$$

$$= \left(\alpha^{(q^{mn}-1)/(q^m-1)}\right)^{(q^m-1)/(q-1)}$$

$$= \alpha^{(q^{mn}-1)/(q-1)} = N_{E/K}(\alpha). \qquad \square$$

If $\{\alpha_1, \ldots, \alpha_m\}$ is a basis of the finite field F over a subfield K, the question arises as to the calculation of the coefficients $c_j(\alpha) \in K$, $1 \le j \le m$, in the unique representation

$$\alpha = c_1(\alpha)\alpha_1 + \cdots + c_m(\alpha)\alpha_m \qquad (2.4)$$

of an element $\alpha \in F$. We note that $c_j : \alpha \mapsto c_j(\alpha)$ is a linear transformation from F into K, and thus, according to Theorem 2.24, there exists a $\beta_j \in F$ such that $c_j(\alpha) = \mathrm{Tr}_{F/K}(\beta_j \alpha)$ for all $\alpha \in F$. Putting $\alpha = \alpha_i$, $1 \le i \le m$, we see that $\mathrm{Tr}_{F/K}(\beta_j \alpha_i) = 0$ for $i \ne j$ and 1 for $i = j$. Furthermore, $\{\beta_1, \ldots, \beta_m\}$ is again a basis of F over K, for if

$$d_1 \beta_1 + \cdots + d_m \beta_m = 0 \quad \text{with } d_i \in K \quad \text{for } 1 \le i \le m,$$

then by multiplying by a fixed α_i and applying the trace function $\mathrm{Tr}_{F/K}$, one shows that $d_i = 0$.

2.30.   Definition.   Let K be a finite field and F a finite extension of K. Then two bases $\{\alpha_1, \ldots, \alpha_m\}$ and $\{\beta_1, \ldots, \beta_m\}$ of F over K are said to be *dual* (or *complementary*) bases if for $1 \le i, j \le m$ we have

$$\mathrm{Tr}_{F/K}(\alpha_i \beta_j) = \begin{cases} 0 & \text{for } i \ne j, \\ 1 & \text{for } i = j. \end{cases}$$

In the discussion above we have shown that for any basis $\{\alpha_1, \ldots, \alpha_m\}$ of F over K there exists a dual basis $\{\beta_1, \ldots, \beta_m\}$. The dual basis is, in fact, uniquely determined since its definition implies that the coefficients $c_j(\alpha)$, $1 \le j \le m$, in (2.4) are given by $c_j(\alpha) = \mathrm{Tr}_{F/K}(\beta_j \alpha)$ for all $\alpha \in F$, and by Theorem 2.24 the element $\beta_j \in F$ is uniquely determined by the linear transformation c_j.

2.31.   Example.   Let $\alpha \in \mathbb{F}_8$ be a root of the irreducible polynomial $x^3 + x^2 + 1$ in $\mathbb{F}_2[x]$. Then $\{\alpha, \alpha^2, 1 + \alpha + \alpha^2\}$ is a basis of $\mathbb{F}_8$ over $\mathbb{F}_2$. One checks easily that its uniquely determined dual basis is again $\{\alpha, \alpha^2, 1 + \alpha + \alpha^2\}$. Such a basis that is its own dual basis is called a *self-dual basis*. The element $\alpha^5 \in \mathbb{F}_8$ can be uniquely represented in the form $\alpha^5 = c_1 \alpha + c_2 \alpha^2 +$

$c_3(1 + \alpha + \alpha^2)$ with $c_1, c_2, c_3 \in \mathbb{F}_2$, and the coefficients are given by

$$c_1 = \mathrm{Tr}_{\mathbb{F}_8}(\alpha \cdot \alpha^5) = 0,$$

$$c_2 = \mathrm{Tr}_{\mathbb{F}_8}(\alpha^2 \cdot \alpha^5) = 1,$$

$$c_3 = \mathrm{Tr}_{\mathbb{F}_8}((1 + \alpha + \alpha^2)\alpha^5) = 1,$$

so that $\alpha^5 = \alpha^2 + (1 + \alpha + \alpha^2)$. $\square$

The number of distinct bases of F over K is rather large (see Exercise 2.37), but there are two special types of bases of particular importance. The first is a *polynomial basis* $\{1, \alpha, \alpha^2, \ldots, \alpha^{m-1}\}$, made up of the powers of a defining element α of F over K. The element α is often taken to be a primitive element of F (compare with Theorem 2.10). Another type of basis is a normal basis defined by a suitable element of F.

2.32.   Definition.   Let $K = \mathbb{F}_q$ and $F = \mathbb{F}_{q^m}$. Then a basis of F over K of the form $\{\alpha, \alpha^q, \ldots, \alpha^{q^{m-1}}\}$, consisting of a suitable element $\alpha \in F$ and its conjugates with respect to K, is called a *normal basis* of F over K.

The basis $\{\alpha, \alpha^2, 1 + \alpha + \alpha^2\}$ of $\mathbb{F}_8$ over $\mathbb{F}_2$ discussed in Example 2.31 is a normal basis of $\mathbb{F}_8$ over $\mathbb{F}_2$ since $1 + \alpha + \alpha^2 = \alpha^4$. We shall show that a normal basis exists in the general case as well. The proof depends on two lemmas, one on a kind of linear independence property of certain group homomorphisms and one on linear operators.

2.33.   Lemma (Artin Lemma).   *Let* $\psi_1, \ldots, \psi_m$ *be distinct homomorphisms from a group G into the multiplicative group F^* of an arbitrary field F, and let* $a_1, \ldots, a_m$ *be elements of F that are not all 0. Then for some $g \in G$ we have*

$$a_1 \psi_1(g) + \cdots + a_m \psi_m(g) \neq 0.$$

Proof.   We proceed by induction on m. The case $m = 1$ being trivial, we assume that $m > 1$ and that the statement is shown for any $m - 1$ distinct homomorphisms. Now take $\psi_1, \ldots, \psi_m$ and $a_1, \ldots, a_m$ as in the lemma. If $a_1 = 0$, the induction hypothesis immediately yields the desired result. Thus let $a_1 \neq 0$. Suppose we had

$$a_1 \psi_1(g) + \cdots + a_m \psi_m(g) = 0 \quad \text{for all } g \in G. \tag{2.5}$$

Since $\psi_1 \neq \psi_m$, there exists $h \in G$ with $\psi_1(h) \neq \psi_m(h)$. Then, replacing g by hg in (2.5), we get

$$a_1 \psi_1(h) \psi_1(g) + \cdots + a_m \psi_m(h) \psi_m(g) = 0 \quad \text{for all } g \in G.$$

After multiplication by $\psi_m(h)^{-1}$ we obtain

$$b_1 \psi_1(g) + \cdots + b_{m-1} \psi_{m-1}(g) + a_m \psi_m(g) = 0 \quad \text{for all } g \in G,$$

where $b_i = a_i \psi_i(h) \psi_m(h)^{-1}$ for $1 \leqslant i \leqslant m - 1$. By subtracting this identity from (2.5), we arrive at

$$c_1\psi_1(g) + \cdots + c_{m-1}\psi_{m-1}(g) = 0 \quad \text{for all } g \in G,$$

where $c_i = a_i - b_i$ for $1 \leqslant i \leqslant m - 1$. But $c_1 = a_1 - a_1\psi_1(h)\psi_m(h)^{-1} \neq 0$, and we have a contradiction to the induction hypothesis. $\square$

We recall a few concepts and facts from linear algebra. If T is a linear operator on the finite-dimensional vector space V over the (arbitrary) field K, then a polynomial $f(x) = a_n x^n + \cdots + a_1 x + a_0 \in K[x]$ is said to *annihilate* T if $a_n T^n + \cdots + a_1 T + a_0 I = 0$, where I is the identity operator and 0 the zero operator on V. The uniquely determined monic polynomial of least positive degree with this property is called the *minimal polynomial* for T. It divides any other polynomial in $K[x]$ annihilating T. In particular, the minimal polynomial for T divides the *characteristic polynomial* $g(x)$ for T (Cayley-Hamilton theorem), which is given by $g(x) = \det(xI - T)$ and is a monic polynomial of degree equal to the dimension of V. A vector $\alpha \in V$ is called a *cyclic vector* for T if the vectors $T^k\alpha$, $k = 0, 1, \ldots$, span V. The following is a standard result from linear algebra.

2.34. Lemma. *Let T be a linear operator on the finite-dimensional vector space V. Then T has a cyclic vector if and only if the characteristic and minimal polynomials for T are identical.*

2.35. Theorem (Normal Basis Theorem). *For any finite field K and any finite extension F of K, there exists a normal basis of F over K.*

Proof. Let $K = \mathbb{F}_q$ and $F = \mathbb{F}_{q^m}$ with $m \geqslant 2$. From Theorem 2.21 and the remarks following it, we know that the distinct automorphisms of F over K are given by $\varepsilon, \sigma, \sigma^2, \ldots, \sigma^{m-1}$, where ε is the identity mapping on F, $\sigma(\alpha) = \alpha^q$ for $\alpha \in F$, and a power σ^j refers to the j-fold composition of σ with itself. Because of $\sigma(\alpha + \beta) = \sigma(\alpha) + \sigma(\beta)$ and $\sigma(c\alpha) = \sigma(c)\sigma(\alpha) = c\sigma(\alpha)$ for $\alpha, \beta \in F$ and $c \in K$, the mapping σ may also be considered as a linear operator on the vector space F over K. Since $\sigma^m = \varepsilon$, the polynomial $x^m - 1 \in K[x]$ annihilates σ. Lemma 2.33, applied to $\varepsilon, \sigma, \sigma^2, \ldots, \sigma^{m-1}$ viewed as endomorphisms of F^*, shows that no nonzero polynomial in $K[x]$ of degree less than m annihilates σ. Consequently, $x^m - 1$ is the minimal polynomial for the linear operator σ. Since the characteristic polynomial for σ is a monic polynomial of degree m that is divisible by the minimal polynomial for σ, it follows that the characteristic polynomial for σ is also given by $x^m - 1$. Lemma 2.34 implies then the existence of an element $\alpha \in F$ such that $\alpha, \sigma(\alpha), \sigma^2(\alpha), \ldots$ span F. By dropping repeated elements, we see that $\alpha, \sigma(\alpha), \sigma^2(\alpha), \ldots, \sigma^{m-1}(\alpha)$ span F and thus form a basis of F over K. Since this basis consists of α and its conjugates with respect to K, it is a normal basis of F over K. $\square$

An alternative proof of the normal basis theorem will be provided in Chapter 3, Section 4, by using so-called linearized polynomials.

We introduce an expression that allows us to decide whether a given set of elements forms a basis of an extension field.

2.36.   Definition.   Let K be a finite field and F an extension of K of degree m over K. Then the *discriminant* $\Delta_{F/K}(\alpha_1,\ldots,\alpha_m)$ of the elements $\alpha_1,\ldots,\alpha_m$ $\in F$ is defined by the determinant of order m given by

$$\Delta_{F/K}(\alpha_1,\ldots,\alpha_m) = \begin{vmatrix} \mathrm{Tr}_{F/K}(\alpha_1\alpha_1) & \mathrm{Tr}_{F/K}(\alpha_1\alpha_2) & \cdots & \mathrm{Tr}_{F/K}(\alpha_1\alpha_m) \\ \mathrm{Tr}_{F/K}(\alpha_2\alpha_1) & \mathrm{Tr}_{F/K}(\alpha_2\alpha_2) & \cdots & \mathrm{Tr}_{F/K}(\alpha_2\alpha_m) \\ \vdots & \vdots & & \vdots \\ \mathrm{Tr}_{F/K}(\alpha_m\alpha_1) & \mathrm{Tr}_{F/K}(\alpha_m\alpha_2) & \cdots & \mathrm{Tr}_{F/K}(\alpha_m\alpha_m) \end{vmatrix}$$

It follows from the definition that $\Delta_{F/K}(\alpha_1,\ldots,\alpha_m)$ is always an element of K. The following simple characterization of bases can now be given.

2.37.   Theorem.   *Let K be a finite field, F an extension of K of degree m over K, and $\alpha_1,\ldots,\alpha_m \in F$. Then $\{\alpha_1,\ldots,\alpha_m\}$ is a basis of F over K if and only if $\Delta_{F/K}(\alpha_1,\ldots,\alpha_m) \neq 0$.*

Proof.   Let $\{\alpha_1,\ldots,\alpha_m\}$ be a basis of F over K. We prove that $\Delta_{F/K}(\alpha_1,\ldots,\alpha_m) \neq 0$ by showing that the row vectors of the determinant defining $\Delta_{F/K}(\alpha_1,\ldots,\alpha_m)$ are linearly independent. For suppose that

$$c_1\mathrm{Tr}_{F/K}(\alpha_1\alpha_j) + \cdots + c_m\mathrm{Tr}_{F/K}(\alpha_m\alpha_j) = 0 \quad \text{for } 1 \leqslant j \leqslant m,$$

where $c_1,\ldots,c_m \in K$. Then with $\beta = c_1\alpha_1 + \cdots + c_m\alpha_m$ we get $\mathrm{Tr}_{F/K}(\beta\alpha_j)$ $= 0$ for $1 \leqslant j \leqslant m$, and since $\alpha_1,\ldots,\alpha_m$ span F, it follows that $\mathrm{Tr}_{F/K}(\beta\alpha) = 0$ for all $\alpha \in F$. However, this is only possible if $\beta = 0$, and then $c_1\alpha_1 + \cdots + c_m\alpha_m = 0$ implies $c_1 = \cdots = c_m = 0$.

Conversely, suppose that $\Delta_{F/K}(\alpha_1,\ldots,\alpha_m) \neq 0$ and $c_1\alpha_1 + \cdots + c_m\alpha_m = 0$ for some $c_1,\ldots,c_m \in K$. Then

$$c_1\alpha_1\alpha_j + \cdots + c_m\alpha_m\alpha_j = 0 \quad \text{for } 1 \leqslant j \leqslant m,$$

and by applying the trace function we get

$$c_1\mathrm{Tr}_{F/K}(\alpha_1\alpha_j) + \cdots + c_m\mathrm{Tr}_{F/K}(\alpha_m\alpha_j) = 0 \quad \text{for } 1 \leqslant j \leqslant m.$$

But since the row vectors of the determinant defining $\Delta_{F/K}(\alpha_1,\ldots,\alpha_m)$ are linearly independent, it follows that $c_1 = \cdots = c_m = 0$. Therefore, $\alpha_1,\ldots,\alpha_m$ are linearly independent over K. $\square$

There is another determinant of order m that serves the same purpose as the discriminant $\Delta_{F/K}(\alpha_1,\ldots,\alpha_m)$. The entries of this determinant are, however, elements of the extension field F. For $\alpha_1,\ldots,\alpha_m \in F$, let

A be the $m \times m$ matrix whose entry in the ith row and jth column is $\alpha_j^{q^{i-1}}$, where q is the number of elements of K. If A^{T} denotes the transpose of A, then a simple calculation shows that $A^{\mathrm{T}}A = B$, where B is the $m \times m$ matrix whose entry in the ith row and jth column is $\mathrm{Tr}_{F/K}(\alpha_i \alpha_j)$. By taking determinants, we obtain

$$\Delta_{F/K}(\alpha_1, \ldots, \alpha_m) = \det(A)^2.$$

The following result is now implied by Theorem 2.37.

2.38.   Corollary.   *Let $\alpha_1, \ldots, \alpha_m \in \mathbb{F}_{q^m}$. Then $\{\alpha_1, \ldots, \alpha_m\}$ is a basis of $\mathbb{F}_{q^m}$ over $\mathbb{F}_q$ if and only if*

$$\begin{vmatrix} \alpha_1 & \alpha_2 & \cdots & \alpha_m \\ \alpha_1^q & \alpha_2^q & \cdots & \alpha_m^q \\ \vdots & \vdots & & \vdots \\ \alpha_1^{q^{m-1}} & \alpha_2^{q^{m-1}} & \cdots & \alpha_m^{q^{m-1}} \end{vmatrix} \neq 0.$$

From the criterion above we are led to a relatively simple way of checking whether a given element gives rise to a normal basis.

2.39.   Theorem.   *For $\alpha \in \mathbb{F}_{q^m}$, $\{\alpha, \alpha^q, \alpha^{q^2}, \ldots, \alpha^{q^{m-1}}\}$ is a normal basis of $\mathbb{F}_{q^m}$ over $\mathbb{F}_q$ if and only if the polynomials $x^m - 1$ and $\alpha x^{m-1} + \alpha^q x^{m-2} + \cdots + \alpha^{q^{m-2}} x + \alpha^{q^{m-1}}$ in $\mathbb{F}_{q^m}[x]$ are relatively prime.*

Proof.   When $\alpha_1 = \alpha$, $\alpha_2 = \alpha^q, \ldots, \alpha_m = \alpha^{q^{m-1}}$, the determinant in Corollary 2.38 becomes

$$\pm \begin{vmatrix} \alpha & \alpha^q & \alpha^{q^2} & \cdots & \alpha^{q^{m-1}} \\ \alpha^{q^{m-1}} & \alpha & \alpha^q & \cdots & \alpha^{q^{m-2}} \\ \alpha^{q^{m-2}} & \alpha^{q^{m-1}} & \alpha & \cdots & \alpha^{q^{m-3}} \\ \vdots & \vdots & \vdots & & \vdots \\ \alpha^q & \alpha^{q^2} & \alpha^{q^3} & \cdots & \alpha \end{vmatrix} \qquad (2.6)$$

after a suitable permutation of the rows. Now consider the resultant $R(f, g)$ of the polynomials $f(x) = x^m - 1$ and $g(x) = \alpha x^{m-1} + \alpha^q x^{m-2} + \cdots + \alpha^{q^{m-2}} x + \alpha^{q^{m-1}}$ of formal degree m resp. $m - 1$, which is given by a determinant of order $2m - 1$ in accordance with Definition 1.93. In this determinant, add the $(m+1)$st column to the first column, the $(m+2)$nd column to the second column, and so on, finally adding the $(2m-1)$st column to the $(m-1)$st column. The resulting determinant factorizes into the determinant of the diagonal matrix of order $m - 1$ with entries -1 along the main diagonal and the determinant in (2.6). Therefore, $R(f, g)$ is, apart from the sign, equal to the determinant in (2.6). The statement of the theorem follows then from Corollary 2.38 and the fact that $R(f, g) \neq 0$ if and only if f and g are relatively prime. $\qquad\square$

In connection with the preceding discussion, we mention without proof the following refinement of the normal basis theorem.

2.40   Theorem.   *For any finite extension F of a finite field K there exists a normal basis of F over K that consists of primitive elements of F.*

4.   ROOTS OF UNITY AND CYCLOTOMIC POLYNOMIALS

In this section we investigate the splitting field of the polynomial $x^n - 1$ over an arbitrary field K, where n is a positive integer. At the same time we obtain a generalization of the concept of a root of unity, well known for complex numbers.

2.41.   Definition.   Let n be a positive integer. The splitting field of $x^n - 1$ over a field K is called the *nth cyclotomic field* over K and denoted by $K^{(n)}$. The roots of $x^n - 1$ in $K^{(n)}$ are called the *nth roots of unity* over K and the set of all these roots is denoted by $E^{(n)}$.

A special case of this general definition is obtained if K is the field of rational numbers. Then $K^{(n)}$ is a subfield of the field of complex numbers and the nth roots of unity have their known geometric interpretation as the vertices of a regular polygon with n vertices on the unit circle in the complex plane.

For our purposes, the most important case is that of a finite field K. The basic properties of roots of unity can, however, be established without using this restriction. The structure of $E^{(n)}$ is determined by the relation of n to the characteristic of K, as the following theorem shows. When we refer to the characteristic p of K in this discussion, we permit the case $p = 0$ as well.

2.42.   Theorem.   *Let n be a positive integer and K a field of characteristic p. Then:*

(i)   *If p does not divide n, then $E^{(n)}$ is a cyclic group of order n with respect to multiplication in $K^{(n)}$.*

(ii)   *If p divides n, write $n = mp^e$ with positive integers m and e and m not divisible by p. Then $K^{(n)} = K^{(m)}$, $E^{(n)} = E^{(m)}$, and the roots of $x^n - 1$ in $K^{(n)}$ are the m elements of $E^{(m)}$, each attained with multiplicity p^e.*

Proof.   (i)   The case $n = 1$ is trivial. For $n \geqslant 2$, $x^n - 1$ and its derivative nx^{n-1} have no common roots, as nx^{n-1} only has the root 0 in $K^{(n)}$. Therefore, by Theorem 1.68, $x^n - 1$ cannot have multiple roots, and hence $E^{(n)}$ has n elements. Now if $\zeta, \eta \in E^{(n)}$, then $(\zeta\eta^{-1})^n = \zeta^n(\eta^n)^{-1} = 1$, thus

$\zeta\eta^{-1} \in E^{(n)}$. It follows that $E^{(n)}$ is a multiplicative group. Let $n = p_1^{e_1} p_2^{e_2} \cdots p_t^{e_t}$ be the prime factor decomposition of n. Then one shows by the same argument as in the proof of Theorem 2.8 that for each i, $1 \leqslant i \leqslant t$, there exists an element $\alpha_i \in E^{(n)}$ that is not a root of the polynomial $x^{n/p_i} - 1$, that $\beta_i = \alpha_i^{n/p_i^{e_i}}$ has order $p_i^{e_i}$, and that $E^{(n)}$ is a cyclic group with generator $\beta = \beta_1 \beta_2 \cdots \beta_t$.

(ii) This follows immediately from $x^n - 1 = x^{mp^e} - 1 = (x^m - 1)^{p^e}$ and part (i). $\square$

2.43.   Definition.   Let K be a field of characteristic p and n a positive integer not divisible by p. Then a generator of the cyclic group $E^{(n)}$ is called a *primitive nth root of unity* over K.

By Theorem 1.15(v) we know that under the conditions of Definition 2.43 there are exactly $\phi(n)$ different primitive nth roots of unity over K. If ζ is one of them, then all primitive nth roots of unity over K are given by ζ^s, where $1 \leqslant s \leqslant n$ and $\gcd(s, n) = 1$. The polynomial whose roots are precisely the primitive nth roots of unity over K is of great interest.

2.44.   Definition.   Let K be a field of characteristic p, n a positive integer not divisible by p, and ζ a primitive nth root of unity over K. Then the polynomial

$$Q_n(x) = \prod_{\substack{s=1 \\ \gcd(s,n)=1}}^{n} (x - \zeta^s)$$

is called the *nth cyclotomic polynomial* over K.

The polynomial $Q_n(x)$ is clearly independent of the choice of ζ. The degree of $Q_n(x)$ is $\phi(n)$ and its coefficients obviously belong to the nth cyclotomic field over K. A simple argument will show that they are actually contained in the prime subfield of K. We use the product symbol $\prod_{d|n}$ to denote a product extended over all positive divisors d of a positive integer n.

2.45.   Theorem.   *Let K be a field of characteristic p and n a positive integer not divisible by p. Then:*

(i) $x^n - 1 = \prod_{d|n} Q_d(x)$;

(ii)   *the coefficients of $Q_n(x)$ belong to the prime subfield of K, and to $\mathbb{Z}$ if the prime subfield of K is the field of rational numbers.*

Proof.   (i) Each nth root of unity over K is a primitive dth root of unity over K for exactly one positive divisor d of n. In detail, if ζ is a primitive nth root of unity over K and ζ^s is an arbitrary nth root of unity over K, then $d = n/\gcd(s, n)$; that is, d is the order of ζ^s in $E^{(n)}$. Since

$$x^n - 1 = \prod_{s=1}^{n} (x - \zeta^s),$$

the formula in (i) is obtained by collecting those factors $(x - \zeta^s)$ for which ζ^s is a primitive dth root of unity over K.

(ii) This is proved by induction on n. Note that $Q_n(x)$ is a monic polynomial. For $n = 1$ we have $Q_1(x) = x - 1$, and the claim is obviously valid. Now let $n > 1$ and suppose the proposition is true for all $Q_d(x)$ with $1 \leqslant d < n$. Then we have by (i), $Q_n(x) = (x^n - 1)/f(x)$, where $f(x) = \prod_{d \mid n, d < n} Q_d(x)$. The induction hypothesis implies that $f(x)$ is a polynomial with coefficients in the prime subfield of K or in $\mathbb{Z}$ in case the characteristic of K is 0. Using long division with $x^n - 1$ and the monic polynomial $f(x)$, we see that the coefficients of $Q_n(x)$ belong to the prime subfield of K or to $\mathbb{Z}$, respectively. $\square$

2.46.   Example.   Let r be a prime and $k \in \mathbb{N}$. Then

$$Q_{r^k}(x) = 1 + x^{r^{k-1}} + x^{2r^{k-1}} + \cdots + x^{(r-1)r^{k-1}}$$

since

$$Q_{r^k}(x) = \frac{x^{r^k} - 1}{Q_1(x)Q_r(x) \cdots Q_{r^{k-1}}(x)} = \frac{x^{r^k} - 1}{x^{r^{k-1}} - 1}$$

by Theorem 2.45(i). For $k = 1$ we simply have $Q_r(x) = 1 + x + x^2 + \cdots + x^{r-1}$. $\square$

An explicit expression for the nth cyclotomic polynomial generalizing the formula for $Q_{r^k}(x)$ in Example 2.46 will be given in Chapter 3, Section 2. For applications to finite fields it is useful to know some properties of cyclotomic fields.

2.47.   Theorem.   *The cyclotomic field $K^{(n)}$ is a simple algebraic extension of K. Moreover:*

(i)   *If $K = \mathbb{Q}$, then the cyclotomic polynomial Q_n is irreducible over K and $[K^{(n)} : K] = \phi(n)$.*

(ii)   *If $K = \mathbb{F}_q$ with $\gcd(q, n) = 1$, then Q_n factors into $\phi(n)/d$ distinct monic irreducible polynomials in $K[x]$ of the same degree d, $K^{(n)}$ is the splitting field of any such irreducible factor over K, and $[K^{(n)} : K] = d$, where d is the least positive integer such that $q^d \equiv 1 \bmod n$.*

Proof.   If there exists a primitive nth root of unity ζ over K, it is clear that $K^{(n)} = K(\zeta)$. Otherwise, we have the situation described in Theorem 2.42(ii), then $K^{(n)} = K^{(m)}$ and the result follows again. As to the remaining statements, we prove only (ii), the important case for our purposes. Let η be a primitive nth root of unity over $\mathbb{F}_q$. Then $\eta \in \mathbb{F}_{q^k}$ if and only if $\eta^{q^k} = \eta$, and the latter identity is equivalent to $q^k \equiv 1 \bmod n$. The smallest positive integer for which this holds is $k = d$, and so η is in $\mathbb{F}_{q^d}$, but in no proper subfield thereof. Thus the minimal polynomial of η over $\mathbb{F}_q$ has degree d, and since η is an arbitrary root of Q_n, the desired results follow. $\square$

2.48.   Example. Let $K = \mathbb{F}_{11}$ and $Q_{12}(x) = x^4 - x^2 + 1 \in \mathbb{F}_{11}[x]$. In the notation of Theorem 2.47(ii) we have $d = 2$. In detail, $Q_{12}(x)$ factors in the form $Q_{12}(x) = (x^2 + 5x + 1)(x^2 - 5x + 1)$, with both factors being irreducible in $\mathbb{F}_{11}[x]$. The cyclotomic field $K^{(12)}$ is equal to $\mathbb{F}_{121}$. $\square$

A further connection between cyclotomic fields and finite fields is given by the following theorem.

2.49.   Theorem.   *The finite field $\mathbb{F}_q$ is the $(q-1)$st cyclotomic field over any one of its subfields.*

Proof.   The polynomial $x^{q-1} - 1$ splits in $\mathbb{F}_q$ since its roots are exactly all nonzero elements of $\mathbb{F}_q$. Obviously, the polynomial cannot split in any proper subfield of $\mathbb{F}_q$, so that $\mathbb{F}_q$ is the splitting field of $x^{q-1} - 1$ over any one of its subfields. $\square$

Since $\mathbb{F}_q^*$ is a cyclic group of order $q - 1$ by Theorem 2.8, there will exist, for any positive divisor n of $q - 1$, a cyclic subgroup $\{1, \alpha, \ldots, \alpha^{n-1}\}$ of $\mathbb{F}_q^*$ of order n (see Theorem 1.15(iii)). All elements of this subgroup are nth roots of unity over any subfield of $\mathbb{F}_q$ and the generating element α is a primitive nth root of unity over any subfield of $\mathbb{F}_q$.

We conclude this section with a lemma we shall need later on.

2.50.   Lemma.   *If d is a divisor of the positive integer n with $1 \leqslant d < n$, then $Q_n(x)$ divides $(x^n - 1)/(x^d - 1)$ whenever $Q_n(x)$ is defined.*

Proof.   From Theorem 2.45(i) we know that $Q_n(x)$ divides

$$x^n - 1 = (x^d - 1) \cdot \frac{x^n - 1}{x^d - 1}.$$

Since d is a proper divisor of n, the polynomials $Q_n(x)$ and $x^d - 1$ have no common root, hence $\gcd(Q_n(x), x^d - 1) = 1$ and the proposition is true. $\square$

5.   REPRESENTATION OF ELEMENTS OF FINITE FIELDS

In this section we describe three different ways of representing the elements of a finite field $\mathbb{F}_q$ with $q = p^n$ elements, where p is the characteristic of $\mathbb{F}_q$.

The first method is based on principles expounded in Chapter 1, Section 4, and in the present chapter. We note that $\mathbb{F}_q$ is a simple algebraic extension of $\mathbb{F}_p$ by Theorem 2.10. In fact, if f is an irreducible polynomial in $\mathbb{F}_p[x]$ of degree n, then f has a root α in $\mathbb{F}_q$ according to Theorem 2.14, and so $\mathbb{F}_q = \mathbb{F}_p(\alpha)$. Then, by Theorem 1.86, every element of $\mathbb{F}_q$ can be uniquely expressed as a polynomial in α over $\mathbb{F}_p$ of degree less than n. We may also view $\mathbb{F}_q$ as the residue class ring $\mathbb{F}_p[x]/(f)$.

2.51. Example. To represent the elements of $\mathbb{F}_9$ in this way, we regard $\mathbb{F}_9$ as a simple algebraic extension of $\mathbb{F}_3$ of degree 2, which is obtained by adjunction of a root α of an irreducible quadratic polynomial over $\mathbb{F}_3$, say $f(x) = x^2 + 1 \in \mathbb{F}_3[x]$. Thus $f(\alpha) = \alpha^2 + 1 = 0$ in $\mathbb{F}_9$, and the nine elements of $\mathbb{F}_9$ are given in the form $a_0 + a_1\alpha$ with $a_0, a_1 \in \mathbb{F}_3$. In detail, $\mathbb{F}_9 = \{0, 1, 2, \alpha, 1 + \alpha, 2 + \alpha, 2\alpha, 1 + 2\alpha, 2 + 2\alpha\}$. The operation tables for $\mathbb{F}_9$ may be constructed as in Example 1.62, with α playing the role of the residue class $[x]$. $\square$

If we use Theorems 2.47 and 2.49, we get another possibility of expressing the elements of $\mathbb{F}_q$. Since $\mathbb{F}_q$ is the $(q-1)$st cyclotomic field over $\mathbb{F}_p$, we can construct it by finding the decomposition of the $(q-1)$st cyclotomic polynomial $Q_{q-1} \in \mathbb{F}_p[x]$ into irreducible factors in $\mathbb{F}_p[x]$, which are all of the same degree. A root of any one of these factors is then a primitive $(q-1)$st root of unity over $\mathbb{F}_p$ and therefore a primitive element of $\mathbb{F}_q$. Thus, $\mathbb{F}_q$ consists of 0 and appropriate powers of that primitive element.

2.52. Example. To apply this to the construction of $\mathbb{F}_9$, we note that $\mathbb{F}_9 = \mathbb{F}_3^{(8)}$, the eighth cyclotomic field over $\mathbb{F}_3$. Now $Q_8(x) = x^4 + 1 \in \mathbb{F}_3[x]$ by Example 2.46, and

$$Q_8(x) = (x^2 + x + 2)(x^2 + 2x + 2)$$

is the decomposition of Q_8 into irreducible factors in $\mathbb{F}_3[x]$. Let ζ be a root of $x^2 + x + 2$; then ζ is a primitive eighth root of unity over $\mathbb{F}_3$. Thus, all nonzero elements of $\mathbb{F}_9$ can be expressed as powers of ζ, and so $\mathbb{F}_9 = \{0, \zeta, \zeta^2, \zeta^3, \zeta^4, \zeta^5, \zeta^6, \zeta^7, \zeta^8\}$. We may arrange the nonzero elements of $\mathbb{F}_9$ in a so-called *index table*, where we list the elements ζ^i according to their exponents i. In order to establish the connection with the representation in Example 2.51, we observe that $x^2 + x + 2 \in \mathbb{F}_3[x]$ has $\zeta = 1 + \alpha$ as a root, where $\alpha^2 + 1 = 0$ as in Example 2.51. Therefore, the index table for $\mathbb{F}_9$ may be written as follows:

i	ζ^i	i	ζ^i
1	$1 + \alpha$	5	$2 + 2\alpha$
2	2α	6	α
3	$1 + 2\alpha$	7	$2 + \alpha$
4	2	8	1

We see that we obtain, of course, the same elements as in Example 2.51, just in a different order. $\square$

A third possibility of representing the elements of $\mathbb{F}_q$ is given by means of matrices. In general, the *companion matrix* of a monic polynomial

$f(x) = a_0 + a_1 x + \cdots + a_{n-1} x^{n-1} + x^n$ of positive degree n over a field is defined to be the $n \times n$ matrix

$$A = \begin{pmatrix} 0 & 0 & 0 & \cdots & 0 & -a_0 \\ 1 & 0 & 0 & \cdots & 0 & -a_1 \\ 0 & 1 & 0 & \cdots & 0 & -a_2 \\ \vdots & \vdots & \vdots & & \vdots & \vdots \\ 0 & 0 & 0 & \cdots & 1 & -a_{n-1} \end{pmatrix}$$

It is well known in linear algebra that A satisfies the equation $f(A) = 0$; that is, $a_0 I + a_1 A + \cdots + a_{n-1} A^{n-1} + A^n = 0$, where I is the $n \times n$ identity matrix.

Thus, if A is the companion matrix of a monic irreducible polynomial f over $\mathbb{F}_p$ of degree n, then $f(A) = 0$, and therefore A can play the role of a root of f. The polynomials in A over $\mathbb{F}_p$ of degree less than n yield a representation of the elements of $\mathbb{F}_q$.

2.53. Example. As in Example 2.51, let $f(x) = x^2 + 1 \in \mathbb{F}_3[x]$. The companion matrix of f is

$$A = \begin{pmatrix} 0 & 2 \\ 1 & 0 \end{pmatrix}.$$

The field $\mathbb{F}_9$ can then be represented in the form $\mathbb{F}_9 = \{0, I, 2I, A, I + A, 2I + A, 2A, I + 2A, 2I + 2A\}$. Explicitly:

$$0 = \begin{pmatrix} 0 & 0 \\ 0 & 0 \end{pmatrix}, \quad I = \begin{pmatrix} 1 & 0 \\ 0 & 1 \end{pmatrix}, \quad 2I = \begin{pmatrix} 2 & 0 \\ 0 & 2 \end{pmatrix}, \quad A = \begin{pmatrix} 0 & 2 \\ 1 & 0 \end{pmatrix},$$

$$I + A = \begin{pmatrix} 1 & 2 \\ 1 & 1 \end{pmatrix}, \quad 2I + A = \begin{pmatrix} 2 & 2 \\ 1 & 2 \end{pmatrix}, \quad 2A = \begin{pmatrix} 0 & 1 \\ 2 & 0 \end{pmatrix},$$

$$I + 2A = \begin{pmatrix} 1 & 1 \\ 2 & 1 \end{pmatrix}, \quad 2I + 2A = \begin{pmatrix} 2 & 1 \\ 2 & 2 \end{pmatrix}.$$

With $\mathbb{F}_9$ given in this way, calculations in this finite field are then carried out by the usual rules of matrix algebra. For instance,

$$(2I + A)(I + 2A) = \begin{pmatrix} 2 & 2 \\ 1 & 2 \end{pmatrix}\begin{pmatrix} 1 & 1 \\ 2 & 1 \end{pmatrix} = \begin{pmatrix} 0 & 1 \\ 2 & 0 \end{pmatrix} = 2A. \qquad \square$$

In the same way, the method based on the factorization of the cyclotomic polynomial Q_{q-1} in $\mathbb{F}_p[x]$ can be adapted to yield a representation of the elements of $\mathbb{F}_q$ in terms of matrices.

2.54. Example. As in Example 2.52, let $h(x) = x^2 + x + 2 \in \mathbb{F}_3[x]$ be an irreducible factor of the cyclotomic polynomial $Q_8 \in \mathbb{F}_3[x]$. The companion matrix of h is

$$C = \begin{pmatrix} 0 & 1 \\ 1 & 2 \end{pmatrix}.$$

The field $\mathbb{F}_9$ can then be represented in the form

$$\mathbb{F}_9 = \{0, C, C^2, C^3, C^4, C^5, C^6, C^7, C^8\}.$$

Explicitly:

$$0 = \begin{pmatrix} 0 & 0 \\ 0 & 0 \end{pmatrix}, \quad C = \begin{pmatrix} 0 & 1 \\ 1 & 2 \end{pmatrix}, \quad C^2 = \begin{pmatrix} 1 & 2 \\ 2 & 2 \end{pmatrix},$$

$$C^3 = \begin{pmatrix} 2 & 2 \\ 2 & 0 \end{pmatrix}, \quad C^4 = \begin{pmatrix} 2 & 0 \\ 0 & 2 \end{pmatrix}, \quad C^5 = \begin{pmatrix} 0 & 2 \\ 2 & 1 \end{pmatrix},$$

$$C^6 = \begin{pmatrix} 2 & 1 \\ 1 & 1 \end{pmatrix}, \quad C^7 = \begin{pmatrix} 1 & 1 \\ 1 & 0 \end{pmatrix}, \quad C^8 = \begin{pmatrix} 1 & 0 \\ 0 & 1 \end{pmatrix}.$$

Calculations proceed by the rules of matrix algebra. For instance,

$$C^6 + C = \begin{pmatrix} 2 & 1 \\ 1 & 1 \end{pmatrix} + \begin{pmatrix} 0 & 1 \\ 1 & 2 \end{pmatrix} = \begin{pmatrix} 2 & 2 \\ 2 & 0 \end{pmatrix} = C^3. \qquad \square$$

6. WEDDERBURN'S THEOREM[1]

All results for finite fields are at the same time also true for all finite division rings by a famous theorem due to Wedderburn. This theorem states that in a finite ring in which all the field properties except commutativity of multiplication are assumed (i.e., in a finite division ring), the multiplication must also be commutative. Basically, the first proof we present of the theorem considers a subring of the finite divison ring that is a field and establishes a numerical relation between the multiplicative group of the field and the multiplicative group of the whole division ring. Using this relation and information about cyclotomic polynomials, one obtains a contra-diction—unless the field is all of the division ring. Before we prove Wedderburn's theorem in detail, we mention some general principles that will be employed.

Let D be a division ring and F a subring that is a field (later on, we will express this more briefly by saying that F is a subfield of D). Then D can be viewed as a (left) vector space over F (compare with the discussion of the analogous situation for fields in Chapter 1, Section 4). If $F = \mathbb{F}_q$ and D is of finite dimension n over $\mathbb{F}_q$, then D has q^n elements. We shall write D^* for the multiplicative group of nonzero elements of D.

For a group G and a nonempty subset S of G, we defined the normalizer $N(S)$ of S in G in Definition 1.24. If S is a singleton $\{b\}$, we may also refer to $N(\{b\})$ as the normalizer of the element b in G. From Theorem

---

[1] This section can be omitted without losing necessary information for the following chapters.

1.25 we infer that if G is finite, then the number of elements in the conjugacy class of b is given by $|G|/|N(\langle b \rangle)|$.

2.55.   Theorem (Wedderburn's Theorem).   *Every finite division ring is a field.*

First Proof.   Let D be a finite division ring and let $Z = \{z \in D : zd = dz$ for all $d \in D\}$ be the *center* of D. We omit the obvious verification that Z is a field. Thus $Z = \mathbb{F}_q$ for some prime power q. Now D is a vector space over Z of finite dimension n, and so D has q^n elements. We shall show that $D = Z$, or, equivalently, that $n = 1$.

Let us suppose, on the contrary, that $n > 1$. Now let $a \in D$ and define $N_a = \{b \in D : ab = ba\}$. Then N_a is a division ring and N_a contains Z. Thus N_a has q^r elements, where $1 \leqslant r \leqslant n$. We wish to show that r divides n. Since N_a^* is a subgroup of D^*, we know that $q^r - 1$ divides $q^n - 1$. If $n = rm + t$ with $0 \leqslant t < r$, then $q^n - 1 = q^{rm}q^t - 1 = q^t(q^{rm} - 1) + (q^t - 1)$. Now $q^r - 1$ divides $q^n - 1$ and also $q^{rm} - 1$, thus it follows that $q^r - 1$ divides $q^t - 1$. But $q^t - 1 < q^r - 1$, and so we must have $t = 0$. This implies that r divides n.

We consider now the class equation for the group D^* (see Theorem 1.27). The center of D^* is Z^*, which has order $q - 1$. For $a \in D^*$, the normalizer of a in D^* is exactly N_a^*. Therefore, a conjugacy class in D^* containing more than one member has $(q^n - 1)/(q^r - 1)$ elements, where r is a divisor of n with $1 \leqslant r < n$. Hence the class equation becomes

$$q^n - 1 = q - 1 + \sum_{i=1}^{k} \frac{q^n - 1}{q^{r_i} - 1}, \tag{2.7}$$

where $r_1, \ldots, r_k$ are (not necessarily distinct) divisors of n with $1 \leqslant r_i < n$ for $1 \leqslant i \leqslant k$.

Now let Q_n be the nth cyclotomic polynomial over the field of rational numbers. Then $Q_n(q)$ is an integer by Theorem 2.45(ii). Furthermore, Lemma 2.50 implies that $Q_n(q)$ divides $(q^n - 1)/(q^{r_i} - 1)$ for $1 \leqslant i \leqslant k$. We conclude then from (2.7) that $Q_n(q)$ divides $q - 1$. However, this will lead to a contradiction. By definition, we have

$$Q_n(x) = \prod_{\substack{s=1 \\ \gcd(s,n)=1}}^{n} (x - \zeta^s),$$

where the complex number ζ is a primitive nth root of unity over the field of rationals. Therefore, as complex numbers,

$$|Q_n(q)| = \prod_{\substack{s=1 \\ \gcd(s,n)=1}}^{n} |q - \zeta^s| > \prod_{\substack{s=1 \\ \gcd(s,n)=1}}^{n} (q - 1) \geqslant q - 1$$

since $n > 1$ and $q \geqslant 2$. This inequality is incompatible with the statement that $Q_n(q)$ divides $q-1$. Hence we must have $n=1$ and $D=Z$, and the theorem is proved. $\square$

Before we start with the second proof of Wedderburn's theorem, we establish some preparatory results. Let D be a finite division ring with center Z, and let F denote a *maximal subfield* of D; that is, F is a subfield of D such that the only subfield of D containing F is F itself. Then F is an extension of Z, for if there were an element $z \in Z$ with $z \notin F$, we could adjoin z to F and obtain a subfield of D properly containing F. From Theorem 2.10 we know that $F = Z(\xi)$, where $\xi \in F^*$ is a root of a monic irreducible polynomial $f \in Z[x]$.

If we view D as a vector space over F, then for each $a \in D$ the assignment $T_a(d) = da$ for $d \in D$ defines a linear operator T_a on this vector space. We consider now the linear operator T_ξ. If d is an eigenvector of T_ξ, then for some $\lambda \in F^*$ we have $d\xi = \lambda d$. This implies $d\xi d^{-1} = \lambda$ and hence $dF^*d^{-1} = F^*$, thus $d \in N(F^*)$, the normalizer of F^* in the group D^*. Conversely, if $d \in N(F^*)$, then $d\xi d^{-1} = \lambda$ for some $\lambda \in F^*$, and so d is an eigenvector of T_ξ. This proves the following result.

   2.56. Lemma.   *An element $d \in D^*$ is an eigenvector of T_ξ if and only if $d \in N(F^*)$.*

Let λ be an eigenvalue of T_ξ with eigenvector d, then $d\xi = \lambda d$. It follows that $0 = df(\xi) = f(\lambda)d$, hence λ must be a root of f. If d_0 is another eigenvector corresponding to the eigenvalue λ, then $d_0 d^{-1} \lambda d d_0^{-1} = \lambda$, and so the element $b = d_0 d^{-1}$ commutes with λ and, consequently, with every element of $F = Z(\lambda)$. Let P be the set of all polynomial expressions in b with coefficients in F. Then it is easily checked that P forms a finite integral domain, and so P is a finite field by Theorem 1.31. But P contains F, and thus $P = F$ by the maximality of F. In particular, we have $b \in F$, and since $d_0 = bd$, we conclude that every eigenspace of T_ξ has dimension 1. We use now the following result from linear algebra.

   2.57. Lemma.   *Let T be a linear operator on the finite-dimensional vector space V over the field K. Then V has a basis consisting of eigenvectors of T if and only if the minimal polynomial for T splits in K into distinct monic linear factors.*

Since $f(\xi) = 0$, the polynomial f annihilates the linear operator T_ξ. Furthermore, f splits in F into distinct monic linear factors by Theorem 2.14. The minimal polynomial for T_ξ divides f, and so it also splits in F into distinct monic linear factors. It follows then from Lemma 2.57 that D has a basis as a vector space over F consisting of eigenvectors of T_ξ. Since every eigenspace of T_ξ has dimension 1, the dimension m of D over F is equal to the number of distinct eigenvalues of T_ξ. Let $\xi = \xi_1, \xi_2, \ldots, \xi_m$ be the distinct eigenvalues of T_ξ and let $1 = d_1, d_2, \ldots, d_m$ be corresponding eigenvectors.

Because $N(F^*)$ is closed under multiplication, it follows from Lemma 2.56 that $d_i d_j$ must correspond to an eigenvalue ξ_k, say, and hence $d_i d_j \xi = \xi_k d_i d_j$. Using $d_j \xi = \xi_j d_j$, we obtain $d_i \xi_j = \xi_k d_i$, or $d_i \xi_j d_i^{-1} = \xi_k$. This shows that for each i, $1 \leqslant i \leqslant m$, the mapping that takes ξ_j to $d_i \xi_j d_i^{-1}$ permutes the eigenvalues among themselves. Consequently, the coefficients of $g(x) = (x - \xi_1) \cdots (x - \xi_m)$ commute with the eigenvectors $d_1, d_2, \ldots, d_m$ of T_ξ. Since the coefficients of g obviously belong to F and thus commute with all the elements of F, they commute with all the elements of D, since these can be written as linear combinations of $d_1, d_2, \ldots, d_m$ with coefficients in F. Thus the coefficients of g are elements of the center Z of D. Since $g(\xi) = 0$, Lemma 2.12 implies that f divides g. On the other hand, we have already observed that every eigenvalue of T_ξ must be a root of f, and so $f = g$. It follows that $[F : Z] = [Z(\xi) : Z] = \deg(f) = m$. Now m is also the dimension of D over F, and so the argument in the proof of Theorem 1.84 shows that D is of dimension m^2 over Z. Since the latter dimension is independent of F, we conclude that every maximal subfield of D has the same degree over Z. We state this result in the following equivalent form.

2.58.   Lemma.   *All maximal subfields of D have the same order.*

Second Proof of Theorem 2.55.   Let D be a finite division ring, and let Z, $F = Z(\xi)$, and $f \in Z[x]$ be as above. Let E be an arbitrary maximal subfield of D. Then, by Lemma 2.58, E and F have the same order, say q. In view of Lemma 2.4, both E and F are splitting fields of $x^q - x$ over Z. It follows then from Theorem 1.91 that there exists an isomorphism from F onto E that keeps the elements of Z fixed. The image $\eta \in E^*$ of ξ under this isomorphism is therefore a root of f in E, and so $E = Z(\eta)$. Consider the linear operator T_η on the vector space D over F. Since $f(\eta) = 0$, the polynomial f annihilates T_η. But f splits in F, and so there exists a root $\lambda \in F$ of f that is an eigenvalue of T_η. For a corresponding eigenvector d we have then $d\eta = \lambda d$, and this implies $E^* = d^{-1}F^*d$. Thus, E^* is a conjugate of the subgroup F^* of D^*.

For an arbitrary $c \in D^*$, the set of polynomial expressions in c with coefficients in Z forms a finite integral domain, and thus a finite field by Theorem 1.31. Hence, any element of D^* is contained in some subfield of D, and so in some maximal subfield of D. From what we have already shown, it follows that any element of D^* belongs to some conjugate of F^*. By Theorem 1.25, the number of distinct conjugates of F^* is given by $|D^*|/|N(F^*)|$, and so it is at most $|D^*|/|F^*|$. Since each conjugate of F^* contains the identity element of D^*, the union of the conjugates of F^* has at most

$$\frac{|D^*|}{|F^*|}(|F^*| - 1) + 1 = |D^*| - \frac{|D^*|}{|F^*|} + 1$$

elements. This number is less than $|D^*|$ except when $D^* = F^*$. Hence $D = F$, and D is a field. $\square$

EXERCISES

2.1. Prove that $x^2 + 1$ is irreducible over $\mathbb{F}_{11}$ and show directly that $\mathbb{F}_{11}[x]/(x^2 + 1)$ has 121 elements. Prove also that $x^2 + x + 4$ is irreducible over $\mathbb{F}_{11}$ and show that $\mathbb{F}_{11}[x]/(x^2 + 1)$ is isomorphic to $\mathbb{F}_{11}[x]/(x^2 + x + 4)$.

2.2. Show that the sum of all elements of a finite field is 0, except for $\mathbb{F}_2$.

2.3. Let a, b be elements of $\mathbb{F}_{2^n}$, n odd. Show that $a^2 + ab + b^2 = 0$ implies $a = b = 0$.

2.4. Determine all primitive elements of $\mathbb{F}_7$.

2.5. Determine all primitive elements of $\mathbb{F}_{17}$.

2.6. Determine all primitive elements of $\mathbb{F}_9$.

2.7. Write all elements of $\mathbb{F}_{25}$ as linear combinations of basis elements over $\mathbb{F}_5$. Then find a primitive element β of $\mathbb{F}_{25}$ and determine for each $\alpha \in \mathbb{F}_{25}^*$ the least nonnegative integer n such that $\alpha = \beta^n$.

2.8. If the elements of $\mathbb{F}_q^*$ are represented as powers of a fixed primitive element $b \in \mathbb{F}_q$, then addition in $\mathbb{F}_q$ is facilitated by the introduction of *Jacobi's logarithm* $L(n)$ defined by the equation $1 + b^n = b^{L(n)}$, where the case $b^n = -1$ is excluded. Show that we have then $b^m + b^n = b^{m+L(n-m)}$ whenever L is defined. Construct a table of Jacobi's logarithm for $\mathbb{F}_9$ and $\mathbb{F}_{17}$.

2.9. Prove: for any field F, every finite subgroup of the multiplicative group F^* is cyclic.

2.10. Let F be any field. If F^* is cyclic, show that F is finite.

2.11. Prove: if F is a finite field, then $H \cup \{0\}$ is a subfield of F for every subgroup H of the multiplicative group F^* if and only if the order of F^* is either 1 or a prime number of the form $2^p - 1$ with a prime p.

2.12. For every finite field $\mathbb{F}_q$ of characteristic p, show that there exists exactly one pth root for each element of $\mathbb{F}_q$.

2.13. For a finite field $\mathbb{F}_q$ with q odd, show that an element $a \in \mathbb{F}_q^*$ has a square root in $\mathbb{F}_q$ if and only if $a^{(q-1)/2} = 1$.

2.14. Prove that for given $k \in \mathbb{N}$ the element $a \in \mathbb{F}_q^*$ is the kth power of some element of $\mathbb{F}_q$ if and only if $a^{(q-1)/d} = 1$, where $d = \gcd(q - 1, k)$.

2.15. Prove: every element of $\mathbb{F}_q$ is the kth power of some element of $\mathbb{F}_q$ if and only if $\gcd(q - 1, k) = 1$.

2.16. Let k be a positive divisor of $q - 1$ and $a \in \mathbb{F}_q$ be such that the equation $x^k = a$ has no solution in $\mathbb{F}_q$. Prove that the same equation has a solution in $\mathbb{F}_{q^m}$ if m is divisible by k, and that the converse holds for a prime number k.

2.17. Prove that $f(x)^q = f(x^q)$ for $f \in \mathbb{F}_q[x]$.

2.18. Show that any quadratic polynomial in $\mathbb{F}_q[x]$ splits over $\mathbb{F}_{q^2}$ into linear factors.

2.19. Show that for $a \in \mathbb{F}_q$ and $n \in \mathbb{N}$ the polynomial $x^{q^n} - x + na$ is divisible by $x^q - x + a$ over $\mathbb{F}_q$.

2.20. Find all automorphisms of a finite field.

2.21. If F is a field and $\Psi : F \to F$ is the mapping defined by $\Psi(a) = a^{-1}$ if $a \neq 0$, $\Psi(a) = 0$ if $a = 0$, show that Ψ is an automorphism of F if and only if F has at most four elements.

2.22. Prove: if p is a prime and n a positive integer, then n divides $\phi(p^n - 1)$. (*Hint:*   Use Corollary 2.19.)

2.23. Let $\mathbb{F}_q$ be a finite field of characteristic p. Prove that $f \in \mathbb{F}_q[x]$ satisfies $f'(x) = 0$ if and only if f is the pth power of some polynomial in $\mathbb{F}_q[x]$.

2.24. Let F be a finite extension of the finite field K with $[F:K] = m$ and let $f(x) = x^d + b_{d-1}x^{d-1} + \cdots + b_0 \in K[x]$ be the minimal polynomial of $\alpha \in F$ over K. Prove that $\mathrm{Tr}_{F/K}(\alpha) = -(m/d)b_{d-1}$ and $N_{F/K}(\alpha) = (-1)^m b_0^{m/d}$.

2.25. Let F be a finite extension of the finite field K and $\alpha \in F$. The mapping $L: \beta \in F \mapsto \alpha\beta \in F$ is a linear transformation of F, considered as a vector space over K. Prove that the characteristic polynomial $g(x)$ of α over K is equal to the characteristic polynomial of the linear transformation L; that is, $g(x) = \det(xI - L)$, where I is the identity transformation.

2.26. Consider the same situation as in Exercise 2.25. Prove that $\mathrm{Tr}_{F/K}(\alpha)$ is equal to the trace of the linear transformation L and that $N_{F/K}(\alpha) = \det(L)$.

2.27. Prove properties (i) and (ii) of Theorem 2.23 by using the interpretation of $\mathrm{Tr}_{F/K}(\alpha)$ obtained in Exercise 2.26.

2.28. Prove properties (i) and (iii) of Theorem 2.28 by using the interpretation of $N_{F/K}(\alpha)$ obtained in Exercise 2.26.

2.29. Let F be a finite extension of the finite field K of characteristic p. Prove that $\mathrm{Tr}_{F/K}(\alpha^{p^n}) = (\mathrm{Tr}_{F/K}(\alpha))^{p^n}$ for all $\alpha \in F$ and $n \in \mathbb{N}$.

2.30. Give an alternative proof of Theorem 2.25 by viewing F as a vector space over K and showing by dimension arguments that the kernel of the linear transformation $\mathrm{Tr}_{F/K}$ is equal to the range of the linear operator L on F defined by $L(\beta) = \beta^q - \beta$ for $\beta \in F$.

2.31. Give an alternative proof of the necessity of the condition in Theorem 2.25 by showing that if $\alpha \in F$ with $\mathrm{Tr}_{F/K}(\alpha) = 0$, $\gamma \in F$ with $\mathrm{Tr}_{F/K}(\gamma) = -1$, and $\delta_j = \alpha + \alpha^q + \cdots + \alpha^{q^{j-1}}$, then

$$\beta = \sum_{j=1}^{[F:K]} \delta_j \gamma^{q^{j-1}}$$

satisfies $\beta^q - \beta = \alpha$.

2.32. Let F be a finite extension of $K = \mathbb{F}_q$ and $\alpha = \beta^q - \beta$ for some $\beta \in F$. Prove that $\alpha = \gamma^q - \gamma$ with $\gamma \in F$ if and only if $\beta - \gamma \in K$.

2.33. Let F be a finite extension of $K = \mathbb{F}_q$. Prove that for $\alpha \in F$ we have $N_{F/K}(\alpha) = 1$ if and only if $\alpha = \beta^{q-1}$ for some $\beta \in F^*$.

2.34. Prove $\sum_{j=0}^{m-1} x^{q^j} - c = \prod (x - \alpha)$ for all $c \in K = \mathbb{F}_q$, where the product is extended over all $\alpha \in F = \mathbb{F}_{q^m}$ with $\mathrm{Tr}_{F/K}(\alpha) = c$.

2.35. Prove

$$x^{q^m} - x = \prod_{c \in \mathbb{F}_q} \left(\sum_{j=0}^{m-1} x^{q^j} - c \right)$$

for any $m \in \mathbb{N}$.

2 36. Consider $\mathbb{F}_{q^m}$ as a vector space over $\mathbb{F}_q$ and prove that for every linear operator L on $\mathbb{F}_{q^m}$ there exists a uniquely determined m-tuple $(\alpha_0, \alpha_1, \ldots, \alpha_{m-1})$ of elements of $\mathbb{F}_{q^m}$ such that

$$L(\beta) = \alpha_0 \beta + \alpha_1 \beta^q + \cdots + \alpha_{m-1} \beta^{q^{m-1}} \text{ for all } \beta \in \mathbb{F}_{q^m}.$$

2.37. Prove that if the order of basis elements is taken into account, then the number of different bases of $\mathbb{F}_{q^m}$ over $\mathbb{F}_q$ is

$$(q^m - 1)(q^m - q)(q^m - q^2) \cdots (q^m - q^{m-1}).$$

2.38. Prove: if $\{\alpha_1, \ldots, \alpha_m\}$ is a basis of $F = \mathbb{F}_{q^m}$ over $K = \mathbb{F}_q$, then $\mathrm{Tr}_{F/K}(\alpha_i) \neq 0$ for at least one i, $1 \leq i \leq m$.

2.39. Prove that there exists a normal basis $\{\alpha, \alpha^q, \ldots, \alpha^{q^{m-1}}\}$ of $F = \mathbb{F}_{q^m}$ over $K = \mathbb{F}_q$ with $\mathrm{Tr}_{F/K}(\alpha) = 1$.

2.40. Let K be a finite field, $F = K(\alpha)$ a finite simple extension of degree n, and $f \in K[x]$ the minimal polynomial of α over K. Let

$$\frac{f(x)}{x - \alpha} = \beta_0 + \beta_1 x + \cdots + \beta_{n-1} x^{n-1} \in F[x] \quad \text{and} \quad \gamma = f'(\alpha).$$

Prove that the dual basis of $\{1, \alpha, \ldots, \alpha^{n-1}\}$ is $\{\beta_0 \gamma^{-1}, \beta_1 \gamma^{-1}, \ldots, \beta_{n-1} \gamma^{-1}\}$.

2.41. Show that there is a self-dual normal basis of $\mathbb{F}_4$ over $\mathbb{F}_2$, but no self-dual normal basis of $\mathbb{F}_{16}$ over $\mathbb{F}_2$ (see Example 2.31 for the definition of a self-dual basis).

2.42. Construct a self-dual basis of $\mathbb{F}_{16}$ over $\mathbb{F}_2$ (see Example 2.31 for the definition of a self-dual basis).

2.43. Prove that the dual basis of a normal basis of $\mathbb{F}_{q^m}$ over $\mathbb{F}_q$ is again a normal basis of $\mathbb{F}_{q^m}$ over $\mathbb{F}_q$.

2.44. Let F be an extension of the finite field K with basis $\{\alpha_1, \ldots, \alpha_m\}$ over K. Let $\beta_1, \ldots, \beta_m \in F$ with $\beta_i = \sum_{j=1}^{m} b_{ij} \alpha_j$ for $1 \leq i \leq m$ and $b_{ij} \in K$. Let B be the $m \times m$ matrix whose (i, j) entry is b_{ij}. Prove that $\Delta_{F/K}(\beta_1, \ldots, \beta_m) = \det(B)^2 \Delta_{F/K}(\alpha_1, \ldots, \alpha_m)$.

2.45.   Let $K = \mathbb{F}_q$ and $F = \mathbb{F}_{q^m}$. Prove that for $\alpha \in F$ we have
$$\Delta_{F/K}(1, \alpha, \ldots, \alpha^{m-1}) = \prod_{0 \leqslant i < j \leqslant m-1} (\alpha^{q^i} - \alpha^{q^j})^2$$

2.46.   Prove that for $\alpha \in F = \mathbb{F}_{q^m}$ with $m \geqslant 2$ and $K = \mathbb{F}_q$ the discriminant $\Delta_{F/K}(1, \alpha, \ldots, \alpha^{m-1})$ is equal to the discriminant of the characteristic polynomial of α over K.

2.47.   Determine the primitive 4th and 8th roots of unity in $\mathbb{F}_9$.

2.48.   Determine the primitive 9th roots of unity in $\mathbb{F}_{19}$.

2.49.   Let ζ be an nth root of unity over a field K. Prove that $1 + \zeta + \zeta^2 + \cdots + \zeta^{n-1} = 0$ or n according as $\zeta \neq 1$ or $\zeta = 1$.

2.50.   For $n \geqslant 2$ let $\zeta_1, \ldots, \zeta_n$ be all the (not necessarily distinct) nth roots of unity over an arbitrary field K. Prove that $\zeta_1^k + \cdots + \zeta_n^k = n$ for $k = 0$ and $\zeta_1^k + \cdots + \zeta_n^k = 0$ for $k = 1, 2, \ldots, n-1$.

2.51.   For an arbitrary field K and an odd positive integer n, show that $K^{(2n)} = K^{(n)}$.

2.52.   Let K be an arbitrary field. Prove that the cyclotomic field $K^{(d)}$ is a subfield of $K^{(n)}$ for any positive divisor d of $n \in \mathbb{N}$. Determine the minimal polynomial over $K^{(4)}$ of a root of unity that can serve as a defining element of $K^{(12)}$ over $K^{(4)}$.

2.53.   Prove that for p prime the $p - 1$ primitive pth roots of unity over $\mathbb{Q}$ are linearly independent over $\mathbb{Q}$ and therefore form a basis of $\mathbb{Q}^{(p)}$ over $\mathbb{Q}$.

2.54.   Let K be an arbitrary field and $n \geqslant 2$. Prove that the polynomial $x^{n-1} + x^{n-2} + \cdots + x + 1$ is irreducible over K only if n is a prime number.

2.55.   Find the least prime p such that $x^{22} + x^{21} + \cdots + x + 1$ is irreducible over $\mathbb{F}_p$.

2.56.   Find the ten least primes p such that $x^{p-1} + x^{p-2} + \cdots + x + 1$ is irreducible over $\mathbb{F}_2$.

2.57.   Prove the following properties of cyclotomic polynomials over a field for which the polynomials exist:

(a) $Q_{mp}(x) = Q_m(x^p)/Q_m(x)$ if p is prime and $m \in \mathbb{N}$ is not divisible by p;

(b) $Q_{mp}(x) = Q_m(x^p)$ for all $m \in \mathbb{N}$ divisible by the prime p;

(c) $Q_{mp^k}(x) = Q_{mp}(x^{p^{k-1}})$ if p is a prime and $m, k \in \mathbb{N}$ are arbitrary;

(d) $Q_{2n}(x) = Q_n(-x)$ if $n \geqslant 3$ and n odd;

(e) $Q_n(0) = 1$ if $n \geqslant 2$;

(f) $Q_n(x^{-1})x^{\phi(n)} = Q_n(x)$ if $n \geqslant 2$;

(g)

$$Q_n(1) = \begin{cases} 0 & \text{if } n = 1, \\ p & \text{if } n \text{ is a power of the prime } p, \\ 1 & \text{if } n \text{ has at least two distinct prime factors;} \end{cases}$$

(h)

$$Q_n(-1) = \begin{cases} 0 & \text{if } n = 2, \\ -2 & \text{if } n = 1, \\ p & \text{if } n \text{ is 2 times a power of the prime } p, \\ 1 & \text{otherwise.} \end{cases}$$

2.58.   Give the matrix representation for the elements of $\mathbb{F}_8$ using the irreducible polynomial $x^3 + x + 1$ over $\mathbb{F}_2$.

2.59.   Let ζ be a primitive element of $F = \mathbb{F}_{16}$ with $\zeta^4 + \zeta + 1 = 0$. For $k \geqslant 0$ write $\zeta^k = \sum_{m=0}^{3} a_{km} \zeta^m$ with $a_{km} \in \mathbb{F}_2$, and let M_k be the 4×4 matrix whose (i, j) entry is $a_{k+i-1, j-1}$. Show that the 15 matrices M_k, $0 \leqslant k \leqslant 14$, and the 4×4 zero matrix form a field (with respect to addition and matrix multiplication over $\mathbb{F}_2$) which is isomorphic to F. For $0 \leqslant k \leqslant 14$ prove that $\operatorname{Tr}_F(\zeta^k) = $ trace of $M_k = a_{k3}$.

Chapter 3

Polynomials over Finite Fields

The theory of polynomials over finite fields is important for investigating the algebraic structure of finite fields as well as for many applications. Above all, irreducible polynomials—the prime elements of the polynomial ring over a finite field—are indispensable for constructing finite fields and computing with the elements of a finite field.

Section 1 introduces the notion of the order of a polynomial. An important fact is the connection between minimal polynomials of primitive elements (so-called primitive polynomials) and polynomials of the highest possible order for a given degree. Results about irreducible polynomials going beyond those discussed in the previous chapters are presented in Section 2. The next section is devoted to constructive aspects of irreducibility and deals also with the problem of calculating the minimal polynomial of an element in an extension field.

Certain special types of polynomials are discussed in the last two sections. Linearized polynomials are singled out by the property that all the exponents occurring in them are powers of the characteristic. The remarkable theory of these polynomials enables us, in particular, to give an alternative proof of the normal basis theorem. Binomials and trinomials —that is, two-term and three-term polynomials—form another class of polynomials for which special results of considerable interest can be established. We remark that another useful collection of polynomials— namely, that of cyclotomic polynomials—was already considered in Chapter

2, Section 4, and that some additional information on cyclotomic polynomials is contained in Section 2 of the present chapter.

1. ORDER OF POLYNOMIALS AND PRIMITIVE POLYNOMIALS

Besides the degree, there is another important integer attached to a nonzero polynomial over a finite field, namely its order. The definition of the order of a polynomial is based on the following result.

3.1. Lemma. *Let* $f \in \mathbb{F}_q[x]$ *be a polynomial of degree* $m \geqslant 1$ *with* $f(0) \neq 0$. *Then there exists a positive integer* $e \leqslant q^m - 1$ *such that* $f(x)$ *divides* $x^e - 1$.

Proof. The residue class ring $\mathbb{F}_q[x]/(f)$ contains $q^m - 1$ nonzero residue classes. The q^m residue classes $x^j + (f)$, $j = 0, 1, \ldots, q^m - 1$, are all nonzero, and so there exist integers r and s with $0 \leqslant r < s \leqslant q^m - 1$ such that $x^s \equiv x^r \bmod f(x)$. Since x and $f(x)$ are relatively prime, it follows that $x^{s-r} \equiv 1 \bmod f(x)$; that is, $f(x)$ divides $x^{s-r} - 1$ and $0 < s - r \leqslant q^m - 1$. $\square$

Since a nonzero constant polynomial divides $x - 1$, these polynomials can be included in the following definition.

3.2. Definition. Let $f \in \mathbb{F}_q[x]$ be a nonzero polynomial. If $f(0) \neq 0$, then the least positive integer e for which $f(x)$ divides $x^e - 1$ is called the *order* of f and denoted by $\mathrm{ord}(f) = \mathrm{ord}(f(x))$. If $f(0) = 0$, then $f(x) = x^h g(x)$, where $h \in \mathbb{N}$ and $g \in \mathbb{F}_q[x]$ with $g(0) \neq 0$ are uniquely determined; $\mathrm{ord}(f)$ is then defined to be $\mathrm{ord}(g)$.

The order of the polynomial f is sometimes also called the *period* of f or the *exponent* of f. The order of an irreducible polynomial f can be characterized in the following alternative fashion.

3.3. Theorem. *Let* $f \in \mathbb{F}_q[x]$ *be an irreducible polynomial over* $\mathbb{F}_q$ *of degree* m *and with* $f(0) \neq 0$. *Then* $\mathrm{ord}(f)$ *is equal to the order of any root of* f *in the multiplicative group* $\mathbb{F}_{q^m}^*$.

Proof. According to Corollary 2.15, $\mathbb{F}_{q^m}$ is the splitting field of f over $\mathbb{F}_q$. The roots of f have the same order in the group $\mathbb{F}_{q^m}^*$ by Theorem 2.18. Let $\alpha \in \mathbb{F}_{q^m}^*$ be any root of f. Then we obtain from Lemma 2.12 that we have $\alpha^e = 1$ if and only if $f(x)$ divides $x^e - 1$. The result follows now from the definitions of $\mathrm{ord}(f)$ and the order of α in the group $\mathbb{F}_{q^m}^*$. $\square$

3.4. Corollary. *If* $f \in \mathbb{F}_q[x]$ *is an irreducible polynomial over* $\mathbb{F}_q$ *of degree* m, *then* $\mathrm{ord}(f)$ *divides* $q^m - 1$.

Proof. If $f(x) = cx$ with $c \in \mathbb{F}_q^*$, then $\mathrm{ord}(f) = 1$ and the result is trivial. Otherwise, the result follows from Theorem 3.3 and the fact that $\mathbb{F}_{q^m}^*$ is a group of order $q^m - 1$.   □

For reducible polynomials the result of Corollary 3.4 need not be valid (see Example 3.10). There is another interpretation of $\mathrm{ord}(f)$ based on associating a square matrix to f in a canonical fashion and considering the order of this matrix in a certain group of matrices (see Lemma 6.26).

Theorem 3.3 leads to a formula for the number of monic irreducible polynomials of given degree and given order. We use again ϕ to denote Euler's function introduced in Theorem 1.15(iv). The following terminology will be convenient: if n is a positive integer and the integer b is relatively prime to n, then the least positive integer k for which $b^k \equiv 1 \bmod n$ is called the *multiplicative order* of b modulo n.

3.5.   Theorem.   *The number of monic irreducible polynomials in $\mathbb{F}_q[x]$ of degree m and order e is equal to $\phi(e)/m$ if $e \geqslant 2$ and m is the multiplicative order of q modulo e, equal to 2 if $m = e = 1$, and equal to 0 in all other cases. In particular, the degree of an irreducible polynomial in $\mathbb{F}_q[x]$ of order e must be equal to the multiplicative order of q modulo e.*

Proof.   Let f be an irreducible polynomial in $\mathbb{F}_q[x]$ with $f(0) \neq 0$. Then, according to Theorem 3.3, we have $\mathrm{ord}(f) = e$ if and only if all roots of f are primitive eth roots of unity over $\mathbb{F}_q$. In other words, we have $\mathrm{ord}(f) = e$ if and only if f divides the cyclotomic polynomial Q_e. By Theorem 2.47(ii), any monic irreducible factor of Q_e has the same degree m, the least positive integer such that $q^m \equiv 1 \bmod e$, and the number of such factors is given by $\phi(e)/m$. For $m = e = 1$, we also have to take into account the monic irreducible polynomial $f(x) = x$.   □

Values of $\mathrm{ord}(f)$ are available in tabulated form, at least for irreducible polynomials f (see Chapter 10, Section 2). Since any polynomial of positive degree can be written as a product of irreducible polynomials, the computation of orders of polynomials can be achieved if one knows how to determine the order of a power of an irreducible polynomial and the order of the product of pairwise relatively prime polynomials. The subsequent discussion is devoted to these questions.

3.6   Lemma.   *Let c be a positive integer. Then the polynomial $f \in \mathbb{F}_q[x]$ with $f(0) \neq 0$ divides $x^c - 1$ if and only if $\mathrm{ord}(f)$ divides c.*

Proof.   If $e = \mathrm{ord}(f)$ divides c, then $f(x)$ divides $x^e - 1$ and $x^e - 1$ divides $x^c - 1$, so that $f(x)$ divides $x^c - 1$. Conversely, if $f(x)$ divides $x^c - 1$, we have $c \geqslant e$, so that we can write $c = me + r$ with $m \in \mathbb{N}$ and $0 \leqslant r < e$. Since $x^c - 1 = (x^{me} - 1)x^r + (x^r - 1)$, it follows that $f(x)$ divides $x^r - 1$, which is only possible for $r = 0$. Therefore, e divides c.   □

3.7.   Corollary.   *If e_1 and e_2 are positive integers, then the greatest common divisor of $x^{e_1} - 1$ and $x^{e_2} - 1$ in $\mathbb{F}_q[x]$ is $x^d - 1$, where d is the greatest common divisor of e_1 and e_2.*

Proof.   Let $f(x)$ be the (monic) greatest common divisor of $x^{e_1} - 1$ and $x^{e_2} - 1$. Since $x^d - 1$ is a common divisor of $x^{e_i} - 1$, $i = 1, 2$, it follows that $x^d - 1$ divides $f(x)$. On the other hand, $f(x)$ is a common divisor of $x^{e_i} - 1$, $i = 1, 2$, and so Lemma 3.6 implies that ord(f) divides e_1 and e_2. Consequently, ord(f) divides d, and hence $f(x)$ divides $x^d - 1$ by Lemma 3.6. Altogether, we have shown that $f(x) = x^d - 1$.   □

Since powers of x are factored out in advance when determining the order of a polynomial, we need not consider powers of the irreducible polynomials $g(x)$ with $g(0) = 0$.

3.8.   Theorem.   *Let $g \in \mathbb{F}_q[x]$ be irreducible over $\mathbb{F}_q$ with $g(0) \neq 0$ and ord(g) $= e$, and let $f = g^b$ with a positive integer b. Let t be the smallest integer with $p^t \geq b$, where p is the characteristic of $\mathbb{F}_q$. Then ord(f) $= ep^t$.*

Proof.   Setting $c = $ord($f$) and noting that the divisibility of $x^c - 1$ by $f(x)$ implies the divisibility of $x^c - 1$ by $g(x)$, we obtain that e divides c by Lemma 3.6. Furthermore, $g(x)$ divides $x^e - 1$; therefore, $f(x)$ divides $(x^e - 1)^b$ and, *a fortiori*, it divides $(x^e - 1)^{p^t} = x^{ep^t} - 1$. Thus according to Lemma 3.6, c divides ep^t. It follows from what we have shown so far that c is of the form $c = ep^u$ with $0 \leq u \leq t$. We note now that $x^e - 1$ has only simple roots, since e is not a multiple of p because of Corollary 3.4. Therefore, all the roots of $x^{ep^u} - 1 = (x^e - 1)^{p^u}$ have multiplicity p^u. But $g(x)^b$ divides $x^{ep^u} - 1$, whence $p^u \geq b$ by comparing multiplicities of roots, and so $u \geq t$. Thus we get $u = t$ and $c = ep^t$.   □

3.9.   Theorem.   *Let $g_1, \ldots, g_k$ be pairwise relatively prime nonzero polynomials over $\mathbb{F}_q$, and let $f = g_1 \cdots g_k$. Then ord(f) is equal to the least common multiple of ord(g_1), ..., ord(g_k).*

Proof.   It is easily seen that it suffices to consider the case where $g_i(0) \neq 0$ for $1 \leq i \leq k$. Set $e = $ord($f$) and $e_i = $ord($g_i$) for $1 \leq i \leq k$, and let $c = \text{lcm}(e_1, \ldots, e_k)$. Then each $g_i(x)$, $1 \leq i \leq k$, divides $x^{e_i} - 1$, and so $g_i(x)$ divides $x^c - 1$. Because of the pairwise relative primality of the polynomials $g_1, \ldots, g_k$, we obtain that $f(x)$ divides $x^c - 1$. An application of Lemma 3.6 shows that e divides c. On the other hand, $f(x)$ divides $x^e - 1$, and so each $g_i(x)$, $1 \leq i \leq k$, divides $x^e - 1$. Again by Lemma 3.6, it follows that each e_i, $1 \leq i \leq k$, divides e, and therefore c divides e. Thus we conclude that $e = c$.   □

By using the same argument as above, one may, in fact, show that the order of the least common multiple of finitely many nonzero polynomials is equal to the least common multiple of the orders of the polynomials.

3.10. Example. Let us compute the order of $f(x) = x^{10} + x^9 + x^3 + x^2 + 1 \in \mathbb{F}_2[x]$. The canonical factorization of $f(x)$ over $\mathbb{F}_2$ is given by $f(x) = (x^2 + x + 1)^3(x^4 + x + 1)$. Since $\operatorname{ord}(x^2 + x + 1) = 3$, we get $\operatorname{ord}((x^2 + x + 1)^3) = 12$ by Theorem 3.8. Furthermore, $\operatorname{ord}(x^4 + x + 1) = 15$, and so Theorem 3.9 implies that $\operatorname{ord}(f)$ is equal to the least common multiple of 12 and 15; that is, $\operatorname{ord}(f) = 60$. Note that $\operatorname{ord}(f)$ does not divide $2^{10} - 1$, which shows that Corollary 3.4 need not hold for reducible polynomials.   □

On the basis of the information provided above, one arrives then at the following general formula for the order of a polynomial. It suffices to consider polynomials of positive degree and with nonzero constant term.

3.11. Theorem. *Let $\mathbb{F}_q$ be a finite field of characteristic p, and let $f \in \mathbb{F}_q[x]$ be a polynomial of positive degree and with $f(0) \neq 0$. Let $f = af_1^{b_1} \cdots f_k^{b_k}$, where $a \in \mathbb{F}_q$, $b_1, \ldots, b_k \in \mathbb{N}$, and $f_1, \ldots, f_k$ are distinct monic irreducible polynomials in $\mathbb{F}_q[x]$, be the canonical factorization of f in $\mathbb{F}_q[x]$. Then $\operatorname{ord}(f) = ep^t$, where e is the least common multiple of $\operatorname{ord}(f_1), \ldots, \operatorname{ord}(f_k)$ and t is the smallest integer with $p^t \geqslant \max(b_1, \ldots, b_k)$.*

A *method of determining the order* of an irreducible polynomial f in $\mathbb{F}_q[x]$ with $f(0) \neq 0$ is based on the observation that the order e of f is the least positive integer such that $x^e \equiv 1 \bmod f(x)$. Furthermore, by Corollary 3.4, e divides $q^m - 1$, where $m = \deg(f)$. Assuming $q^m > 2$, we start from the prime factor decomposition

$$q^m - 1 = \prod_{j=1}^{s} p_j^{r_j}.$$

For $1 \leqslant j \leqslant s$ we calculate the residues of $x^{(q^m - 1)/p_j} \bmod f(x)$. This is accomplished by multiplying together a suitable combination of the residues of $x, x^q, x^{q^2}, \ldots, x^{q^{m-1}} \bmod f(x)$. If $x^{(q^m - 1)/p_j} \not\equiv 1 \bmod f(x)$, then e is a multiple of $p_j^{r_j}$. If $x^{(q^m - 1)/p_j} \equiv 1 \bmod f(x)$, then e is not a multiple of $p_j^{r_j}$. In the latter case we check to see whether e is a multiple of $p_j^{r_j - 1}, p_j^{r_j - 2}, \ldots, p_j$ by calculating the residues of

$$x^{(q^m - 1)/p_j^2}, x^{(q^m - 1)/p_j^3}, \ldots, x^{(q^m - 1)/p_j^{r_j}} \bmod f(x).$$

This computation is repeated for each prime factor of $q^m - 1$.

A key step in the method above is the factorization of the integer $q^m - 1$. There exist extensive tables for the complete factorization of numbers of this form, especially for the case $q = 2$.

We compare now the orders of polynomials obtained from each other by simple algebraic transformations. The following is a typical example.

3.12. Definition. Let

$$f(x) = a_n x^n + a_{n-1} x^{n-1} + \cdots + a_1 x + a_0 \in \mathbb{F}_q[x]$$

with $a_n \neq 0$. Then the *reciprocal polynomial f^** of f is defined by

$$f^*(x) = x^n f\left(\frac{1}{x}\right) = a_0 x^n + a_1 x^{n-1} + \cdots + a_{n-1}x + a_n.$$

3.13.   Theorem.   *Let f be a nonzero polynomial in $\mathbb{F}_q[x]$ and f^* its reciprocal polynomial. Then $\mathrm{ord}(f) = \mathrm{ord}(f^*)$.*

Proof.   First consider the case $f(0) \neq 0$. Then the result follows from the fact that $f(x)$ divides $x^e - 1$ if and only if $f^*(x)$ does. If $f(0) = 0$, write $f(x) = x^h g(x)$ with $h \in \mathbb{N}$ and $g \in \mathbb{F}_q[x]$ satisfying $g(0) \neq 0$. Then from what we have already shown it follows that $\mathrm{ord}(f) = \mathrm{ord}(g) = \mathrm{ord}(g^*) = \mathrm{ord}(f^*)$, where the last identity is valid since $g^* = f^*$.   □

There is also a close relationship between the orders of $f(x)$ and $f(-x)$. Since $f(x) = f(-x)$ for a field of characteristic 2, it suffices to consider finite fields of odd characteristic.

3.14.   Theorem.   *For odd q, let $f \in \mathbb{F}_q[x]$ be a polynomial of positive degree with $f(0) \neq 0$. Let e and E be the orders of $f(x)$ and $f(-x)$, respectively. Then $E = e$ if e is a multiple of 4 and $E = 2e$ if e is odd. If e is twice an odd number, then $E = e/2$ if all irreducible factors of f have even order and $E = e$ otherwise.*

Proof.   Since $\mathrm{ord}(f(x)) = e$, $f(x)$ divides $x^{2e} - 1$, and so $f(-x)$ divides $(-x)^{2e} - 1 = x^{2e} - 1$. Thus E divides $2e$ by Lemma 3.6. By the same argument, e divides $2E$, and so E can only be $2e$, e, or $e/2$. If e is a multiple of 4, then both e and E are even. Since $f(x)$ divides $x^e - 1$, $f(-x)$ divides $(-x)^e - 1 = x^e - 1$, and so E divides e. Similarly, e divides E, and thus it follows that $E = e$. If e is odd, then $f(-x)$ divides $(-x)^e - 1 = -x^e - 1$ and so $x^e + 1$. But then $f(-x)$ cannot divide $x^e - 1$, and so we must have $E = 2e$.

In the remaining case we have $e = 2h$ with an odd integer h. Let f be a power of an irreducible polynomial in $\mathbb{F}_q[x]$. Then $f(x)$ divides $(x^h - 1)(x^h + 1)$ and $f(x)$ does not divide $x^h - 1$ since $\mathrm{ord}(f) = 2h$. But $x^h - 1$ and $x^h + 1$ are relatively prime, and this implies that $f(x)$ divides $x^h + 1$. Consequently, $f(-x)$ divides $(-x)^h + 1 = -x^h + 1$ and so $x^h - 1$. It follows that $E = e/2$. Note that by Theorem 3.8 the power of an irreducible polynomial has even order if and only if the irreducible polynomial itself has even order.

For general f we have a factorization $f = g_1 \cdots g_k$, where each g_i is a power of an irreducible polynomial and $g_1, \ldots, g_k$ are pairwise relatively prime. Furthermore, $2h = \mathrm{lcm}(\mathrm{ord}(g_1), \ldots, \mathrm{ord}(g_k))$ according to Theorem 3.9. We arrange the g_i in such a way that $\mathrm{ord}(g_i) = 2h_i$ for $1 \leq i \leq m$ and $\mathrm{ord}(g_i) = h_i$ for $m + 1 \leq i \leq k$, where the h_i are odd integers with $\mathrm{lcm}(h_1, \ldots, h_k) = h$. By what we have already shown, we get $\mathrm{ord}(g_i(-x)) = h_i$ for $1 \leq i \leq m$ and $\mathrm{ord}(g_i(-x)) = 2h_i$ for $m + 1 \leq i \leq k$. Then Theorem 3.9 yields

$$E = \mathrm{lcm}(h_1, \ldots, h_m, 2h_{m+1}, \ldots, 2h_k),$$

and so $E = h = e/2$ if $m = k$ and $E = 2h = e$ if $m < k$. These formulas are equivalent to those given in the last part of the theorem.   □

It follows from Lemma 3.1 and Definition 3.2 that the order of a polynomial of degree $m \geqslant 1$ over $\mathbb{F}_q$ is at most $q^m - 1$. This bound is attained for an important class of polynomials—namely, so-called primitive polynomials. The definition of a primitive polynomial is based on the notion of primitive element introduced in Definition 2.9.

3.15. Definition. A polynomial $f \in \mathbb{F}_q[x]$ of degree $m \geqslant 1$ is called a *primitive* polynomial over $\mathbb{F}_q$ if it is the minimal polynomial over $\mathbb{F}_q$ of a primitive element of $\mathbb{F}_{q^m}$.

Thus, a primitive polynomial over $\mathbb{F}_q$ of degree m may be described as a monic polynomial that is irreducible over $\mathbb{F}_q$ and has a root $\alpha \in \mathbb{F}_{q^m}$ that generates the multiplicative group of $\mathbb{F}_{q^m}$. Primitive polynomials can also be characterized as follows.

3.16. Theorem. *A polynomial* $f \in \mathbb{F}_q[x]$ *of degree m is a primitive polynomial over* $\mathbb{F}_q$ *if and only if f is monic, $f(0) \neq 0$, and* $\mathrm{ord}(f) = q^m - 1$.

Proof. If f is primitive over $\mathbb{F}_q$, then f is monic and $f(0) \neq 0$. Since f is irreducible over $\mathbb{F}_q$, we get $\mathrm{ord}(f) = q^m - 1$ from Theorem 3.3 and the fact that f has a primitive element of $\mathbb{F}_{q^m}$ as a root.

Conversely, the property $\mathrm{ord}(f) = q^m - 1$ implies that $m \geqslant 1$. Next, we claim that f is irreducible over $\mathbb{F}_q$. Suppose f were reducible over $\mathbb{F}_q$. Then f is either a power of an irreducible polynomial or it can be written as a product of two relatively prime polynomials of positive degree. In the first case, we have $f = g^b$ with $g \in \mathbb{F}_q[x]$ irreducible over $\mathbb{F}_q$, $g(0) \neq 0$, and $b \geqslant 2$. Then, according to Theorem 3.8, $\mathrm{ord}(f)$ is divisible by the characteristic of $\mathbb{F}_q$, but $q^m - 1$ is not, a contradiction. In the second case, we have $f = g_1 g_2$ with relatively prime monic polynomials $g_1, g_2 \in \mathbb{F}_q[x]$ of positive degree m_1 and m_2, respectively. If $e_i = \mathrm{ord}(g_i)$ for $i = 1, 2$, then $\mathrm{ord}(f) \leqslant e_1 e_2$ by Theorem 3.9. Furthermore, $e_i \leqslant q^{m_i} - 1$ for $i = 1, 2$ by Lemma 3.1, hence

$$\mathrm{ord}(f) \leqslant (q^{m_1} - 1)(q^{m_2} - 1) < q^{m_1 + m_2} - 1 = q^m - 1,$$

a contradiction. Therefore, f is irreducible over $\mathbb{F}_q$, and it follows then from Theorem 3.3 that f is a primitive polynomial over $\mathbb{F}_q$.   □

We remark that the condition $f(0) \neq 0$ in the theorem above is only needed to rule out the non-primitive polynomial $f(x) = x$ in case $q = 2$ and $m = 1$. Still another characterization of primitive polynomials is based on the following auxiliary result.

3.17. Lemma. *Let $f \in \mathbb{F}_q[x]$ be a polynomial of positive degree with $f(0) \neq 0$. Let r be the least positive integer for which x^r is congruent* mod $f(x)$ *to some element of $\mathbb{F}_q$, so that $x^r \equiv a \bmod f(x)$ with a uniquely determined*

$a \in \mathbb{F}_q^*$. Then ord$(f) = hr$, where h is the order of a in the multiplicative group $\mathbb{F}_q^*$.

Proof. Put $e = $ ord(f). Since $x^e \equiv 1 \bmod f(x)$, we must have $e \geqslant r$. Thus we can write $e = sr + t$ with $s \in \mathbb{N}$ and $0 \leqslant t < r$. Now

$$1 \equiv x^e \equiv x^{sr+t} \equiv a^s x^t \bmod f(x), \qquad (3.1)$$

thus $x^t \equiv a^{-s} \bmod f(x)$, and because of the definition of r this is only possible if $t = 0$. The congruence (3.1) yields then $a^s \equiv 1 \bmod f(x)$, thus $a^s = 1$, and so $s \geqslant h$ and $e \geqslant hr$. On the other hand, $x^{hr} \equiv a^h \equiv 1 \bmod f(x)$, and so $e = hr$. □

3.18. Theorem. *The monic polynomial $f \in \mathbb{F}_q[x]$ of degree $m \geqslant 1$ is a primitive polynomial over $\mathbb{F}_q$ if and only if $(-1)^m f(0)$ is a primitive element of $\mathbb{F}_q$ and the least positive integer r for which x^r is congruent $\bmod f(x)$ to some element of $\mathbb{F}_q$ is $r = (q^m - 1)/(q - 1)$. In case f is primitive over $\mathbb{F}_q$, we have $x^r \equiv (-1)^m f(0) \bmod f(x)$.*

Proof. If f is primitive over $\mathbb{F}_q$, then f has a root $\alpha \in \mathbb{F}_{q^m}$, which is a primitive element of $\mathbb{F}_{q^m}$. By calculating the norm $N_{\mathbb{F}_{q^m}/\mathbb{F}_q}(\alpha)$ both by Definition 2.27 and by (2.3) and observing that f is the characteristic polynomial of α over $\mathbb{F}_q$, we arrive at the identity

$$(-1)^m f(0) = \alpha^{(q^m - 1)/(q-1)}. \qquad (3.2)$$

It follows that the order of $(-1)^m f(0)$ in $\mathbb{F}_q^*$ is $q - 1$; that is, $(-1)^m f(0)$ is a primitive element of $\mathbb{F}_q$. Since f is the minimal polynomial of α over $\mathbb{F}_q$, the identity (3.2) implies that

$$x^{(q^m - 1)/(q-1)} \equiv (-1)^m f(0) \bmod f(x),$$

and so $r \leqslant (q^m - 1)/(q - 1)$. But Theorem 3.16 and Lemma 3.17 yield $q^m - 1 = $ ord$(f) \leqslant (q - 1)r$, thus $r = (q^m - 1)/(q - 1)$.

Conversely, suppose the conditions of the theorem are satisfied. It follows from $r = (q^m - 1)/(q - 1)$ and Lemma 3.17 that ord(f) is relatively prime to q. Then Theorem 3.11 shows that f has a factorization of the form $f = f_1 \cdots f_k$, where the f_i are distinct monic irreducible polynomials over $\mathbb{F}_q$. If $m_i = \deg(f_i)$, then ord(f_i) divides $q^{m_i} - 1$ for $1 \leqslant i \leqslant k$ according to Corollary 3.4. Now $q^{m_i} - 1$ divides

$$d = (q^{m_1} - 1) \cdots (q^{m_k} - 1)/(q-1)^{k-1},$$

thus ord(f_i) divides d for $1 \leqslant i \leqslant k$. It follows from Lemma 3.6 that $f_i(x)$ divides $x^d - 1$ for $1 \leqslant i \leqslant k$, and so $f(x)$ divides $x^d - 1$. If $k \geqslant 2$, then

$$d < (q^{m_1 + \cdots + m_k} - 1)/(q-1) = (q^m - 1)/(q-1) = r,$$

a contradiction to the definition of r. Thus $k = 1$ and f is irreducible over $\mathbb{F}_q$.

If $\beta \in \mathbb{F}_{q^m}$ is a root of f, then the argument leading to (3.2) shows that $\beta^r = (-1)^m f(0)$, and so $x^r \equiv (-1)^m f(0) \bmod f(x)$. Since the order of $(-1)^m f(0)$ in $\mathbb{F}_q^*$ is $q-1$, it follows from Lemma 3.17 that $\mathrm{ord}(f) = q^m - 1$, so that f is primitive over $\mathbb{F}_q$ by Theorem 3.16. $\square$

3.19. Example. Consider the polynomial $f(x) = x^4 + x^3 + x^2 + 2x + 2 \in \mathbb{F}_3[x]$. Since f is irreducible over $\mathbb{F}_3$, one can use the method outlined after Theorem 3.11 to show that $\mathrm{ord}(f) = 80 = 3^4 - 1$. Consequently, f is primitive over $\mathbb{F}_3$ by Theorem 3.16. We have $x^{40} \equiv 2 \bmod f(x)$ in accordance with Theorem 3.18. $\square$

2. IRREDUCIBLE POLYNOMIALS

We recall that a polynomial $f \in \mathbb{F}_q[x]$ is *irreducible* over $\mathbb{F}_q$ if f has positive degree and every factorization of f in $\mathbb{F}_q[x]$ must involve a constant polynomial (see Definition 1.57). Elementary properties of irreducible polynomials over $\mathbb{F}_q$ were discussed in Chapter 2, Section 2.

3.20. Theorem. *For every finite field $\mathbb{F}_q$ and every $n \in \mathbb{N}$, the product of all monic irreducible polynomials over $\mathbb{F}_q$ whose degrees divide n is equal to $x^{q^n} - x$.*

Proof. According to Lemma 2.13, the monic irreducible polynomials over $\mathbb{F}_q$ occurring in the canonical factorization of $g(x) = x^{q^n} - x$ in $\mathbb{F}_q[x]$ are precisely those whose degrees divide n. Since $g'(x) = -1$, Theorem 1.68 implies that g has no multiple roots in its splitting field over $\mathbb{F}_q$, and so each monic irreducible polynomial over $\mathbb{F}_q$ whose degree divides n occurs exactly once in the canonical factorization of g in $\mathbb{F}_q[x]$. $\square$

3.21. Corollary. *If $N_q(d)$ is the number of monic irreducible polynomials in $\mathbb{F}_q[x]$ of degree d, then*

$$q^n = \sum_{d \mid n} d N_q(d) \quad \text{for all } n \in \mathbb{N}, \tag{3.3}$$

where the sum is extended over all positive divisors d of n.

Proof. The identity (3.3) follows from Theorem 3.20 by comparing the degree of $g(x) = x^{q^n} - x$ with the total degree of the canonical factorization of $g(x)$. $\square$

With a little elementary number theory we can derive from (3.3) an explicit formula for the number of monic irreducible polynomials in $\mathbb{F}_q[x]$ of fixed degree. We need an arithmetic function, called the Moebius function, which is defined as follows.

3.22. **Definition.** The *Moebius function* μ is the function on $\mathbb{N}$ defined by

$$\mu(n) = \begin{cases} 1 & \text{if } n=1, \\ (-1)^k & \text{if } n \text{ is the product of } k \text{ distinct primes,} \\ 0 & \text{if } n \text{ is divisible by the square of a prime.} \end{cases}$$

As in (3.3), we use the summation symbol $\sum_{d \mid n}$ to denote a sum extended over all positive divisors d of $n \in \mathbb{N}$. A similar convention applies to the product symbol $\prod_{d \mid n}$.

3.23. **Lemma.** *For $n \in \mathbb{N}$ the Moebius function μ satisfies*

$$\sum_{d \mid n} \mu(d) = \begin{cases} 1 & \text{if } n=1, \\ 0 & \text{if } n>1. \end{cases}$$

Proof. For $n>1$ we have to take into account only those positive divisors d of n for which $\mu(d) \neq 0$—that is, for which $d=1$ or d is a product of distinct primes. Thus, if $p_1, p_2, \ldots, p_k$ are the distinct prime divisors of n, we get

$$\sum_{d \mid n} \mu(d) = \mu(1) + \sum_{i=1}^{k} \mu(p_i) + \sum_{1 \leq i_1 < i_2 \leq k} \mu(p_{i_1} p_{i_2}) + \cdots + \mu(p_1 p_2 \cdots p_k)$$

$$= 1 + \binom{k}{1}(-1) + \binom{k}{2}(-1)^2 + \cdots + \binom{k}{k}(-1)^k$$

$$= (1+(-1))^k = 0.$$

The case $n=1$ is trivial. $\qquad\qquad\qquad\qquad\qquad\qquad\qquad\qquad\qquad\square$

3.24. **Theorem** (Moebius Inversion Formula)

(i) Additive case: *Let h and H be two functions from $\mathbb{N}$ into an additively written abelian group G. Then*

$$H(n) = \sum_{d \mid n} h(d) \quad \text{for all } n \in \mathbb{N} \tag{3.4}$$

if and only if

$$h(n) = \sum_{d \mid n} \mu\left(\frac{n}{d}\right) H(d) = \sum_{d \mid n} \mu(d) H\left(\frac{n}{d}\right) \quad \text{for all } n \in \mathbb{N}. \tag{3.5}$$

(ii) Multiplicative case: *Let h and H be two functions from $\mathbb{N}$ into a multiplicatively written abelian group G. Then*

$$H(n) = \prod_{d \mid n} h(d) \quad \text{for all } n \in \mathbb{N} \tag{3.6}$$

if and only if

$$h(n) = \prod_{d \mid n} H(d)^{\mu(n/d)} = \prod_{d \mid n} H\left(\frac{n}{d}\right)^{\mu(d)} \quad \text{for all } n \in \mathbb{N}. \tag{3.7}$$

Proof. Assuming (3.4) and using Lemma 3.23, we get

$$\sum_{d\mid n} \mu\left(\frac{n}{d}\right) H(d) = \sum_{d\mid n} \mu(d) H\left(\frac{n}{d}\right) = \sum_{d\mid n} \mu(d) \sum_{c\mid n/d} h(c)$$

$$= \sum_{c\mid n} \sum_{d\mid n/c} \mu(d) h(c) = \sum_{c\mid n} h(c) \sum_{d\mid n/c} \mu(d) = h(n)$$

for all $n \in \mathbb{N}$. The converse is derived by a similar calculation. The proof of part (ii) follows immediately from the proof of part (i) if we replace the sums by products and the multiples by powers.   □

3.25.   Theorem.   *The number $N_q(n)$ of monic irreducible polynomials in $\mathbb{F}_q[x]$ of degree n is given by*

$$N_q(n) = \frac{1}{n} \sum_{d\mid n} \mu\left(\frac{n}{d}\right) q^d = \frac{1}{n} \sum_{d\mid n} \mu(d) q^{n/d}.$$

Proof.   We apply the additive case of the Moebius inversion formula to the group $G = \mathbb{Z}$, the additive group of integers. Let $h(n) = nN_q(n)$ and $H(n) = q^n$ for all $n \in \mathbb{N}$. Then (3.4) is satisfied because of the identity (3.3), and so (3.5) already gives the desired formula.   □

3.26.   Example.   The number of monic irreducible polynomials in $\mathbb{F}_q[x]$ of degree 20 is given by

$$N_q(20) = \tfrac{1}{20}\left(\mu(1)q^{20} + \mu(2)q^{10} + \mu(4)q^5 + \mu(5)q^4 + \mu(10)q^2 + \mu(20)q\right)$$

$$= \tfrac{1}{20}\left(q^{20} - q^{10} - q^4 + q^2\right).$$ □

It should be noted that the formula in Theorem 3.25 shows again that for every finite field $\mathbb{F}_q$ and every $n \in \mathbb{N}$ there exists an irreducible polynomial in $\mathbb{F}_q[x]$ of degree n (compare with Corollary 2.11). Namely, using $\mu(1) = 1$ and $\mu(d) \geq -1$ for all $d \in \mathbb{N}$, a crude estimate yields

$$N_q(n) \geq \frac{1}{n}\left(q^n - q^{n-1} - q^{n-2} - \cdots - q\right) = \frac{1}{n}\left(q^n - \frac{q^n - q}{q-1}\right) > 0.$$

As another application of the Moebius inversion formula, we establish an explicit formula for the nth cyclotomic polynomial Q_n.

3.27.   Theorem.   *For a field K of characteristic p and $n \in \mathbb{N}$ not divisible by p, the nth cyclotomic polynomial Q_n over K satisfies*

$$Q_n(x) = \prod_{d\mid n} (x^d - 1)^{\mu(n/d)} = \prod_{d\mid n} (x^{n/d} - 1)^{\mu(d)}.$$

Proof.   We apply the multiplicative case of the Moebius inversion formula to the multiplicative group G of nonzero rational functions over K. Let $h(n) = Q_n(x)$ and $H(n) = x^n - 1$ for all $n \in \mathbb{N}$. Then Theorem 2.45(i) shows that (3.6) is satisfied, and so (3.7) yields the desired result.   □

3.28.   **Example.**   For fields K over which Q_{12} is defined, we have

$$Q_{12}(x) = \prod_{d \mid 12} (x^{12/d} - 1)^{\mu(d)}$$

$$= (x^{12} - 1)^{\mu(1)}(x^6 - 1)^{\mu(2)}(x^4 - 1)^{\mu(3)}(x^3 - 1)^{\mu(4)}$$

$$(x^2 - 1)^{\mu(6)}(x - 1)^{\mu(12)}$$

$$= \frac{(x^{12} - 1)(x^2 - 1)}{(x^6 - 1)(x^4 - 1)} = x^4 - x^2 + 1. \qquad \square$$

The explicit formula in Theorem 3.27 can be used to establish the basic properties of cyclotomic polynomials (compare with Exercise 3.35).

In Theorem 3.25 we determined the *number* of monic irreducible polynomials in $\mathbb{F}_q[x]$ of fixed degree. We present now a formula for the *product* of all monic irreducible polynomials in $\mathbb{F}_q[x]$ of fixed degree.

3.29.   **Theorem.**   *The product $I(q, n; x)$ of all monic irreducible polynomials in $\mathbb{F}_q[x]$ of degree n is given by*

$$I(q, n; x) = \prod_{d \mid n} (x^{q^d} - x)^{\mu(n/d)} = \prod_{d \mid n} (x^{q^{n/d}} - x)^{\mu(d)}$$

Proof.   It follows from Theorem 3.20 that

$$x^{q^n} - x = \prod_{d \mid n} I(q, d; x).$$

We apply the multiplicative case of the Moebius inversion formula to the multiplicative group G of nonzero rational functions over $\mathbb{F}_q$, putting $h(n) = I(q, n; x)$ and $H(n) = x^{q^n} - x$ for all $n \in \mathbb{N}$, and we obtain the desired formula. $\qquad \square$

3.30.   **Example.**   For $q = 2$, $n = 4$ we get

$$I(2, 4; x) = (x^{16} - x)^{\mu(1)}(x^4 - x)^{\mu(2)}(x^2 - x)^{\mu(4)}$$

$$= \frac{x^{16} - x}{x^4 - x} = \frac{x^{15} - 1}{x^3 - 1}$$

$$= x^{12} + x^9 + x^6 + x^3 + 1. \qquad \square$$

All monic irreducible polynomials in $\mathbb{F}_q[x]$ of degree n can be determined by factoring $I(q, n; x)$. For this purpose it is advantageous to have $I(q, n; x)$ available in a partially factored form. This is achieved by the following result.

3.31.   **Theorem.**   *Let $I(q, n; x)$ be as in Theorem 3.29. Then for $n > 1$ we have*

$$I(q, n; x) = \prod_m Q_m(x), \qquad (3.8)$$

where the product is extended over all positive divisors m of $q^n - 1$ for which n is the multiplicative order of q modulo m, and where $Q_m(x)$ is the mth cyclotomic polynomial over $\mathbb{F}_q$.

Proof. For $n > 1$ let S be the set of elements of $\mathbb{F}_{q^n}$ that are of degree n over $\mathbb{F}_q$. Then every $\alpha \in S$ has a minimal polynomial over $\mathbb{F}_q$ of degree n and is thus a root of $I(q, n; x)$. On the other hand, if β is a root of $I(q, n; x)$, then β is a root of some monic irreducible polynomial in $\mathbb{F}_q[x]$ of degree n, which implies that $\beta \in S$. Therefore,

$$I(q, n; x) = \prod_{\alpha \in S} (x - \alpha).$$

If $\alpha \in S$, then $\alpha \in \mathbb{F}_{q^n}^*$, and so the order of α in that multiplicative group is a divisor of $q^n - 1$. We note that $\gamma \in \mathbb{F}_{q^n}^*$ is an element of a proper subfield $\mathbb{F}_{q^d}$ of $\mathbb{F}_{q^n}$ if and only if $\gamma^{q^d} = \gamma$ —that is, if and only if the order of γ divides $q^d - 1$. Thus, the order m of an element α of S must be such that n is the least positive integer with $q^n \equiv 1 \bmod m$ —that is, such that n is the multiplicative order of q modulo m. For a positive divisor m of $q^n - 1$ with this property, let S_m be the set of elements of S of order m. Then S is the disjoint union of the subsets S_m, so that we can write

$$I(q, n; x) = \prod_m \prod_{\alpha \in S_m} (x - \alpha).$$

Now S_m contains exactly all elements of $\mathbb{F}_{q^n}^*$ of order m. In other words, S_m is the set of primitive mth roots of unity over $\mathbb{F}_q$. From the definition of cyclotomic polynomials (see Definition 2.44), it follows that

$$\prod_{\alpha \in S_m} (x - \alpha) = Q_m(x),$$

and so (3.8) is established. $\square$

3.32. Example. We determine all (monic) irreducible polynomials in $\mathbb{F}_2[x]$ of degree 4. The identity (3.8) yields $I(2, 4; x) = Q_5(x)Q_{15}(x)$. By Theorem 2.47(ii), $Q_5(x) = x^4 + x^3 + x^2 + x + 1$ is irreducible in $\mathbb{F}_2[x]$. By the same theorem, $Q_{15}(x)$ factors into two irreducible polynomials in $\mathbb{F}_2[x]$ of degree 4. Since $Q_5(x + 1) = x^4 + x^3 + 1$ is irreducible in $\mathbb{F}_2[x]$, this polynomial must divide $Q_{15}(x)$, and so

$$Q_{15}(x) = x^8 + x^7 + x^5 + x^4 + x^3 + x + 1 = (x^4 + x^3 + 1)(x^4 + x + 1).$$

Therefore, the irreducible polynomials in $\mathbb{F}_2[x]$ of degree 4 are $x^4 + x^3 + x^2 + x + 1$, $x^4 + x^3 + 1$, and $x^4 + x + 1$. $\square$

Irreducible polynomials often arise as minimal polynomials of elements of an extension field. Minimal polynomials were introduced in Definition 1.81 and their fundamental properties established in Theorem 1.82. With special reference to finite fields, we summarize now the most useful facts about minimal polynomials.

3.33.   Theorem.   *Let α be an element of the extension field $\mathbb{F}_{q^m}$ of $\mathbb{F}_q$. Suppose that the degree of α over $\mathbb{F}_q$ is d and that $g \in \mathbb{F}_q[x]$ is the minimal polynomial of α over $\mathbb{F}_q$. Then:*

   (i)   *g is irreducible over $\mathbb{F}_q$ and its degree d divides m.*
  (ii)   *A polynomial $f \in \mathbb{F}_q[x]$ satisfies $f(\alpha) = 0$ if and only if g divides f.*
 (iii)   *If f is a monic irreducible polynomial in $\mathbb{F}_q[x]$ with $f(\alpha) = 0$, then $f = g$.*
 (iv)   *$g(x)$ divides $x^{q^d} - x$ and $x^{q^m} - x$.*
  (v)   *The roots of g are $\alpha, \alpha^q, \ldots, \alpha^{q^{d-1}}$, and g is the minimal polynomial over $\mathbb{F}_q$ of all these elements.*
 (vi)   *If $\alpha \neq 0$, then $\mathrm{ord}(g)$ is equal to the order of α in the multiplicative group $\mathbb{F}_{q^m}^*$.*
(vii)   *g is a primitive polynomial over $\mathbb{F}_q$ if and only if α is of order $q^d - 1$ in $\mathbb{F}_{q^m}^*$.*

Proof. (i)   The first part follows from Theorem 1.82(i) and the second part from Theorem 1.86.
   (ii)   This follows from Theorem 1.82(ii).
  (iii)   This is an immediate consequence of (ii).
  (iv)   This follows from (i) and Lemma 2.13.
   (v)   The first part follows from (i) and Theorem 2.14 and the second part from (iii).
  (vi)   Since $\alpha \in \mathbb{F}_{q^d}^*$ and $\mathbb{F}_{q^d}^*$ is a subgroup of $\mathbb{F}_{q^m}^*$, the result is contained in Theorem 3.3.
  (vii)   If g is primitive over $\mathbb{F}_q$, then $\mathrm{ord}(g) = q^d - 1$, and so α is of order $q^d - 1$ in $\mathbb{F}_{q^m}^*$ because of (vi). Conversely, if α is of order $q^d - 1$ in $\mathbb{F}_{q^m}^*$ and so in $\mathbb{F}_{q^d}^*$, then α is a primitive element of $\mathbb{F}_{q^d}$, and therefore g is primitive over $\mathbb{F}_q$ by Definition 3.15.   □

3.   CONSTRUCTION OF IRREDUCIBLE POLYNOMIALS

We first describe a general principle of obtaining new irreducible polynomials from known ones. It depends on an auxiliary result from number theory. We recall that if n is a positive integer and the integer b is relatively prime to n, then the least positive integer k for which $b^k \equiv 1 \bmod n$ is called the *multiplicative order* of b modulo n. We note that this multiplicative order divides any other positive integer h for which $b^h \equiv 1 \bmod n$.

3.34.   Lemma.   *Let $s \geqslant 2$ and $e \geqslant 2$ be relatively prime integers and let m be the multiplicative order of s modulo e. Let $t \geqslant 2$ be an integer whose prime factors divide e but not $(s^m - 1)/e$. Assume also that $s^m \equiv 1 \bmod 4$ if $t \equiv 0 \bmod 4$. Then the multiplicative order of s modulo et is equal to mt.*

Proof. We proceed by induction on the number of prime factors of t, each counted with its multiplicity. First, let t be a prime number. Writing $d = (s^m - 1)/e$, we have $s^m = 1 + de$, and so

$$s^{mt} = (1 + de)^t$$

$$= 1 + \binom{t}{1} de + \binom{t}{2} d^2 e^2 + \cdots + \binom{t}{t-1} d^{t-1} e^{t-1} + d^t e^t.$$

In the last expression, each term except the first and the last is divisible by et because of a property of binomial coefficients noted in the proof of Theorem 1.46. Furthermore, the last term is divisible by et since t divides e. Therefore, $s^{mt} \equiv 1 \bmod et$, and so the multiplicative order k of s modulo et divides mt. Also, $s^k \equiv 1 \bmod et$ implies $s^k \equiv 1 \bmod e$, and so k is divisible by m. Since t is a prime number, k can only be m or mt. If $k = m$, then $s^m \equiv 1 \bmod et$, hence $de \equiv 0 \bmod et$ and t divides d, a contradiction. Thus we must have $k = mt$.

Now suppose that t has at least two prime factors and write $t = rt_0$, where r is a prime factor of t. By what we have already shown, the multiplicative order of s modulo er is equal to mr. If we can prove that each prime factor of t_0 divides er but not $d_0 = (s^{mr} - 1)/er$, then the induction hypothesis applied to t_0 yields that the multiplicative order of s modulo $ert_0 = et$ is equal to $mrt_0 = mt$. Let r_0 be a prime factor of t_0. Since every prime factor of t divides e, it is trivial that r_0 divides er. We write again $d = (s^m - 1)/e$. We have $s^{mr} - 1 = c(s^m - 1)$ with $c = s^{m(r-1)} + \cdots + s^m + 1$, thus $d_0 = c(s^m - 1)/er = cd/r$. Furthermore, since $s^m \equiv 1 \bmod e$ and r divides e, we get $s^m \equiv 1 \bmod r$, and so $c \equiv r \equiv 0 \bmod r$. Thus c/r is an integer. Since r_0 does not divide d, it suffices to demonstrate that r_0 does not divide c/r in order to prove that r_0 does not divide $d_0 = cd/r$. We note that $s^m \equiv 1 \bmod r_0$, and so $c \equiv r \bmod r_0$. If $r_0 \neq r$, then $c/r \equiv 1 \bmod r_0$, thus r_0 does not divide c/r. Now let $r_0 = r$. Then $s^m \equiv 1 + br \bmod r^2$ for some $b \in \mathbb{Z}$, hence $s^{mj} \equiv (1 + br)^j \equiv 1 + jbr \bmod r^2$ for all $j \geqslant 0$, and thus

$$c \equiv r + br \sum_{j=0}^{r-1} j \equiv r + br \frac{r(r-1)}{2} \bmod r^2.$$

It follows that

$$\frac{c}{r} \equiv 1 + b \frac{r(r-1)}{2} \bmod r.$$

If r is odd, then $c/r \equiv 1 \bmod r$, so that $r_0 = r$ does not divide c/r. In the remaining case we have $r_0 = r = 2$. Then $t \equiv 0 \bmod 4$, and so $s^m \equiv 1 \bmod 4$ by hypothesis. Since $c = s^m + 1$ in this case, we get $c \equiv 2 \bmod 4$, and thus $c/r = c/2 \equiv 1 \bmod 2$. It follows again that r_0 does not divide c/r. $\square$

3.35.   Theorem.   *Let $f_1(x), f_2(x), \ldots, f_N(x)$ be all the distinct monic irreducible polynomials in $\mathbb{F}_q[x]$ of degree m and order e, and let $t \geqslant 2$ be an* *integer whose prime factors divide e but not $(q^m - 1)/e$. Assume also that $q^m \equiv 1 \bmod 4$ if $t \equiv 0 \bmod 4$. Then $f_1(x^t), f_2(x^t),\ldots,f_N(x^t)$ are all the distinct monic irreducible polynomials in $\mathbb{F}_q[x]$ of degree mt and order et.*

Proof. The condition on e implies $e \geqslant 2$. According to Theorem 3.5, monic irreducible polynomials in $\mathbb{F}_q[x]$ of degree m and order $e \geqslant 2$ exist only if m is the multiplicative order of q modulo e, and then $N = \phi(e)/m$. By Lemma 3.34, the multiplicative order of q modulo et is equal to mt, and since $\phi(et)/mt = \phi(e)/m$ by the formula in Exercise 1.4, part (c), it follows that the number of monic irreducible polynomials in $\mathbb{F}_q[x]$ of degree mt and order et is also equal to N. Therefore, it remains to show that each of the polynomials $f_j(x^t)$, $1 \leqslant j \leqslant N$, is irreducible in $\mathbb{F}_q[x]$ and of order et. Since the roots of each $f_j(x)$ are primitive eth roots of unity over $\mathbb{F}_q$ by Theorem 3.3, it follows that $f_j(x)$ divides the cyclotomic polynomial $Q_e(x)$ over $\mathbb{F}_q$. Then $f_j(x^t)$ divides $Q_e(x^t)$, and repeated use of the property enunciated in Exercise 2.57, part (b), shows that $Q_e(x^t) = Q_{et}(x)$. Thus $f_j(x^t)$ divides $Q_{et}(x)$. According to Theorem 2.47(ii), the degree of each irreducible factor of $Q_{et}(x)$ in $\mathbb{F}_q[x]$ is equal to the multiplicative order of q modulo et, which is mt. Since $f_j(x^t)$ has degree mt, it follows that $f_j(x^t)$ is irreducible in $\mathbb{F}_q[x]$. Furthermore, since $f_j(x^t)$ divides $Q_{et}(x)$, the order of $f_j(x^t)$ is et.   □

3.36. Example. The irreducible polynomials in $\mathbb{F}_2[x]$ of degree 4 and order 15 are $x^4 + x + 1$ and $x^4 + x^3 + 1$. Then the irreducible polynomials in $\mathbb{F}_2[x]$ of degree 12 and order 45 are $x^{12} + x^3 + 1$ and $x^{12} + x^9 + 1$. The irreducible polynomials in $\mathbb{F}_2[x]$ of degree 60 and order 225 are $x^{60} + x^{15} + 1$ and $x^{60} + x^{45} + 1$. The irreducible polynomials in $\mathbb{F}_2[x]$ of degree 100 and order 375 are $x^{100} + x^{25} + 1$ and $x^{100} + x^{75} + 1$.   □

The case in which $t \equiv 0 \bmod 4$ and $q^m \equiv -1 \bmod 4$ is not covered in Theorem 3.35. Here we must have $q \equiv -1 \bmod 4$ and m odd. The result referring to this case is somewhat more complicated than Theorem 3.35.

3.37. Theorem. *Let $f_1(x), f_2(x),\ldots,f_N(x)$ be all the distinct monic irreducible polynomials in $\mathbb{F}_q[x]$ of odd degree m and of order e. Let $q = 2^a u - 1$, $t = 2^b v$ with $a, b \geqslant 2$, where u and v are odd and all prime factors of t divide e but not $(q^m - 1)/e$. Let k be the smaller of a and b. Then each of the polynomials $f_j(x^t)$ factors as a product of 2^{k-1} monic irreducible polynomials $g_{ij}(x)$ in $\mathbb{F}_q[x]$ of degree $mt2^{1-k}$. The $2^{k-1}N$ polynomials $g_{ij}(x)$ are all the distinct monic irreducible polynomials in $\mathbb{F}_q[x]$ of degree $mt2^{1-k}$ and order et.*

Proof. If $v \geqslant 3$, then Theorem 3.35 implies that $f_1(x^v), f_2(x^v),\ldots,f_N(x^v)$ are all the distinct monic irreducible polynomials in $\mathbb{F}_q[x]$ of odd degree mv and of order ev. Thus we will be done once the special case $t = 2^b$ is settled.

Let now $t = 2^b$, and note that as in the proof of Theorem 3.35 we obtain that m is the multiplicative order of q modulo e, $N = \phi(e)/m$, and each $f_j(x^t)$ divides $Q_{et}(x)$. By Theorem 2.47(ii), $Q_{et}(x)$ factors into distinct monic irreducible polynomials in $\mathbb{F}_q[x]$ of degree d, where d is the multiplicative order of q modulo et. Since $q^d \equiv 1 \bmod et$, we have $q^d \equiv 1 \bmod e$, and so m divides d. Consider first the case $a \geq b$. Then $q^{2m} - 1 = (q^m - 1)(q^m + 1)$, and the first factor is divisible by e, whereas the second factor is divisible by t since $q \equiv -1 \bmod 2^a$ implies $q \equiv -1 \bmod t$, and thus $q^m \equiv (-1)^m \equiv -1 \bmod t$. Altogether, we get $q^{2m} \equiv 1 \bmod et$, and so d can only be m or $2m$. If $d = m$, then $q^m \equiv 1 \bmod et$, hence $q^m \equiv 1 \bmod t$, a contradiction. Thus $d = 2m = m2^{b-k+1}$ since $k = b$ in this case.

Now consider the case $a < b$. We prove by induction on h that

$$q^{m2^h} \equiv 1 + w2^{a+h} \bmod 2^{a+h+1} \quad \text{for all } h \in \mathbb{N},\qquad(3.9)$$

where w is odd. For $h = 1$ we get

$$q^{2m} = (2^a u - 1)^{2m}$$

$$= 1 - 2^{a+1} um + \sum_{n=2}^{2m} \binom{2m}{n}(-1)^{2m-n} 2^{na} u^n \equiv 1 + w2^{a+1} \bmod 2^{a+2}$$

with $w = -um$. If (3.9) is shown for some $h \in \mathbb{N}$, then

$$q^{m2^h} = 1 + w2^{a+h} + c2^{a+h+1} \quad \text{for some } c \in \mathbb{Z}.$$

It follows that

$$q^{m2^{h+1}} = \left(1 + w2^{a+h} + c2^{a+h+1}\right)^2 \equiv 1 + w2^{a+h+1} \bmod 2^{a+h+2},$$

and so the proof of (3.9) is complete. Applying (3.9) with $h = b - a + 1$, we get $q^{m2^{b-a+1}} \equiv 1 \bmod 2^{b+1}$. Furthermore, $q^m \equiv 1 \bmod e$ implies $q^{m2^{b-a+1}} \equiv 1 \bmod e$, and so $q^{m2^{b-a+1}} \equiv 1 \bmod L$, where L is the least common multiple of 2^{b+1} and e. Now e is even since all prime factors of t divide e, but also $e \not\equiv 0 \bmod 4$ since $q^m \equiv 1 \bmod e$ and $q^m \equiv -1 \bmod 4$. Therefore, $L = e2^b = et$, and thus $q^{m2^{b-a+1}} \equiv 1 \bmod et$. On the other hand, using (3.9) with $h = b - a$ we get

$$q^{m2^{b-a}} \equiv 1 + w2^b \not\equiv 1 \bmod 2^{b+1},$$

which implies $q^{m2^{b-a}} \not\equiv 1 \bmod et$. Consequently, we must have $d = m2^{b-a+1} = m2^{b-k+1}$ since $k = a$ in this case. Therefore, the formula $d = m2^{b-k+1} = mt2^{1-k}$ is valid in both cases.

Since $Q_{et}(x)$ factors into distinct monic irreducible polynomials in $\mathbb{F}_q[x]$ of degree $mt2^{1-k}$, each $f_j(x^t)$ factors into such polynomials. By comparing degrees, the number of factors is found to be 2^{k-1}. Since each irreducible factor $g_{ij}(x)$ of $f_j(x^t)$ divides $Q_{et}(x)$, each $g_{ij}(x)$ is of order et. The various polynomials $g_{ij}(x)$, $1 \leq i \leq 2^{k-1}$, $1 \leq j \leq N$, are distinct, for otherwise one such polynomial, say $g(x)$, would divide $f_{j_1}(x^t)$ and $f_{j_2}(x^t)$ for $j_1 \neq j_2$, and then any root β of $g(x)$ would lead to a common root β^t of $f_{j_1}(x)$ and $f_{j_2}(x)$, a contradiction. By Theorem 3.5, the number of monic irreducible polynomials in $\mathbb{F}_q[x]$ of degree $mt2^{1-k}$ and order et is $\phi(et)/mt2^{1-k} = 2^{k-1}\phi(et)/mt = 2^{k-1}\phi(e)/m = 2^{k-1}N$, and so the $g_{ij}(x)$ yield all such polynomials.   □

We will show how, from a given irreducible polynomial of order e, all the irreducible polynomials whose orders divide e may be obtained. Since in all cases $g(x) = x$ will be among the latter polynomials, we only consider polynomials g with $g(0) \neq 0$. Let f be a monic irreducible polynomial in $\mathbb{F}_q[x]$ of degree m and order e and with $f(0) \neq 0$. Let $\alpha \in \mathbb{F}_{q^m}$ be a root of f, and for every $t \in \mathbb{N}$ let $g_t \in \mathbb{F}_q[x]$ be the minimal polynomial of α^t over $\mathbb{F}_q$. Let $T = \{t_1, t_2, \ldots, t_n\}$ be a set of positive integers such that for each $t \in \mathbb{N}$ there exists a uniquely determined i, $1 \leqslant i \leqslant n$, with $t \equiv t_i q^b \bmod e$ for some integer $b \geqslant 0$. Such a set T can, for instance, be constructed as follows. Put $t_1 = 1$ and, when $t_1, t_2, \ldots, t_{j-1}$ have been constructed, let t_j be the least positive integer such that $t_j \not\equiv t_i q^b \bmod e$ for $1 \leqslant i < j$ and all integers $b \geqslant 0$. This procedure stops after finitely many steps.

With the notation introduced above, we have then the following general result.

3.38.   Theorem.   *The polynomials $g_{t_1}, g_{t_2}, \ldots, g_{t_n}$ are all the distinct monic irreducible polynomials in $\mathbb{F}_q[x]$ whose orders divide e and whose constant terms are nonzero.*

Proof.   Each g_{t_i} is monic and irreducible in $\mathbb{F}_q[x]$ by definition and satisfies $g_{t_i}(0) \neq 0$. Furthermore, since g_{t_i} has the root α^{t_i} whose order in the group $\mathbb{F}_{q^m}^*$ divides the order of α, it follows from Theorem 3.3 that $\mathrm{ord}(g_{t_i})$ divides e.

Let g be an arbitrary monic irreducible polynomial in $\mathbb{F}_q[x]$ of order d dividing e and with $g(0) \neq 0$. If β is a root of g, then $\beta^d = 1$ implies $\beta^e = 1$, and so β is an eth root of unity over $\mathbb{F}_q$. Since α is a primitive eth root of unity over $\mathbb{F}_q$, it follows from Theorem 2.42(i) that $\beta = \alpha^t$ for some $t \in \mathbb{N}$. Then the definition of the set T implies that $t \equiv t_i q^b \bmod e$ for some i, $1 \leqslant i \leqslant n$, and some $b \geqslant 0$. Hence $\beta = \alpha^t = (\alpha^{t_i})^{q^b}$, and so β is a root of g_{t_i} because of Theorem 2.14. Since g is the minimal polynomial of β over $\mathbb{F}_q$, it follows from Theorem 3.33(iii) that $g = g_{t_i}$.

It remains to show that the polynomials g_{t_i}, $1 \leqslant i \leqslant n$, are distinct. Suppose $g_{t_i} = g_{t_j}$ for $i \neq j$. Then α^{t_i} and α^{t_j} are roots of g_{t_i}, and so $\alpha^{t_j} = (\alpha^{t_i})^{q^b}$ for some $b \geqslant 0$. This implies $t_j \equiv t_i q^b \bmod e$, but since we also have $t_j \equiv t_j q^0 \bmod e$, we obtain a contradiction to the definition of the set T.   □

The minimal polynomial g_t of $\alpha^t \in \mathbb{F}_{q^m}$ over $\mathbb{F}_q$ is usually calculated by means of the characteristic polynomial f_t of $\alpha^t \in \mathbb{F}_{q^m}$ over $\mathbb{F}_q$. From the discussion following Definition 2.22 we know that $f_t = g_t^r$, where $r = m/k$ and k is the degree of g_t. Since g_t is irreducible in $\mathbb{F}_q[x]$, k is the multiplicative order of q modulo $d = \mathrm{ord}(g_t)$, and d is equal to the order of

α^t in the group $\mathbb{F}_{q^m}^*$, which is $e/\gcd(t, e)$ by Theorem 1.15(ii). Therefore d, and so k and r, can be determined easily.

Several methods are known for calculating f_t. One of them is based on a useful relationship between f_t and the given polynomial f.

3.39.   Theorem.   *Let f be a monic irreducible polynomial in $\mathbb{F}_q[x]$ of degree m. Let $\alpha \in \mathbb{F}_{q^m}$ be a root of f, and for $t \in \mathbb{N}$ let f_t be the characteristic polynomial of $\alpha^t \in \mathbb{F}_{q^m}$ over $\mathbb{F}_q$. Then*

$$f_t(x^t) = (-1)^{m(t+1)} \prod_{j=1}^{t} f(\omega_j x),$$

where $\omega_1, \ldots, \omega_t$ are the t th roots of unity over $\mathbb{F}_q$ counted according to multiplicity.

Proof.   Let $\alpha = \alpha_1, \alpha_2, \ldots, \alpha_m$ be all the roots of f. Then $\alpha_1^t, \alpha_2^t, \ldots, \alpha_m^t$ are the roots of f_t counted according to multiplicity. Thus

$$f_t(x^t) = \prod_{i=1}^{m} (x^t - \alpha_i^t)$$

$$= \prod_{i=1}^{m} \prod_{j=1}^{t} (x - \alpha_i \omega_j)$$

$$= \prod_{i=1}^{m} \prod_{j=1}^{t} \omega_j (\omega_j^{-1} x - \alpha_i).$$

A comparison of coefficients in the identity

$$x^t - 1 = \prod_{j=1}^{t} (x - \omega_j)$$

shows that

$$\prod_{j=1}^{t} \omega_j = (-1)^{t+1},$$

and so

$$f_t(x^t) = (-1)^{m(t+1)} \prod_{j=1}^{t} \prod_{i=1}^{m} (\omega_j^{-1} x - \alpha_i)$$

$$= (-1)^{m(t+1)} \prod_{j=1}^{t} f(\omega_j^{-1} x) = (-1)^{m(t+1)} \prod_{j=1}^{t} f(\omega_j x)$$

since $\omega_1^{-1}, \ldots, \omega_t^{-1}$ run exactly through all t th roots of unity over $\mathbb{F}_q$.   □

3.40.   Example.   Consider the irreducible polynomial $f(x) = x^4 + x + 1$ in $\mathbb{F}_2[x]$. To calculate f_3, we note that the third roots of unity over $\mathbb{F}_2$ are 1, ω, and ω^2, where ω is a root of $x^2 + x + 1$ in $\mathbb{F}_4$. Then

$$f_3(x^3) = (-1)^{16} f(x) f(\omega x) f(\omega^2 x)$$
$$= (x^4 + x + 1)(\omega x^4 + \omega x + 1)(\omega^2 x^4 + \omega^2 x + 1)$$
$$= x^{12} + x^9 + x^6 + x^3 + 1,$$

so that $f_3(x) = x^4 + x^3 + x^2 + x + 1$.   □

Another method of calculating f_t is based on matrix theory. Let $f(x) = x^m - a_{m-1} x^{m-1} - \cdots - a_1 x - a_0$ and let A be the *companion matrix* of f, which is defined to be the $m \times m$ matrix

$$A = \begin{pmatrix} 0 & 0 & \cdots & 0 & a_0 \\ 1 & 0 & \cdots & 0 & a_1 \\ 0 & 1 & \cdots & 0 & a_2 \\ \vdots & \vdots & & \vdots & \vdots \\ 0 & 0 & \cdots & 1 & a_{m-1} \end{pmatrix}$$

Then f is the *characteristic polynomial* of A in the sense of linear algebra; that is, $f(x) = \det(xI - A)$ with I being the $m \times m$ identity matrix over $\mathbb{F}_q$. For each $t \in \mathbb{N}$, f_t is the characteristic polynomial of A^t, the tth power of A. Thus, by calculating the powers of A one obtains the polynomials f_t.

3.41. Example. It is of interest to determine which polynomials f_t are irreducible in $\mathbb{F}_q[x]$. From the discussion prior to Theorem 3.39 it follows immediately that f_t is irreducible in $\mathbb{F}_q[x]$ if and only if $k = m$, that is, if and only if m is the multiplicative order of q modulo $d = e/\gcd(t, e)$. Consider, for instance, the case $q = 2$, $m = 6$, $e = 63$. Since the multiplicative order of q modulo a divisor of e must be a divisor of m, the only possibilities for the multiplicative order apart from m are $k = 1, 2, 3$. Then $q^k - 1 = 1, 3, 7$, and $q^k \equiv 1 \bmod d$ is only possible when $d = 1, 3, 7$. Thus f_t is reducible in $\mathbb{F}_2[x]$ precisely if $\gcd(t, 63) = 9, 21, 63$. Since it suffices to consider values of t with $1 \leqslant t \leqslant 63$, it follows that f_t is irreducible in $\mathbb{F}_2[x]$ except when $t = 9, 18, 21, 27, 36, 42, 45, 54, 63$.   □

In practice, irreducible polynomials often arise as minimal polynomials of elements in an extension field. If in the discussion above we let f be a primitive polynomial over $\mathbb{F}_q$, so that $e = q^m - 1$, then the powers of α run through all nonzero elements of $\mathbb{F}_{q^m}$. Therefore, the methods outlined above can be used to calculate the minimal polynomial over $\mathbb{F}_q$ of each element of $\mathbb{F}_{q^m}^*$.

A straightforward *method of determining minimal polynomials* is the following one. Let θ be a defining element of $\mathbb{F}_{q^m}$ over $\mathbb{F}_q$, so that $\{1, \theta, \ldots, \theta^{m-1}\}$ is a basis of $\mathbb{F}_{q^m}$ over $\mathbb{F}_q$. In order to find the minimal polynomial g of $\beta \in \mathbb{F}_{q^m}^*$ over $\mathbb{F}_q$, we express the powers $\beta^0, \beta^1, \ldots, \beta^m$ in terms of the basis elements. Let

$$\beta^{i-1} = \sum_{j=1}^{m} b_{ij}\theta^{j-1} \quad \text{for } 1 \leqslant i \leqslant m+1.$$

We write g in the form $g(x) = c_m x^m + \cdots + c_1 x + c_0$. We want g to be the monic polynomial of least positive degree with $g(\beta) = 0$. The condition $g(\beta) = c_m \beta^m + \cdots + c_1 \beta + c_0 = 0$ leads to the homogeneous system of linear equations

$$\sum_{i=1}^{m+1} c_{i-1} b_{ij} = 0 \quad \text{for } 1 \leqslant j \leqslant m \tag{3.10}$$

with unknowns $c_0, c_1, \ldots, c_m$. Let B be the matrix of coefficients of the system—that is, B is the $(m+1) \times m$ matrix whose (i, j) entry is b_{ij}—and let r be the rank of B. Then the dimension of the space of solutions of the system is $s = m + 1 - r$, and since $1 \leqslant r \leqslant m$, we have $1 \leqslant s \leqslant m$. Therefore, we can prescribe values for s of the unknowns $c_0, c_1, \ldots, c_m$, and then the remaining ones are uniquely determined. If $s = 1$, we set $c_m = 1$, and if $s > 1$, we set $c_m = c_{m-1} = \cdots = c_{m-s+2} = 0$ and $c_{m-s+1} = 1$.

3.42. Example. Let $\theta \in \mathbb{F}_{64}$ be a root of the irreducible polynomial $x^6 + x + 1$ in $\mathbb{F}_2[x]$. For $\beta = \theta^3 + \theta^4$ we have

$$\begin{aligned}
\beta^0 &= 1 \\
\beta^1 &= \qquad\qquad\quad \theta^3 + \theta^4 \\
\beta^2 &= 1 + \theta + \theta^2 + \theta^3 \\
\beta^3 &= \quad\ \theta + \theta^2 + \theta^3 \\
\beta^4 &= \quad\ \theta + \theta^2 \qquad\ + \theta^4 \\
\beta^5 &= 1 \qquad\qquad\quad + \theta^3 + \theta^4 \\
\beta^6 &= 1 + \theta + \theta^2 \qquad\ + \theta^4
\end{aligned}$$

Therefore, the matrix B is given by

$$B = \begin{pmatrix}
1 & 0 & 0 & 0 & 0 & 0 \\
0 & 0 & 0 & 1 & 1 & 0 \\
1 & 1 & 1 & 1 & 0 & 0 \\
0 & 1 & 1 & 1 & 0 & 0 \\
0 & 1 & 1 & 0 & 1 & 0 \\
1 & 0 & 0 & 1 & 1 & 0 \\
1 & 1 & 1 & 0 & 1 & 0
\end{pmatrix}$$

and its rank is $r = 3$. Hence $s = m + 1 - r = 4$, so that we set $c_6 = c_5 = c_4 = 0$, $c_3 = 1$. The remaining coefficients are determined from (3.10), and this yields $c_2 = 1, c_1 = 0, c_0 = 1$. Consequently, the minimal polynomial of β over $\mathbb{F}_2$ is $g(x) = x^3 + x^2 + 1$.   □

Still another *method of determining minimal polynomials* is based on Theorem 3.33(v). If we wish to find the minimal polynomial g of $\beta \in \mathbb{F}_{q^m}$

over $\mathbb{F}_q$, we calculate the powers $\beta, \beta^q, \beta^{q^2}, \dots$ until we find the least positive integer d for which $\beta^{q^d} = \beta$. This integer d is the degree of g, and g itself is given by

$$g(x) = (x - \beta)(x - \beta^q) \cdots (x - \beta^{q^{d-1}}).$$

The elements $\beta, \beta^q, \dots, \beta^{q^{d-1}}$ are the distinct conjugates of β with respect to $\mathbb{F}_q$, and g is the minimal polynomial over $\mathbb{F}_q$ of all these elements.

3.43. Example. We compute the minimal polynomials over $\mathbb{F}_2$ of all elements of $\mathbb{F}_{16}$. Let $\theta \in \mathbb{F}_{16}$ be a root of the primitive polynomial $x^4 + x + 1$ over $\mathbb{F}_2$, so that every nonzero element of $\mathbb{F}_{16}$ can be written as a power of θ. We have the following index table for $\mathbb{F}_{16}$:

i	θ^i		i	θ^i
0	1		8	$1 + \theta^2$
1	θ		9	$\theta + \theta^3$
2	θ^2		10	$1 + \theta + \theta^2$
3	θ^3		11	$\theta + \theta^2 + \theta^3$
4	$1 + \theta$		12	$1 + \theta + \theta^2 + \theta^3$
5	$\theta + \theta^2$		13	$1 + \theta^2 + \theta^3$
6	$\theta^2 + \theta^3$		14	$1 + \theta^3$
7	$1 + \theta + \theta^3$			

The minimal polynomials of the elements β of $\mathbb{F}_{16}$ over $\mathbb{F}_2$ are:

$\beta = 0$: $g_1(x) = x$.

$\beta = 1$: $g_2(x) = x + 1$.

$\beta = \theta$:   The distinct conjugates of θ with respect to $\mathbb{F}_2$ are $\theta, \theta^2, \theta^4, \theta^8$, and the minimal polynomial is

$$g_3(x) = (x - \theta)(x - \theta^2)(x - \theta^4)(x - \theta^8)$$
$$= x^4 + x + 1.$$

$\beta = \theta^3$:   The distinct conjugates of θ^3 with respect to $\mathbb{F}_2$ are $\theta^3, \theta^6, \theta^{12}, \theta^{24} = \theta^9$, and the minimal polynomial is

$$g_4(x) = (x - \theta^3)(x - \theta^6)(x - \theta^9)(x - \theta^{12})$$
$$= x^4 + x^3 + x^2 + x + 1.$$

$\beta = \theta^5$:   Since $\beta^4 = \beta$, the distinct conjugates of this element with respect to $\mathbb{F}_2$ are θ^5, θ^{10}, and the minimal polynomial is

$$g_5(x) = (x - \theta^5)(x - \theta^{10}) = x^2 + x + 1.$$

$\beta = \theta^7$:   The distinct conjugates of θ^7 with respect to $\mathbb{F}_2$ are $\theta^7, \theta^{14}, \theta^{28} = \theta^{13}, \theta^{56} = \theta^{11}$, and the minimal polynomial is

$$g_6(x) = (x - \theta^7)(x - \theta^{11})(x - \theta^{13})(x - \theta^{14})$$
$$= x^4 + x^3 + 1.$$

These elements, together with their conjugates with respect to $\mathbb{F}_2$, exhaust $\mathbb{F}_{16}$. $\qquad\qquad\qquad\qquad\qquad\qquad\qquad\qquad\qquad\qquad\qquad\qquad\qquad\square$

An important problem is that of the *determination of primitive polynomials*. One approach is based on the fact that the product of all primitive polynomials over $\mathbb{F}_q$ of degree m is equal to the cyclotomic polynomial Q_e with $e = q^m - 1$ (see Theorem 2.47(ii) and Exercise 3.42). Therefore, all primitive polynomials over $\mathbb{F}_q$ of degree m can be determined by applying one of the factorization algorithms in Chapter 4 to the cyclotomic polynomial Q_e.

Another method depends on constructing a primitive element of $\mathbb{F}_{q^m}$ and then determining the minimal polynomial of this element over $\mathbb{F}_q$ by the methods described above. To find a primitive element of $\mathbb{F}_{q^m}$, one starts from the order $q^m - 1$ of such an element in the group $\mathbb{F}_{q^m}^*$ and factors it in the form $q^m - 1 = h_1 \cdots h_k$, where the positive integers $h_1, \ldots, h_k$ are pairwise relatively prime. If for each $i, 1 \le i \le k$, one can find an element $\alpha_i \in \mathbb{F}_{q^m}^*$ of order h_i, then the product $\alpha_1 \cdots \alpha_k$ has order $q^m - 1$ and is thus a primitive element of $\mathbb{F}_{q^m}$.

3.44.   Example.   We determine a primitive polynomial over $\mathbb{F}_3$ of degree 4. Since $3^4 - 1 = 16 \cdot 5$, we first construct two elements of $\mathbb{F}_{81}^*$ of order 16 and 5, respectively. The elements of order 16 are the roots of the cyclotomic polynomial $Q_{16}(x) = x^8 + 1 \in \mathbb{F}_3[x]$. Since the multiplicative order of 3 modulo 16 is 4, Q_{16} factors into two monic irreducible polynomials in $\mathbb{F}_3[x]$ of degree 4. Now

$$x^8 + 1 = (x^4 - 1)^2 - x^4$$
$$= (x^4 - 1 + x^2)(x^4 - 1 - x^2),$$

and so $f(x) = x^4 - x^2 - 1$ is irreducible over $\mathbb{F}_3$ and with a root θ of f we have $\mathbb{F}_{81} = \mathbb{F}_3(\theta)$. Furthermore, θ is an element of $\mathbb{F}_{81}^*$ of order 16. To find an element α of order 5, we write $\alpha = a + b\theta + c\theta^2 + d\theta^3$ with $a, b, c, d \in \mathbb{F}_3$, and since we must have $\alpha^{10} = 1$, we get

$$1 = \alpha^9 \alpha = (a + b\theta^9 + c\theta^{18} + d\theta^{27})(a + b\theta + c\theta^2 + d\theta^3)$$
$$= (a - b\theta + c\theta^2 - d\theta^3)(a + b\theta + c\theta^2 + d\theta^3)$$
$$= (a + c\theta^2)^2 - (b\theta + d\theta^3)^2 = a^2 + (2ac - b^2)\theta^2 + (c^2 - 2bd)\theta^4 - d^2\theta^6$$
$$= a^2 + c^2 - d^2 + bd + (c^2 + d^2 - b^2 - ac + bd)\theta^2.$$

A comparison of coefficients yields

$$a^2 + c^2 - d^2 + bd = 1, \qquad c^2 + d^2 - b^2 - ac + bd = 0.$$

Setting $a = d = 0$, we get $b^2 = c^2 = 1$. Take $b = c = 1$, and then it is easily checked that $\alpha = \theta + \theta^2$ has order 5. Therefore, $\zeta = \theta\alpha = \theta^2 + \theta^3$ has order 80 and is thus a primitive element of $\mathbb{F}_{81}$. The minimal polynomial g of ζ

over $\mathbb{F}_3$ is

$$
\begin{aligned}
g(x) &= (x - \zeta)(x - \zeta^3)(x - \zeta^9)(x - \zeta^{27}) \\
&= (x - \theta^2 - \theta^3)(x - 1 + \theta + \theta^2)(x - \theta^2 + \theta^3)(x - 1 - \theta + \theta^2) \\
&= x^4 + x^3 + x^2 - x - 1,
\end{aligned}
$$

and we have thus obtained a primitive polynomial over $\mathbb{F}_3$ of degree 4. □

3.45. Example. We determine a primitive polynomial over $\mathbb{F}_2$ of degree 6. Since $2^6 - 1 = 9 \cdot 7$, we first construct two elements of $\mathbb{F}_{64}^*$ of order 9 and 7, respectively. The multiplicative order of 2 modulo 9 is 6, and so the cyclotomic polynomial $Q_9(x) = x^6 + x^3 + 1$ is irreducible over $\mathbb{F}_2$. A root θ of Q_9 has order 9 and $\mathbb{F}_{64} = \mathbb{F}_2(\theta)$. An element $\alpha \in \mathbb{F}_{64}^*$ of order 7 satisfies $\alpha^8 = \alpha$, thus writing $\alpha = \sum_{i=0}^{5} a_i \theta^i$ with $a_i \in \mathbb{F}_2$, $0 \leqslant i \leqslant 5$, we get

$$
\begin{aligned}
\sum_{i=0}^{5} a_i \theta^i &= \left(\sum_{i=0}^{5} a_i \theta^i \right)^8 \\
&= \sum_{i=0}^{5} a_i \theta^{8i} \\
&= a_0 + a_1 \theta^8 + a_2 \theta^7 + a_3 \theta^6 + a_4 \theta^5 + a_5 \theta^4 \\
&= a_0 + a_3 + a_2 \theta + a_1 \theta^2 + a_3 \theta^3 + (a_2 + a_5)\theta^4 + (a_1 + a_4)\theta^5,
\end{aligned}
$$

and a comparison of coefficients yields $a_3 = 0$, $a_1 = a_2$, $a_4 = a_2 + a_5$. Choose $a_0 = a_3 = a_4 = 0$, $a_1 = a_2 = a_5 = 1$, so that $\alpha = \theta + \theta^2 + \theta^5$ is an element of order 7. Thus, $\zeta = \theta\alpha = 1 + \theta^2$ is a primitive element of $\mathbb{F}_{64}$. Then $\zeta^2 = 1 + \theta^4$, $\zeta^3 = \theta^2 + \theta^3 + \theta^4$, $\zeta^4 = 1 + \theta^2 + \theta^5$, $\zeta^5 = 1 + \theta + \theta^5$, $\zeta^6 = 1 + \theta^2 + \theta^3 + \theta^4 + \theta^5$ An application of the method in Example 3.42 yields the minimal polynomial $g(x) = x^6 + x^4 + x^3 + x + 1$ of ζ over $\mathbb{F}_2$ and thus a primitive polynomial over $\mathbb{F}_2$ of degree 6. □

If a primitive polynomial g over $\mathbb{F}_q$ of degree m is known, all other such primitive polynomials can be obtained by considering a root θ of g in $\mathbb{F}_{q^m}$ and determining the minimal polynomials over $\mathbb{F}_q$ of all elements θ^t, where t runs through all positive integers $\leqslant q^m - 1$ that are relatively prime to $q^m - 1$. The calculation of these minimal polynomials is carried out by the methods described earlier in this section.

It is useful to be able to decide whether an irreducible polynomial over a finite field remains irreducible over a certain finite extension field. The following results address themselves to this question.

3.46. Theorem. *Let f be an irreducible polynomial over $\mathbb{F}_q$ of degree n and let $k \in \mathbb{N}$. Then f factors into d irreducible polynomials in $\mathbb{F}_{q^k}[x]$ of the same degree n/d, where $d = \gcd(k, n)$.*

Proof. Since the case $f(0) = 0$ is trivial, we can assume $f(0) \neq 0$. Let g be an irreducible factor of f in $\mathbb{F}_{q^k}[x]$. If $\mathrm{ord}(f) = e$, then also $\mathrm{ord}(g) = e$ by Theorem 3.3 since the roots of g are also roots of f. By Theorem 3.5 the multiplicative order of q modulo e is n and the degree of g is equal to the multiplicative order of q^k modulo e. The powers q^j, $j = 0, 1, \ldots$, considered modulo e, form a cyclic group of order n. Thus it follows from Theorem 1.15(ii) that the multiplicative order of q^k modulo e is n/d, and so the degree of g is n/d.   □

3.47. Corollary. *An irreducible polynomial over $\mathbb{F}_q$ of degree n remains irreducible over $\mathbb{F}_{q^k}$ if and only if k and n are relatively prime.*

Proof. This is an immediate consequence of Theorem 3.46.   □

3.48. Example. Consider the primitive polynomial $g(x) = x^6 + x^4 + x^3 + x + 1$ over $\mathbb{F}_2$ from Example 3.45 as a polynomial over $\mathbb{F}_{16}$. Then, in the notation of Theorem 3.46, we have $n = 6$, $k = 4$, and thus $d = 2$. Therefore, g factors in $\mathbb{F}_{16}[x]$ into two irreducible cubic polynomials. Using the notation of Example 3.45, let g_1 be the factor that has $\zeta = 1 + \theta^2$ as a root. The other roots of g_1 must be the conjugates ζ^{16} and $\zeta^{256} = \zeta^4$ with respect to $\mathbb{F}_{16}$. Since these elements are also conjugates with respect to $\mathbb{F}_4$, it follows that g_1 is actually in $\mathbb{F}_4[x]$. Now $\beta = \zeta^{21}$ is a primitive third root of unity over $\mathbb{F}_2$, and so $\mathbb{F}_4 = \langle 0, 1, \beta, \beta^2 \rangle$. Furthermore,

$$g_1(x) = (x - \zeta)(x - \zeta^4)(x - \zeta^{16})$$

$$= x^3 + (\zeta + \zeta^4 + \zeta^{16})x^2 + (\zeta^5 + \zeta^{17} + \zeta^{20})x + \zeta^{21}.$$

We have $\zeta^4 = 1 + \theta^2 + \theta^5$, $\zeta^{16} = 1 + \theta^5$, and so $\zeta + \zeta^4 + \zeta^{16} = 1$. Similarly, we obtain $\zeta^5 + \zeta^{17} + \zeta^{20} = 1$, so that $g_1(x) = x^3 + x^2 + x + \beta$. By dividing g by g_1 we get the second factor and thus the factorization

$$g(x) = (x^3 + x^2 + x + \beta)(x^3 + x^2 + x + \beta^2)$$

in $\mathbb{F}_4[x]$, and hence in $\mathbb{F}_{16}[x]$. The two factors of g are primitive polynomials over $\mathbb{F}_4$, but not over $\mathbb{F}_{16}$. By Corollary 3.47, the polynomial g remains irreducible over certain other extension fields of $\mathbb{F}_2$, such as $\mathbb{F}_{32}$ and $\mathbb{F}_{128}$.   □

4. LINEARIZED POLYNOMIALS

Both in theory and in applications the special class of polynomials to be introduced below is of importance. A useful feature of these polynomials is the structure of the set of roots that facilitates the determination of the roots. Let q, as usual, denote a prime power.

3.49.   Definition.   A polynomial of the form

$$L(x) = \sum_{i=0}^{n} \alpha_i x^{q^i}$$

with coefficients in an extension field $\mathbb{F}_{q^m}$ of $\mathbb{F}_q$ is called a *q-polynomial* over $\mathbb{F}_{q^m}$.

If the value of q is fixed once and for all or is clear from the context, it is also customary to speak of a *linearized polynomial*. This terminology stems from the following property of linearized polynomials. If F is an arbitrary extension field of $\mathbb{F}_{q^m}$ and $L(x)$ is a linearized polynomial (i.e., a q-polynomial) over $\mathbb{F}_{q^m}$, then

$$L(\beta + \gamma) = L(\beta) + L(\gamma) \quad \text{for all } \beta, \gamma \in F, \tag{3.11}$$

$$L(c\beta) = cL(\beta) \quad \text{for all } c \in \mathbb{F}_q \text{ and all } \beta \in F. \tag{3.12}$$

The identity (3.11) follows immediately from Theorem 1.46 and (3.12) follows from the fact that $c^{q^i} = c$ for $c \in \mathbb{F}_q$ and $i \geq 0$. Thus, if F is considered as a vector space over $\mathbb{F}_q$, then the linearized polynomial $L(x)$ induces a linear operator on F.

The special character of the set of roots of a linearized polynomial is shown by the following result.

3.50.   Theorem.   *Let $L(x)$ be a nonzero q-polynomial over $\mathbb{F}_{q^m}$ and let the extension field $\mathbb{F}_{q^s}$ of $\mathbb{F}_{q^m}$ contain all the roots of $L(x)$. Then each root of $L(x)$ has the same multiplicity, which is either 1 or a power of q, and the roots form a linear subspace of $\mathbb{F}_{q^s}$, where $\mathbb{F}_{q^s}$ is regarded as a vector space over $\mathbb{F}_q$.*

Proof.   It follows from (3.11) and (3.12) that any linear combination of roots with coefficients in $\mathbb{F}_q$ is again a root, and so the roots of $L(x)$ form a linear subspace of $\mathbb{F}_{q^s}$. If

$$L(x) = \sum_{i=0}^{n} \alpha_i x^{q^i},$$

then $L'(x) = \alpha_0$, so that $L(x)$ has only simple roots in case $\alpha_0 \neq 0$. Otherwise, we have $\alpha_0 = \alpha_1 = \cdots = \alpha_{k-1} = 0$, but $\alpha_k \neq 0$ for some $k \geq 1$, and then

$$L(x) = \sum_{i=k}^{n} \alpha_i x^{q^i} = \sum_{i=k}^{n} \alpha_i^{q^{mk}} x^{q^i} = \left(\sum_{i=k}^{n} \alpha_i^{q^{(m-1)k}} x^{q^{i-k}} \right)^{q^k},$$

which is the q^k th power of a linearized polynomial having only simple roots. In this case, each root of $L(x)$ has multiplicity q^k. $\square$

There is also a partial converse of Theorem 3.50, which is given by Theorem 3.52. It depends on a result about certain determinants which extends Corollary 2.38.

3.51.   Lemma.   *Let* $\beta_1, \beta_2, \ldots, \beta_n$ *be elements of* $\mathbb{F}_{q^m}$. *Then*

$$\begin{vmatrix} \beta_1 & \beta_1^q & \beta_1^{q^2} & \cdots & \beta_1^{q^{n-1}} \\ \beta_2 & \beta_2^q & \beta_2^{q^2} & \cdots & \beta_2^{q^{n-1}} \\ \vdots & \vdots & \vdots & & \vdots \\ \beta_n & \beta_n^q & \beta_n^{q^2} & \cdots & \beta_n^{q^{n-1}} \end{vmatrix} = \beta_1 \prod_{j=1}^{n-1} \prod_{c_1, \ldots, c_j \in \mathbb{F}_q} \left(\beta_{j+1} - \sum_{k=1}^{j} c_k \beta_k \right),$$

(3.13)

and so the determinant is $\neq 0$ *if and only if* $\beta_1, \beta_2, \ldots, \beta_n$ *are linearly independent over* $\mathbb{F}_q$.

Proof.   Let D_n be the determinant on the left-hand side of (3.13). We prove (3.13) by induction on n and note that the formula is trivial for $n = 1$ if the empty product on the right-hand side is interpreted as 1. Suppose the formula is shown for some $n \geqslant 1$. Consider the polynomial

$$D(x) = \begin{vmatrix} \beta_1 & \beta_1^q & \cdots & \beta_1^{q^{n-1}} & \beta_1^{q^n} \\ \beta_2 & \beta_2^q & \cdots & \beta_2^{q^{n-1}} & \beta_2^{q^n} \\ \vdots & \vdots & & \vdots & \vdots \\ \beta_n & \beta_n^q & \cdots & \beta_n^{q^{n-1}} & \beta_n^{q^n} \\ x & x^q & \cdots & x^{q^{n-1}} & x^{q^n} \end{vmatrix}.$$

By expansion along the last row we get

$$D(x) = D_n x^{q^n} + \sum_{i=0}^{n-1} \alpha_i x^{q^i}$$

with $\alpha_i \in \mathbb{F}_{q^m}$ for $0 \leqslant i \leqslant n-1$. Assume first that $\beta_1, \ldots, \beta_n$ are linearly independent over $\mathbb{F}_q$. We have $D(\beta_k) = 0$ for $1 \leqslant k \leqslant n$, and since $D(x)$ is a q-polynomial over $\mathbb{F}_{q^m}$, all linear combinations $c_1 \beta_1 + \cdots + c_n \beta_n$ with $c_k \in \mathbb{F}_q$ for $1 \leqslant k \leqslant n$ are roots of $D(x)$. Thus $D(x)$ has q^n distinct roots, so that we obtain a factorization

$$D(x) = D_n \prod_{c_1, \ldots, c_n \in \mathbb{F}_q} \left(x - \sum_{k=1}^{n} c_k \beta_k \right).$$

(3.14)

If $\beta_1, \ldots, \beta_n$ are linearly dependent over $\mathbb{F}_q$, then $D_n = 0$ and $\sum_{k=1}^{n} b_k \beta_k = 0$ for some $b_1, \ldots, b_n \in \mathbb{F}_q$, not all of which are 0. It follows that

$$\sum_{k=1}^{n} b_k \beta_k^{q^j} = \left(\sum_{k=1}^{n} b_k \beta_k \right)^{q^j} = 0 \quad \text{for } j = 0, 1, \ldots, n,$$

and so the first n row vectors in the determinant defining $D(x)$ are linearly

4. Linearized Polynomials

dependent over $\mathbb{F}_q$. Thus $D(x) = 0$, and the identity (3.14) is satisfied in all cases. Consequently,

$$D_{n+1} = D(\beta_{n+1}) = D_n \prod_{c_1,\ldots,c_n \in \mathbb{F}_q} \left(\beta_{n+1} - \sum_{k=1}^n c_k \beta_k \right),$$

and (3.13) is established. $\qquad\square$

3.52. Theorem. *Let U be a linear subspace of $\mathbb{F}_{q^m}$, considered as a vector space over $\mathbb{F}_q$. Then for any nonnegative integer k the polynomial*

$$L(x) = \prod_{\beta \in U} (x - \beta)^{q^k}$$

is a q-polynomial over $\mathbb{F}_{q^m}$.

Proof. Since the q^kth power of a q-polynomial over $\mathbb{F}_{q^m}$ is again such a polynomial, it suffices to consider the case $k = 0$. Let $\{\beta_1, \ldots, \beta_n\}$ be a basis of U over $\mathbb{F}_q$. Then the determinant D_n on the left-hand side of (3.13) is $\neq 0$ by Lemma 3.51, and so

$$L(x) = \prod_{\beta \in U} (x - \beta)$$

$$= \prod_{c_1, \ldots, c_n \in \mathbb{F}_q} \left(x - \sum_{k=1}^n c_k \beta_k \right) = D_n^{-1} D(x)$$

by (3.14), which shows already that $L(x)$ is a q-polynomial over $\mathbb{F}_{q^m}$. $\qquad\square$

The properties of linearized polynomials lead to the following *method of determining roots* of such polynomials. Let

$$L(x) = \sum_{i=0}^n \alpha_i x^{q^i}$$

be a q-polynomial over $\mathbb{F}_{q^m}$, and suppose we want to find all roots of $L(x)$ in the finite extension F of $\mathbb{F}_{q^m}$. As we noted above, the mapping L: $\beta \in F \mapsto L(\beta) \in F$ is a linear operator on the vector space F over $\mathbb{F}_q$. Therefore, L can be represented by a matrix over $\mathbb{F}_q$. Specifically, let $\{\beta_1, \ldots, \beta_s\}$ be a basis of F over $\mathbb{F}_q$, so that every $\beta \in F$ can be written in the form

$$\beta = \sum_{j=1}^s c_j \beta_j \quad \text{with } c_j \in \mathbb{F}_q \quad \text{for } 1 \leqslant j \leqslant s;$$

then

$$L(\beta) = \sum_{j=1}^s c_j L(\beta_j).$$

Now let

$$L(\beta_j) = \sum_{k=1}^{s} b_{jk}\beta_k \quad \text{for } 1 \leqslant j \leqslant s,$$

where $b_{jk} \in \mathbb{F}_q$ for $1 \leqslant j, k \leqslant s$, and let B be the $s \times s$ matrix over $\mathbb{F}_q$ whose (j, k) entry is b_{jk}. Then, if

$$(c_1, \ldots, c_s) B = (d_1, \ldots, d_s),$$

we have

$$L(\beta) = \sum_{k=1}^{s} d_k \beta_k.$$

Therefore, the equation $L(\beta) = 0$ is equivalent to

$$(c_1, \ldots, c_s) B = (0, \ldots, 0). \tag{3.15}$$

This is a homogeneous system of s linear equations for $c_1, \ldots, c_s$. If r is the rank of the matrix B, then (3.15) has q^{s-r} solution vectors $(c_1, \ldots, c_s)$. Each solution vector $(c_1, \ldots, c_s)$ yields a root $\beta = \sum_{j=1}^{s} c_j \beta_j$ of $L(x)$ in F. Thus, the problem of finding the roots of $L(x)$ in F is reduced to the easier problem of solving a homogeneous system of linear equations.

3.53. Example. Consider the linearized polynomial $L(x) = x^9 - x^3 - \alpha x$ $\in \mathbb{F}_9[x]$, where α is a root of the primitive polynomial $x^2 + x - 1$ over $\mathbb{F}_3$. In order to find the roots of $L(x)$ in $\mathbb{F}_{81}$, we choose the basis $\{1, \zeta, \zeta^2, \zeta^3\}$ of $\mathbb{F}_{81}$ over $\mathbb{F}_3$, where ζ is a root of the primitive polynomial $x^4 + x^3 + x^2 - x - 1$ over $\mathbb{F}_3$ (compare with Example 3.44). Because of the orders involved, we must have $\alpha = \zeta^{10j}$ with $j = 1, 3, 5,$ or 7, and since $\zeta^{20} + \zeta^{10} - 1 = 0$, we can take $\alpha = \zeta^{10} = -1 + \zeta + \zeta^2 - \zeta^3$. Next, we calculate

$$L(1) = -\alpha = 1 - \zeta - \zeta^2 + \zeta^3,$$

$$L(\zeta) = \zeta^9 - \zeta^3 - \alpha\zeta = -\zeta - \zeta^2 - \zeta^3,$$

$$L(\zeta^2) = \zeta^{18} - \zeta^6 - \alpha\zeta^2 = -1 + \zeta^3,$$

$$L(\zeta^3) = \zeta^{27} - \zeta^9 - \alpha\zeta^3 = 1 - \zeta^3,$$

and so we get

$$B = \begin{pmatrix} 1 & -1 & -1 & 1 \\ 0 & -1 & -1 & -1 \\ -1 & 0 & 0 & 1 \\ 1 & 0 & 0 & -1 \end{pmatrix}.$$

The system (3.15) has two linearly independent solutions, such as $(0, 0, 1, 1)$ and $(-1, 1, 0, 1)$. All solutions of (3.15) are obtained by forming all linear combinations of these two vectors with coefficients in $\mathbb{F}_3$. The roots of $L(x)$ in $\mathbb{F}_{81}$ are then $\theta_1 = 0$, $\theta_2 = \zeta^2 + \zeta^3$, $\theta_3 = -\zeta^2 - \zeta^3$, $\theta_4 = -1 + \zeta + \zeta^3$,

$\theta_5 = 1 - \zeta - \zeta^3,\quad \theta_6 = -1 + \zeta + \zeta^2 - \zeta^3,\quad \theta_7 = 1 - \zeta - \zeta^2 + \zeta^3,\quad \theta_8 = 1 - \zeta + \zeta^2,$
$\theta_9 = -1 + \zeta - \zeta^2.$ $\square$

This method of finding roots can also be applied to a somewhat more general class of polynomials—namely, affine polynomials.

3.54. Definition. A polynomial of the form $A(x) = L(x) - \alpha$, where $L(x)$ is a q-polynomial over $\mathbb{F}_{q^m}$ and $\alpha \in \mathbb{F}_{q^m}$, is called an *affine q-polynomial* over $\mathbb{F}_{q^m}$.

An element $\beta \in F$ is a root of $A(x)$ if and only if $L(\beta) = \alpha$. In the notation of (3.15), the equation $L(\beta) = \alpha$ is equivalent to

$$(c_1, \ldots, c_s)B = (d_1, \ldots, d_s), \tag{3.16}$$

where $\alpha = \sum_{k=1}^{s} d_k \beta_k$. The system (3.16) of linear equations is solved for $c_1, \ldots, c_s$, and each solution vector $(c_1, \ldots, c_s)$ yields a root $\beta = \sum_{j=1}^{s} c_j \beta_j$ of $A(x)$ in F.

The fact that roots are easier to determine for affine polynomials suggests the following *method of finding the roots* of an arbitrary polynomial $f(x)$ over $\mathbb{F}_{q^m}$ of positive degree in an extension field F of $\mathbb{F}_{q^m}$. First determine a nonzero affine q-polynomial $A(x)$ over $\mathbb{F}_{q^m}$ that is divisible by $f(x)$—that is, a so-called *affine multiple* of $f(x)$. Next, obtain all the roots of $A(x)$ in F by the method described above. Since the roots of $f(x)$ in F must be among the roots of $A(x)$ in F, it suffices then to calculate $f(\beta)$ for all roots β of $A(x)$ in F in order to locate the roots of $f(x)$ in F.

The only point that remains to be settled is how to determine an affine multiple $A(x)$ of $f(x)$. This can be achieved as follows. Let $n \geqslant 1$ be the degree of $f(x)$. For $i = 0, 1, \ldots, n-1$, calculate the unique polynomial $r_i(x)$ of degree $\leqslant n-1$ with $x^{q^i} \equiv r_i(x) \bmod f(x)$. Then determine elements $\alpha_i \in \mathbb{F}_{q^m}$, not all 0, such that $\sum_{i=0}^{n-1} \alpha_i r_i(x)$ is a constant polynomial. This involves $n-1$ conditions concerning the vanishing of the coefficients of x^j, $1 \leqslant j \leqslant n-1$, and thus leads to a homogeneous system of $n-1$ linear equations for the n unknowns $\alpha_0, \alpha_1, \ldots, \alpha_{n-1}$. Such a system always has a nontrivial solution. Once a nontrivial solution has been fixed, we have $\sum_{i=0}^{n-1} \alpha_i r_i(x) = \alpha$ for some $\alpha \in \mathbb{F}_{q^m}$. It follows that

$$\sum_{i=0}^{n-1} \alpha_i x^{q^i} \equiv \sum_{i=0}^{n-1} \alpha_i r_i(x) \equiv \alpha \bmod f(x),$$

and so

$$A(x) = \sum_{i=0}^{n-1} \alpha_i x^{q^i} - \alpha$$

is a nonzero affine q-polynomial over $\mathbb{F}_{q^m}$ divisible by $f(x)$. It is clear that we may take $A(x)$ to be a monic polynomial.

3.55. Example. Let $f(x) = x^4 + \theta^2 x^3 + \theta x^2 + x + \theta \in \mathbb{F}_4[x]$, where θ is a root of $x^2 + x + 1 \in \mathbb{F}_2[x]$. We want to find the roots of $f(x)$ in $\mathbb{F}_{64}$. We first determine an affine multiple $A(x)$ of $f(x)$ by using the method described above with $q = 2$. Modulo $f(x)$ we have $x \equiv x = r_0(x)$, $x^2 \equiv x^2 = r_1(x)$, $x^4 \equiv \theta^2 x^3 + \theta x^2 + x + \theta = r_2(x)$, $x^8 \equiv \theta x^3 + \theta x^2 + x + \theta = r_3(x)$. The condition that $\alpha_0 r_0(x) + \alpha_1 r_1(x) + \alpha_2 r_2(x) + \alpha_3 r_3(x)$ should be a constant polynomial leads to the system

$$
\begin{aligned}
\alpha_0 \quad + \quad \alpha_2 + \alpha_3 &= 0 \\
\alpha_1 + \quad \theta\alpha_2 + \theta\alpha_3 &= 0 \\
\theta^2\alpha_2 + \theta\alpha_3 &= 0.
\end{aligned}
$$

We choose $\alpha_3 = 1$ and then obtain $\alpha_2 = \theta^2$, $\alpha_1 = \theta^2$, $\alpha_0 = \theta$. Furthermore,

$$\alpha = \alpha_0 r_0(x) + \alpha_1 r_1(x) + \alpha_2 r_2(x) + \alpha_3 r_3(x) = \theta^2,$$

and so

$$A(x) = \alpha_3 x^8 + \alpha_2 x^4 + \alpha_1 x^2 + \alpha_0 x - \alpha = x^8 + \theta^2 x^4 + \theta^2 x^2 + \theta x + \theta^2.$$

Next, we calculate the roots of $A(x)$ in $\mathbb{F}_{64}$. We have to solve the equation $L(x) = \theta^2$ with the 2-polynomial $L(x) = x^8 + \theta^2 x^4 + \theta^2 x^2 + \theta x$ over $\mathbb{F}_4$. Let ζ be a root of the primitive polynomial $x^6 + x + 1$ over $\mathbb{F}_2$. Then $\{1, \zeta, \zeta^2, \zeta^3, \zeta^4, \zeta^5\}$ is a basis of $\mathbb{F}_{64}$ over $\mathbb{F}_2$. Since θ is a primitive third root of unity over $\mathbb{F}_2$, we can take $\theta = \zeta^{21} = 1 + \zeta + \zeta^3 + \zeta^4 + \zeta^5$. Using $\theta^2 = \theta + 1 = \zeta + \zeta^3 + \zeta^4 + \zeta^5$, we obtain

$$
\begin{aligned}
L(1) &= \zeta && + \zeta^3 && + \zeta^4 && + \zeta^5 \\
L(\zeta) &= \zeta + \zeta^2 && && && + \zeta^5 \\
L(\zeta^2) &= && \zeta^2 + \zeta^3 && + \zeta^4 && + \zeta^5 \\
L(\zeta^3) &= \zeta && + \zeta^3 && + \zeta^4 && \\
L(\zeta^4) &= && && && \zeta^5 \\
L(\zeta^5) &= && \zeta^2 + \zeta^3 && + \zeta^4 &&
\end{aligned}
$$

Thus the matrix B in (3.16) is given by

$$
B = \begin{pmatrix}
0 & 1 & 0 & 1 & 1 & 1 \\
0 & 1 & 1 & 0 & 0 & 1 \\
0 & 0 & 1 & 1 & 1 & 1 \\
0 & 1 & 0 & 1 & 1 & 0 \\
0 & 0 & 0 & 0 & 0 & 1 \\
0 & 0 & 1 & 1 & 1 & 0
\end{pmatrix}
$$

From the representation for θ^2 given above it follows that the vector $(d_1, \ldots, d_s)$ in (3.16) is equal to $(0, 1, 0, 1, 1, 1)$. The general solution of the system (3.16) is then

$$(1, 0, 0, 0, 0, 0) + a_1(0, 1, 1, 1, 0, 0) + a_2(1, 1, 1, 0, 1, 0) + a_3(1, 1, 0, 0, 0, 1)$$

with $a_1, a_2, a_3 \in \mathbb{F}_2$. Thus the roots of $A(x)$ in $\mathbb{F}_{64}$ are $\eta_1 = 1$, $\eta_2 = \zeta + \zeta^5$, $\eta_3 = \zeta + \zeta^2 + \zeta^4$, $\eta_4 = 1 + \zeta^2 + \zeta^4 + \zeta^5$, $\eta_5 = 1 + \zeta + \zeta^2 + \zeta^3$, $\eta_6 = \zeta^2 + \zeta^3 + \zeta^5$, $\eta_7 = \zeta^3 + \zeta^4$, $\eta_8 = 1 + \zeta + \zeta^3 + \zeta^4 + \zeta^5 = \theta$. By calculating $f(\eta_j)$ for $j = 1, 2, \ldots, 8$, we find that the roots of $f(x)$ in $\mathbb{F}_{64}$ are $\eta_3, \eta_5, \eta_7, \eta_8$. □

The method of determining the roots of an affine polynomial shows, in particular, that these roots form an affine subspace—that is, a translate of a linear subspace. This can also be deduced from abstract principles, together with a statement concerning multiplicities.

3.56.   Theorem.   *Let $A(x)$ be an affine q-polynomial over $\mathbb{F}_{q^m}$ of positive degree and let the extension field $\mathbb{F}_{q^s}$ of $\mathbb{F}_{q^m}$ contain all the roots of $A(x)$. Then each root of $A(x)$ has the same multiplicity, which is either 1 or a power of q, and the roots form an affine subspace of $\mathbb{F}_{q^s}$, where $\mathbb{F}_{q^s}$ is regarded as a vector space over $\mathbb{F}_q$.*

Proof.   The result about the multiplicities is shown in the same way as in the proof of Theorem 3.50. Now let $A(x) = L(x) - \alpha$, where $L(x)$ is a q-polynomial over $\mathbb{F}_{q^m}$, and let β be a fixed root of $A(x)$. Then $\gamma \in \mathbb{F}_{q^s}$ is a root of $A(x)$ if and only if $L(\gamma) = \alpha = L(\beta)$ if and only if $L(\gamma - \beta) = 0$ if and only if $\gamma \in \beta + U$, where U is the linear subspace of $\mathbb{F}_{q^s}$ consisting of the roots of $L(x)$. Thus the roots of $A(x)$ form an affine subspace of $\mathbb{F}_{q^s}$. □

3.57.   Theorem.   *Let T be an affine subspace of $\mathbb{F}_{q^m}$, considered as a vector space over $\mathbb{F}_q$. Then for any nonnegative integer k the polynomial*

$$A(x) = \prod_{\gamma \in T} (x - \gamma)^{q^k}$$

is an affine q-polynomial over $\mathbb{F}_{q^m}$.

Proof.   Let $T = \eta + U$, where U is a linear subspace of $\mathbb{F}_{q^m}$. Then

$$L(x) = \prod_{\beta \in U} (x - \beta)^{q^k}$$

is a q-polynomial over $\mathbb{F}_{q^m}$ according to Theorem 3.52. Furthermore,

$$A(x) = \prod_{\gamma \in T} (x - \gamma)^{q^k} = \prod_{\beta \in U} (x - \eta - \beta)^{q^k} = L(x - \eta),$$

and $L(x - \eta)$ is easily seen to be an affine q-polynomial over $\mathbb{F}_{q^m}$. □

The ordinary product of linearized polynomials need not be a linearized polynomial. However, the composition $L_1(L_2(x))$ of two q-polynomials $L_1(x), L_2(x)$ over $\mathbb{F}_{q^m}$ is again a q-polynomial. Instead of the word composition (or substitution) we use the phrase "symbolic multiplication." Thus, we define *symbolic multiplication* by

$$L_1(x) \otimes L_2(x) = L_1(L_2(x)).$$

If we consider only q-polynomials over $\mathbb{F}_q$, then a simple investigation shows that symbolic multiplication is commutative, associative, and distributive (with respect to ordinary addition). In fact, the set of q-polynomials over $\mathbb{F}_q$ forms an integral domain under the operations of symbolic multiplication and ordinary addition. The operation of symbolic multiplication can be related to the conventional arithmetic of polynomials by means of the following notion.

3.58. Definition. The polynomials

$$l(x) = \sum_{i=0}^{n} \alpha_i x^i \quad \text{and} \quad L(x) = \sum_{i=0}^{n} \alpha_i x^{q^i}$$

over $\mathbb{F}_{q^m}$ are called q-*associates* of each other. More specifically, $l(x)$ is the *conventional q-associate* of $L(x)$ and $L(x)$ is the *linearized q-associate* of $l(x)$.

    3.59. Lemma. *Let $L_1(x)$ and $L_2(x)$ be q-polynomials over $\mathbb{F}_q$ with conventional q-associates $l_1(x)$ and $l_2(x)$. Then $l(x) = l_1(x)l_2(x)$ and $L(x) = L_1(x) \otimes L_2(x)$ are q-associates of each other.*

    Proof. The equations

$$l(x) = \sum_i a_i x^i = \sum_j b_j x^j \sum_k c_k x^k = l_1(x)l_2(x)$$

and

$$L(x) = \sum_i a_i x^{q^i} = \sum_j b_j \left(\sum_k c_k x^{q^k} \right)^{q^j} = \sum_j b_j \sum_k c_k x^{q^{j+k}} = L_1(x) \otimes L_2(x)$$

are each true if and only if

$$a_i = \sum_{j+k=i} b_j c_k \quad \text{for every } i. \qquad \square$$

If $L_1(x)$ and $L(x)$ are q-polynomials over $\mathbb{F}_q$, we say that $L_1(x)$ *symbolically divides* $L(x)$ (or that $L(x)$ is *symbolically divisible* by $L_1(x)$) if $L(x) = L_1(x) \otimes L_2(x)$ for some q-polynomial $L_2(x)$ over $\mathbb{F}_q$. The following criterion is then an immediate consequence of Lemma 3.59.

    3.60. Corollary. *Let $L_1(x)$ and $L(x)$ be q-polynomials over $\mathbb{F}_q$ with conventional q-associates $l_1(x)$ and $l(x)$. Then $L_1(x)$ symbolically divides $L(x)$ if and only if $l_1(x)$ divides $l(x)$.*

3.61. Example. Let $L(x)$ be a q-polynomial over $\mathbb{F}_q$ that symbolically divides $x^{q^m} - x$ for some $m \in \mathbb{N}$. Then there exists a q-polynomial $L_1(x)$ over $\mathbb{F}_q$ such that

$$x^{q^m} - x = L(x) \otimes L_1(x) = L_1(x) \otimes L(x) = L_1(L(x)). \qquad (3.17)$$

This can be applied as follows. Let α be a fixed element of $\mathbb{F}_{q^m}$. Then the affine polynomial $L(x) - \alpha$ has at least one root in $\mathbb{F}_{q^m}$ if and only if $L_1(\alpha) = 0$, and if $L_1(\alpha) = 0$, then actually all the roots of $L(x) - \alpha$ are in $\mathbb{F}_{q^m}$. For if $\beta \in \mathbb{F}_{q^m}$ is a root of $L(x) - \alpha$, then $L(\beta) = \alpha$, and substituting x by β in (3.17) yields $L_1(\alpha) = \beta^{q^m} - \beta = 0$. Conversely, suppose $L_1(\alpha) = 0$ and let γ be a root of $L(x) - \alpha$ in some extension field of $\mathbb{F}_{q^m}$; then $L(\gamma) = \alpha$, and substituting x by γ in (3.17) yields $\gamma^{q^m} - \gamma = L_1(\alpha) = 0$, so that $\gamma \in \mathbb{F}_{q^m}$. The polynomial $L_1(x)$ can be calculated by letting $l(x)$ be the conventional q-associate of $L(x)$, determining $l_1(x) = (x^m - 1)/l(x)$, and then taking $L_1(x)$ to be the linearized q-associate of $l_1(x)$. This application contains Theorem 2.25 as a special case, as one sees easily by choosing $L(x) = x^q - x$. $\square$

It is an important fact that although symbolic multiplication and ordinary multiplication are quite different operations, the divisibility concepts for linearized polynomials based on these operations are equivalent.

3.62. Theorem. *Let $L_1(x)$ and $L(x)$ be q-polynomials over $\mathbb{F}_q$ with conventional q-associates $l_1(x)$ and $l(x)$. Then the following properties are equivalent: (i) $L_1(x)$ symbolically divides $L(x)$; (ii) $L_1(x)$ divides $L(x)$ in the ordinary sense; (iii) $l_1(x)$ divides $l(x)$.*

Proof. Since the equivalence of (i) and (iii) has been established in Corollary 3.60, it suffices to show the equivalence of (i) and (ii). If $L_1(x)$ symbolically divides $L(x)$, then

$$L(x) = L_1(x) \otimes L_2(x) = L_2(x) \otimes L_1(x) = L_2(L_1(x))$$

for some q-polynomial $L_2(x)$ over $\mathbb{F}_q$. Let

$$L_2(x) = \sum_{i=0}^{n} a_i x^{q^i},$$

then

$$L(x) = a_0 L_1(x) + a_1 L_1(x)^q + \cdots + a_n L_1(x)^{q^n},$$

and so $L_1(x)$ divides $L(x)$ in the ordinary sense. Conversely, suppose $L_1(x)$ divides $L(x)$ in the ordinary sense, where we can assume that $L_1(x)$ is nonzero. Using the division algorithm, we write $l(x) = k(x) l_1(x) + r(x)$, where $\deg(r(x)) < \deg(l_1(x))$, and turning to linearized q-associates we get in an obvious notation $L(x) = K(x) \otimes L_1(x) + R(x)$. By what we have already shown, $L_1(x)$ divides $K(x) \otimes L_1(x)$ in the ordinary sense, and so $L_1(x)$ divides $R(x)$ in the ordinary sense. But since $\deg(R(x)) < \deg(L_1(x))$, $R(x)$ must be the zero polynomial, and this proves that $L_1(x)$ symbolically divides $L(x)$. $\square$

This result can be used to establish an interesting relationship between an irreducible polynomial and the irreducible factors of its linearized q-associate.

3.63.   Theorem.   *Let $f(x)$ be irreducible in $\mathbb{F}_q[x]$ and let $F(x)$ be its linearized q-associate. Then the degree of every irreducible factor of $F(x)/x$ in $\mathbb{F}_q[x]$ is equal to ord($f(x)$).*

Proof.   Since the case $f(0) = 0$ is trivial, we can assume $f(0) \neq 0$. Put $e = \text{ord}(f(x))$ and let $h(x) \in \mathbb{F}_q[x]$ be an irreducible factor of $F(x)/x$ of degree d. Then $f(x)$ divides $x^e - 1$, and so by Theorem 3.62 $F(x)$ divides $x^{q^e} - x$. It follows that $h(x)$ divides $x^{q^e} - x$, hence d divides e by Theorem 3.20. By the division algorithm, we can write $x^d - 1 = g(x)f(x) + r(x)$ with $g(x), r(x) \in \mathbb{F}_q[x]$ and $\deg(r(x)) < \deg(f(x))$. Turning to linearized q-associates, we get

$$x^{q^d} - x = G(x) \otimes F(x) + R(x),$$

and since $h(x)$ divides $x^{q^d} - x$ and $G(x) \otimes F(x)$, it follows that $h(x)$ divides $R(x)$. If $r(x)$ is not the zero polynomial, then $r(x)$ and $f(x)$ are relatively prime, and so by Theorem 1.55 there exist polynomials $s(x), k(x) \in \mathbb{F}_q[x]$ with

$$s(x)r(x) + k(x)f(x) = 1.$$

Turning to linearized q-associates, we get

$$S(x) \otimes R(x) + K(x) \otimes F(x) = x.$$

Since $h(x)$ divides $R(x)$ and $F(x)$, it follows that $h(x)$ divides x, which is impossible. Thus $r(x)$ is the zero polynomial, so that $f(x)$ divides $x^d - 1$, and therefore e divides d by Lemma 3.6. Altogether, we have shown $d = e$. $\square$

We say that a q-polynomial $L(x)$ over $\mathbb{F}_q$ of degree > 1 is *symbolically irreducible* over $\mathbb{F}_q$ if the only symbolic decompositions $L(x) = L_1(x) \otimes L_2(x)$ with q-polynomials $L_1(x), L_2(x)$ over $\mathbb{F}_q$ are those for which one of the factors has degree 1. A symbolically irreducible polynomial is always reducible in the ordinary sense since any linearized polynomial of degree > 1 has the nontrivial factor x. By using Lemma 3.59, one shows immediately that the q-polynomial $L(x)$ is symbolically irreducible over $\mathbb{F}_q$ if and only if its conventional q-associate $l(x)$ is irreducible over $\mathbb{F}_q$.

Every q-polynomial $L(x)$ over $\mathbb{F}_q$ of degree > 1 has a *symbolic factorization* into symbolically irreducible polynomials over $\mathbb{F}_q$ and this factorization is essentially unique, in the sense that all other symbolic factorizations are obtained by rearranging factors and by multiplying factors by nonzero elements of $\mathbb{F}_q$. Using the correspondence between linearized polynomials and their conventional q-associates, one sees that the symbolic factorization of $L(x)$ is obtained by writing down the canonical factorization in $\mathbb{F}_q[x]$ of its conventional q-associate $l(x)$ and then turning to linearized q-associates.

3.64.   Example.   Consider the 2-polynomial $L(x) = x^{16} + x^8 + x^2 + x$ over $\mathbb{F}_2$. Its conventional 2-associate $l(x) = x^4 + x^3 + x + 1$ has the canonical factorization $l(x) = (x^2 + x + 1)(x + 1)^2$ in $\mathbb{F}_2[x]$. Thus,

$$L(x) = (x^4 + x^2 + x) \otimes (x^2 + x) \otimes (x^2 + x)$$

is the symbolic factorization of $L(x)$ into symbolically irreducible polynomials over $\mathbb{F}_2$.   □

For two or more q-polynomials over $\mathbb{F}_q$, not all of them 0, we may define their *greatest common symbolic divisor* to be the monic q-polynomial over $\mathbb{F}_q$ of highest degree that symbolically divides all of them. In order to compare this notion with that of the ordinary greatest common divisor, we note first that the roots of the greatest common divisor are exactly the common roots of the given q-polynomials. Since the intersection of linear subspaces is another linear subspace, it follows that the roots of the greatest common divisor form a linear subspace of some extension field $\mathbb{F}_{q^m}$, considered as a vector space over $\mathbb{F}_q$. Furthermore, by applying the first part of Theorem 3.50 to the given q-polynomials, we conclude that each root of the greatest common divisor has the same multiplicity, which is either 1 or a power of q. Therefore, Theorem 3.52 implies that the greatest common divisor is a q-polynomial. It follows then from Theorem 3.62 that *the greatest common divisor and the greatest common symbolic divisor are identical*. An efficient way of calculating the greatest common (symbolic) divisor of q-polynomials over $\mathbb{F}_q$ is to consider the conventional q-associates and determine their greatest common divisor; then the linearized q-associate of this greatest common divisor is the greatest common (symbolic) divisor of the given q-polynomials.

By Theorem 3.50 the roots of a nonzero q-polynomial over $\mathbb{F}_q$ form a vector space over $\mathbb{F}_q$. The roots have the additional property that the qth power of a root is again a root. A finite-dimensional vector space M over $\mathbb{F}_q$ that is contained in some extension field of $\mathbb{F}_q$ and has the property that the qth power of every element of M is again in M is called a *q-modulus*. On the basis of this concept we can establish the following criterion.

3.65.   Theorem.   *The monic polynomial $L(x)$ is a q-polynomial over $\mathbb{F}_q$ if and only if each root of $L(x)$ has the same multiplicity, which is either 1 or a power of q, and the roots form a q-modulus.*

Proof.   The necessity of the conditions follows from Theorem 3.50 and the remarks above. Conversely, the given conditions and Theorem 3.52 imply that $L(x)$ is a q-polynomial over some extension field of $\mathbb{F}_q$. If M is the q-modulus consisting of the roots of $L(x)$, then

$$L(x) = \prod_{\beta \in M} (x - \beta)^{q^k}$$

for some nonnegative integer k. Since $M = \{\beta^q : \beta \in M\}$, we obtain

$$L(x)^q = \prod_{\beta \in M} (x^q - \beta^q)^{q^k} = \prod_{\beta \in M} (x^q - \beta)^{q^k} = L(x^q).$$

If

$$L(x) = \sum_{i=0}^{n} \alpha_i x^{q^i},$$

then

$$\sum_{i=0}^{n} \alpha_i^q x^{q^{i+1}} = L(x)^q = L(x^q) = \sum_{i=0}^{n} \alpha_i x^{q^{i+1}},$$

so that for $0 \leqslant i \leqslant n$ we have $\alpha_i^q = \alpha_i$ and thus $\alpha_i \in \mathbb{F}_q$. Therefore, $L(x)$ is a q-polynomial over $\mathbb{F}_q$. $\square$

Any q-polynomial over $\mathbb{F}_q$ of degree q is symbolically irreducible over $\mathbb{F}_q$. For q-polynomials of degree $> q$, the notion of q-modulus can be used to characterize symbolically irreducible polynomials.

3.66.   Theorem.   *The q-polynomial $L(x)$ over $\mathbb{F}_q$ of degree $> q$ is symbolically irreducible over $\mathbb{F}_q$ if and only if $L(x)$ has simple roots and the q-modulus M consisting of the roots of $L(x)$ contains no q-modulus other than $\{0\}$ and M itself.*

Proof.   Suppose $L(x)$ is symbolically irreducible over $\mathbb{F}_q$. If $L(x)$ had multiple roots, then Theorem 3.65 would imply that we could write $L(x) = L_1(x)^q$ with a q-polynomial $L_1(x)$ over $\mathbb{F}_q$ of degree > 1. But then $L(x) = x^q \otimes L_1(x)$, a contradiction to the symbolic irreducibility of $L(x)$. Thus $L(x)$ has only simple roots. Furthermore, if N is a q-modulus contained in M, then Theorem 3.65 shows that $L_2(x) = \prod_{\beta \in N}(x - \beta)$ is a q-polynomial over $\mathbb{F}_q$. Since $L_2(x)$ divides $L(x)$ in the ordinary sense, it symbolically divides $L(x)$ by Theorem 3.62. But $L(x)$ is symbolically irreducible over $\mathbb{F}_q$, and so $\deg(L_2(x))$ must be either 1 or $\deg(L(x))$; that is, N is either $\{0\}$ or M.

To prove the sufficiency of the condition, suppose that $L(x) = L_1(x) \otimes L_2(x)$ is a symbolic decomposition with q-polynomials $L_1(x), L_2(x)$ over $\mathbb{F}_q$. Then $L_1(x)$ symbolically divides $L(x)$, and so it divides $L(x)$ in the ordinary sense by Theorem 3.62. It follows that $L_1(x)$ has simple roots and that the q-modulus N consisting of the roots of $L_1(x)$ is contained in M. Consequently, N is either $\{0\}$ or M, and so $\deg(L_1(x))$ is either 1 or $\deg(L(x))$. Thus, either $L_1(x)$ or $L_2(x)$ is of degree 1, which means that $L(x)$ is symbolically irreducible over $\mathbb{F}_q$. $\square$

3.67.   Definition.   Let $L(x)$ be a nonzero q-polynomial over $\mathbb{F}_{q^m}$. A root ζ of $L(x)$ is called a *q-primitive root* over $\mathbb{F}_{q^m}$ if it is not a root of any nonzero q-polynomial over $\mathbb{F}_{q^m}$ of lower degree.

This concept may also be viewed as follows. Let $g(x)$ be the minimal polynomial of ζ over $\mathbb{F}_{q^m}$. Then ζ is a q-primitive root of $L(x)$ over $\mathbb{F}_{q^m}$ if and only if $g(x)$ divides $L(x)$ and $g(x)$ does not divide any nonzero q-polynomial over $\mathbb{F}_{q^m}$ of lower degree.

Given an element ζ of a finite extension field of $\mathbb{F}_{q^m}$, one can always find a nonzero q-polynomial over $\mathbb{F}_{q^m}$ for which ζ is a q-primitive root over $\mathbb{F}_{q^m}$. To see this, we proceed as in the construction of an affine multiple. Let $g(x)$ be the minimal polynomial of ζ over $\mathbb{F}_{q^m}$, let n be the degree of $g(x)$, and calculate for $i = 0, 1, \ldots, n$ the unique polynomial $r_i(x)$ of degree $\leqslant n - 1$ with $x^{q^i} \equiv r_i(x) \bmod g(x)$. Then determine elements $\alpha_i \in \mathbb{F}_{q^m}$, not all 0, such that $\sum_{i=0}^{n} \alpha_i r_i(x) = 0$. This involves n conditions concerning the vanishing of the coefficients of x^j, $0 \leqslant j \leqslant n - 1$, and thus leads to a homogeneous system of n linear equations for the $n + 1$ unknowns $\alpha_0, \alpha_1, \ldots, \alpha_n$. Such a system always has a nontrivial solution, and with such a solution we get

$$L(x) = \sum_{i=0}^{n} \alpha_i x^{q^i} \equiv \sum_{i=0}^{n} \alpha_i r_i(x) \equiv 0 \bmod g(x),$$

so that $L(x)$ is a nonzero q-polynomial over $\mathbb{F}_{q^m}$ divisible by $g(x)$. By choosing the α_i in such a way that $L(x)$ is monic and of the lowest possible degree, one finds that ζ is a q-primitive root of $L(x)$ over $\mathbb{F}_{q^m}$. It is easily seen that this monic q-polynomial $L(x)$ over $\mathbb{F}_{q^m}$ of least positive degree that is divisible by $g(x)$ is uniquely determined; it is called the *minimal q-polynomial* of ζ over $\mathbb{F}_{q^m}$.

3.68.   Theorem.   *Let ζ be an element of a finite extension field of $\mathbb{F}_{q^m}$ and let $M(x)$ be its minimal q-polynomial over $\mathbb{F}_{q^m}$. Then a q-polynomial $K(x)$ over $\mathbb{F}_{q^m}$ has ζ as a root if and only if $K(x) = L(x) \otimes M(x)$ for some q-polynomial $L(x)$ over $\mathbb{F}_{q^m}$. In particular, for the case $m = 1$ this means that $K(x)$ has ζ as a root if and only if $K(x)$ is symbolically divisible by $M(x)$.*

Proof.   If $K(x) = L(x) \otimes M(x) = L(M(x))$, it follows immediately that $K(\zeta) = 0$. Conversely, let

$$M(x) = \sum_{j=0}^{t} \gamma_j x^{q^j} \quad \text{with } \gamma_t = 1$$

and suppose

$$K(x) = \sum_{h=0}^{r} \alpha_h x^{q^h} \quad \text{with } r \geqslant t$$

has ζ as a root. Put $s = r - t$ and $\gamma_j = 0$ for $j < 0$, and consider the following system of $s + 1$ linear equations in the $s + 1$ unknowns $\beta_0, \beta_1, \ldots, \beta_s$:

$$\beta_0 + \gamma_{t-1}^q \beta_1 + \gamma_{t-2}^{q^2} \beta_2 + \cdots + \gamma_{t-s}^{q^s} \beta_s = \alpha_t$$

$$\beta_1 + \gamma_{t-1}^{q^2} \beta_2 + \cdots + \gamma_{t-s+1}^{q^s} \beta_s = \alpha_{t+1}$$

$$\vdots \qquad \vdots \qquad \vdots$$

$$\beta_{s-1} + \gamma_{t-1}^{q^s} \beta_s = \alpha_{r-1}$$

$$\beta_s = \alpha_r.$$

It is clear that this system has a unique solution involving elements $\beta_0, \beta_1, \ldots, \beta_s$ of $\mathbb{F}_{q^m}$. With

$$L(x) = \sum_{i=0}^{s} \beta_i x^{q^i} \quad \text{and} \quad R(x) = K(x) - L(M(x))$$

we get

$$R(x) = \sum_{h=0}^{r} \alpha_h x^{q^h} - \sum_{i=0}^{s} \beta_i \left(\sum_{j=0}^{t} \gamma_j x^{q^j} \right)^{q^i}$$

$$= \sum_{h=0}^{r} \alpha_h x^{q^h} - \sum_{i=0}^{s} \beta_i \sum_{j=0}^{t} \gamma_j^{q^i} x^{q^{i+j}}$$

$$= \sum_{h=0}^{r} \alpha_h x^{q^h} - \sum_{h=0}^{r} \left(\sum_{i=0}^{s} \gamma_{h-i}^{q^i} \beta_i \right) x^{q^h}$$

$$= \sum_{h=0}^{r} \left(\alpha_h - \sum_{i=0}^{s} \gamma_{h-i}^{q^i} \beta_i \right) x^{q^h}$$

It follows from the system above that $R(x)$ has degree $< q^t$. But since $R(\zeta) = K(\zeta) - L(M(\zeta)) = 0$, the definition of $M(x)$ implies that $R(x)$ is the zero polynomial. Therefore, we have $K(x) = L(M(x)) = L(x) \otimes M(x)$. □

We consider now the problem of determining the number N_L of q-primitive roots over $\mathbb{F}_q$ of a nonzero q-polynomial $L(x)$ over $\mathbb{F}_q$. If $L(x)$ has multiple roots, then by Theorem 3.65 we can write $L(x) = L_1(x)^q$ with a q-polynomial $L_1(x)$ over $\mathbb{F}_q$. Since every root of $L(x)$ is then also a root of $L_1(x)$, we have $N_L = 0$. Thus we can assume that $L(x)$ has only simple roots. If $L(x)$ has degree 1, it is obvious that $N_L = 1$. If $L(x)$ has degree $q^n > 1$ and is monic (without loss of generality), let

$$L(x) = \underbrace{L_1(x) \otimes \cdots \otimes L_1(x)}_{e_1} \otimes \cdots \otimes \underbrace{L_r(x) \otimes \cdots \otimes L_r(x)}_{e_r}$$

be the symbolic factorization of $L(x)$ with distinct monic symbolically irreducible polynomials $L_i(x)$ over $\mathbb{F}_q$. We obtain N_L by subtracting from the total number q^n of roots the number of roots of $L(x)$ that are already roots of some nonzero q-polynomial over $\mathbb{F}_q$ of degree $< q^n$. If ζ is a root of $L(x)$ of the latter kind and $M(x)$ is the minimal q-polynomial of ζ over $\mathbb{F}_q$, then $\deg(M(x)) < q^n$ and $M(x)$ symbolically divides $L(x)$ by Theorem 3.68. It follows that $M(x)$ symbolically divides one of the polynomials $K_i(x)$, $1 \le i \le r$, obtained from the symbolic factorization of $L(x)$ by omitting the symbolic factor $L_i(x)$, in which case $K_i(\zeta) = 0$ by Theorem 3.68. Since every root of $K_i(x)$ is automatically a root of $L(x)$, it follows that N_L is q^n minus the number of ζ that are roots of some $K_i(x)$. If q^{n_i} is the degree of $L_i(x)$, then the degree, and thus the number of roots, of $K_i(x)$ is q^{n-n_i}. If $i_1, \ldots, i_s$ are distinct subscripts, then the number of common roots of $K_{i_1}(x), \ldots, K_{i_s}(x)$ is equal to the degree of the greatest common divisor, which is the same as the degree of the greatest common symbolic divisor (see the discussion following Example 3.64). Using symbolic factorizations, one finds that this degree is equal to

$$q^{n - n_{i_1} - \cdots - n_{i_s}}.$$

Altogether, the inclusion-exclusion principle of combinatorics yields

$$N_L = q^n - \sum_{i=1}^{r} q^{n-n_i} + \sum_{1 \le i < j \le r} q^{n-n_i-n_j} \mp \cdots + (-1)^r q^{n-n_1-\cdots-n_r}$$
$$= q^n(1 - q^{-n_1}) \cdots (1 - q^{-n_r}).$$

This expression can also be interpreted in a different way. Let $l(x)$ be the conventional q-associate of $L(x)$. Then

$$l(x) = l_1(x)^{e_1} \cdots l_r(x)^{e_r}$$

is the canonical factorization of $l(x)$ in $\mathbb{F}_q[x]$, where $l_i(x)$ is the conventional q-associate of $L_i(x)$. We define an analog of Euler's ϕ-function (see Exercise 1.4) for nonzero $f \in \mathbb{F}_q[x]$ by letting $\Phi_q(f(x)) = \Phi_q(f)$ denote the number of polynomials in $\mathbb{F}_q[x]$ that are of smaller degree than f as well as relatively prime to f. The following result will then imply the identity $N_L = \Phi_q(l(x))$ for the case under consideration.

3.69. Lemma. *The function* Φ_q *defined for nonzero polynomials in* $\mathbb{F}_q[x]$ *has the following properties*:

  (i) $\Phi_q(f) = 1$ *if* $\deg(f) = 0$;
  (ii) $\Phi_q(fg) = \Phi_q(f)\Phi_q(g)$ *whenever* f *and* g *are relatively prime*;
  (iii) *if* $\deg(f) = n \ge 1$, *then*

$$\Phi_q(f) = q^n(1 - q^{-n_1}) \cdots (1 - q^{-n_r}),$$

  where the n_i *are the degrees of the distinct monic irreducible polynomials appearing in the canonical factorization of* f *in* $\mathbb{F}_q[x]$.

Proof. Property (i) is trivial. For property (ii), let $\Phi_q(f) = s$ and $\Phi_q(g) = t$, and let $f_1, \ldots, f_s$ resp. $g_1, \ldots, g_t$ be the polynomials counted by $\Phi_q(f)$ resp. $\Phi_q(g)$. If $h \in \mathbb{F}_q[x]$ is a polynomial with $\deg(h) < \deg(fg)$ and $\gcd(fg, h) = 1$, then $\gcd(f, h) = \gcd(g, h) = 1$, and so $h \equiv f_i \bmod f$, $h \equiv g_j \bmod g$ for a unique ordered pair (i, j) with $1 \leqslant i \leqslant s$, $1 \leqslant j \leqslant t$. On the other hand, given an ordered pair (i, j), the Chinese remainder theorem for $\mathbb{F}_q[x]$ (see Exercise 1.37) shows that there exists a unique $h \in \mathbb{F}_q[x]$ with $h \equiv f_i \bmod f$, $h \equiv g_j \bmod g$, and $\deg(h) < \deg(fg)$. This h satisfies $\gcd(f, h) = \gcd(g, h) = 1$, and so $\gcd(fg, h) = 1$. Therefore, there is a one-to-one correspondence between the st ordered pairs (i, j) and the polynomials $h \in \mathbb{F}_q[x]$ with $\deg(h) < \deg(fg)$ and $\gcd(fg, h) = 1$. Consequently, $\Phi_q(fg) = st = \Phi_q(f)\Phi_q(g)$.

For an irreducible polynomial b in $\mathbb{F}_q[x]$ of degree m and a positive integer e, we can calculate $\Phi_q(b^e)$ directly. The polynomials $h \in \mathbb{F}_q[x]$ with $\deg(h) < \deg(b^e) = em$ that are not relatively prime to b^e are exactly those divisible by b, and they are thus of the form $h = gb$ with $\deg(g) < em - m$. Since there are q^{em-m} different choices for g, we get $\Phi_q(b^e) = q^{em} - q^{em-m} = q^{em}(1 - q^{-m})$. Property (iii) follows now from property (ii). $\square$

3.70. Theorem. *Let $L(x)$ be a nonzero q-polynomial over $\mathbb{F}_q$ with conventional q-associate $l(x)$. Then the number N_L of q-primitive roots of $L(x)$ over $\mathbb{F}_q$ is given by $N_L = 0$ if $L(x)$ has multiple roots and by $N_L = \Phi_q(l(x))$ if $L(x)$ has simple roots.*

Proof. This follows from Lemma 3.69 and the discussion preceding it. $\square$

3.71. Corollary. *Every nonzero q-polynomial over $\mathbb{F}_q$ with simple roots has at least one q-primitive root over $\mathbb{F}_q$.*

Earlier in this section we introduced the notion of a q-modulus. The results about q-primitive roots can be used to construct a special type of basis for a q-modulus.

3.72. Theorem. *Let M be a q-modulus of dimension $m \geqslant 1$ over $\mathbb{F}_q$. Then there exists an element $\zeta \in M$ such that $\{\zeta, \zeta^q, \zeta^{q^2}, \ldots, \zeta^{q^{m-1}}\}$ is a basis of M over $\mathbb{F}_q$.*

Proof. According to Theorem 3.65, $L(x) = \prod_{\beta \in M}(x - \beta)$ is a q-polynomial over $\mathbb{F}_q$. By Corollary 3.71, $L(x)$ has a q-primitive root ζ over $\mathbb{F}_q$. Then $\zeta, \zeta^q, \zeta^{q^2}, \ldots, \zeta^{q^{m-1}}$ are elements of M. If these elements were linearly dependent over $\mathbb{F}_q$, then ζ would be a root of a nonzero q-polynomial over $\mathbb{F}_q$ of degree less than $q^m = \deg(L(x))$, a contradiction to the definition of a q-primitive root of $L(x)$ over $\mathbb{F}_q$. Therefore, these m elements are linearly independent over $\mathbb{F}_q$, and so they form a basis of M over $\mathbb{F}_q$. $\square$

3.73. Theorem. *In $\mathbb{F}_{q^m}$ there exist exactly $\Phi_q(x^m-1)$ elements ζ such that $\{\zeta, \zeta^q, \zeta^{q^2}, \ldots, \zeta^{q^{m-1}}\}$ is a basis of $\mathbb{F}_{q^m}$ over $\mathbb{F}_q$.*

Proof. Since $\mathbb{F}_{q^m}$ can be viewed as a q-modulus, the argument in the proof of Theorem 3.72 applies. Here

$$L(x) = \prod_{\beta \in \mathbb{F}_{q^m}} (x-\beta) = x^{q^m} - x$$

by Lemma 2.4, and every q-primitive root of $L(x)$ over $\mathbb{F}_q$ yields a basis of the desired type. On the other hand, if $\zeta \in \mathbb{F}_{q^m}$ is not a q-primitive root of $L(x)$ over $\mathbb{F}_q$, then $\zeta, \zeta^q, \zeta^{q^2}, \ldots, \zeta^{q^{m-1}}$ are linearly dependent over $\mathbb{F}_q$, and so they do not form a basis of $\mathbb{F}_{q^m}$ over $\mathbb{F}_q$. Consequently, the number of $\zeta \in \mathbb{F}_{q^m}$ such that $\{\zeta, \zeta^q, \zeta^{q^2}, \ldots, \zeta^{q^{m-1}}\}$ is a basis of $\mathbb{F}_{q^m}$ over $\mathbb{F}_q$ is equal to the number of q-primitive roots of $L(x)$ over $\mathbb{F}_q$, which is given by $\Phi_q(x^m-1)$ according to Theorem 3.70. □

This result provides a refinement of the normal basis theorem (compare with Definition 2.32 and Theorem 2.35). Since each of the elements $\zeta, \zeta^q, \zeta^{q^2}, \ldots, \zeta^{q^{m-1}}$ generates the same normal basis of $\mathbb{F}_{q^m}$ over $\mathbb{F}_q$, the number of different normal bases of $\mathbb{F}_{q^m}$ over $\mathbb{F}_q$ is given by $(1/m)\Phi_q(x^m-1)$.

3.74. Example. We calculate the number of different normal bases of $\mathbb{F}_{64}$ over $\mathbb{F}_2$. Since $64 = 2^6$, this number is given by $\frac{1}{6}\Phi_2(x^6-1)$. From the canonical factorization

$$x^6 - 1 = (x+1)^2(x^2+x+1)^2$$

in $\mathbb{F}_2[x]$ and Lemma 3.69(iii) it follows that

$$\Phi_2(x^6-1) = 2^6\left(1-\tfrac{1}{2}\right)\left(1-\tfrac{1}{4}\right) = 24,$$

and so there are four different normal bases of $\mathbb{F}_{64}$ over $\mathbb{F}_2$. □

5. BINOMIALS AND TRINOMIALS

A *binomial* is a polynomial with two nonzero terms, one of them being the constant term. Irreducible binomials can be characterized explicitly. For this purpose it suffices to consider nonlinear, monic binomials.

3.75. Theorem. *Let $t \geqslant 2$ be an integer and $a \in \mathbb{F}_q^*$. Then the binomial $x^t - a$ is irreducible in $\mathbb{F}_q[x]$ if and only if the following two conditions are satisfied: (i) each prime factor of t divides the order e of a in $\mathbb{F}_q^*$, but not $(q-1)/e$; (ii) $q \equiv 1 \bmod 4$ if $t \equiv 0 \bmod 4$.*

(i) $p \mid t \Leftrightarrow \operatorname{ord}_p(e) > 0$ + $\operatorname{ord}_p(e) = \operatorname{ord}_p(q-1)$.

Proof. Suppose (i) and (ii) are satisfied. Then we note that $f(x) = x - a$ is an irreducible polynomial in $\mathbb{F}_q[x]$ of order e, and so $f(x^t) = x^t - a$ is irreducible in $\mathbb{F}_q[x]$ by Theorem 3.35.

Suppose (i) is violated. Then there exists a prime factor r of t that either divides $(q-1)/e$ or does not divide e. In the first case, we have $rs = (q-1)/e$ for some $s \in \mathbb{N}$. The subgroup of $\mathbb{F}_q^*$ consisting of rth powers has order $(q-1)/r = es$ and thus contains the subgroup of order e of $\mathbb{F}_q^*$ generated by a. In particular, $a = b^r$ for some $b \in \mathbb{F}_q^*$, and so $x^t - a = x^{t_1 r} - b^r$ has the factor $x^{t_1} - b$. In the remaining case, r divides neither $(q-1)/e$ nor e, and so r does not divide $q - 1$. Then $r_1 r \equiv 1 \bmod(q-1)$ for some $r_1 \in \mathbb{N}$, and thus $x^t - a = x^{t_1 r} - a^{r_1 r}$ has the factor $x^{t_1} - a^{r_1}$.

Suppose (i) is satisfied and (ii) is violated. Then $t = 4t_2$ for some $t_2 \in \mathbb{N}$ and $q \not\equiv 1 \bmod 4$. But (i) implies that e is even, and since e divides $q - 1$, q must be odd. Hence $q \equiv 3 \bmod 4$. The fact that $x^t - a$ is reducible in $\mathbb{F}_q[x]$ is then a consequence of Theorem 3.37. This can also be seen directly as follows. First we note that the information on e and q yields $e \equiv 2 \bmod 4$. Moreover, $a^{e/2} = -1$, and so $x^t - a = x^t + a^{(e/2)+1} = x^t + a^d$, where $d = (e/2)+1$ is even. Now

$$a^d = 4(2^{-1}a^{d/2})^2$$
$$= 4(2^{-1}a^{d/2})^{q+1} = 4c^4 \quad \text{with } c = (2^{-1}a^{d/2})^{(q+1)/4},$$

and this leads to the decomposition

$$x^t - a = x^{4t_2} + 4c^4$$
$$= (x^{2t_2} + 2cx^{t_2} + 2c^2)(x^{2t_2} - 2cx^{t_2} + 2c^2). \qquad \square$$

If $q \equiv 3 \bmod 4$, we can write q in the form $q = 2^A u - 1$ with $A \geq 2$ and u odd. Suppose condition (i) in Theorem 3.75 is satisfied and t is divisible by 2^A. We write $t = Bv$ with $B = 2^{A-1}$ and v even. Then $k = A$ in Theorem 3.37, so that with $f(x) = x - a$ the polynomial $f(x^t) = x^t - a$ factors as a product of B monic irreducible polynomials in $\mathbb{F}_q[x]$ of degree $t/B = v$. These irreducible factors can be determined explicitly. We note that as in the last part of the proof of Theorem 3.75, $d = (e/2)+1$ is even. Since $\gcd(2B, q-1) = 2$, there exists $r \in \mathbb{N}$ with $2Br \equiv d \bmod(q-1)$. Setting $b = a^r \in \mathbb{F}_q$, we get then the following canonical factorization.

3.76. Theorem. *With the conditions and the notation introduced above, let*

$$F(x) = \sum_{i=0}^{B/2} \frac{(B-i-1)!B}{i!(B-2i)!} x^{B-2i} \in \mathbb{F}_q[x].$$

Then the roots $c_1, \ldots, c_B$ of $F(x)$ are all in $\mathbb{F}_q$, and in $\mathbb{F}_q[x]$ we have the *canonical factorization*

$$x^t - a = \prod_{j=1}^{B} \left(x^v - bc_j x^{v/2} - b^2 \right).$$

Proof. For a nonzero element γ in an extension field of $\mathbb{F}_q$ we have

$$(x - \gamma)(x + \gamma^{-1}) = x^2 - \beta x - 1 \text{ with } \beta = \gamma - \gamma^{-1}.$$

Using the statement and the notation of Waring's formula (see Theorem 1.76), we get

$$s_B(x_1, x_2) = x_1^B + x_2^B$$

$$= \sum_{\substack{i_1 + 2i_2 = B \\ i_1, i_2 \geqslant 0}} (-1)^{i_2} \frac{(i_1 + i_2 - 1)!B}{i_1! i_2!} \sigma_1(x_1, x_2)^{i_1} \sigma_2(x_1, x_2)^{i_2}$$

$$= \sum_{i_2 = 0}^{B/2} (-1)^{i_2} \frac{(B - i_2 - 1)!B}{(B - 2i_2)! i_2!} (x_1 + x_2)^{B - 2i_2} (x_1 x_2)^{i_2}.$$

Putting $x_1 = \gamma$, $x_2 = -\gamma^{-1}$, we obtain

$$\gamma^B + \gamma^{-B} = \sum_{i=0}^{B/2} (-1)^i \frac{(B - i - 1)!B}{i!(B - 2i)!} \beta^{B-2i}(-1)^i = F(\beta).$$

If c_j is a root of $F(x)$ in some extension field of $\mathbb{F}_q$ and γ_j is such that $\gamma_j - \gamma_j^{-1} = c_j$, then $\gamma_j^B + \gamma_j^{-B} = F(c_j) = 0$, and so $\gamma_j^{2B} = -1$. Since $q + 1 = 2Bu$ with u odd, we get $\gamma_j^{q+1} = -1$, hence $\gamma_j^q = -\gamma_j^{-1}$. Then

$$c_j^q = \left(\gamma_j - \gamma_j^{-1} \right)^q = \gamma_j^q - \gamma_j^{-q} = -\gamma_j^{-1} + \gamma_j = c_j,$$

and so $c_j \in \mathbb{F}_q$. Since $F(x)$ is monic, we have

$$F(x) = \prod_{j=1}^{B} (x - c_j),$$

hence

$$\gamma^B + \gamma^{-B} = F(\beta) = \prod_{j=1}^{B} (\beta - c_j) = \prod_{j=1}^{B} \left(\gamma - \gamma^{-1} - c_j \right).$$

It follows that

$$\gamma^{2B} + 1 = \prod_{j=1}^{B} \left(\gamma^2 - c_j \gamma - 1 \right).$$

Since this identity holds for any element γ of any extension field of $\mathbb{F}_q$ (also for $\gamma = 0$), we get the polynomial identity

$$x^{2B} + 1 = \prod_{j=1}^{B} \left(x^2 - c_j x - 1 \right).$$

By substituting $b^{-1}x^{v/2}$ for x and multiplying by b^{2B}, we get a factorization of $x^{Bv} + b^{2B} = x^t + a^{2Br} = x^t + a^d = x^t - a$ (compare with the final portion of the proof of Theorem 3.75 for the last step). The resulting factors are irreducible in $\mathbb{F}_q[x]$ because we know already that the canonical factorization of $x^t - a$ involves B irreducible polynomials in $\mathbb{F}_q[x]$ of degree v (see the discussion preceding Theorem 3.76). $\square$

3.77.   Example.   We factor the binomial $x^{24} - 3$ in $\mathbb{F}_7[x]$. Here $q = 2^3 - 1$, so that $A = 3$, $B = 4$, and $v = 6$. Furthermore, the element $a = 3$ is of order $e = 6$ in $\mathbb{F}_7^*$, and so condition (i) in Theorem 3.75 is satisfied and Theorem 3.76 can be applied. We have $d = 4$, and a solution of the congruence $8r \equiv 4 \bmod 6$ is given by $r = 2$. Therefore, $b = a^2 = 2$. Furthermore, $F(x) = x^4 + 4x^2 + 2$ has the roots ± 1 and ± 3 in $\mathbb{F}_7$. Thus $x^{24} - 3 = (x^6 - 2x^3 - 4)(x^6 + 2x^3 - 4)(x^6 + x^3 - 4)(x^6 - x^3 - 4)$ is the canonical factorization in $\mathbb{F}_7[x]$. $\square$

A *trinomial* is a polynomial with three nonzero terms, one of them being the constant term. We first consider trinomials that are also affine polynomials.

3.78.   Theorem.   *Let $a \in \mathbb{F}_q$ and let p be the characteristic of $\mathbb{F}_q$. Then the trinomial $x^p - x - a$ is irreducible in $\mathbb{F}_q[x]$ if and only if it has no root in $\mathbb{F}_q$.*

Proof. If β is a root of $x^p - x - a$ in some extension field of $\mathbb{F}_q$, then by the proof of Theorem 3.56 the set of roots of $x^p - x - a$ is $\beta + U$, where U is the set of roots of the linearized polynomial $x^p - x$. But $U = \mathbb{F}_p$, and so

$$x^p - x - a = \prod_{b \in \mathbb{F}_p} (x - \beta - b).$$

Suppose now that $x^p - x - a$ has a factor $g \in \mathbb{F}_q[x]$ with $1 \leqslant r = \deg(g) < p$ and g monic. Then

$$g(x) = \prod_{i=1}^{r} (x - \beta - b_i)$$

for certain $b_i \in \mathbb{F}_p$. A comparison of the coefficients of x^{r-1} shows that $r\beta + b_1 + \cdots + b_r$ is an element of $\mathbb{F}_q$. Since r has a multiplicative inverse in $\mathbb{F}_q$, it follows that $\beta \in \mathbb{F}_q$. Thus we have shown that if $x^p - x - a$ factors nontrivially in $\mathbb{F}_q[x]$, then it has a root in $\mathbb{F}_q$. The converse is trivial. $\square$

3.79.   Corollary.   *With the notation of Theorem 3.78, the trinomial $x^p - x - a$ is irreducible in $\mathbb{F}_q[x]$ if and only if $\mathrm{Tr}_{\mathbb{F}_q}(a) \neq 0$.*

Proof.   By Theorem 2.25, $x^p - x - a$ has a root in $\mathbb{F}_q$ if and only if the absolute trace $\mathrm{Tr}_{\mathbb{F}_q}(a)$ is 0. The rest follows from Theorem 3.78. $\square$

Since for $b \in \mathbb{F}_q^*$ the polynomial $f(x)$ is irreducible over $\mathbb{F}_q$ if and only if $f(bx)$ is irreducible over $\mathbb{F}_q$, the criteria above hold also for trinomials of the form $b^p x^p - bx - a$.

If we consider more general trinomials of the above type for which the degree is a higher power of the characteristic, then these criteria need not be valid any longer. In fact, the following decomposition formula can be established.

3.80. Theorem. *For $x^q - x - a$ with a being an element of the subfield $K = \mathbb{F}_r$ of $F = \mathbb{F}_q$, we have the decomposition*

$$x^q - x - a = \prod_{j=1}^{q/r} \left(x^r - x - \beta_j \right) \tag{3.18}$$

in $\mathbb{F}_q[x]$, where the β_j are the distinct elements of $\mathbb{F}_q$ with $\operatorname{Tr}_{F/K}(\beta_j) = a$.

Proof. For a given β_j, let γ be a root of $x^r - x - \beta_j$ in some extension field of $\mathbb{F}_q$. Then $\gamma^r - \gamma = \beta_j$, and also

$$a = \operatorname{Tr}_{F/K}(\beta_j)$$

$$= \operatorname{Tr}_{F/K}(\gamma^r - \gamma)$$

$$= (\gamma^r - \gamma) + (\gamma^r - \gamma)^r + (\gamma^r - \gamma)^{r^2} + \cdots + (\gamma^r - \gamma)^{q/r} = \gamma^q - \gamma,$$

so that γ is a root of $x^q - x - a$. Since $x^r - x - \beta_j$ has only simple roots, $x^r - x - \beta_j$ divides $x^q - x - a$. Now the polynomials $x^r - x - \beta_j$, $1 \leqslant j \leqslant q/r$, are pairwise relatively prime, and so the polynomial on the right-hand side of (3.18) divides $x^q - x - a$. A comparison of degrees and of leading coefficients shows that the two sides of (3.18) are identical. □

3.81. Example. Consider $x^9 - x - 1$ in $\mathbb{F}_9[x]$. Viewing $\mathbb{F}_9$ as $\mathbb{F}_3(\alpha)$, where α is a root of the irreducible polynomial $x^2 - x - 1$ in $\mathbb{F}_3[x]$, we find that the elements of $\mathbb{F}_9$ with absolute trace equal to 1 are $-1, \alpha, 1 - \alpha$. Thus (3.18) yields the decomposition

$$x^9 - x - 1 = (x^3 - x + 1)(x^3 - x - \alpha)(x^3 - x - 1 + \alpha).$$

Since all three factors are irreducible in $\mathbb{F}_9[x]$, we have also obtained the canonical factorization of $x^9 - x - 1$ in $\mathbb{F}_9[x]$. □

The information about irreducible trinomials can be applied to the construction of new irreducible polynomials from given ones.

3.82. Theorem. *Let $f(x) = x^m + a_{m-1}x^{m-1} + \cdots + a_0$ be an irreducible polynomial over the finite field $\mathbb{F}_q$ of characteristic p and let $b \in \mathbb{F}_q$. Then the polynomial $f(x^p - x - b)$ is irreducible over $\mathbb{F}_q$ if and only if the absolute trace $\operatorname{Tr}_{\mathbb{F}_q}(mb - a_{m-1})$ is $\neq 0$.*

Proof. Suppose $\text{Tr}_{\mathbb{F}_q}(mb - a_{m-1}) \neq 0$. Put $K = \mathbb{F}_q$ and let F be the splitting field of f over K. If $\alpha \in F$ is a root of f, then, according to Theorem 2.14, all the roots of f are given by $\alpha, \alpha^q, \ldots, \alpha^{q^{m-1}}$ and $F = K(\alpha)$. Furthermore, $\text{Tr}_{F/K}(\alpha) = -a_{m-1}$ by (2.2), and using Theorem 2.26 we get

$$\text{Tr}_F(\alpha + b) = \text{Tr}_K\big(\text{Tr}_{F/K}(\alpha + b)\big) = \text{Tr}_K(-a_{m-1} + mb) \neq 0.$$

By Corollary 3.79, the trinomial $x^p - x - (\alpha + b)$ is irreducible over F. Thus $[F(\beta): F] = p$, where β is a root of $x^p - x - (\alpha + b)$. It follows from Theorem 1.84 that

$$[F(\beta): K] = [F(\beta): F][F: K] = pm.$$

Now $\alpha = \beta^p - \beta - b$, so that $\alpha \in K(\beta)$ and $K(\beta) = K(\alpha, \beta) = F(\beta)$. Hence $[K(\beta): K] = pm$ and the minimal polynomial of β over K has degree pm. But $f(\beta^p - \beta - b) = f(\alpha) = 0$, and so β is a root of the monic polynomial $f(x^p - x - b) \in K[x]$ of degree pm. Theorem 3.33(ii) shows that $f(x^p - x - b)$ is the minimal polynomial of β over K. By Theorem 3.33(i), $f(x^p - x - b)$ is irreducible over $K = \mathbb{F}_q$.

If $\text{Tr}_{\mathbb{F}_q}(mb - a_{m-1}) = 0$, then $x^p - x - (\alpha + b)$ is reducible over F, and so $[F(\beta): F] < p$ for any root β of $x^p - x - (\alpha + b)$. The same arguments as above show that β is a root of $f(x^p - x - b)$ and that $[F(\beta): K] < pm$, hence $f(x^p - x - b)$ is reducible over $K = \mathbb{F}_q$. $\square$

For certain types of reducible trinomials we can establish the form of the canonical factorization. The hypothesis for this result involves the irreducibility of a binomial, which can be checked by Theorem 3.75.

3.83. Theorem. *Let* $f(x) = x^r - ax - b \in \mathbb{F}_q[x]$, *where* $r > 2$ *is a power of the characteristic of* $\mathbb{F}_q$, *and suppose that the binomial* $x^{r-1} - a$ *is irreducible over* $\mathbb{F}_q$. *Then* $f(x)$ *is the product of a linear polynomial and an irreducible polynomial over* $\mathbb{F}_q$ *of degree* $r - 1$.

Proof. Since $f'(x) = -a \neq 0$, $f(x)$ has only simple roots. If p is the characteristic of $\mathbb{F}_q$, then $f(x)$ is an affine p-polynomial over $\mathbb{F}_q$. Hence, Theorem 3.56 shows that the difference γ of two distinct roots of $f(x)$ is a root of the p-polynomial $x^r - ax$, and so a root of $x^{r-1} - a$. From $r - 1 > 1$ and the hypothesis about this binomial, it follows that γ is not an element of $\mathbb{F}_q$, and so there exists a root α of $f(x)$ that is not an element of $\mathbb{F}_q$. Then $\alpha^q \neq \alpha$ is also a root of $f(x)$ and, by what we have already shown, $\alpha^q - \alpha$ is a root of the irreducible polynomial $x^{r-1} - a$ over $\mathbb{F}_q$, so that $[\mathbb{F}_q(\alpha^q - \alpha): \mathbb{F}_q] = r - 1$. Since $\mathbb{F}_q(\alpha^q - \alpha) \subseteq \mathbb{F}_q(\alpha)$, it follows that $m = [\mathbb{F}_q(\alpha): \mathbb{F}_q]$ is a multiple of $r - 1$. On the other hand, α is a root of the polynomial $f(x)$ of degree r, so that $m \leq r$. Because of $r > 2$, this is only possible if $m = r - 1$. Thus the minimal polynomial of α over $\mathbb{F}_q$ is an irreducible polynomial over $\mathbb{F}_q$ of degree $r - 1$ that divides $f(x)$. The result follows now immediately. $\square$

In the special case of prime fields, one can characterize the primitive polynomials among trinomials of a certain kind.

3.84.   Theorem.   *For a prime p, the trinomial $x^p - x - a \in \mathbb{F}_p[x]$ is a primitive polynomial over $\mathbb{F}_p$ if and only if a is a primitive element of $\mathbb{F}_p$ and* $\mathrm{ord}(x^p - x - 1) = (p^p - 1)/(p - 1)$.

Proof.   Suppose first that $f(x) = x^p - x - a$ is a primitive polynomial over $\mathbb{F}_p$. Then a must be a primitive element of $\mathbb{F}_p$ because of Theorem 3.18. If β is a root of $g(x) = x^p - x - 1$ in some extension field of $\mathbb{F}_p$, then

$$0 = ag(\beta) = a(\beta^p - \beta - 1) = a^p\beta^p - a\beta - a = f(a\beta),$$

and so $\alpha = a\beta$ is a root of $f(x)$. Consequently, we have $\beta^r \neq 1$ for $0 < r < (p^p - 1)/(p - 1)$, for otherwise $\alpha^{r(p-1)} = 1$ with $0 < r(p-1) < p^p - 1$, a contradiction to α being a primitive element of $\mathbb{F}_{p^p}$. On the other hand, $g(x)$ is irreducible over $\mathbb{F}_p$ by Corollary 3.79, and so

$$g(x) = x^p - x - 1 = (x - \beta)(x - \beta^p) \cdots \left(x - \beta^{p^{p-1}}\right).$$

A comparison of the constant terms leads to $\beta^{(p^p - 1)/(p-1)} = 1$, hence $\mathrm{ord}(x^p - x - 1) = (p^p - 1)/(p - 1)$ on account of Theorem 3.3.

Conversely, if the conditions of the theorem are satisfied, then a and β have orders $p - 1$ and $(p^p - 1)/(p - 1)$, respectively, in the multiplicative group $\mathbb{F}_{p^p}^*$. Now

$$(p^p - 1)/(p - 1) = 1 + p + p^2 + \cdots + p^{p-1} \equiv 1 + 1 + 1 + \cdots + 1$$

$$\equiv p \equiv 1 \bmod(p - 1),$$

so that $p - 1$ and $(p^p - 1)/(p - 1)$ are relatively prime. Therefore, $\alpha = a\beta$ has order $(p - 1) \cdot (p^p - 1)/(p - 1) = p^p - 1$ in $\mathbb{F}_{p^p}^*$. Hence α is a primitive element of $\mathbb{F}_{p^p}$ and $f(x)$ is a primitive polynomial over $\mathbb{F}_p$. $\square$

3.85.   Example.   For $p = 5$ we have $(p^p - 1)/(p - 1) = 781 = 11 \cdot 71$. From the proof of Theorem 3.84 it follows that $x^{781} \equiv 1 \bmod(x^5 - x - 1)$, and since $x^{11} \not\equiv 1 \bmod(x^5 - x - 1)$ and $x^{71} \not\equiv 1 \bmod(x^5 - x - 1)$, we obtain $\mathrm{ord}(x^5 - x - 1) = 781$. Now 2 and 3 are primitive elements of $\mathbb{F}_5$, and so $x^5 - x - 2$ and $x^5 - x - 3$ are primitive polynomials over $\mathbb{F}_5$ by Theorem 3.84. $\square$

For a trinomial $x^2 + x + a$ over a finite field $\mathbb{F}_q$ of odd characteristic, it is easily seen that it is irreducible over $\mathbb{F}_q$ if and only if a is not of the form $a = 4^{-1} - b^2$, $b \in \mathbb{F}_q$. Thus, there are exactly $(q - 1)/2$ choices for $a \in \mathbb{F}_q$ that make $x^2 + x + a$ irreducible over $\mathbb{F}_q$. More generally, the number of $a \in \mathbb{F}_q$ that make $x^n + x + a$ irreducible over $\mathbb{F}_q$ is usually asymptotic to q/n, according to the following result.

3.86. Theorem. *Let* $\mathbb{F}_q$ *be a finite field of characteristic p. For an integer* $n \geqslant 2$ *such that* $2n(n-1)$ *is not divisible by p, let* $T_n(q)$ *denote the number of* $a \in \mathbb{F}_q$ *for which the trinomial* $x^n + x + a$ *is irreducible over* $\mathbb{F}_q$. *Then there is a constant* B_n, *depending only on n, such that*

$$\left| T_n(q) - \frac{q}{n} \right| \leqslant B_n q^{1/2}.$$

We omit the proof, as it depends on an elaborate investigation of certain Galois groups.

In Definition 1.92 we defined the discriminant of a polynomial. The following result gives an explicit formula for the discriminant of a trinomial.

3.87. Theorem. *The discriminant of the trinomial* $x^n + ax^k + b \in \mathbb{F}_q[x]$ *with* $n > k \geqslant 1$ *is given by*

$$D(x^n + ax^k + b) = (-1)^{n(n-1)/2} b^{k-1}$$

$$\cdot \left(n^N b^{N-K} - (-1)^N (n-k)^{N-K} k^K a^N \right)^d,$$

where $d = \gcd(n, k)$, $N = n/d$, $K = k/d$.

EXERCISES

3.1. Determine the order of the polynomial $(x^2 + x + 1)^5(x^3 + x + 1)$ over $\mathbb{F}_2$.

3.2. Determine the order of the polynomial $x^7 - x^6 + x^4 - x^2 + x$ over $\mathbb{F}_3$.

3.3. Determine ord(f) for all monic irreducible polynomials f in $\mathbb{F}_3[x]$ of degree 3.

3.4. Prove that the polynomial $x^8 + x^7 + x^3 + x + 1$ is irreducible over $\mathbb{F}_2$ and determine its order.

3.5. Let $f \in \mathbb{F}_q[x]$ be a polynomial of degree $m \geqslant 1$ with $f(0) \neq 0$ and suppose that the roots $\alpha_1, \ldots, \alpha_m$ of f in the splitting field of f over $\mathbb{F}_q$ are all simple. Prove that ord(f) is equal to the least positive integer e such that $\alpha_i^e = 1$ for $1 \leqslant i \leqslant m$.

3.6. Prove that ord(Q_e) = e for all e for which the cyclotomic polynomial $Q_e \in \mathbb{F}_q[x]$ is defined.

3.7. Let f be irreducible over $\mathbb{F}_q$ with $f(0) \neq 0$. For $e \in \mathbb{N}$ relatively prime to q, prove that ord(f) = e if and only if f divides the cyclotomic polynomial Q_e.

3.8. Let $f \in \mathbb{F}_q[x]$ be as in Exercise 3.5 and let $b \in \mathbb{N}$. Find a general formula showing the relationship between ord(f^b) and ord(f).

3.9. Let $\mathbb{F}_q$ be a finite field of characteristic p, and let $f \in \mathbb{F}_q[x]$ be a polynomial of positive degree with $f(0) \neq 0$. Prove that ord($f(x^p)$) = p ord($f(x)$).

3.10. Let f be an irreducible polynomial in $\mathbb{F}_q[x]$ with $f(0) \neq 0$ and ord(f) = e, and let r be a prime not dividing q. Prove: (i) if r divides e, then every irreducible factor of $f(x^r)$ in $\mathbb{F}_q[x]$ has order er; (ii) if r does not divide e, then one irreducible factor of $f(x^r)$ in $\mathbb{F}_q[x]$ has order e and the other factors have order er.

3.11. Deduce from Exercise 3.10 that if $f \in \mathbb{F}_q[x]$ is a polynomial of positive degree with $f(0) \neq 0$, and if r is a prime not dividing q, then ord($f(x^r)$) = r ord($f(x)$).

3.12. Prove that the reciprocal polynomial of an irreducible polynomial f over $\mathbb{F}_q$ with $f(0) \neq 0$ is again irreducible over $\mathbb{F}_q$.

3.13. A nonzero polynomial $f \in \mathbb{F}_q[x]$ is called *self-reciprocal* if $f = f^*$. Prove that if $f = gh$, where g and h are irreducible in $\mathbb{F}_q[x]$ and f is self-reciprocal, then either (i) $h^* = ag$ with $a \in \mathbb{F}_q^*$; or (ii) $g^* = bg$, $h^* = bh$ with $b = \pm 1$.

3.14. Prove: if f is a self-reciprocal irreducible polynomial in $\mathbb{F}_q[x]$ of degree $m > 1$, then m must be even.

3.15. Prove: if f is a self-reciprocal irreducible polynomial in $\mathbb{F}_q[x]$ of degree > 1 and of order e, then every irreducible polynomial in $\mathbb{F}_q[x]$ of degree > 1 whose order divides e is self-reciprocal.

3.16. Show that $x^6 + x^5 + x^2 + x + 1$ is a primitive polynomial over $\mathbb{F}_2$.

3.17. Show that $x^8 + x^6 + x^5 + x + 1$ is a primitive polynomial over $\mathbb{F}_2$.

3.18. Show that $x^5 - x + 1$ is a primitive polynomial over $\mathbb{F}_3$.

3.19. Let $f \in \mathbb{F}_q[x]$ be monic of degree $m \geq 1$. Prove that f is primitive over $\mathbb{F}_q$ if and only if f is an irreducible factor over $\mathbb{F}_q$ of the cyclotomic polynomial $Q_d \in \mathbb{F}_q[x]$ with $d = q^m - 1$.

3.20. Determine the number of primitive polynomials over $\mathbb{F}_q$ of degree m.

3.21. If $m \in \mathbb{N}$ is not a prime, prove that not every monic irreducible polynomial over $\mathbb{F}_q$ of degree m can be a primitive polynomial over $\mathbb{F}_q$.

3.22. If m is a prime, prove that all monic irreducible polynomials over $\mathbb{F}_q$ of degree m are primitive over $\mathbb{F}_q$ if and only if $q = 2$ and $2^m - 1$ is a prime.

3.23. If f is a primitive polynomial over $\mathbb{F}_q$, prove that $f(0)^{-1}f^*$ is again primitive over $\mathbb{F}_q$.

3.24. Prove that the only self-reciprocal primitive polynomials are $x + 1$ and $x^2 + x + 1$ over $\mathbb{F}_2$ and $x + 1$ over $\mathbb{F}_3$ (see Exercise 3.13 for the definition of a self-reciprocal polynomial).

3.25. Prove: if $f(x)$ is irreducible in $\mathbb{F}_q[x]$, then $f(ax + b)$ is irreducible in $\mathbb{F}_q[x]$ for any $a, b \in \mathbb{F}_q$ with $a \neq 0$.

3.26. Prove that $N_q(n) \leqslant (1/n)(q^n - q)$ with equality if and only if n is prime.

3.27. Prove that

$$N_q(n) \geqslant \frac{1}{n}q^n - \frac{q}{n(q-1)}(q^{n/2}-1).$$

3.28. Give a detailed proof of the fact that (3.5) implies (3.4).

3.29. Prove that the Moebius function μ satisfies $\mu(mn) = \mu(m)\mu(n)$ for all $m, n \in \mathbb{N}$ with $\gcd(m, n) = 1$.

3.30. Prove the identity

$$\sum_{d|n} \frac{\mu(d)}{d} = \frac{\phi(n)}{n} \quad \text{for all } n \in \mathbb{N}.$$

3.31. Prove that $\sum_{d|n}\mu(d)\phi(d) = 0$ for every even integer $n \geqslant 2$.

3.32. Prove the identity $\sum_{d|n}|\mu(d)| = 2^k$, where k is the number of distinct prime factors of $n \in \mathbb{N}$.

3.33. Prove that $N_q(n)$ is divisible by eq provided that $n \geqslant 2$, e is a divisor of $q - 1$, and $\gcd(eq, n) = 1$.

3.34. Calculate the cyclotomic polynomials Q_{12} and Q_{30} from the explicit formula in Theorem 3.27.

3.35. Establish the properties of cyclotomic polynomials listed in Exercise 2.57, Parts (a)–(f), by using the explicit formula in Theorem 3.27.

3.36. Prove that the cyclotomic polynomial Q_n with $\gcd(n, q) = 1$ is irreducible over $\mathbb{F}_q$ if and only if the multiplicative order of q modulo n is $\phi(n)$.

3.37. If Q_n is irreducible over $\mathbb{F}_2$, prove that n must be a prime $\equiv \pm 3 \bmod 8$ or a power of such a prime. Show also that this condition is not sufficient.

3.38. Prove that Q_{15} is reducible over any finite field over which it is defined.

3.39. Prove that for $n \in \mathbb{N}$ there exists an integer b relatively prime to n whose multiplicative order modulo n is $\phi(n)$ if and only if $n = 1, 2, 4$, p^r, or $2p^r$, where p is an odd prime and $r \in \mathbb{N}$.

3.40. Dirichlet's theorem on primes in arithmetic progressions states that any arithmetic progression of integers $b, b + n, \ldots, b + kn, \ldots$ with $n \in \mathbb{N}$ and $\gcd(b, n) = 1$ contains infinitely many primes. Use this theorem to prove the following: the integers $n \in \mathbb{N}$ for which there exists a finite field $\mathbb{F}_q$ with $\gcd(n, q) = 1$ over which the cyclotomic polynomial Q_n is irreducible are exactly given by $n = 1, 2, 4, p^r$, or $2p^r$, where p is an odd prime and $r \in \mathbb{N}$.

3.41. Prove that Q_{19} and Q_{27} are two cyclotomic polynomials over $\mathbb{F}_2$ of the same degree that are both irreducible over $\mathbb{F}_2$.

3.42. If $e \geqslant 2$, $\gcd(e, q) = 1$, and m is the multiplicative order of q modulo e, prove that the product of all monic irreducible polynomials in $\mathbb{F}_q[x]$ of degree m and order e is equal to the cyclotomic polynomial Q_e over $\mathbb{F}_q$.

3.43. Find the factorization of $x^{32} - x$ into irreducible polynomials over $\mathbb{F}_2$.

3.44. Calculate $I(2,6; x)$ from the formula in Theorem 3.29.

3.45. Calculate $I(2,6; x)$ from the formula in Theorem 3.31.

3.46. Prove that

$$I(q, n; x) = \prod_{d \mid n} \left(x^{q^d - 1} - 1 \right)^{\mu(n/d)} \quad \text{for } n > 1.$$

3.47. Prove that over a finite field of odd order q the polynomial $\frac{1}{2}(1 + x^{(q+1)/2} + (1 - x)^{(q+1)/2})$ is the square of a polynomial.

3.48. Determine all irreducible polynomials in $\mathbb{F}_2[x]$ of degree 6 and order 21 and then all irreducible polynomials in $\mathbb{F}_2[x]$ of degree 294 and order 1029.

3.49. Determine all monic irreducible polynomials in $\mathbb{F}_3[x]$ of degree 3 and order 26 and then all monic irreducible polynomials in $\mathbb{F}_3[x]$ of degree 6 and order 104.

3.50. Proceed as in Example 3.41 to determine which polynomials f_t are irreducible in $\mathbb{F}_q[x]$ in the case $q = 5$, $m = 4$, $e = 78$.

3.51. In the notation of Example 3.41, prove that if t is a prime with $t - 1$ dividing $m - 1$, then f_t is irreducible in $\mathbb{F}_2[x]$.

3.52. Given the irreducible polynomial $f(x) = x^3 - x^2 + x + 1$ over $\mathbb{F}_3$, calculate f_2 and f_5 by the matrix-theoretic method.

3.53. Calculate f_2 and f_5 in the previous exercise by using the result of Theorem 3.39.

3.54. Use a root of the primitive polynomial $x^3 - x + 1$ over $\mathbb{F}_3$ to represent all elements of $\mathbb{F}_{27}^*$ and compute the minimal polynomials over $\mathbb{F}_3$ of all elements of $\mathbb{F}_{27}$.

3.55. Let $\theta \in \mathbb{F}_{64}$ be a root of the irreducible polynomial $x^6 + x + 1$ in $\mathbb{F}_2[x]$. Find the minimal polynomial of $\beta = 1 + \theta^2 + \theta^3$ over $\mathbb{F}_2$.

3.56. Let $\theta \in \mathbb{F}_{64}$ be a root of the irreducible polynomial $x^6 + x^4 + x^3 + x + 1$ in $\mathbb{F}_2[x]$. Find the minimal polynomial of $\beta = 1 + \theta + \theta^5$ over $\mathbb{F}_2$.

3.57. Determine all primitive polynomials over $\mathbb{F}_3$ of degree 2.

3.58. Determine all primitive polynomials over $\mathbb{F}_4$ of degree 2.

3.59. Determine a primitive polynomial over $\mathbb{F}_5$ of degree 3.

3.60. Factor the polynomial $g \in \mathbb{F}_3[x]$ from Example 3.44 in $\mathbb{F}_9[x]$ to obtain primitive polynomials over $\mathbb{F}_9$.

3.61. Factor the polynomial $g \in \mathbb{F}_2[x]$ from Example 3.45 in $\mathbb{F}_8[x]$ to obtain primitive polynomials over $\mathbb{F}_8$.

3.62. Find the roots of the following linearized polynomials in their splitting fields:
(a) $L(x) = x^8 + x^4 + x^2 + x \in \mathbb{F}_2[x]$;
(b) $L(x) = x^9 + x \in \mathbb{F}_3[x]$.

3.63. Find the roots of the following polynomials in the indicated fields by

first determining an affine multiple:
(a) $f(x) = x^7 + x^6 + x^3 + x^2 + 1 \in \mathbb{F}_2[x]$ in $\mathbb{F}_{32}$;
(b) $f(x) = x^4 + \theta x^3 - x^2 - (\theta + 1)x + 1 - \theta \in \mathbb{F}_9[x]$ in $\mathbb{F}_{729}$, where θ
     is a root of $x^2 - x - 1 \in \mathbb{F}_3[x]$.

3.64. Prove that for every polynomial f over $\mathbb{F}_{q^m}$ of positive degree there
       exists a nonzero q-polynomial over $\mathbb{F}_{q^m}$ that is divisible by f.

3.65. Prove that the greatest common divisor of two or more nonzero
       q-polynomials over $\mathbb{F}_{q^m}$ is again a q-polynomial, but that their least
       common multiple need not necessarily be a q-polynomial.

3.66. Determine the greatest common divisor of the following linearized
       polynomials:
       (a) $L_1(x) = x^{64} + x^{16} + x^8 + x^4 + x^2 + x \in \mathbb{F}_2[x]$,
            $L_2(x) = x^{32} + x^8 + x^2 + x \in \mathbb{F}_2[x]$;
       (b) $L_1(x) = x^{243} - x^{81} - x^9 + x^3 + x \in \mathbb{F}_3[x]$,
            $L_2(x) = x^{81} + x \in \mathbb{F}_3[x]$.

3.67. Determine the symbolic factorization of the following linearized
       polynomials into symbolically irreducible polynomials over the given
       prime fields:
       (a) $L(x) = x^{32} + x^{16} + x^8 + x^4 + x^2 + x \in \mathbb{F}_2[x]$;
       (b) $L(x) = x^{81} - x^9 - x^3 - x \in \mathbb{F}_3[x]$.

3.68. Prove that the q-polynomial $L_1(x)$ over $\mathbb{F}_{q^m}$ divides the q-polynomial
       $L(x)$ over $\mathbb{F}_{q^m}$ if and only if $L(x) = L_2(x) \otimes L_1(x)$ for some q-poly-
       nomial $L_2(x)$ over $\mathbb{F}_{q^m}$.

3.69. Prove that the greatest common divisor of two or more affine
       q-polynomials over $\mathbb{F}_{q^m}$, not all of them 0, is again an affine q-poly-
       nomial.

3.70. If $A_1(x) = L_1(x) - \alpha_1$ and $A_2(x) = L_2(x) - \alpha_2$ are affine q-polynomi-
       als over $\mathbb{F}_{q^m}$ and $A_1(x)$ divides $A_2(x)$, prove that the q-polynomial
       $L_1(x)$ divides the q-polynomial $L_2(x)$.

3.71. Let $f(x)$ be irreducible in $\mathbb{F}_q[x]$ with $f(0) \neq 0$ and let $F(x)$ be its
       linearized q-associate. Prove that $F(x)/x$ is irreducible in $\mathbb{F}_q[x]$ if
       and only if $f(x)$ is a primitive polynomial over $\mathbb{F}_q$ or a nonzero
       constant multiple of such a polynomial.

3.72. Let ζ be an element of a finite extension field of $\mathbb{F}_{q^m}$. Prove that a
       q-polynomial $K(x)$ over $\mathbb{F}_{q^m}$ has ζ as a root if and only if $K(x)$ is
       divisible by the minimal q-polynomial of ζ over $\mathbb{F}_{q^m}$.

3.73. For a nonzero polynomial $f \in \mathbb{F}_q[x]$, prove that $\sum \Phi_q(g) = q^{\deg(f)}$,
       where the sum is extended over all monic divisors $g \in \mathbb{F}_q[x]$ of f.

3.74. For a nonzero polynomial $f \in \mathbb{F}_q[x]$ and $g \in \mathbb{F}_q[x]$ with $\gcd(f, g) = 1$,
       prove that $g^k \equiv 1 \bmod f$, where $k = \Phi_q(f)$.

3.75. The function μ_q is defined on the set S of nonzero polynomials f over
       $\mathbb{F}_q$ by $\mu_q(f) = 1$ if $\deg(f) = 0$, $\mu_q(f) = 0$ if f has at least one multiple
       root, and $\mu_q(f) = (-1)^k$ if $\deg(f) \geq 1$ and f has only simple roots,
       where k is the number of irreducible factors in the canonical factori-

zation of f in $\mathbb{F}_q[x]$. Let Σ denote a sum extended over all monic divisors $g \in \mathbb{F}_q[x]$ of f. Prove the following properties:

(a) $\sum \mu_q(g) = \begin{cases} 1 & \text{if } \deg(f) = 0, \\ 0 & \text{if } \deg(f) \geqslant 1; \end{cases}$

(b) $\mu_q(fg) = \mu_q(f)\mu_q(g)$ for all $f, g \in S$ with $\gcd(f, g) = 1$;

(c) $\sum q^{\deg(g)} \mu_q(f/g) = \Phi_q(f)$ for all $f \in S$;

(d) if ψ is a mapping from S into an additively written abelian group G with $\psi(cf) = \psi(f)$ for all $c \in \mathbb{F}_q^*$ and $f \in S$, and if $\Psi(f) = \Sigma\psi(g)$ for all $f \in S$, then $\psi(f) = \Sigma\mu_q(f/g)\Psi(g) = \Sigma\mu_q(g)\Psi(f/g)$ for all $f \in S$.

3.76. Prove that the number of different normal bases of $\mathbb{F}_{q^m}$ over $\mathbb{F}_q$ is

$$\frac{1}{m} \prod_{d \mid m} \left(q^{\phi(d)} - 1 \right)$$

provided that $\gcd(m, q) = 1$ and the multiplicative order of q modulo m is $\phi(m)$.

3.77. Refer to Example 2.31 for the definition of a self-dual basis and show that there exists a self-dual normal basis of $\mathbb{F}_{2^m}$ over $\mathbb{F}_2$ whenever m is odd. (*Hint:* Show first that the number of different normal bases of $\mathbb{F}_{2^m}$ over $\mathbb{F}_2$ is odd whenever m is odd.)

3.78. For a prime r and $a \in \mathbb{F}_q$, prove that $x^r - a$ is either irreducible in $\mathbb{F}_q[x]$ or has a root in $\mathbb{F}_q$.

3.79. For an odd prime r, an integer $n \geqslant 1$, and $a \in \mathbb{F}_q$, prove that $x^{r^n} - a$ is irreducible in $\mathbb{F}_q[x]$ if and only if a is not an rth power of an element of $\mathbb{F}_q$.

3.80. Find the canonical factorization of the following binomials over the given prime fields:

(a) $f(x) = x^8 + 1 \in \mathbb{F}_3[x]$;

(b) $f(x) = x^{27} - 4 \in \mathbb{F}_{19}[x]$;

(c) $f(x) = x^{88} - 10 \in \mathbb{F}_{23}[x]$.

3.81. Prove that under the conditions of Theorem 3.76 the roots of the polynomial $F(x)$ introduced there are simple.

3.82. Prove that the resultant of two binomials $x^n - a$ and $x^m - b$ in $\mathbb{F}_q[x]$ is given by $(-1)^n(b^{n/d} - a^{m/d})^d$ with $d = \gcd(n, m)$, where n and m are considered to be the formal degrees of the binomials (compare with Definition 1.93).

3.83. For a nonzero element b of a prime field $\mathbb{F}_p$, prove that the trinomial $x^p - x - b$ is irreducible in $\mathbb{F}_{p^n}[x]$ if and only if n is not divisible by p.

3.84. Prove that any polynomial of the form $x^q - ax - b \in \mathbb{F}_q[x]$ with $a \neq 1$ has a root in $\mathbb{F}_q$.

3.85. Prove: if $x^p - x - a$ is irreducible over the field $\mathbb{F}_q$ of characteristic p

and β is a root of this trinomial in an extension field of $\mathbb{F}_q$, then $x^p - x - a\beta^{p-1}$ is irreducible over $\mathbb{F}_q(\beta)$.

3.86. Prove: if $f(x) = x^m + a_{m-1}x^{m-1} + \cdots + a_0$ is irreducible over the field $\mathbb{F}_q$ of characteristic p and $b \in \mathbb{F}_q$ is such that $\text{Tr}_{\mathbb{F}_q}(mb - a_{m-1}) = 0$, then $f(x^p - x - b)$ is the product of p irreducible polynomials over $\mathbb{F}_q$ of degree m.

3.87. If m and p are distinct primes and the multiplicative order of p modulo m is $m - 1$, prove that $\sum_{i=0}^{m-1}(x^p - x)^i$ is irreducible over $\mathbb{F}_p$.

3.88. Find the canonical factorization of the given polynomial over the indicated field:
   (a) $f(x) = x^8 - \alpha x - 1 \in \mathbb{F}_{64}[x]$, where α satisfies $\alpha^3 = \alpha + 1$;
   (b) $f(x) = x^9 - \alpha x + \alpha \in \mathbb{F}_9[x]$, where α satisfies $\alpha^2 = \alpha + 1$.

3.89. Let $A(x) = L(x) - a \in \mathbb{F}_q[x]$ be an affine p-polynomial of degree $r > 2$, and suppose the p-polynomial $L(x)$ is such that $L(x)/x$ is irreducible over $\mathbb{F}_q$. Prove that $A(x)$ is the product of a linear polynomial and an irreducible polynomial over $\mathbb{F}_q$ of degree $r - 1$.

3.90. Prove: the trinomial $x^n + ax^k + b \in \mathbb{F}_q[x]$, $n > k \geqslant 1$, q even, has multiple roots if and only if n and k are both even.

3.91. Prove that the degree of every irreducible factor of $x^{2^n} + x + 1$ in $\mathbb{F}_2[x]$ divides $2n$.

3.92. Prove that the degree of every irreducible factor of $x^{2^n+1} + x + 1$ in $\mathbb{F}_2[x]$ divides $3n$.

3.93. Recall the notion of a self-reciprocal polynomial defined in Exercise 3.13. Prove that if $f \in \mathbb{F}_2[x]$ is a self-reciprocal polynomial of positive degree, then f divides a trinomial in $\mathbb{F}_2[x]$ only if $\text{ord}(f)$ is a multiple of 3. Prove also that the converse holds if f is irreducible over $\mathbb{F}_2$.

3.94. Prove that for odd $d \in \mathbb{N}$ the cyclotomic polynomial $Q_d \in \mathbb{F}_2[x]$ divides a trinomial in $\mathbb{F}_2[x]$ if and only if d is a multiple of 3.

3.95. Let $f(x) = x^n + ax^k + b \in \mathbb{F}_q[x]$, $n > k \geqslant 1$, be a trinomial and let $m \in \mathbb{N}$ be a multiple of $\text{ord}(f)$. Prove that $f(x)$ divides the trinomial $g(x) = x^{m-k} + b^{-1}x^{n-k} + ab^{-1}$.

3.96. Prove that the trinomial $x^{2n} + x^n + 1$ is irreducible over $\mathbb{F}_2$ if and only if $n = 3^k$ for some nonnegative integer k.

3.97. Prove that the trinomial $x^{4n} + x^n + 1$ is irreducible over $\mathbb{F}_2$ if and only if $n = 3^k 5^m$ for some nonnegative integers k and m.

Chapter 4

Factorization of Polynomials

Any nonconstant polynomial over a field can be expressed as a product of irreducible polynomials. In the case of finite fields, some reasonably efficient algorithms can be devised for the actual calculation of the irreducible factors of a given polynomial of positive degree.

The availability of feasible factorization algorithms for polynomials over finite fields is important for coding theory and for the study of linear recurrence relations in finite fields. Beyond the realm of finite fields, there are various computational problems in algebra and number theory that depend in one way or another on the factorization of polynomials over finite fields. We mention the factorization of polynomials over the ring of integers, the determination of the decomposition of rational primes in algebraic number fields, the calculation of the Galois group of an equation over the rationals, and the construction of field extensions.

We shall present several algorithms for the factorization of polynomials over finite fields. The decision on the choice of algorithm for a specific factorization problem usually depends on whether the underlying finite field is "small" or "large." In Section 1 we describe those algorithms that are better adapted to "small" finite fields and in the next section those that work better for "large" finite fields. Some of these algorithms reduce the problem of factoring polynomials to that of finding the roots of certain other polynomials. Therefore, Section 3 is devoted to the discussion of the latter problem from the computational viewpoint.

1. FACTORIZATION OVER SMALL FINITE FIELDS

Any polynomial $f \in \mathbb{F}_q[x]$ of positive degree has a canonical factorization in $\mathbb{F}_q[x]$ by Theorem 1.59. For the discussion of factorization algorithms it will suffice to consider only monic polynomials. Our goal is thus to express a monic polynomial $f \in \mathbb{F}_q[x]$ of positive degree in the form

$$f = f_1^{e_1} \cdots f_k^{e_k}, \tag{4.1}$$

where $f_1, \ldots, f_k$ are distinct monic irreducible polynomials in $\mathbb{F}_q[x]$ and $e_1, \ldots, e_k$ are positive integers.

First we simplify our task by showing that the problem can be reduced to that of factoring a polynomial *with no repeated factors*, which means that the exponents $e_1, \ldots, e_k$ in (4.1) are all equal to 1 (or, equivalently, that the polynomial has no multiple roots). To this end, we calculate

$$d(x) = \gcd(f(x), f'(x)),$$

the greatest common divisor of $f(x)$ and its derivative, by the Euclidean algorithm.

If $d(x) = 1$, then we know that $f(x)$ has no repeated factors because of Theorem 1.68. If $d(x) = f(x)$, we must have $f'(x) = 0$. Hence $f(x) = g(x)^p$, where $g(x)$ is a suitable polynomial in $\mathbb{F}_q[x]$ and p is the characteristic of $\mathbb{F}_q$. If necessary, the reduction process can be continued by applying the method to $g(x)$.

If $d(x) \neq 1$ and $d(x) \neq f(x)$, then $d(x)$ is a nontrivial factor of $f(x)$ and $f(x)/d(x)$ has no repeated factors. The factorization of $f(x)$ is achieved by factoring $d(x)$ and $f(x)/d(x)$ separately. In case $d(x)$ still has repeated factors, further applications of the reduction process will have to be carried out.

By applying this process sufficiently often, the original problem is reduced to that of factoring a certain number of polynomials with no repeated factors. The canonical factorizations of these polynomials lead directly to the canonical factorization of the original polynomial. Therefore, we may restrict the attention to polynomials with no repeated factors. The following theorem is crucial.

4.1.   Theorem.   *If $f \in \mathbb{F}_q[x]$ is monic and $h \in \mathbb{F}_q[x]$ is such that $h^q \equiv h \bmod f$, then*

$$f(x) = \prod_{c \in \mathbb{F}_q} \gcd(f(x), h(x) - c). \tag{4.2}$$

Proof.   Each greatest common divisor on the right-hand side of (4.2) divides $f(x)$. Since the polynomials $h(x) - c$, $c \in \mathbb{F}_q$, are pairwise relatively prime, so are the greatest common divisors with $f(x)$, and thus the product of these greatest common divisors divides $f(x)$. On the other hand, $f(x)$

divides

$$h(x)^q - h(x) = \prod_{c \in F_q} (h(x) - c),$$

and so $f(x)$ divides the right-hand side of (4.2). Thus, the two sides of (4.2) are monic polynomials that divide each other, and therefore they must be equal.   □

In general, (4.2) does not yield the complete factorization of f since $\gcd(f(x), h(x) - c)$ may be reducible in $F_q[x]$. If $h(x) \equiv c \bmod f(x)$ for some $c \in F_q$, then Theorem 4.1 gives a trivial factorization of f and therefore is of no use. However, if h is such that Theorem 4.1 yields a nontrivial factorization of f, we say that h is an *f-reducing polynomial*. Any h with $h^q \equiv h \bmod f$ and $0 < \deg(h) < \deg(f)$ is obviously f-reducing. In order to obtain factorization algorithms on the basis of Theorem 4.1, we have to find methods of constructing f-reducing polynomials. It should be clear at this stage already that since the factorization provided by (4.2) depends on the calculation of q greatest common divisors, a direct application of this formula will only be feasible for small finite fields F_q.

The first method of constructing f-reducing polynomials makes use of the Chinese remainder theorem for polynomials (see Exercise 1.37). Let us assume that f has no repeated factors, so that $f = f_1 \cdots f_k$ is a product of distinct monic irreducible polynomials over F_q. If $(c_1, \ldots, c_k)$ is any k-tuple of elements of F_q, the Chinese remainder theorem implies that there is a unique $h \in F_q[x]$ with $h(x) \equiv c_i \bmod f_i(x)$ for $1 \leqslant i \leqslant k$ and $\deg(h) < \deg(f)$. The polynomial $h(x)$ satisfies the condition

$$h(x)^q \equiv c_i^q = c_i \equiv h(x) \bmod f_i(x) \quad \text{for } 1 \leqslant i \leqslant k,$$

and therefore

$$h^q \equiv h \bmod f, \qquad \deg(h) < \deg(f). \tag{4.3}$$

On the other hand, if h is a solution of (4.3), then the identity

$$h(x)^q - h(x) = \prod_{c \in F_q} (h(x) - c)$$

implies that every irreducible factor of f divides one of the polynomials $h(x) - c$. Thus, all solutions of (4.3) satisfy $h(x) \equiv c_i \bmod f_i(x)$, $1 \leqslant i \leqslant k$, for some k-tuple $(c_1, \ldots, c_k)$ of elements of F_q. Consequently, there are exactly q^k solutions of (4.3).

We find these solutions by reducing (4.3) to a system of linear equations. With $n = \deg(f)$ we construct the $n \times n$ matrix $B = (b_{ij})$, $0 \leqslant i, j \leqslant n - 1$, by calculating the powers $x^{iq} \bmod f(x)$. Specifically, let

$$x^{iq} \equiv \sum_{j=0}^{n-1} b_{ij} x^j \bmod f(x) \quad \text{for } 0 \leqslant i \leqslant n - 1. \tag{4.4}$$

Then $h(x) = a_0 + a_1 x + \cdots + a_{n-1} x^{n-1} \in \mathbb{F}_q[x]$ is a solution of (4.3) if and only if

$$(a_0, a_1, \ldots, a_{n-1}) B = (a_0, a_1, \ldots, a_{n-1}). \qquad (4.5)$$

This follows from the fact that (4.5) holds if and only if

$$
\begin{aligned}
h(x) &= \sum_{j=0}^{n-1} a_j x^j \\
&= \sum_{j=0}^{n-1} \sum_{i=0}^{n-1} a_i b_{ij} x^j \\
&\equiv \sum_{i=0}^{n-1} a_i x^{iq} = h(x)^q \bmod f(x).
\end{aligned}
$$

The system (4.5) may be written in the equivalent form

$$(a_0, a_1, \ldots, a_{n-1})(B - I) = (0, 0, \ldots, 0), \qquad (4.6)$$

where I is the $n \times n$ identity matrix over $\mathbb{F}_q$. By the considerations above, the system (4.6) has q^k solutions. Thus, *the dimension of the null space of the matrix $B - I$ is k, the number of distinct monic irreducible factors of f, and the rank of $B - I$ is $n - k$.*

Since the constant polynomial $h_1(x) = 1$ is always a solution of (4.3), the vector $(1, 0, \ldots, 0)$ is always a solution of (4.6), as can also be checked directly. There will exist polynomials $h_2(x), \ldots, h_k(x)$ of degree $\leqslant n - 1$ such that the vectors corresponding to $h_1(x), h_2(x), \ldots, h_k(x)$ form a basis for the null space of $B - I$. The polynomials $h_2(x), \ldots, h_k(x)$ have positive degree and are thus f-reducing.

In this approach, an important role is played by the determination of the rank r of the matrix $B - I$. We have $r = n - k$ as noted above, so that once the rank r is found, we know that *the number of distinct monic irreducible factors of f is given by $n - r$.* On the basis of this information we can then decide when the factorization procedure can be stopped. The rank of $B - I$ can be determined by using row and column operations to reduce the matrix to echelon form. However, since we also want to solve the system (4.6), it is advisable to use only column operations because they leave the null space invariant. Thus, we are allowed to multiply any column of the matrix $B - I$ by a nonzero element of $\mathbb{F}_q$ and to add any multiple of one of its columns to a different column. The rank r is the number of nonzero columns in the column echelon form.

Having found r, we form $k = n - r$. If $k = 1$, we know that f is irreducible over $\mathbb{F}_q$ and the procedure terminates. In this case, the only solutions of (4.3) are the constant polynomials and the null space of $B - I$ contains only the vectors of the form $(c, 0, \ldots, 0)$ with $c \in \mathbb{F}_q$. If $k \geqslant 2$, we take the f-reducing basis polynomial $h_2(x)$ and calculate

$\gcd(f(x), h_2(x)-c)$ for all $c \in \mathbb{F}_q$. The result will be a nontrivial factorization of $f(x)$ afforded by (4.2). If the use of $h_2(x)$ does not succeed in splitting $f(x)$ into k factors, we calculate $\gcd(g(x), h_3(x)-c)$ for all $c \in \mathbb{F}_q$ and all nontrivial factors $g(x)$ found so far. This procedure is continued until k factors of $f(x)$ are obtained.

The process described above must eventually yield all the factors. For if we consider two distinct monic irreducible factors of $f(x)$, say $f_1(x)$ and $f_2(x)$, then by the argument following (4.3) there exist elements c_{j1}, $c_{j2} \in \mathbb{F}_q$ such that $h_j(x) \equiv c_{j1} \bmod f_1(x)$, $h_j(x) \equiv c_{j2} \bmod f_2(x)$ for $1 \leqslant j \leqslant k$. Suppose we had $c_{j1} = c_{j2}$ for $1 \leqslant j \leqslant k$. Then, since any solution $h(x)$ of (4.3) is a linear combination of $h_1(x), \ldots, h_k(x)$ with coefficients in $\mathbb{F}_q$, there would exist for any such $h(x)$ an element $c \in \mathbb{F}_q$ with $h(x) \equiv c \bmod f_1(x)$, $h(x) \equiv c \bmod f_2(x)$. But the argument leading to (4.3) shows, in particular, that there is a solution $h(x)$ of (4.3) with $h(x) \equiv 0 \bmod f_1(x)$, $h(x) \equiv 1 \bmod f_2(x)$. This contradiction proves that $c_{j1} \neq c_{j2}$ for some j with $1 \leqslant j \leqslant k$ (in fact, since $h_1(x) = 1$, we will have $j \geqslant 2$). Therefore, $h_j(x) - c_{j1}$ will be divisible by $f_1(x)$, but not by $f_2(x)$. Hence any two distinct monic irreducible factors of $f(x)$ will be separated by some $h_j(x)$.

This factorization algorithm based on determining f-reducing polynomials by solving the system (4.6) is called *Berlekamp's algorithm*.

4.2. Example. Factor $f(x) = x^8 + x^6 + x^4 + x^3 + 1$ over $\mathbb{F}_2$ by Berlekamp's algorithm. Since $\gcd(f(x), f'(x)) = 1$, $f(x)$ has no repeated factors. We have to compute $x^{iq} \bmod f(x)$ for $q = 2$ and $0 \leqslant i \leqslant 7$. This yields the following congruences $\bmod f(x)$:

$$
\begin{aligned}
x^0 &\equiv 1 \\
x^2 &\equiv x^2 \\
x^4 &\equiv x^4 \\
x^6 &\equiv x^6 \\
x^8 &\equiv 1 + x^3 + x^4 + x^6 \\
x^{10} &\equiv 1 + x^2 + x^3 + x^4 + x^5 \\
x^{12} &\equiv x^2 + x^4 + x^5 + x^6 + x^7 \\
x^{14} &\equiv 1 + x + x^3 + x^4 + x^5
\end{aligned}
$$

Therefore, the 8×8 matrix B is given by

$$
B = \begin{pmatrix}
1 & 0 & 0 & 0 & 0 & 0 & 0 & 0 \\
0 & 0 & 1 & 0 & 0 & 0 & 0 & 0 \\
0 & 0 & 0 & 0 & 1 & 0 & 0 & 0 \\
0 & 0 & 0 & 0 & 0 & 0 & 1 & 0 \\
1 & 0 & 0 & 1 & 1 & 0 & 1 & 0 \\
1 & 0 & 1 & 1 & 1 & 1 & 0 & 0 \\
0 & 0 & 1 & 0 & 1 & 1 & 1 & 1 \\
1 & 1 & 0 & 1 & 1 & 1 & 0 & 0
\end{pmatrix}
$$

and $B - I$ is given by

$$B - I = \begin{pmatrix} 0 & 0 & 0 & 0 & 0 & 0 & 0 & 0 \\ 0 & 1 & 1 & 0 & 0 & 0 & 0 & 0 \\ 0 & 0 & 1 & 0 & 1 & 0 & 0 & 0 \\ 0 & 0 & 0 & 1 & 0 & 0 & 1 & 0 \\ 1 & 0 & 0 & 1 & 0 & 0 & 1 & 0 \\ 1 & 0 & 1 & 1 & 1 & 0 & 0 & 0 \\ 0 & 0 & 1 & 0 & 1 & 1 & 0 & 1 \\ 1 & 1 & 0 & 1 & 1 & 1 & 0 & 1 \end{pmatrix}.$$

The matrix $B - I$ has rank 6, and the two vectors $(1,0,0,0,0,0,0,0)$ and $(0,1,1,0,0,1,1,1)$ form a basis of the null space of $B - I$. The corresponding polynomials are $h_1(x) = 1$ and $h_2(x) = x + x^2 + x^5 + x^6 + x^7$. We calculate $\gcd(f(x), h_2(x) - c)$ for $c \in \mathbb{F}_2$ by the Euclidean algorithm and obtain $\gcd(f(x), h_2(x)) = x^6 + x^5 + x^4 + x + 1$, $\gcd(f(x), h_2(x) - 1) = x^2 + x + 1$. The desired canonical factorization is therefore

$$f(x) = (x^6 + x^5 + x^4 + x + 1)(x^2 + x + 1). \qquad \square$$

A second method of obtaining f-reducing polynomials is based on the explicit construction of a family of polynomials among which at least one f-reducing polynomial can be found. Let f be again a monic polynomial of degree n with no repeated factors. Let $f = f_1 \cdots f_k$ be its canonical factorization in $\mathbb{F}_q[x]$ with $\deg(f_j) = n_j$ for $1 \leqslant j \leqslant k$. If N is the least positive integer with $x^{q^N} \equiv x \bmod f(x)$, then it follows from Theorem 3.20 that $N = \mathrm{lcm}(n_1, \ldots, n_k)$, and it is also easily seen that N is the degree of the splitting field F of f over $\mathbb{F}_q$. Let the polynomial $T \in \mathbb{F}_q[x]$ be given by $T(x) = x + x^q + x^{q^2} + \cdots + x^{q^{N-1}}$ and define $T_i(x) = T(x^i)$ for $i = 0, 1, \ldots$. The following result guarantees that in the case of interest, namely, when f is reducible, there are f-reducing polynomials among the T_i.

4.3. Theorem. *If f is reducible in $\mathbb{F}_q[x]$, then at least one of the polynomials T_i, $1 \leqslant i \leqslant n - 1$, is f-reducing.*

Proof. It is immediate that any polynomial T_i satisfies $T_i^q \equiv T_i \bmod f$. Suppose now that for all T_i, $1 \leqslant i \leqslant n - 1$, the factorization of f afforded by (4.2) were trivial. This means that there exist elements $c_1, \ldots, c_{n-1} \in \mathbb{F}_q$ such that $T_i(x) \equiv c_i \bmod f(x)$ for $1 \leqslant i \leqslant n - 1$. With $c_0 = N$, viewed as an element of $\mathbb{F}_q$, we get $T(x^i) \equiv c_i \bmod f(x)$ for $0 \leqslant i \leqslant n - 1$. For any

$$g(x) = \sum_{i=1}^{n-1} a_i x^i \in \mathbb{F}_q[x]$$

of degree less than n we have then

$$T(g(x)) = T\left(\sum_{i=0}^{n-1} a_i x^i \right) = \sum_{i=0}^{n-1} a_i T(x^i) \equiv \sum_{i=0}^{n-1} a_i c_i \bmod f(x).$$

Putting

$$c(g) = \sum_{i=0}^{n-1} a_i c_i \in \mathbb{F}_q,$$

we obtain

$$T(g(x)) \equiv c(g) \bmod f_j(x) \quad \text{for } 1 \leqslant j \leqslant k. \tag{4.7}$$

Since $N = \text{lcm}(n_1, \ldots, n_k)$, at least one of the integers N/n_j, say N/n_1, is not divisible by the characteristic of $\mathbb{F}_q$. Let θ_1 be a root of f_1 in the splitting field F_1 of f_1 over $\mathbb{F}_q$. Because of Theorem 2.23(iii) there exists $g_1 \in \mathbb{F}_q[x]$ with

$$\text{Tr}_{F_1/\mathbb{F}_q}(g_1(\theta_1)) = 1. \tag{4.8}$$

Since $k \geqslant 2$ by assumption, we can apply the Chinese remainder theorem to obtain a polynomial $g \in \mathbb{F}_q[x]$ of degree $< n$ with

$$g \equiv g_1 \bmod f_1, \, g \equiv 0 \bmod f_2. \tag{4.9}$$

From (4.8) and (4.9) we deduce that

$$\text{Tr}_{F_1/\mathbb{F}_q}(g(\theta_1)) = 1,$$

and Theorems 2.23(iv) and 2.26 imply that

$$\text{Tr}_{F/\mathbb{F}_q}(g(\theta_1)) = N/n_1.$$

Because of the definitions of the trace and of the element θ_1, it follows that

$$T(g(x)) \equiv N/n_1 \bmod f_1(x).$$

However, the second congruence in (4.9) leads to $T(g(x)) \equiv 0 \bmod f_2(x)$, and since $N/n_1 \neq 0$ as an element of $\mathbb{F}_q$, we get a contradiction to (4.7). Therefore, at least one of the T_i, $1 \leqslant i \leqslant n-1$, is f-reducing. $\square$

4.4. Example. Factor $f(x) = x^{17} + x^{14} + x^{13} + x^{12} + x^{11} + x^{10} + x^9 + x^8 + x^7 + x^5 + x^4 + x + 1$ over $\mathbb{F}_2$. We have $\gcd(f(x), f'(x)) = x^{10} + x^8 + 1$, and so $f_0(x) = f(x)/\gcd(f(x), f'(x)) = x^7 + x^5 + x^4 + x + 1$ has no repeated factors. We factor f_0 by finding an f_0-reducing polynomial of the type described above. To this end, we calculate the powers $x, x^2, x^4, \ldots \bmod f_0(x)$ until we obtain the least positive integer N with $x^{2^N} \equiv x \bmod f_0(x)$. We simplify the notation by identifying a polynomial $\sum_{i=0}^{n-1} a_i x^i$ with the n-tuple $a_0 a_1 \cdots a_{n-1}$ of its coefficients, so that, for instance, $f_0(x) = 11001101$. The calculation of the required powers of $x \bmod f_0(x)$ is facilitated by the observation that squaring a polynomial $a_0 a_1 \cdots a_6 \bmod f_0(x)$ is the same as multiplying the vector $a_0 a_1 \cdots a_6$ by the 7×7 matrix of even powers

$x^0, x^2, \ldots, x^{12} \bmod f_0(x)$. This matrix is obtained from

$$
\begin{array}{cccccccc}
x^0 \equiv 1 & 0 & 0 & 0 & 0 & 0 & 0 \\
x^2 \equiv 0 & 0 & 1 & 0 & 0 & 0 & 0 \\
x^4 \equiv 0 & 0 & 0 & 0 & 1 & 0 & 0 \\
x^6 \equiv 0 & 0 & 0 & 0 & 0 & 0 & 1 \\
x^8 \equiv 0 & 1 & 1 & 0 & 0 & 1 & 1 \\
x^{10} \equiv 1 & 0 & 1 & 1 & 0 & 0 & 1 \\
x^{12} \equiv 0 & 1 & 0 & 0 & 1 & 0 & 1
\end{array}
$$

where all the congruences are mod $f_0(x)$. Therefore we get mod $f_0(x)$:

$$
\begin{array}{cccccccc}
x \equiv 0 & 1 & 0 & 0 & 0 & 0 & 0 \\
x^2 \equiv 0 & 0 & 1 & 0 & 0 & 0 & 0 \\
x^4 \equiv 0 & 0 & 0 & 0 & 1 & 0 & 0 \\
x^8 \equiv 0 & 1 & 1 & 0 & 0 & 1 & 1 \\
x^{16} \equiv 1 & 1 & 0 & 1 & 0 & 0 & 0 \\
x^{32} \equiv 1 & 0 & 1 & 0 & 0 & 0 & 1 \\
x^{64} \equiv 1 & 1 & 0 & 0 & 0 & 0 & 1 \\
x^{128} \equiv 1 & 1 & 1 & 0 & 1 & 0 & 1 \\
x^{256} \equiv 1 & 0 & 0 & 0 & 0 & 1 & 0 \\
x^{512} \equiv 0 & 0 & 1 & 1 & 0 & 0 & 1 \\
x^{1024} \equiv 0 & 1 & 0 & 0 & 0 & 0 & 0
\end{array}
$$

Thus $N = 10$ and

$$
T_1(x) = \sum_{j=0}^{9} x^{2^j} \equiv 1 \quad 1 \quad 1 \quad 0 \quad 0 \quad 0 \quad 1 \bmod f_0(x).
$$

Since $T_1(x)$ is not congruent to a constant mod $f_0(x)$, $T_1(x)$ is f_0-reducing. We have

$$
\gcd(f_0(x), T_1(x)) = \gcd(1\ 1\ 0\ 0\ 1\ 1\ 0\ 1, 1\ 1\ 1\ 0\ 0\ 0\ 1)
$$
$$
= x^5 + x^4 + x^3 + x^2 + 1,
$$
$$
\gcd(f_0(x), T_1(x) - 1) = \gcd(1\ 1\ 0\ 0\ 1\ 1\ 0\ 1, 0\ 1\ 1\ 0\ 0\ 0\ 1)
$$
$$
= x^2 + x + 1,
$$

and so

$$
f_0(x) = (x^5 + x^4 + x^3 + x^2 + 1)(x^2 + x + 1).
$$

The second factor is obviously irreducible in $\mathbb{F}_2[x]$. Since $N = 10$ is the least common multiple of the degrees of the irreducible factors of $f_0(x)$, any nontrivial factorization of the first factor would lead to a value of N different from 10, so that the first factor is also irreducible in $\mathbb{F}_2[x]$.

It remains to factor $\gcd(f(x),\ f'(x)) = x^{10} + x^8 + 1$. We have $x^{10} + x^8 + 1 = (x^5 + x^4 + 1)^2$, and by checking whether $x^5 + x^4 + 1$ is divisible by one of the irreducible factors of $f_0(x)$, we find that $x^5 + x^4 + 1 = (x^3 + x + 1)(x^2 + x + 1)$, with $x^3 + x + 1$ irreducible in $\mathbb{F}_2[x]$. Hence

$$f(x) = (x^5 + x^4 + x^3 + x^2 + 1)(x^3 + x + 1)^2(x^2 + x + 1)^3$$

is the canonical factorization of $f(x)$ in $\mathbb{F}_2[x]$.   □

It should be noted that, in general, the f-reducing polynomials T_i do not yield the complete factorization of f since the T_i are not able to separate those irreducible factors f_j for which N/n_j is divisible by the characteristic of $\mathbb{F}_q$. In practice, however, one calculates the first f-reducing T_i and then calculates new T_i for each of the resulting factors. In this way, one eventually obtains the complete factorization of f.

It is, however, possible to construct a related set of polynomials R_i that are capable of separating all the irreducible factors of f at once. We assume, without loss of generality, that $f(0) \neq 0$. Let $\mathrm{ord}(f(x)) = e$, so that $f(x)$ divides $x^e - 1$. Since f has no repeated factors, e and q are relatively prime by Corollary 3.4 and Theorem 3.9. For each $i \geq 0$ let m_i be the least positive integer with

$$x^{iq^{m_i}} \equiv x^i \bmod f(x). \tag{4.10}$$

Then we define

$$R_i(x) = x^i + x^{iq} + x^{iq^2} + \cdots + x^{iq^{m_i-1}}$$

Since (4.10) is equivalent to

$$iq^{m_i} \equiv i \bmod e, \tag{4.11}$$

which is in turn equivalent to $q^{m_i} \equiv 1 \bmod(e/\gcd(e, i))$, it follows that m_i can also be described as the multiplicative order of q modulo $e/\gcd(e, i)$. A comparison with the definition of $T_i(x)$ shows that

$$T_i(x) \equiv \frac{N}{m_i} R_i(x) \bmod f(x).$$

It is clear that $R_i^q \equiv R_i \bmod f$ for all i, so that the R_i can be used in (4.2) in place of h. We prove now the claim about the R_i made above.

4.5. Theorem. *Let f be monic and reducible in $\mathbb{F}_q[x]$ with no repeated factors, and suppose that $f(0) \neq 0$ and $\mathrm{ord}(f) = e$. Then, if all the polynomials R_i, $1 \leq i \leq e - 1$, are used in (4.2), they will separate all irreducible factors of f.*

Proof. Let $h(x) = \sum_{i=0}^{e-1} a_i x^i \in \mathbb{F}_q[x]$ be a solution of $h(x)^q \equiv h(x) \bmod(x^e - 1)$. If we interpret subscripts mod e, then $h(x) \equiv \sum_{i=0}^{e-1} a_{iq} x^{iq} \bmod(x^e - 1)$ since iq, $i = 0, 1, \ldots, e - 1$, runs through all residues mod e as q and e are relatively prime. Since $h(x)^q = \sum_{i=0}^{e-1} a_i x^{iq}$, we get

$$\sum_{i=0}^{e-1} a_i x^{iq} \equiv \sum_{i=0}^{e-1} a_{iq} x^{iq} \bmod(x^e - 1).$$

By considering the exponents mod e, it follows that corresponding coefficients are identical. Thus $a_i = a_{iq}$ for all i, and so $a_i = a_{iq} = a_{iq^2} = \cdots$ for all i. Since m_i is the least positive integer for which (4.11) holds, we obtain

$$h(x) \equiv \sum_{i \in J} a_i R_i(x) \bmod(x^e - 1),$$

where the set J contains exactly one representative from each equivalence class of residues mod e determined by the equivalence relation $\sim$ which is defined by $i_1 \sim i_2$ if and only if $i_1 \equiv i_2 q^t \bmod e$ for some $t \geqslant 0$. Thus, for suitable $b_i \in \mathbb{F}_q$ we have

$$h(x) \equiv \sum_{i=0}^{e-1} b_i R_i(x) \bmod(x^e - 1). \tag{4.12}$$

Let now $f_1(x)$ and $f_2(x)$ be two distinct monic irreducible factors of $f(x)$, and so of $x^e - 1$. By the argument leading to (4.3), there is a solution $h(x) \in \mathbb{F}_q[x]$ of $h(x)^q \equiv h(x) \bmod(x^e - 1)$, $\deg(h(x)) < e$, with

$$h(x) \equiv 0 \bmod f_1(x), \qquad h(x) \equiv 1 \bmod f_2(x). \tag{4.13}$$

Since $R_i^q \equiv R_i \bmod f$, the argument subsequent to (4.3) shows that there exist elements $c_{i1}, c_{i2} \in \mathbb{F}_q$ with $R_i(x) \equiv c_{i1} \bmod f_1(x)$, $R_i(x) \equiv c_{i2} \bmod f_2(x)$ for $0 \leqslant i \leqslant e - 1$. If we had $c_{i1} = c_{i2}$ for $0 \leqslant i \leqslant e - 1$, then it would follow from (4.12) that $h(x) \equiv c \bmod f_1(x)$, $h(x) \equiv c \bmod f_2(x)$ for some $c \in \mathbb{F}_q$, a contradiction to (4.13). Thus $c_{i1} \neq c_{i2}$ for some i with $0 \leqslant i \leqslant e - 1$, and since $R_0(x) = 1$, we must have $i \geqslant 1$. Then $R_i(x) - c_{i1}$ will be divisible by $f_1(x)$, but not by $f_2(x)$. Hence the use of this $R_i(x)$ in (4.2) will separate $f_1(x)$ from $f_2(x)$. □

The argument in the proof of Theorem 4.5 shows, of course, that the polynomials R_i, with i running through the nonzero elements of the set J, are already separating all irreducible factors of f. However, the determination of the set J depends on knowing the order e, and a direct calculation of e (i.e., one that does not have recourse to the canonical factorization of f) will be lengthy in most cases.

This problem does not arise in the special cases $f(x) = x^e - 1$ and $f(x) = Q_e(x)$, the eth cyclotomic polynomial, since it is trivial that $\mathrm{ord}(x^e - 1) = \mathrm{ord}(Q_e(x)) = e$. The polynomials R_i are, in fact, well suited for factoring these binomials and cyclotomic polynomials.

4.6. Example. We determine the canonical factorization of the cyclo-tomic polynomial $Q_{52}(x)$ in $\mathbb{F}_3[x]$. According to Theorem 3.27 we have

$$Q_{52}(x) = \frac{(x^{52}-1)(x^2-1)}{(x^{26}-1)(x^4-1)}$$

$$= x^{24} - x^{22} + x^{20} - x^{18} + x^{16} - x^{14} + x^{12}$$

$$- x^{10} + x^8 - x^6 + x^4 - x^2 + 1.$$

Now $R_1(x) = x + x^3 + x^9 + x^{27} + x^{81} + x^{243}$, and since $x^{26} \equiv -1 \bmod Q_{52}(x)$, we get $R_1(x) \equiv 0 \bmod Q_{52}(x)$, so that R_1 is not Q_{52}-reduc-ing. With $R_2(x) = x^2 + x^6 + x^{18}$ we get

$$\gcd(Q_{52}(x), R_2(x)) = x^6 - x^2 + 1,$$

$$\gcd(Q_{52}(x), R_2(x)+1) = x^6 + x^4 - x^2 + 1,$$

$$\gcd(Q_{52}(x), R_2(x)-1) = x^{12} + x^{10} - x^8 + x^6 + x^4 + x^2 + 1 = g(x),$$

say, so that (4.2) yields

$$Q_{52}(x) = (x^6 - x^2 + 1)(x^6 + x^4 - x^2 + 1)g(x).$$

By Theorem 2.47(ii), $Q_{52}(x)$ is the product of four irreducible factors in $\mathbb{F}_3[x]$ of degree 6. Thus, it remains to factor $g(x)$. Since $R_3(x) = x^3 + x^9 + x^{27} + x^{81} + x^{243} + x^{729} \equiv 0 \bmod Q_{52}(x)$, we next use $R_4(x) = x^4 + x^{12} + x^{36}$. We note that $x^{12} \equiv -x^{10} + x^8 - x^6 - x^4 - x^2 - 1 \bmod g(x)$, $x^{36} \equiv -x^{10} \bmod g(x)$, and so

$$R_4(x) \equiv x^{10} + x^8 - x^6 - x^2 - 1 \bmod g(x).$$

Therefore,

$$\gcd(g(x), R_4(x)) = \gcd(g(x), x^{10} + x^8 - x^6 - x^2 - 1) = 1,$$

$$\gcd(g(x), R_4(x)+1) = \gcd(g(x), x^{10} + x^8 - x^6 - x^2) = x^6 - x^4 + x^2 + 1,$$

$$\gcd(g(x), R_4(x)-1) = \gcd(g(x), x^{10} + x^8 - x^6 - x^2 + 1) = x^6 - x^4 + 1.$$

Thus,

$$Q_{52}(x) = (x^6 - x^2 + 1)(x^6 + x^4 - x^2 + 1)(x^6 - x^4 + x^2 + 1)(x^6 - x^4 + 1)$$

is the desired canonical factorization.   □

2. FACTORIZATION OVER LARGE FINITE FIELDS

If $\mathbb{F}_q$ is a finite field with a large number q of elements, the practical implementation of the methods in the previous section will become more difficult. We may still be able to find an f-reducing polynomial with a reasonable effort, but a direct application of the basic formula (4.2) will be problematic since it requires the calculation of q greatest common divisors. Thus, to make the use of f-reducing polynomials feasible for large finite fields, it is imperative that we devise ways of reducing the number of elements $c \in \mathbb{F}_q$ for which the greatest common divisor in (4.2) needs to be calculated. We note that in the context of factorization we consider q to be "large" if q is (substantially) bigger than the degree of the polynomial to be factored.

Let f again be a monic polynomial in $\mathbb{F}_q[x]$ with no repeated factors, let $\deg(f) = n$, and let k be the number of distinct monic irreducible factors of f. Suppose that $h \in \mathbb{F}_q[x]$ satisfies $h^q \equiv h \bmod f$ and $0 < \deg(h) < n$, so that h is f-reducing. Since the various greatest common divisors in (4.2) are pairwise relatively prime, it is clear that at most k of these greatest common divisors will be $\neq 1$. The problem is to find an *a priori* characterization of those $c \in \mathbb{F}_q$ for which $\gcd(f(x), h(x) - c) \neq 1$.

One such characterization can be obtained by using the theory of resultants (see Definition 1.93 and the remarks following it). Let $R(f(x), h(x) - c)$ be the resultant of $f(x)$ and $h(x) - c$, where the degrees of the two polynomials are taken as the formal degrees in the definition of the resultant. Then $\gcd(f(x), h(x) - c) \neq 1$ if and only if $R(f(x), h(x) - c) = 0$. We are thus led to consider

$$F(y) = R(f(x), h(x) - y),$$

which, from the representation of the resultant as a determinant, is seen to be a polynomial in y of degree $\leq n$. Then we have $\gcd(f(x), h(x) - c) \neq 1$ if and only if c is a root of $F(y)$ in $\mathbb{F}_q$.

The polynomial $F(y)$ may be calculated from the definition, which involves the evaluation of a determinant of order $\leq 2n - 1$ whose entries are either elements of $\mathbb{F}_q$ or linear polynomials in y. In many cases it will, however, be preferable to use the following method. Choose $n + 1$ distinct elements $c_0, c_1, \ldots, c_n \in \mathbb{F}_q$ and calculate the resultants $r_i = R(f(x), h(x) - c_i)$ for $0 \leq i \leq n$. Then the unique polynomial $F(y)$ of degree $\leq n$ with $F(c_i) = r_i$ for $0 \leq i \leq n$ is obtained from the Lagrange interpolation formula (see Theorem 1.71). This method has the advantage that if any of the r_i are 0, we automatically get roots of the polynomial $F(y)$ in $\mathbb{F}_q$. At any rate, the question of isolating the elements $c \in \mathbb{F}_q$ with $\gcd(f(x), h(x) - c) \neq 1$ is now reduced to that of finding the roots of a polynomial in $\mathbb{F}_q$. Computational methods for dealing with this problem will be discussed in the next section.

4.7. Example. Factor $f(x) = x^6 - 3x^5 + 5x^4 - 9x^3 - 5x^2 + 6x + 7$ over $\mathbb{F}_{23}$. Since $\gcd(f(x), f'(x)) = 1$, $f(x)$ has no repeated factors. We proceed by Berlekamp's algorithm and calculate $x^{23i} \bmod f(x)$ for $0 \leq i \leq 5$. This yields the 6×6 matrix

$$B = \begin{pmatrix} 1 & 0 & 0 & 0 & 0 & 0 \\ 5 & 0 & -1 & 8 & -3 & -10 \\ -10 & 10 & 10 & 0 & 1 & -9 \\ 0 & 7 & 9 & -8 & 10 & -11 \\ 11 & 0 & -4 & 7 & 7 & 2 \\ -3 & 0 & -10 & 9 & 2 & -9 \end{pmatrix},$$

and thus $B - I$ is given by

$$B - I = \begin{pmatrix} 0 & 0 & 0 & 0 & 0 & 0 \\ 5 & -1 & -1 & 8 & -3 & -10 \\ -10 & 10 & 9 & 0 & 1 & -9 \\ 0 & 7 & 9 & -9 & 10 & -11 \\ 11 & 0 & -4 & 7 & 6 & 2 \\ -3 & 0 & -10 & 9 & 2 & -10 \end{pmatrix}.$$

Reduction to column echelon form shows that $B - I$ has rank $r = 3$, so that f has $k = 6 - r = 3$ distinct monic irreducible factors in $\mathbb{F}_{23}[x]$. A basis for the null space of $B - I$ is given by the vectors $h_1 = (1,0,0,0,0,0)$, $h_2 = (0,4,2,1,0,0)$, $h_3 = (0,-2,9,0,1,1)$, which correspond to the polynomials $h_1(x) = 1$, $h_2(x) = x^3 + 2x^2 + 4x$, $h_3(x) = x^5 + x^4 + 9x^2 - 2x$. We take the f-reducing polynomial $h_2(x)$ and consider

$$F(y) = R(f(x), h_2(x) - y)$$

$$= \begin{vmatrix} 1 & -3 & 5 & -9 & -5 & 6 & 7 & 0 & 0 \\ 0 & 1 & -3 & 5 & -9 & -5 & 6 & 7 & 0 \\ 0 & 0 & 1 & -3 & 5 & -9 & -5 & 6 & 7 \\ 1 & 2 & 4 & -y & 0 & 0 & 0 & 0 & 0 \\ 0 & 1 & 2 & 4 & -y & 0 & 0 & 0 & 0 \\ 0 & 0 & 1 & 2 & 4 & -y & 0 & 0 & 0 \\ 0 & 0 & 0 & 1 & 2 & 4 & -y & 0 & 0 \\ 0 & 0 & 0 & 0 & 1 \cdot & 2 & 4 & -y & 0 \\ 0 & 0 & 0 & 0 & 0 & 1 & 2 & 4 & -y \end{vmatrix}.$$

In this case a direct computation of $F(y)$ is feasible, and we obtain $F(y) = y^6 + 4y^5 + 3y^4 - 7y^3 + 10y^2 + 11y + 7$. Since f has three distinct monic irreducible factors in $\mathbb{F}_{23}[x]$, the polynomial F can have at most three roots in $\mathbb{F}_{23}$. By using either the methods to be discussed in the next section or trial and error, one determines the roots of F in $\mathbb{F}_{23}$ to be -3, 2, and 6. Furthermore,

$$\gcd(f(x), h_2(x) + 3) = x - 4,$$
$$\gcd(f(x), h_2(x) - 2) = x^2 - x + 7,$$
$$\gcd(f(x), h_2(x) - 6) = x^3 + 2x^2 + 4x - 6,$$

so that

$$f(x) = (x-4)(x^2 - x + 7)(x^3 + 2x^2 + 4x - 6)$$

is the canonical factorization of $f(x)$ in $\mathbb{F}_{23}[x]$.   □

Another method of characterizing the elements $c \in \mathbb{F}_q$ for which the greatest common divisors in (4.2) need to be calculated is based on the following considerations. With the notation as above, let C be the set of all $c \in \mathbb{F}_q$ such that $\gcd(f(x), h(x) - c) \neq 1$. Then (4.2) implies

$$f(x) = \prod_{c \in C} \gcd(f(x), h(x) - c), \qquad (4.14)$$

and so $f(x)$ divides $\prod_{c \in C}(h(x) - c)$. We introduce the polynomial

$$G(y) = \prod_{c \in C} (y - c).$$

Then $f(x)$ divides $G(h(x))$ and the polynomial $G(y)$ may be characterized as follows.

4.8.   Theorem.   *Among all the polynomials $g \in \mathbb{F}_q[y]$ such that $f(x)$ divides $g(h(x))$, the polynomial $G(y)$ is the unique monic polynomial of least degree.*

Proof.   We have already shown that the monic polynomial $G(y)$ is such that $f(x)$ divides $G(h(x))$. It is easily seen that the polynomials $g \in \mathbb{F}_q[y]$ with $f(x)$ dividing $g(h(x))$ form a nonzero ideal of $\mathbb{F}_q[y]$. By Theorem 1.54, this ideal is a principal ideal generated by a uniquely determined monic polynomial $G_0 \in \mathbb{F}_q[y]$. It follows that $G_0(y)$ divides $G(y)$, and so

$$G_0(y) = \prod_{c \in C_1} (y - c)$$

for some subset C_1 of C. Furthermore, $f(x)$ divides $G_0(h(x)) = \prod_{c \in C_1}(h(x) - c)$, and hence

$$f(x) = \prod_{c \in C_1} \gcd(f(x), h(x) - c).$$

A comparison with (4.14) shows that $C_1 = C$. Therefore $G_0(y) = G(y)$, and the theorem follows.   □

This result is applied in the following manner. Let m be the number of elements of the set C. Then we write

$$G(y) = \prod_{c \in C} (y - c) = \sum_{j=0}^{m} b_j y^j$$

with coefficients $b_j \in \mathbb{F}_q$. Now $f(x)$ divides $G(h(x))$, so that we have

$$\sum_{j=0}^{m} b_j h(x)^j \equiv 0 \bmod f(x).$$

Since $b_m = 1$, this may be viewed as a nontrivial linear dependence relation over $\mathbb{F}_q$ of the residues of $1, h(x), h(x)^2, \ldots, h(x)^m \bmod f(x)$. Theorem 4.8 says that with the normalization $b_m = 1$ this linear dependence relation is unique, and that the residues of $1, h(x), h(x)^2, \ldots, h(x)^{m-1} \bmod f(x)$ are linearly independent over $\mathbb{F}_q$. The bound $m \leqslant k$ follows from (4.14).

The polynomial G can thus be determined by calculating the residues mod $f(x)$ of $1, h(x), h(x)^2, \ldots$ until we find the smallest power of $h(x)$ that is linearly dependent (over $\mathbb{F}_q$) on its predecessors. The coefficients of this first linear dependence relation, in the normalized form, are the coefficients of G. We know that we need not go beyond $h(x)^k$ to find this linear dependence relation, and k can be obtained from Berlekamp's algorithm. The elements of C are now precisely the roots of the polynomial G. This method of reducing the problem of finding the elements of C to that of calculating the roots of a polynomial in $\mathbb{F}_q$ is called the *Zassenhaus algorithm*.

4.9. Example. Consider again the polynomial $f \in \mathbb{F}_{23}[x]$ from Example 4.7. From Berlekamp's algorithm we obtained $k = 3$ and the f-reducing polynomial $h(x) = x^3 + 2x^2 + 4x \in \mathbb{F}_{23}[x]$. We apply the Zassenhaus algorithm in order to determine the elements $c \in \mathbb{F}_{23}$ for which $\gcd(f(x), h(x) - c) \neq 1$. We have

$$h(x) \equiv \qquad\qquad x^3 + 2x^2 + 4x \qquad \bmod f(x),$$

$$h(x)^2 \equiv 7x^5 + 7x^4 + 2x^3 - 2x^2 - 6x - 7 \bmod f(x),$$

and so it is clear that $h(x)^2$ is not linearly dependent on 1 and $h(x)$. Therefore, $h(x)^3$ must be the smallest power of $h(x)$ that is linearly dependent on its predecessors. We have

$$h(x)^3 \equiv -11x^5 - 11x^4 - x^3 - 9x^2 - 5x - 2 \bmod f(x),$$

and the linear dependence relation is

$$h(x)^3 - 5h(x)^2 + 11h(x) - 10 \equiv 0 \bmod f(x),$$

so that $G(y) = y^3 - 5y^2 + 11y - 10$. By using either the methods to be discussed in the next section or trial and error, one determines the roots of G to be -3, 2, and 6. The canonical factorization of f in $\mathbb{F}_{23}[x]$ is then obtained as in the last part of Example 4.7. $\square$

A method that is conceptually more complicated, but of great theoretical interest, is based on the use of matrices of polynomials. By a *matrix of polynomials* we mean here a matrix whose entries are elements of $\mathbb{F}_q[x]$.

4.10.   Definition.   A square matrix of polynomials is called *nonsingular* if its determinant is a nonzero polynomial, and it is called *unimodular* if its determinant is a nonzero element of $\mathbb{F}_q$.

4.11.   Definition.   Two square matrices P and Q of polynomials are said to be *equivalent* if there exists a unimodular matrix U of polynomials and a nonsingular matrix E with entries in $\mathbb{F}_q$ such that $P = UQE$.

It is easily verified that this notion of equivalence is an equivalence relation, in the sense that it is reflexive, symmetric, and transitive.

We have seen in Section 1 that there are polynomials $h_2, \ldots, h_k \in \mathbb{F}_q[x]$ with $0 < \deg(h_i) < \deg(f)$ for $2 \leqslant i \leqslant k$, which together with $h_1 = 1$ are solutions of $h^q \equiv h \bmod f$ that are linearly independent over $\mathbb{F}_q$. Clearly, the polynomials h_i may be taken to be monic. The following theorem is fundamental.

4.12.   Theorem.   *Let $f = f_1 \cdots f_k$, where $f_1, \ldots, f_k$ are distinct monic irreducible polynomials in $\mathbb{F}_q[x]$, and let $h_2, \ldots, h_k \in \mathbb{F}_q[x]$ be monic polynomials with $0 < \deg(h_i) < \deg(f)$ for $2 \leqslant i \leqslant k$, which together with $h_1 = 1$ are solutions of $h^q \equiv h \bmod f$ that are linearly independent over $\mathbb{F}_q$. Then the diagonal matrix of polynomials*

$$D = \begin{pmatrix} f_1 & 0 & 0 & \cdots & 0 \\ 0 & f_2 & 0 & \cdots & 0 \\ 0 & 0 & f_3 & \cdots & 0 \\ \vdots & \vdots & \vdots & \ddots & \vdots \\ 0 & 0 & 0 & \cdots & f_k \end{pmatrix}$$

is equivalent to the matrix of polynomials

$$A = \begin{pmatrix} f & 0 & 0 & \cdots & 0 \\ h_2 & -1 & 0 & \cdots & 0 \\ h_3 & 0 & -1 & \cdots & 0 \\ \vdots & \vdots & \vdots & \ddots & \vdots \\ h_k & 0 & 0 & \cdots & -1 \end{pmatrix}.$$

Proof.   By the argument following (4.3) we have $h_i(x) \equiv e_{ij} \bmod f_j(x)$ with $e_{ij} \in \mathbb{F}_q$ for $1 \leqslant i, j \leqslant k$. Let E be the $k \times k$ matrix whose (i, j) entry is e_{ij}. We show first that E is nonsingular. Otherwise, there would exist elements $d_1, \ldots, d_k \in \mathbb{F}_q$, not all zero, such that

$$\sum_{i=1}^{k} d_i e_{ij} = 0 \quad \text{for } 1 \leqslant j \leqslant k.$$

This implies that

$$\sum_{i=1}^{k} d_i h_i \equiv 0 \bmod f_j \quad \text{for } 1 \leqslant j \leqslant k,$$

and so $\sum_{i=1}^{k} d_i h_i \equiv 0 \bmod f$. Since $\deg(h_i) < \deg(f)$ for $1 \leqslant i \leqslant k$, it follows that $\sum_{i=1}^{k} d_i h_i = 0$, a contradiction to the linear independence of $h_1, \ldots, h_k$.

Next we note that AE is a nonsingular matrix of polynomials. Thus we can write $D = (D(AE)^{-1})AE$, so that the theorem is established once we have shown that $U = D(AE)^{-1}$ is a unimodular matrix of polynomials.

Let $b_{ij} \in \mathbb{F}_q[x]$ be the (i, j) entry of AE. Then $b_{1j} = fe_{1j} = f \equiv 0 \bmod f_j$ for $1 \leqslant j \leqslant k$, and for $2 \leqslant i \leqslant k$ we have $b_{ij} = h_i e_{1j} - e_{ij} = h_i - e_{ij} \equiv 0 \bmod f_j$ for $1 \leqslant j \leqslant k$, so that

$$b_{ij} \equiv 0 \bmod f_j \quad \text{for } 1 \leqslant i, j \leqslant k. \tag{4.15}$$

Now

$$(AE)^{-1} = \frac{1}{\det(AE)} (B_{ij})_{1 \leqslant i, j \leqslant k} = \frac{(-1)^{k-1}}{\det(E)f} (B_{ij})_{1 \leqslant i, j \leqslant k},$$

where B_{ij} is the cofactor of the (j, i) entry in AE, and

$$U = D(AE)^{-1} = \frac{(-1)^{k-1}}{\det(E)f} (f_i B_{ij})_{1 \leqslant i, j \leqslant k}.$$

Since (4.15) implies that $B_{ij} \equiv 0 \bmod(f/f_i)$, it follows that each entry of U is a polynomial over $\mathbb{F}_q$. Furthermore,

$$\det(U) = \frac{\det(D)}{\det(AE)} = \frac{(-1)^{k-1}}{\det(E)},$$

which is a nonzero element of $\mathbb{F}_q$. Thus, U is a unimodular matrix of polynomials. $\square$

Theorem 4.12 leads to the theoretical possibility of determining the irreducible factors of f by diagonalizing the matrix A. The number k as well as the entries $h_2, \ldots, h_k$ in the first column of A can be obtained with relative ease by Berlekamp's algorithm. The algorithm that achieves the diagonalization of A is, however, quite complicated.

The *diagonalization algorithm* is based on the use of the following *elementary operations*: (i) permute any pair of rows (columns); (ii) multiply any row (column) by an element of $\mathbb{F}_q^*$; (iii) multiply some row (column) by a monomial (element of $\mathbb{F}_q$) and add the result to any other row (column).

The elementary row operations may be performed by multiplying the original matrix from the left by an appropriate unimodular matrix of polynomials, whereas the elementary column operations may be performed by multiplying the original matrix from the right by an appropriate nonsingular matrix with entries in $\mathbb{F}_q$. Therefore, the new matrix obtained by any of these elementary operations is equivalent to the original matrix.

One can show that A is equivalent to a matrix R of polynomials with the property that for each row of R the degree of the diagonal entry is greater than the degrees of the other entries in the row. The matrix R can be computed from A by performing at most $(2\Delta + k - 1)(k - 1)$ elementary operations, where $\Delta = \deg(h_2) + \cdots + \deg(h_k)$.

We note that the diagonal entries of R can be permuted by carrying out suitable row and column permutations. We can thus obtain a matrix S that, in addition to the property of R stated above, satisfies $\deg(s_{ii}) \geq \deg(s_{jj})$ for $1 \leq i \leq j \leq k$, where the s_{ii} are the diagonal entries of S. By multiplying the rows of S by appropriate elements of $\mathbb{F}_q^*$, if necessary, we may assume that the s_{ii} are monic polynomials. A matrix S of polynomials with all these properties is called a *normalized matrix*.

The diagonal entries of the matrix D in Theorem 4.12 may also be arranged in such a way that $\deg(f_i) \geq \deg(f_j)$ for $1 \leq i \leq j \leq k$. The resulting equivalent matrix, which we again call D, is then diagonal and normalized. Using the fact that the normalized matrix S is equivalent to D, one can then show that $\deg(s_{ii}) = \deg(f_i)$ for $1 \leq i \leq k$. Thus, one can read off the degrees of the various irreducible factors of f from the diagonal entries of S. Furthermore, if d is a positive integer which occurs as the degree of some s_{ii}, and if $S^{(d)}$ is the square submatrix of S whose main diagonal contains exactly all s_{ii} of degree d, then one can prove that the determinant of $S^{(d)}$ is equal to the determinant of the corresponding submatrix of D. Thus $\det(S^{(d)}) = g_d$, where g_d is the product of all f_i of degree d. In this way we are led to the partial factorization

$$f = \prod_d g_d, \tag{4.16}$$

where the product is over all positive integers d that occur as the degree of some f_i.

In summary, we see that the matrix S can be used to obtain the following information about the distinct monic irreducible factors of f: the degrees of these factors, the number of these factors of given degree, and the product of all these factors of given degree. If the f_i have distinct degrees, or, equivalently, if the s_{ii} have distinct degrees, then (4.16) represents already the canonical factorization of f in $\mathbb{F}_q[x]$.

If (4.16) is not yet the canonical factorization, then one can proceed in various ways. An obvious option is the application of one of the methods discussed earlier to factor the polynomials g_d. One can also continue with the diagonalization algorithm in order to obtain the diagonal matrix D equivalent to the normalized matrix S.

For the latter purpose, we assume as above that D is put in normalized form. In addition to the properties mentioned above, it is then also true that each of the submatrices $S^{(d)}$ is equivalent to the corresponding submatrix $D^{(d)}$ of D. It is therefore sufficient to diagonalize each of the submatrices $S^{(d)}$ separately. By the equivalence of $S^{(d)}$ and $D^{(d)}$ we have $S^{(d)} = UD^{(d)}E$ for some unimodular matrix U of polynomials and some nonsingular matrix E with entries in $\mathbb{F}_q$. We may then write

$$S^{(d)} = S_0^{(d)} + S_1^{(d)}x + \cdots + S_d^{(d)}x^d,$$

$$U = U_0 + U_1 x + \cdots + U_m x^m,$$

$$D^{(d)} = D_0^{(d)} + D_1^{(d)}x + \cdots + D_d^{(d)}x^d,$$

where the $S_r^{(d)}$, $D_r^{(d)}$, and U_t, $0 \leqslant r \leqslant d$, $0 \leqslant t \leqslant m$, are matrices with entries in $\mathbb{F}_q$, $U_m \neq 0$, and $S_d^{(d)} = D_d^{(d)} = I$, the identity matrix of appropriate order. A comparison of the matrix coefficients of the highest powers of x on both sides of the equation $S^{(d)} = UD^{(d)}E$ yields $I = U_m IE$ and $m = 0$. Thus, $U = U_0 = E^{-1}$ and hence $S^{(d)} = E^{-1}D^{(d)}E$.

Comparing the matrix coefficients of like powers of x in the last identity gives $S_r^{(d)} = E^{-1}D_r^{(d)}E$ for $0 \leqslant r \leqslant d$. Consequently, $S_r^{(d)}$ and $D_r^{(d)}$ have the same characteristic polynomial and eigenvalues, and since $D_r^{(d)}$ is diagonal, its eigenvalues are exactly its diagonal entries. Therefore, the latter can be determined by finding the roots of the characteristic polynomial of $S_r^{(d)}$, which must all be in $\mathbb{F}_q$. As in the earlier methods, we have thus again reduced the factorization problem to that of finding the roots of certain polynomials in $\mathbb{F}_q$.

The partial factorization (4.16) can also be obtained by an entirely different method. To this end, we extend the definition of g_d by letting g_i, $i \geqslant 1$, be the product of all monic irreducible polynomials in $\mathbb{F}_q[x]$ of degree i that divide f. In particular, $g_i(x) = 1$ in case f has no irreducible factor in $\mathbb{F}_q[x]$ of degree i. We can thus write

$$f = \prod_{i \geqslant 1} g_i$$

It is trivial that only those i with $i \leqslant \deg(f)$ need to be considered. We calculate now recursively the polynomials $r_0(x)$, $r_1(x), \ldots$ and $F_0(x)$, $F_1(x), \ldots$ as well as $d_1(x), d_2(x), \ldots$. We start with

$$r_0(x) = x, \quad F_0(x) = f(x),$$

and for $i \geq 1$ we use the formulas

$$r_i(x) \equiv r_{i-1}(x)^q \bmod F_{i-1}(x), \deg(r_i) < \deg(F_{i-1}),$$
$$d_i(x) = \gcd(F_{i-1}(x), r_i(x) - x),$$
$$F_i(x) = F_{i-1}(x)/d_i(x).$$

The algorithm can be stopped when $d_i(x) = F_{i-1}(x)$.

4.13.   Theorem.   *With the notation above, we have $d_i(x) = g_i(x)$ for all $i \geq 1$.*

Proof.   Using the fact that F_i divides F_{i-1}, a straightforward induction shows that

$$r_i(x) \equiv x^{q^i} \bmod F_{i-1}(x) \quad \text{for all } i \geq 1. \tag{4.17}$$

We prove now by induction that

$$F_{i-1} = \prod_{j \geq i} g_j \quad \text{and} \quad d_i = g_i \quad \text{for all } i \geq 1. \tag{4.18}$$

For $i = 1$ the first identity holds since $F_0 = f$. As to the second identity, we have

$$d_1(x) = \gcd(F_0(x), r_1(x) - x) = \gcd(f(x), x^q - x)$$

by (4.17), and since $x^q - x$ is the product of all monic linear polynomials in $\mathbb{F}_q[x]$, it follows that d_1 is the product of all monic linear polynomials in $\mathbb{F}_q[x]$ dividing f, and hence $d_1 = g_1$. Now assume that (4.18) is shown for some $i \geq 1$. Then

$$F_i = F_{i-1}/d_i = F_{i-1}/g_i = \prod_{j \geq i+1} g_j, \tag{4.19}$$

which proves the first identity in (4.18) for $i + 1$. Furthermore,

$$d_{i+1}(x) = \gcd(F_i(x), r_{i+1}(x) - x) = \gcd\left(F_i(x), x^{q^{i+1}} - x\right)$$

by (4.17). According to Theorem 3.20, $x^{q^{i+1}} - x$ is the product of all monic irreducible polynomials in $\mathbb{F}_q[x]$ whose degrees divide $i + 1$. Consequently, d_{i+1} is the product of all monic irreducible polynomials in $\mathbb{F}_q[x]$ that divide F_i and whose degrees divide $i + 1$. It follows then from (4.19) that $d_{i+1} = g_{i+1}$.   □

In the algorithm above, the most complicated step from the viewpoint of calculation is that of obtaining r_i by computing the qth power of $r_{i-1} \bmod F_{i-1}$. A common technique of cutting down the amount of calculation somewhat is based on computing first the residues mod F_{i-1} of $r_{i-1}, r_{i-1}^2, r_{i-1}^4, \ldots, r_{i-1}^{2^e}$ by repeated squaring and reduction mod F_{i-1}, where 2^e is the largest power of 2 that is $\leq q$, and then multiplying together an appropriate combination of these residues mod F_{i-1} to obtain the residue of

$r_{i-1}^q \bmod F_{i-1}$. For instance, to get the residue of $r_{i-1}^{23} \bmod F_{i-1}$, one would multiply together the residues of r_{i-1}^{16}, r_{i-1}^4, r_{i-1}^2, and $r_{i-1} \bmod F_{i-1}$.

Instead of working with the repeated squaring technique, we could employ the matrix B from Berlekamp's algorithm in Section 1 to calculate r_i from r_{i-1}. We write $n = \deg(f)$ and

$$r_{i-1}(x) = \sum_{j=0}^{n-1} r_{i-1}^{(j)} x^j,$$

and define $(s_i^{(0)}, s_i^{(1)}, \ldots, s_i^{(n-1)}) \in \mathbb{F}_q^n$ by the matrix identity

$$\left(s_i^{(0)}, s_i^{(1)}, \ldots, s_i^{(n-1)}\right) = \left(r_{i-1}^{(0)}, r_{i-1}^{(1)}, \ldots, r_{i-1}^{(n-1)}\right) B, \qquad (4.20)$$

where B is the $n \times n$ matrix in (4.5). With

$$s_i(x) = \sum_{j=0}^{n-1} s_i^{(j)} x^j \qquad (4.21)$$

we get then $r_{i-1}(x)^q \equiv s_i(x) \bmod f(x)$, hence $r_{i-1}(x)^q \equiv s_i(x) \bmod F_{i-1}(x)$, and thus

$$r_i(x) \equiv s_i(x) \bmod F_{i-1}(x)$$

Therefore, once the matrix B has been calculated, we compute r_i from r_{i-1} in each step by reduction mod F_{i-1} of the polynomial s_i obtained from (4.20) and (4.21).

4.14. Example. We consider $f(x) = x^6 - 3x^5 + 5x^4 - 9x^3 - 5x^2 + 6x + 7 \in \mathbb{F}_{23}[x]$ as in Example 4.7. Then

$$B = \begin{pmatrix} 1 & 0 & 0 & 0 & 0 & 0 \\ 5 & 0 & -1 & 8 & -3 & -10 \\ -10 & 10 & 10 & 0 & 1 & -9 \\ 0 & 7 & 9 & -8 & 10 & -11 \\ 11 & 0 & -4 & 7 & 7 & 2 \\ -3 & 0 & -10 & 9 & 2 & -9 \end{pmatrix}$$

We start the algorithm with $r_0(x) = x$, $F_0(x) = f(x)$. From (4.20) and (4.21) we get $s_1(x) = -10x^5 - 3x^4 + 8x^3 - x^2 + 5$, and reduction mod $F_0(x)$ yields $r_1(x) = s_1(x)$. By Theorem 4.13 we have $g_1(x) = d_1(x) = \gcd(F_0(x), r_1(x) - x) = x - 4$. Furthermore, $F_1(x) = F_0(x)/d_1(x) = x^5 + x^4 + 9x^3 + 4x^2 + 11x + 4$.

In the second iteration, we use again (4.20) and (4.21) to obtain $s_2(x) = 5x^5 - 8x^4 + 9x^3 - 10x^2 - 11$, and reduction mod $F_1(x)$ leads to $r_2(x) = 10x^4 + 10x^3 - 7x^2 - 9x - 8$. By Theorem 4.13 we have $g_2(x) = d_2(x) = \gcd(F_1(x), r_2(x) - x) = x^2 - x + 7$. Furthermore, $F_2(x) = F_1(x)/d_2(x) = x^3 + 2x^2 + 4x - 6$. But, according to the first part of (4.18), all irreducible factors of $F_2(x)$ have degree ≥ 3, so that $F_2(x)$ itself must be irreducible in $\mathbb{F}_{23}[x]$ and $g_3(x) = F_2(x)$. Thus, we arrive at the partial factorization

$$f(x) = (x - 4)(x^2 - x + 7)(x^3 + 2x^2 + 4x - 6),$$

which, in this case, is already the canonical factorization of $f(x)$ in $\mathbb{F}_{23}[x]$. □

3. CALCULATION OF ROOTS OF POLYNOMIALS

We have seen in the preceding section that the problem of determining the canonical factorization of a polynomial can often be reduced to that of finding the roots of an auxiliary polynomial in a finite field. The calculation of roots of a polynomial is, of course, a matter of independent interest as well.

In general, one will be interested in determining the roots of a polynomial in an extension of the field from which the coefficients are taken. However, it suffices to consider the situation in which we are asked to find the roots of a polynomial $f \in \mathbb{F}_q[x]$ of positive degree in $\mathbb{F}_q$, since a polynomial over a subfield can always be viewed as a polynomial over $\mathbb{F}_q$.

It is clear that every factorization algorithm is, in particular, a root-finding algorithm since the roots of f in $\mathbb{F}_q$ can be read off from the linear factors that occur in the canonical factorization of f in $\mathbb{F}_q[x]$. Thus, the algorithms presented in the earlier sections of this chapter can also be used for the determination of roots. However, these algorithms will often not be the most efficient procedures for the more specialized task of calculating roots. Therefore, we shall discuss methods that are better suited to this particular purpose.

As a first step, one may isolate that part of f which contains the roots of f in $\mathbb{F}_q$. This is achieved by calculating $\gcd(f(x), x^q - x)$. Since $x^q - x$ is the product of all monic linear polynomials in $\mathbb{F}_q[x]$, this greatest common divisor is the product of all monic linear polynomials over $\mathbb{F}_q$ dividing f, and so its roots are precisely the roots of f in $\mathbb{F}_q$. Therefore, we may assume, without loss of generality, that the polynomial for which we want to find the roots in $\mathbb{F}_q$ is a product of distinct monic linear polynomials over $\mathbb{F}_q$.

A useful method of finding roots of polynomials was already discussed in Chapter 3, Section 4. It is based on the determination of an affine multiple of the given polynomial. See Example 3.55 for an illustration of this method.

In order to arrive at other methods, we consider first the case of a prime field $\mathbb{F}_p$. As we have seen above, it suffices to deal with polynomials of the form

$$f(x) = \prod_{i=1}^{n} (x - c_i),$$

where $c_1,\ldots,c_n$ are distinct elements of $\mathbb{F}_p$. If p is small, then it is feasible to determine the roots of f by trial and error, that is, by simply calculating $f(0), f(1),\ldots,f(p-1)$.

For large p the following method may be employed. For $b \in \mathbb{F}_p$, p odd, we consider

$$f(x-b) = \prod_{i=1}^{n} \bigl(x - (b+c_i)\bigr).$$

We note that $f(x-b)$ divides $x^p - x = x(x^{(p-1)/2} + 1)(x^{(p-1)/2} - 1)$. If x is a factor of $f(x-b)$, then $f(-b) = 0$ and a root of f has been found. If x is not a factor of $f(x-b)$, then we have

$$f(x-b) = \gcd\bigl(f(x-b), x^{(p-1)/2} + 1\bigr)\gcd\bigl(f(x-b), x^{(p-1)/2} - 1\bigr).$$

$$(4.22)$$

The identity (4.22) is now used as follows. We calculate the residue mod $f(x-b)$ of $x^{(p-1)/2}$—for example, by the repeated squaring technique discussed after Theorem 4.13. If $x^{(p-1)/2} \not\equiv \pm 1 \bmod f(x-b)$, then (4.22) yields a nontrivial partial factorization of $f(x-b)$. Replacing x by $x+b$, we get then a nontrivial partial factorization of $f(x)$. In the rather unlikely case where $x^{(p-1)/2} \equiv \pm 1 \bmod f(x-b)$, we try another value of b. Thus, by using, if necessary, several choices for b, we will find either a root of f or a nontrivial partial factorization of f. Continuing this process, we will eventually obtain all the roots of f. It should be noted that, strictly speaking, this is not a deterministic, but a probabilistic root-finding algorithm, as it depends on the random selection of several elements $b \in \mathbb{F}_p$.

4.15. Example. Find the roots of $f(x) = x^6 - 7x^5 + 3x^4 - 7x^3 + 4x^2 - x - 2 \in \mathbb{F}_{17}[x]$ contained in $\mathbb{F}_{17}$. The roots of $f(x)$ in $\mathbb{F}_{17}$ are precisely the roots of $g(x) = \gcd(f(x), x^{17} - x)$ in $\mathbb{F}_{17}$. By the Euclidean algorithm we obtain $g(x) = x^4 + 6x^3 - 5x^2 + 7x - 2$. To find the roots of $g(x)$, we use the algorithm above and first select $b = 0$. A straightforward calculation yields $x^{(p-1)/2} = x^8 \equiv 1 \bmod g(x)$, and so this value of b does not afford a nontrivial partial factorization of $g(x)$. Next we choose $b = 1$. Then $g(x-1) = x^4 + 2x^3 - 3x - 2$ and $x^8 \equiv -4x^3 - 7x^2 + 8x - 5 \bmod g(x-1)$, so that $b = 1$ yields a nontrivial partial factorization of $g(x-1)$. We have

$$\gcd\bigl(g(x-1), x^8 + 1\bigr) = \gcd\bigl(x^4 + 2x^3 - 3x - 2, -4x^3 - 7x^2 + 8x - 4\bigr)$$

$$= x^2 - 7x + 4$$

and

$$\gcd\bigl(g(x-1), x^8 - 1\bigr) = \gcd\bigl(x^4 + 2x^3 - 3x - 2, -4x^3 - 7x^2 + 8x - 6\bigr)$$

$$= x^2 - 8x + 8,$$

hence (4.22) implies

$$g(x - 1) = (x^2 - 7x + 4)(x^2 - 8x + 8),$$

which leads to the partial factorization

$$g(x) = (x^2 - 5x - 2)(x^2 - 6x + 1) = g_1(x)g_2(x),$$

say. In order to factor $g_1(x)$ and $g_2(x)$, we try $b = 2$. We have $g_1(x - 2) = x^2 + 8x - 5$ and $x^8 \equiv -8x + 2 \bmod g_1(x - 2)$. Furthermore,

$$\gcd(g_1(x - 2), x^8 + 1) = \gcd(x^2 + 8x - 5, -8x + 3) = x + 6,$$

and long division yields $g_1(x - 2) = (x + 6)(x + 2)$, so that

$$g_1(x) = (x + 8)(x + 4).$$

Turning to $g_2(x)$, we have $g_2(x - 2) = x^2 + 7x = x(x + 7)$, thus -2 is a root of $g_2(x)$ and

$$g_2(x) = (x + 2)(x - 8).$$

Combining these factorizations, we get

$$g(x) = (x + 8)(x + 4)(x + 2)(x - 8).$$

Therefore, the roots of $g(x)$, and thus of $f(x)$, in $\mathbb{F}_{17}$ are $-8, -4, -2, 8$.   □

Next we discuss a root-finding algorithm for large finite fields $\mathbb{F}_q$ with small characteristic p. As before, it suffices to consider the case where

$$f(x) = \prod_{i=1}^{n} (x - \gamma_i)$$

with distinct elements $\gamma_1, \ldots, \gamma_n \in \mathbb{F}_q$. Let $q = p^m$ and define the polynomial

$$S(x) = \sum_{j=0}^{m-1} x^{p^j}$$

We note that for $\gamma \in \mathbb{F}_q$ we have $S(\gamma) = \mathrm{Tr}_{\mathbb{F}_q}(\gamma) \in \mathbb{F}_p$, where $\mathrm{Tr}_{\mathbb{F}_q}$ is the absolute trace function (see Definition 2.22). Because of Theorem 2.23(iii), the equation $S(\gamma) = c$ has p^{m-1} solutions $\gamma \in \mathbb{F}_q$ for every $c \in \mathbb{F}_p$, and this observation leads to the identity

$$x^q - x = \prod_{c \in \mathbb{F}_p} (S(x) - c). \tag{4.23}$$

Since $f(x)$ divides $x^q - x$, we get

$$\prod_{c \in \mathbb{F}_p} (S(x) - c) \equiv 0 \bmod f(x),$$

and so

$$f(x) = \prod_{c \in \mathbb{F}_p} \gcd(f(x), S(x) - c). \tag{4.24}$$

This yields a partial factorization of $f(x)$ that calls for the calculation of p greatest common divisors. If p is small, this is certainly a feasible method.

It can, however, happen that the factorization in (4.24) is trivial—namely, precisely when $S(x) \equiv c \bmod f(x)$ for some $c \in \mathbb{F}_p$. In this case, other auxiliary polynomials related to $S(x)$ have to be used. Let β be a defining element of $\mathbb{F}_q$ over $\mathbb{F}_p$, so that $\{1, \beta, \beta^2, \dots, \beta^{m-1}\}$ is a basis of $\mathbb{F}_q$ over $\mathbb{F}_p$. For $j = 0, 1, \dots, m-1$ we substitute $\beta^j x$ for x in (4.23) and we get

$$(\beta^j)^q x^q - \beta^j x = \prod_{c \in \mathbb{F}_p} (S(\beta^j x) - c).$$

Since $(\beta^j)^q = \beta^j$, we obtain

$$x^q - x = \beta^{-j} \prod_{e \in \mathbb{F}_p} (S(\beta^j x) - c).$$

This yields the following generalization of (4.24):

$$f(x) = \prod_{c \in \mathbb{F}_p} \gcd(f(x), S(\beta^j x) - c) \quad \text{for } 0 \leqslant j \leqslant m-1. \quad (4.25)$$

We show now that if $n = \deg(f) \geqslant 2$, then there exists at least one j, $0 \leqslant j \leqslant m-1$, for which the partial factorization in (4.25) is nontrivial. For suppose, on the contrary, that all the partial factorizations in (4.25) are trivial. Then for each j, $0 \leqslant j \leqslant m-1$, there exists a $c_j \in \mathbb{F}_p$ with

$$S(\beta^j x) \equiv c_j \bmod f(x).$$

In particular, we get

$$S(\beta^j \gamma_1) = S(\beta^j \gamma_2) = c_j \quad \text{for } 0 \leqslant j \leqslant m-1.$$

By the linearity of the trace it follows that

$$\mathrm{Tr}_{\mathbb{F}_q}((\gamma_1 - \gamma_2)\beta^j) = 0 \quad \text{for } 0 \leqslant j \leqslant m-1$$

and

$$\mathrm{Tr}_{\mathbb{F}_q}((\gamma_1 - \gamma_2)\alpha) = 0 \quad \text{for all } \alpha \in \mathbb{F}_q.$$

Using the second part of Theorem 2.24, we conclude that $\gamma_1 - \gamma_2 = 0$, which is a contradiction. Thus, for at least one j the partial factorization in (4.25) is nontrivial.

The defining element β of $\mathbb{F}_q$ over $\mathbb{F}_p$ used in (4.25) is chosen as a root of a known irreducible polynomial in $\mathbb{F}_p[x]$ of degree m. Once a nontrivial factorization of the form (4.25) has been found, the method is applied to the nontrivial factors by employing other values of j. The argument above shows also that all distinct roots of f can eventually be separated by using all the values of j in (4.25).

4.16.   Example.   Consider $\mathbb{F}_{64} = \mathbb{F}_2(\beta)$, where β is a root of the irreducible polynomial $x^6 + x + 1$ in $\mathbb{F}_2[x]$, and let

$$f(x) = x^4 + (\beta^5 + \beta^4 + \beta^3 + \beta^2)x^3 + (\beta^5 + \beta^4 + \beta^2 + \beta + 1)x^2$$
$$+ (\beta^4 + \beta^3 + \beta)x + \beta^3 + \beta \in \mathbb{F}_{64}[x].$$

Using

$$x^6 \equiv (\beta^5 + \beta + 1)x^3 + (\beta^4 + \beta^3 + \beta^2)x^2 + (\beta^5 + \beta^3 + \beta^2 + 1)x$$
$$+ \beta^5 + \beta^4 + \beta^2 + 1 \bmod f(x),$$

we get the following congruences mod $f(x)$ by repeated squaring:

$$x \equiv \hspace{6cm} x$$
$$x^2 \equiv \hspace{5.5cm} x^2$$
$$x^4 \equiv (\beta^5 + \beta^4 + \beta^3 + \beta^2)x^3 + \quad (\beta^5 + \beta^4 + \beta^2 + \beta + 1)x^2 + (\beta^4 + \beta^3 + \beta)x + \beta^3 + \beta$$
$$x^8 \equiv \quad (\beta^4 + \beta^3 + \beta^2)x^3 + \quad (\beta^5 + \beta + 1)x^2 + (\beta^5 + \beta + 1)x + \beta^5 + \beta^4$$
$$x^{16} \equiv \quad (\beta^5 + \beta^3 + \beta)x^3 + \quad (\beta^3 + \beta)x^2 + \beta^5 x + \beta^4 + \beta^3 + \beta^2 + \beta + 1$$
$$x^{32} \equiv \quad (\beta^5 + \beta^2 + 1)x^3 + (\beta^5 + \beta^4 + \beta^3 + \beta^2 + \beta + 1)x^2 + (\beta^4 + \beta^2)x + \beta^5 + \beta^3$$
$$x^{64} \equiv \hspace{6cm} x$$

Thus, $f(x)$ divides $x^{64} - x$ and so has four distinct roots in $\mathbb{F}_{64}$. We consider now $S(x) = x + x^2 + x^4 + x^8 + x^{16} + x^{32}$. From the congruences above we obtain

$$S(x) \equiv (\beta^5 + \beta^3 + \beta^2 + \beta + 1)x^3 + \beta^5 x^2 + (\beta^3 + \beta^2)x$$
$$+ \beta^3 + \beta^2 + 1 \bmod f(x),$$

and therefore

$$\gcd(f(x), S(x)) = \gcd(f(x), (\beta^5 + \beta^3 + \beta^2 + \beta + 1)x^3 + \beta^5 x^2$$
$$+ (\beta^3 + \beta^2)x + \beta^3 + \beta^2 + 1)$$
$$= x^3 + (\beta^4 + \beta^3 + \beta^2)x^2 + (\beta^5 + \beta^2 + 1)x + \beta^3 + \beta^2 = g(x)$$

say, and

$$\gcd(f(x), S(x) - 1) = \gcd(f(x), (\beta^5 + \beta^3 + \beta^2 + \beta + 1)x^3 + \beta^5 x^2$$
$$+ (\beta^3 + \beta^2)x + \beta^3 + \beta^2) = x + \beta^5.$$

Then (4.24) yields

$$f(x) = g(x)(x + \beta^5). \tag{4.26}$$

To find the roots of $g(x)$, we next use (4.25) with $j = 1$. We have

$$S(\beta x) = \beta x + \beta^2 x^2 + \beta^4 x^4 + \beta^8 x^8 + \beta^{16} x^{16} + \beta^{32} x^{32}$$
$$= \beta x + \beta^2 x^2 + \beta^4 x^4 + (\beta^3 + \beta^2)x^8$$
$$+ (\beta^4 + \beta + 1)x^{16} + (\beta^3 + 1)x^{32},$$

and the congruences above yield

$$S(\beta x) \equiv (\beta^2 + 1)x^3 + (\beta^3 + \beta + 1)x^2 + (\beta^5 + \beta^4 + \beta^3 + \beta^2 + \beta + 1)x$$
$$+ \beta^4 + \beta^2 + \beta \bmod f(x).$$

Since $g(x)$ divides $f(x)$, this congruence holds also mod $g(x)$, and so

$$S(\beta x) \equiv (\beta^2 + 1)x^3 + (\beta^3 + \beta + 1)x^2 + (\beta^5 + \beta^4 + \beta^3 + \beta^2 + \beta + 1)x$$
$$+ \beta^4 + \beta^2 + \beta$$
$$\equiv (\beta^5 + \beta^2)x^2 + \beta^3 x + \beta^5 + \beta^3 + \beta \bmod g(x).$$

Thus,

$$\gcd(g(x), S(\beta x)) = \gcd(g(x), (\beta^5 + \beta^2)x^2 + \beta^3 x + \beta^5 + \beta^3 + \beta)$$
$$= x^2 + (\beta^3 + 1)x + \beta^4 + \beta^3 + \beta^2 + \beta = h(x),$$

say, and

$$\gcd(g(x), S(\beta x) - 1) = \gcd(g(x), (\beta^5 + \beta^2)x^2 + \beta^3 x + \beta^5 + \beta^3 + \beta + 1)$$
$$= x + \beta^4 + \beta^2 + 1.$$

Then (4.25) with $j = 1$ yields

$$g(x) = h(x)(x + \beta^4 + \beta^2 + 1). \tag{4.27}$$

To find the roots of $h(x)$, we use (4.25) with $j = 2$. We have

$$S(\beta^2 x) = \beta^2 x + \beta^4 x^2 + \beta^8 x^4 + \beta^{16} x^8 + \beta^{32} x^{16} + \beta^{64} x^{32}$$
$$= \beta^2 x + \beta^4 x^2 + (\beta^3 + \beta^2)x^4 + (\beta^4 + \beta + 1)x^8$$
$$+ (\beta^3 + 1)x^{16} + \beta x^{32},$$

and a similar calculation as for $S(\beta x)$ yields

$$S(\beta^2 x) \equiv (\beta^5 + \beta^2 + 1)x + \beta^5 + \beta^3 + \beta^2 \bmod h(x).$$

Therefore,

$$\gcd(h(x), S(\beta^2 x)) = \gcd(h(x), (\beta^5 + \beta^2 + 1)x + \beta^5 + \beta^3 + \beta^2)$$
$$= x + \beta + 1$$

and

$$\gcd(h(x), S(\beta^2 x) - 1) = \gcd(h(x), (\beta^5 + \beta^2 + 1)x + \beta^5 + \beta^3 + \beta^2 + 1)$$
$$= x + \beta^3 + \beta,$$

so that from (4.25) with $j = 2$ we get

$$h(x) = (x + \beta + 1)(x + \beta^3 + \beta). \tag{4.28}$$

Combining (4.26), (4.27), and (4.28), we arrive at the factorization

$$f(x) = (x + \beta + 1)(x + \beta^3 + \beta)(x + \beta^4 + \beta^2 + 1)(x + \beta^5),$$

and so the roots of $f(x)$ are $\beta + 1$, $\beta^3 + \beta$, $\beta^4 + \beta^2 + 1$, and β^5.   □

Finally we consider the root-finding problem for large finite fields $\mathbb{F}_q$ with large characteristic p. As we have seen before, it suffices to know how to treat polynomials of the form

$$f(x) = \prod_{i=1}^{n} (x - \gamma_i)$$

with distinct elements $\gamma_1, \ldots, \gamma_n \in \mathbb{F}_q$. To check whether $f(x)$ has this form, we need only verify the congruence $x^q \equiv x \bmod f(x)$ (compare with the first part of Example 4.16). We can assume that q is the least power of p for which this holds. The polynomial $f(x)$ will, of course, be given by its standard representation

$$f(x) = \sum_{j=0}^{n} \alpha_j x^j,$$

where $\alpha_j \in \mathbb{F}_q$ for $0 \leq j \leq n$ and $\alpha_n = 1$.

It will be our first aim to find a nontrivial factor of $f(x)$. To exclude a trivial case, we can assume $n \geq 2$. Let $q = p^m$ and define the polynomials

$$f_k(x) = \sum_{j=0}^{n} \alpha_j^{p^k} x^j \quad \text{for } 0 \leq k \leq m-1, \tag{4.29}$$

so that $f_0(x) = f(x)$ and each $f_k(x)$ is a monic polynomial over $\mathbb{F}_q$. Furthermore,

$$f_k(\gamma_i^{p^k}) = \sum_{j=0}^{n} \alpha_j^{p^k} \gamma_i^{jp^k} = \left(\sum_{j=0}^{n} \alpha_j \gamma_i^j \right)^{p^k} = 0$$

for $1 \leq i \leq n$, $0 \leq k \leq m-1$, and so

$$f_k(x) = \prod_{i=1}^{n} (x - \gamma_i^{p^k}) \quad \text{for } 0 \leq k \leq m-1.$$

We calculate now the polynomial

$$F(x) = \prod_{k=0}^{m-1} f_k(x). \tag{4.30}$$

This is a polynomial over $\mathbb{F}_p$ since

$$F(x) = \prod_{k=0}^{m-1} \prod_{i=1}^{n} (x - \gamma_i^{p^k}) = \prod_{i=1}^{n} \prod_{k=0}^{m-1} (x - \gamma_i^{p^k}) = \prod_{i=1}^{n} F_i(x)^{m/d_i},$$

where $F_i(x)$ is the minimal polynomial of γ_i over $\mathbb{F}_p$ and d_i is its degree (compare with the discussion following Definition 2.22). The $F_i(x)$ are therefore the irreducible factors of $F(x)$ in $\mathbb{F}_p[x]$, but certain $F_i(x)$ could be identical. Thus, the canonical factorization of $F(x)$ in $\mathbb{F}_p[x]$ has the form

$$F(x) = G_1(x) \cdots G_r(x), \tag{4.31}$$

where the $G_t(x)$, $1 \leqslant t \leqslant r$, are powers of the distinct $F_i(x)$. This canonical factorization can be obtained by one of the factorization algorithms in Section 2 of this chapter. Since $f(x) = f_0(x)$ divides $F(x)$, it follows from (4.31) that

$$f(x) = \prod_{t=1}^{r} \gcd(f(x), G_t(x)). \tag{4.32}$$

In most cases, (4.32) will provide a nontrivial partial factorization of $f(x)$. The factorization will be trivial precisely if $\gcd(f(x), G_t(x)) = f(x)$ for some t, $1 \leqslant t \leqslant r$, which is equivalent to $r = 1$ and $f(x)$ dividing $F_1(x)$. A comparison of degrees shows then $n \leqslant d_1 = m$. Furthermore, the roots of $f(x)$ are then all conjugate with respect to $\mathbb{F}_p$. Thus, by labelling the roots of $f(x)$ suitably, we can write

$$\gamma_i = \gamma_1^{p^{b_i}} \quad \text{for} \quad 1 \leqslant i \leqslant n, \text{ with } 0 = b_1 < b_2 < \cdots < b_n < m.$$

We set $b_{n+1} = m$ and

$$d = \min_{1 \leqslant i \leqslant n} (b_{i+1} - b_i).$$

It is clear that $d \leqslant m/n$. The following two possibilities can occur:

(A) $b_{i+1} - b_i > d$ for some i, $1 \leqslant i \leqslant n$;
(B) $b_{i+1} - b_i = d$ for all i, $1 \leqslant i \leqslant n$.

In case (A) we note that the set of roots of $f(x)$ is

$$\left\{ \gamma_1^{p^{b_1}}, \gamma_1^{p^{b_2}}, \ldots, \gamma_1^{p^{b_n}} \right\}$$

and the set of roots of $f_d(x)$ is

$$\left\{ \gamma_1^{p^{b_1+d}}, \gamma_1^{p^{b_2+d}}, \ldots, \gamma_1^{p^{b_n+d}} \right\}.$$

The condition in (A) implies that these two sets of roots are not identical. On the other hand, since $b_{i+1} - b_i = d$ for some i, $1 \leqslant i \leqslant n$, the two sets of roots have a common element. Thus, $\gcd(f(x), f_d(x)) \neq f(x)$ and $\neq 1$; that is, $\gcd(f(x), f_d(x))$ is a nontrivial factor of $f(x)$. We observe also that in this case we have $d < m/n$.

In case (B) a comparison of the sets of roots of $f(x)$ and $f_d(x)$ shows that $f(x) = f_d(x)$, whereas $\gcd(f(x), f_k(x)) = 1$ for $1 \leqslant k < d$. Moreover, we have $d = m/n$, so that n divides m, and also $b_i = d(i-1)$ for $1 \leqslant i \leqslant n$. It follows that

$$\gamma_i = \gamma_1^{p^{d(i-1)}} \quad \text{for } 1 \leqslant i \leqslant n,$$

hence the γ_i are exactly all the conjugates of γ_1 with respect to $\mathbb{F}_{p^d}$. Consequently, $f(x)$ is the minimal polynomial of γ_1 over $\mathbb{F}_{p^d}$ and thus irreducible over $\mathbb{F}_{p^d}$.

Therefore, corresponding to the cases (A) and (B) above we have the following alternatives:

(A) $\gcd(f(x), f_k(x))$ is a nontrivial factor of $f(x)$ for some k, $1 \leqslant k < m/n$;

(B) $\gcd(f(x), f_k(x)) = 1$ for $1 \leqslant k < d = m/n \in \mathbb{N}$ and $f(x) = f_d(x)$ is the minimal polynomial of γ_1 over $\mathbb{F}_{p^d}$.

In alternative (A) our aim of finding a nontrivial factor of $f(x)$ has been achieved.

Further work is needed in alternative (B). Let β again denote a defining element of $\mathbb{F}_q$ over $\mathbb{F}_p$. Then $\mathbb{F}_{p^d}(\beta) = \mathbb{F}_q = \mathbb{F}_{p^m}$, and so β is of degree $m/d = n$ over $\mathbb{F}_{p^d}$. In particular, we have $\beta^j \notin \mathbb{F}_{p^d}$ for $1 \leqslant j \leqslant n - 1$. Now let the coefficients α_j of $f(x)$ be such that $\alpha_{j_0} \neq 0$ for some j_0 with $1 \leqslant j_0 \leqslant n - 1$. Consider

$$\bar{f}(x) = \beta^{-n} f(\beta x), \tag{4.33}$$

which is a monic polynomial of degree n over $\mathbb{F}_q$. Since $\beta^{n-j_0} \notin \mathbb{F}_{p^d}$ and $\alpha_{j_0} \in \mathbb{F}_{p^d}^*$, it follows that the coefficient of x^{j_0} in $\bar{f}(x)$ is not an element of $\mathbb{F}_{p^d}$. Thus $\bar{f}(x)$ is not a polynomial over $\mathbb{F}_{p^d}$, and so the alternative (B) cannot occur if the procedure above is applied to $\bar{f}(x)$. Since $f(x) = \beta^n \bar{f}(\beta^{-1} x)$, any nontrivial factor of $\bar{f}(x)$ yields immediately a nontrivial factor of $f(x)$.

It remains to consider the case where alternative (B) is valid and $\alpha_j = 0$ for $1 \leqslant j \leqslant n - 1$. Then $f(x)$ is the binomial $x^n + \alpha_0 \in \mathbb{F}_{p^d}[x]$. Now n is not a multiple of p, for otherwise we would have $f(x) = (x^{n/p} + \alpha_0^{p^{d-1}})^p$, which would contradict the irreducibility of $f(x)$ over $\mathbb{F}_{p^d}$. We set

$$\bar{f}(x) = \beta^{-n} f(\beta x + 1), \tag{4.34}$$

and then it is easily seen from $\beta^{-1} \notin \mathbb{F}_{p^d}$ that the coefficient of x^{n-1} in $\bar{f}(x)$ is not in $\mathbb{F}_{p^d}$. Thus, the alternative (B) cannot occur if the procedure described above is applied to $\bar{f}(x)$. Since $f(x) = \beta^n \bar{f}(\beta^{-1}(x - 1))$, any nontrivial factor of $\bar{f}(x)$ yields immediately a nontrivial factor of $f(x)$.

This root-finding algorithm is thus carried out as follows. We first form the polynomials $f_k(x)$ according to (4.29) and then the polynomial $F(x) \in \mathbb{F}_p[x]$ according to (4.30). Next, we apply a factorization algorithm to obtain the canonical factorization (4.31) of $F(x)$ in $\mathbb{F}_p[x]$. This leads to the partial factorization of $f(x)$ given by (4.32). Should this factorization be trivial, we calculate $\gcd(f(x), f_k(x))$ for $1 \leqslant k < m/n$. If this also does not produce a nontrivial factor of $f(x)$, we transform $f(x)$ into $\bar{f}(x)$ by either (4.34) or (4.33), depending on whether $f(x)$ is a binomial or not. As we have shown above, an application of the algorithm to $\bar{f}(x)$ is bound to yield a nontrivial factor of $\bar{f}(x)$ and thus of $f(x)$. Once a nontrivial factor of $f(x)$ has been found, the procedure is continued with the resulting factors in place of $f(x)$, until $f(x)$ is split up completely into linear factors.

EXERCISES

4.1. Factor $x^{12} + x^7 + x^5 + x^4 + x^3 + x^2 + 1$ over $\mathbb{F}_2$ by Berlekamp's algorithm.

4.2. Factor $x^7 + x^6 + x^5 - x^3 + x^2 - x - 1$ over $\mathbb{F}_3$ by Berlekamp's algorithm.

4.3. Let $\mathbb{F}_4 = \mathbb{F}_2(\theta)$ and factor $x^5 + \theta x^4 + x^3 + (1 + \theta)x + \theta$ over $\mathbb{F}_4$ by Berlekamp's algorithm.

4.4. Use Berlekamp's algorithm to prove that $x^6 - x^3 - x - 1$ is irreducible in $\mathbb{F}_3[x]$.

4.5. Use Berlekamp's algorithm to determine the number of distinct monic irreducible factors of $x^4 + 1$ in $\mathbb{F}_p[x]$ for all odd primes p.

4.6. Use the polynomials T_i in Section 1 to factor $x^5 + x^4 + 1$ over $\mathbb{F}_2$.

4.7. Determine the splitting field of $x^8 + x^6 + x^5 + x^4 + x^3 + x^2 + 1$ over $\mathbb{F}_2$.

4.8. Determine the splitting field of $x^6 - x^4 - x^2 - x + 1$ over $\mathbb{F}_3$.

4.9. Use the polynomials R_i in Section 1 to factor the polynomial of Exercise 4.1 over $\mathbb{F}_2$.

4.10. Find the canonical factorization of $x^8 + x^6 + x^4 + x^3 + 1$ in $\mathbb{F}_2[x]$ by using the polynomials R_i in Section 1.

4.11. Determine the canonical factorization of the cyclotomic polynomial $Q_{31}(x)$ in $\mathbb{F}_2[x]$.

4.12. Factor $f(x) = x^8 + x^3 + 1$ over $\mathbb{F}_2$ and determine $\mathrm{ord}(f(x))$.

4.13. Factor $f(x) = x^9 + x^8 + x^7 + x^4 + x^3 + x + 1$ over $\mathbb{F}_2$ and determine $\mathrm{ord}(f(x))$.

4.14. Prove in detail that if f is a nonzero polynomial over a field and $d = \gcd(f, f')$, then f/d has no repeated factors. (*Note*: Count nonzero constant polynomials among the polynomials with no repeated factors.)

4.15. Let f be a monic polynomial of positive degree with integer coefficients. Prove that if f has no repeated factors, then there are only finitely many primes p such that f, considered as a polynomial over $\mathbb{F}_p$, has repeated factors.

4.16. Determine the number of monic polynomials in $\mathbb{F}_q[x]$ of degree $n \geqslant 1$ with no repeated factors.

4.17. Let f be a monic polynomial over $\mathbb{F}_q$ and let $g_1, \ldots, g_r$ be nonzero polynomials over $\mathbb{F}_q$ that are pairwise relatively prime. Prove that if f divides $g_1 \cdots g_r$, then $f = \prod_{i=1}^{r} \gcd(f, g_i)$.

4.18. Use Berlekamp's algorithm to prove the following special case of

Theorem 3.75: the binomial $x^t - a$, where t is a prime divisor of $q - 1$ and $a \in \mathbb{F}_q^*$, is irreducible in $\mathbb{F}_q[x]$ if and only if $a^{(q-1)/t} \neq 1$.

4.19. Let f be an irreducible polynomial in $\mathbb{F}_q[x]$ of degree n and define the $n \times n$ matrix $B = (b_{ij})$ by (4.4). Prove that the characteristic polynomial $\det(xI - B)$ of B is equal to $x^n - 1$.

4.20. Let $f = f_1 \cdots f_k$ be a product of k distinct monic irreducible polynomials $f_1, \ldots, f_k$ in $\mathbb{F}_q[x]$ of degree $n_1, \ldots, n_k$, respectively. Put $\deg(f) = n = n_1 + \cdots + n_k$ and define the $n \times n$ matrix $B = (b_{ij})$ by (4.4). Prove that the characteristic polynomial $\det(xI - B)$ of B is equal to $(x^{n_1} - 1) \cdots (x^{n_k} - 1)$.

4.21. In the notation of Section 1, prove that the polynomials T_i do not separate those irreducible factors f_j of f for which N/n_j is divisible by the characteristic of $\mathbb{F}_q$.

4.22. Let $f \in \mathbb{F}_q[x]$ be monic of degree $n \geq 1$. Define $h \in \mathbb{F}_q[x, y]$ by

$$h(x, y) = (y - x)(y - x^q)(y - x^{q^2}) \cdots (y - x^{q^{n-1}}) - f(y)$$

and write

$$h(x, y) = s_{n-1}(x) y^{n-1} + \cdots + s_1(x) y + s_0(x).$$

Prove that f is irreducible over $\mathbb{F}_q$ if and only if f divides s_j for $0 \leq j \leq n - 1$.

4.23. Use the criterion in the preceding exercise to prove that $x^7 + x^6 + x^3 + x^2 + 1$ is reducible over $\mathbb{F}_2$.

4.24. Prove that the quadratic polynomial $f(x) = x^2 + bx + c$ is irreducible over $\mathbb{F}_q$ if and only if $f(x)$ divides $x^q + x + b$.

4.25. Let f be an irreducible polynomial in $\mathbb{F}_q[x]$ of degree m and let λ be a root of f in $\mathbb{F}_{q^m}$. Let g and h be nonzero polynomials in $\mathbb{F}_q[x]$. Prove that $h(x)^m f(g(x)/h(x))$ is irreducible in $\mathbb{F}_q[x]$ if and only if $g(x) - \lambda h(x)$ is irreducible in $\mathbb{F}_{q^m}[x]$.

4.26. Use the method in Example 4.7 to factor $x^4 + 3x^3 + 4x^2 + 2x - 1$ over $\mathbb{F}_{13}$.

4.27. Use the method in Example 4.7 to factor $x^3 - 6x^2 - 8x - 8$ over $\mathbb{F}_{19}$.

4.28. Use the Zassenhaus algorithm to factor $x^4 + 3x^3 + 4x^2 + 2x - 1$ over $\mathbb{F}_{13}$.

4.29. Use the Zassenhaus algorithm to factor $x^3 - 6x^2 - 8x - 8$ over $\mathbb{F}_{19}$.

4.30. Use the Zassenhaus algorithm to factor $x^5 + 3x^4 + 2x^3 - 6x^2 + 5$ over $\mathbb{F}_{17}$.

4.31. Factor $x^4 - 7x^3 + 4x^2 + 2x + 4$ over $\mathbb{F}_{17}$.

4.32. Factor $x^4 - 3x^3 + 4x^2 - 6x - 8$ over $\mathbb{F}_{19}$.

4.33. Prove in detail that equivalence of square matrices of polynomials as defined by Definition 4.11 is reflexive, symmetric, and transitive.

4.34. Use the method in Example 4.14 to factor $x^3 - 6x^2 - 8x - 8$ over $\mathbb{F}_{19}$.

4.35. Use the method in Example 4.14 to factor $x^5 + 3x^4 + 2x^3 - 6x^2 + 5$ over $\mathbb{F}_{17}$.

4.36. Use the method in Example 4.14 to obtain a partial factorization of $x^7 - 2x^6 - 4x^4 + 3x^3 - 5x^2 + 3x + 5$ over $\mathbb{F}_{11}$ and complete the factorization by another method.

4.37. Find the roots of $f(x) = x^5 - x^4 + 2x^3 + x^2 - x - 2 \in \mathbb{F}_5[x]$ contained in $\mathbb{F}_5$.

4.38. Find the roots of $f(x) = x^5 + 6x^4 + 2x^3 - 6x^2 - 5x + 5 \in \mathbb{F}_{13}[x]$ contained in $\mathbb{F}_{13}$.

4.39. Prove that all the roots of $f(x) = x^3 + 8x^2 + 6x - 7 \in \mathbb{F}_{19}[x]$ are contained in $\mathbb{F}_{19}$ and find them.

4.40. Let $\mathbb{F}_{32} = \mathbb{F}_2(\beta)$, where β is a root of the irreducible polynomial $x^5 + x^2 + 1$ over $\mathbb{F}_2$. Prove that all the roots of $f(x) = x^3 + (\beta^4 + \beta^3 + 1)x^2 + \beta^2 x + \beta^4 + \beta^3 + \beta + 1 \in \mathbb{F}_{32}[x]$ are contained in $\mathbb{F}_{32}$ and find them.

4.41. Let $\mathbb{F}_{27} = \mathbb{F}_3(\beta)$, where β is a root of the irreducible polynomial $x^3 - x + 1$ over $\mathbb{F}_3$. Prove that all the roots of $f(x) = x^3 + x^2 - (\beta^2 - \beta + 1)x + \beta^2 - 1 \in \mathbb{F}_{27}[x]$ are contained in $\mathbb{F}_{27}$ and find them.

4.42. Let $\mathbb{F}_{169} = \mathbb{F}_{13}(\beta)$, where β is a root of the irreducible polynomial $x^2 - x - 1$ over $\mathbb{F}_{13}$. Find the roots of $f(x) = x^2 + (3\beta + 1)x + \beta + 5 \in \mathbb{F}_{169}[x]$ contained in $\mathbb{F}_{169}$.

4.43. If the polynomial $f(x - b)$ in (4.22) is quadratic with constant term $c \neq 0$, prove that the factorization in (4.22) is nontrivial if and only if c is not the square of an element of $\mathbb{F}_p$.

4.44. Let β be a defining element of $F = \mathbb{F}_{2^m}$ over $\mathbb{F}_2$. Prove:
   (a) There exists k, $0 \leq k \leq m - 1$, with $\mathrm{Tr}_F(\beta^k) = 1$.
   (b) For each $i = 0, 1, \ldots, m - 1$ there exists an $\alpha_i \in F$ such that

$$\alpha_i^2 + \alpha_i = \begin{cases} \beta^i & \text{if } \mathrm{Tr}_F(\beta^i) = 0, \\ \beta^i + \beta^k & \text{if } \mathrm{Tr}_F(\beta^i) = 1. \end{cases}$$

   (c) If $\gamma = \sum_{i=0}^{m-1} c_i \beta^i$, $c_i \in \mathbb{F}_2$, and $\mathrm{Tr}_F(\gamma) = 0$, then the roots of $x^2 + x + \gamma$ are $\sum_{i=0}^{m-1} c_i \alpha_i$ and $1 + \sum_{i=0}^{m-1} c_i \alpha_i$.

Chapter 5

Exponential Sums

Exponential sums are important tools in number theory for solving problems involving integers—and real numbers in general—that are often intractable by other means. Analogous sums can be considered in the framework of finite fields and turn out to be useful in various applications of finite fields.

A basic role in setting up exponential sums for finite fields is played by special group homomorphisms called characters. It is necessary to distinguish between two types of characters—namely, additive and multiplicative characters—depending on whether reference is made to the additive or the multiplicative group of the finite field. Exponential sums are formed by using the values of one or more characters and possibly combining them with weights or with other function values. If we only sum the values of a single character, we speak of a character sum.

In Section 1 we lay the foundation by first discussing characters of finite abelian groups and then specializing to finite fields. Explicit formulas for additive and multiplicative characters of finite fields can be given. Both types of characters satisfy important orthogonality relations.

Section 2 is devoted to Gaussian sums, which are arguably the most important types of exponential sums for finite fields as they govern the transition from the additive to the multiplicative structure and vice versa. They also appear in many other contexts in algebra and number theory. As an illustration of their usefulness in number theory, we present a proof of the law of quadratic reciprocity based on properties of Gaussian sums.

Exponential sums with the terms of a linear recurring sequence as arguments will be treated in Chapter 6, Section 7. Deep investigations on exponential sums for finite fields have been carried out with the help of algebraic geometry, leading to the famous results of Weil and Deligne, but a presentation of this work would lead far beyond the scope of this book.

1. CHARACTERS

Let G be a finite abelian group (written multiplicatively) of order $|G|$ with identity element 1_G. A *character* χ of G is a homomorphism from G into the multiplicative group U of complex numbers of absolute value 1—that is, a mapping from G into U with $\chi(g_1 g_2) = \chi(g_1)\chi(g_2)$ for all $g_1, g_2 \in G$. Since $\chi(1_G) = \chi(1_G)\chi(1_G)$, we must have $\chi(1_G) = 1$. Furthermore,

$$\left(\chi(g)\right)^{|G|} = \chi\left(g^{|G|}\right) = \chi(1_G) = 1$$

for every $g \in G$, so that the values of χ are $|G|$th roots of unity. We note also that $\chi(g)\chi(g^{-1}) = \chi(gg^{-1}) = \chi(1_G) = 1$, and so $\chi(g^{-1}) = (\chi(g))^{-1} = \overline{\chi(g)}$ for every $g \in G$, where the bar denotes complex conjugation.

Among the characters of G we have the *trivial* character χ_0 defined by $\chi_0(g) = 1$ for all $g \in G$; all other characters of G are called *nontrivial*. With each character χ of G there is associated the *conjugate* character $\overline{\chi}$ defined by $\overline{\chi}(g) = \overline{\chi(g)}$ for all $g \in G$. Given finitely many characters $\chi_1, \ldots, \chi_n$ of G, one can form the product character $\chi_1 \cdots \chi_n$ by setting $(\chi_1 \cdots \chi_n)(g) = \chi_1(g) \cdots \chi_n(g)$ for all $g \in G$. If $\chi_1 = \cdots = \chi_n = \chi$, we write χ^n for $\chi_1 \cdots \chi_n$. It is obvious that the set $G^\wedge$ of characters of G forms an abelian group under this multiplication of characters. Since the values of characters of G can only be $|G|$th roots of unity, $G^\wedge$ is finite.

After briefly considering the special case of à finite cyclic group, we establish some basic facts about characters.

5.1. Example. Let G be a finite cyclic group of order n, and let g be a generator of G. For a fixed integer j, $0 \leqslant j \leqslant n - 1$, the function

$$\chi_j(g^k) = e^{2\pi i j k / n}, \quad k = 0, 1, \ldots, n - 1,$$

defines a character of G. On the other hand, if χ is any character of G, then $\chi(g)$ must be an nth root of unity, say $\chi(g) = e^{2\pi i j / n}$ for some j, $0 \leqslant j \leqslant n - 1$, and it follows that $\chi = \chi_j$. Therefore, $G^\wedge$ consists exactly of the characters $\chi_0, \chi_1, \ldots, \chi_{n-1}$. $\qquad \square$

5.2. Theorem. *Let H be a subgroup of the finite abelian group G and let ψ be a character of H. Then ψ can be extended to a character of G; that is, there exists a character χ of G with $\chi(h) = \psi(h)$ for all $h \in H$.*

Proof. We may suppose that H is a proper subgroup of G. Choose $a \in G$ with $a \notin H$, and let H_1 be the subgroup of G generated by H and a. Let m be the least positive integer for which $a^m \in H$. Then every element $g \in H_1$ can be written uniquely in the form $g = a^j h$ with $0 \leqslant j < m$ and $h \in H$. Define a function ψ_1 on H_1 by $\psi_1(g) = \omega^j \psi(h)$, where ω is a fixed complex number satisfying $\omega^m = \psi(a^m)$. To check that ψ_1 is indeed a character of H_1, let $g_1 = a^k h_1$, $0 \leqslant k < m$, $h_1 \in H$, be another element of H_1. If $j + k < m$, then $\psi_1(gg_1) = \omega^{j+k} \psi(hh_1) = \psi_1(g)\psi_1(g_1)$. If $j + k \geqslant m$, then $gg_1 = a^{j+k-m}(a^m hh_1)$, and so

$$\psi_1(gg_1) = \omega^{j+k-m}\psi(a^m hh_1)$$
$$= \omega^{j+k-m}\psi(a^m)\psi(hh_1) = \omega^{j+k}\psi(hh_1) = \psi_1(g)\psi_1(g_1).$$

It is obvious that $\psi_1(h) = \psi(h)$ for $h \in H$. If $H_1 = G$, then we are done. Otherwise, we can continue the process above until, after finitely many steps, we obtain an extension of ψ to G.   □

5.3.   Corollary.   *For any two distinct elements g_1, $g_2 \in G$ there exists a character χ of G with $\chi(g_1) \neq \chi(g_2)$.*

Proof.   It suffices to show that for $h = g_1 g_2^{-1} \neq 1_G$ there exists a character χ of G with $\chi(h) \neq 1$. This follows, however, from Example 5.1 and Theorem 5.2 by letting H be the cyclic subgroup of G generated by h. □

5.4.   Theorem.   *If χ is a nontrivial character of the finite abelian group G, then*

$$\sum_{g \in G} \chi(g) = 0. \tag{5.1}$$

If $g \in G$ with $g \neq 1_G$, then

$$\sum_{\chi \in G^\wedge} \chi(g) = 0. \tag{5.2}$$

Proof.   Since χ is nontrivial, there exists $h \in G$ with $\chi(h) \neq 1$. Then

$$\chi(h) \sum_{g \in G} \chi(g) = \sum_{g \in G} \chi(hg) = \sum_{g \in G} \chi(g),$$

because if g runs through G, so does hg. Thus we have

$$(\chi(h) - 1) \sum_{g \in G} \chi(g) = 0,$$

which already implies (5.1). For the second part, we note that the function $\hat{g}$ defined by $\hat{g}(\chi) = \chi(g)$ for $\chi \in G^\wedge$ is a character of the finite abelian group $G^\wedge$. This character is nontrivial since, by Corollary 5.3, there exists $\chi \in G^\wedge$ with $\chi(g) \neq \chi(1_G) = 1$. Therefore from (5.1) applied to the group $G^\wedge$,

$$\sum_{\chi \in G^\wedge} \chi(g) = \sum_{\chi \in G^\wedge} \hat{g}(\chi) = 0. \qquad \Box$$

5.5. Theorem. *The number of characters of a finite abelian group G is equal to $|G|$.*

Proof. This follows from

$$|G^\wedge| = \sum_{g \in G} \sum_{\chi \in G^\wedge} \chi(g) = \sum_{\chi \in G^\wedge} \sum_{g \in G} \chi(g) = |G|,$$

where we used (5.2) in the first identity and (5.1) in the last identity. $\qquad \Box$

The statements of Theorems 5.4 and 5.5 can be combined into the *orthogonality relations for characters*. Let χ and ψ be characters of G. Then

$$\frac{1}{|G|} \sum_{g \in G} \chi(g)\overline{\psi(g)} = \begin{cases} 0 & \text{for } \chi \neq \psi, \\ 1 & \text{for } \chi = \psi. \end{cases} \qquad (5.3)$$

The first part follows, of course, by applying (5.1) to the character $\chi\bar{\psi}$; the second part is trivial.

Furthermore, if g and h are elements of G, then

$$\frac{1}{|G|} \sum_{\chi \in G^\wedge} \chi(g)\overline{\chi(h)} = \begin{cases} 0 & \text{for } g \neq h, \\ 1 & \text{for } g = h. \end{cases} \qquad (5.4)$$

Here, the first part is obtained from (5.2) applied to the element gh^{-1}, whereas the second part follows from Theorem 5.5.

Character theory is often used to obtain expressions for the number of solutions of equations in a finite abelian group G. Let f be an arbitrary map from the cartesian product $G^n = G \times \cdots \times G$ (n factors) into G. Then, for fixed $h \in G$, the number $N(h)$ of n-tuples $(g_1,\ldots,g_n) \in G^n$ with $f(g_1,\ldots,g_n) = h$ is given by

$$N(h) = \frac{1}{|G|} \sum_{g_1 \in G} \cdots \sum_{g_n \in G} \sum_{\chi \in G^\wedge} \chi(f(g_1,\ldots,g_n))\overline{\chi(h)}, \qquad (5.5)$$

on account of (5.4).

A character χ of G may be nontrivial on G, but still annihilate a whole subgroup H of G, in the sense that $\chi(h) = 1$ for all $h \in H$. The set of all characters of G annihilating a given subgroup H is called the *annihilator* of H in $G^\wedge$.

5.6. Theorem. *Let H be a subgroup of the finite abelian group G. Then the annihilator of H in $G^\wedge$ is a subgroup of $G^\wedge$ of order $|G|/|H|$.*

Proof. Let A be the annihilator in question. Then it is obvious from the definition that A is a subgroup of $G^\wedge$. Let $\chi \in A$; then $\mu(gH) = \chi(g)$, $g \in G$, is a well-defined character of the factor group G/H. Conversely, if μ

is a character of G/H, then $\chi(g) = \mu(gH)$, $g \in G$, defines a character of G annihilating H. Distinct elements of A correspond to distinct characters of G/H. Therefore, A is in one-to-one correspondence with the character group $(G/H)^\wedge$, and so the order of A is equal to the order of $(G/H)^\wedge$, which is $|G/H| = |G|/|H|$ according to Theorem 5.5. $\square$

In a finite field $\mathbb{F}_q$ there are two finite abelian groups that are of significance—namely, the additive group and the multiplicative group of the field. Therefore, we will have to make an important distinction between the characters pertaining to these two group structures. In both cases, explicit formulas for the characters can be given.

Consider first the *additive group* of $\mathbb{F}_q$. Let p be the characteristic of $\mathbb{F}_q$; then the prime field contained in $\mathbb{F}_q$ is $\mathbb{F}_p$, which we identify with $\mathbb{Z}/(p)$. Let $\mathrm{Tr}: \mathbb{F}_q \to \mathbb{F}_p$ be the absolute trace function from $\mathbb{F}_q$ to $\mathbb{F}_p$ (see Definition 2.22). Then the function χ_1 defined by

$$\chi_1(c) = e^{2\pi i \, \mathrm{Tr}(c)/p} \quad \text{for all } c \in \mathbb{F}_q \tag{5.6}$$

is a character of the additive group of $\mathbb{F}_q$, since for $c_1, c_2 \in \mathbb{F}_q$ we have $\mathrm{Tr}(c_1 + c_2) = \mathrm{Tr}(c_1) + \mathrm{Tr}(c_2)$, and so $\chi_1(c_1 + c_2) = \chi_1(c_1)\chi_1(c_2)$. Instead of "character of the additive group of $\mathbb{F}_q$," we shall henceforth use the term *additive character* of $\mathbb{F}_q$. The character χ_1 in (5.6) will be called the *canonical additive character* of $\mathbb{F}_q$. All additive characters of $\mathbb{F}_q$ can be expressed in terms of χ_1.

5.7. Theorem. *For $b \in \mathbb{F}_q$, the function χ_b with $\chi_b(c) = \chi_1(bc)$ for all $c \in \mathbb{F}_q$ is an additive character of $\mathbb{F}_q$, and every additive character of $\mathbb{F}_q$ is obtained in this way.*

Proof. For $c_1, c_2 \in \mathbb{F}_q$ we have

$$\chi_b(c_1 + c_2) = \chi_1(bc_1 + bc_2)$$
$$= \chi_1(bc_1)\chi_1(bc_2) = \chi_b(c_1)\chi_b(c_2),$$

and the first part is established. Since Tr maps $\mathbb{F}_q$ onto $\mathbb{F}_p$ by Theorem 2.23(iii), χ_1 is a nontrivial character. Therefore, if $a, b \in \mathbb{F}_q$ with $a \neq b$, then

$$\frac{\chi_a(c)}{\chi_b(c)} = \frac{\chi_1(ac)}{\chi_1(bc)} = \chi_1((a-b)c) \neq 1$$

for suitable $c \in \mathbb{F}_q$, and so χ_a and χ_b are distinct characters. Hence, if b runs through $\mathbb{F}_q$, we get q distinct additive characters χ_b. On the other hand, $\mathbb{F}_q$ has exactly q additive characters by Theorem 5.5, and so the list of additive characters of $\mathbb{F}_q$ is already complete. $\square$

By setting $b = 0$ in Theorem 5.7, we obtain the trivial additive character χ_0, for which $\chi_0(c) = 1$ for all $c \in \mathbb{F}_q$.

Let E be a finite extension field of $\mathbb{F}_q$, let χ_1 be the canonical additive character of $\mathbb{F}_q$, and let μ_1 be the canonical additive character of E defined in analogy with (5.6), where Tr is of course replaced by the absolute trace function Tr_E from E to $\mathbb{F}_p$. Then χ_1 and μ_1 are connected by the identity

$$\chi_1\big(\mathrm{Tr}_{E/\mathbb{F}_q}(\beta)\big) = \mu_1(\beta) \quad \text{for all } \beta \in E, \tag{5.7}$$

where $\mathrm{Tr}_{E/\mathbb{F}_q}$ is the trace function from E to $\mathbb{F}_q$. This follows from the transitivity relation

$$\mathrm{Tr}_E(\beta) = \mathrm{Tr}\big(\mathrm{Tr}_{E/\mathbb{F}_q}(\beta)\big) \quad \text{for all } \beta \in E,$$

which was shown in Theorem 2.26.

Characters of the *multiplicative group* $\mathbb{F}_q^*$ of $\mathbb{F}_q$ are called *multiplicative characters* of $\mathbb{F}_q$. Since $\mathbb{F}_q^*$ is a cyclic group of order $q - 1$ by Theorem 2.8, its characters can be easily determined.

5.8.   Theorem.   *Let g be a fixed primitive element of $\mathbb{F}_q$. For each $j = 0, 1, \ldots, q - 2$, the function ψ_j with*

$$\psi_j\big(g^k\big) = e^{2\pi i j k/(q-1)} \quad \text{for } k = 0, 1, \ldots, q - 2$$

defines a multiplicative character of $\mathbb{F}_q$, and every multiplicative character of $\mathbb{F}_q$ is obtained in this way.

Proof.   This follows immediately from Example 5.1.   □

No matter what g is, the character ψ_0 will always represent the trivial multiplicative character, which satisfies $\psi_0(c) = 1$ for all $c \in \mathbb{F}_q^*$.

5.9.   Corollary.   *The group of multiplicative characters of $\mathbb{F}_q$ is cyclic of order $q - 1$ with identity element ψ_0.*

Proof.   Every character ψ_j in Theorem 5.8 with j relatively prime to $q - 1$ is a generator of the group in question.   □

5.10.   Example.   Let q be odd and let η be the real-valued function on $\mathbb{F}_q^*$ with $\eta(c) = 1$ if c is the square of an element of $\mathbb{F}_q^*$ and $\eta(c) = -1$ otherwise. Then η is a multiplicative character of $\mathbb{F}_q$. It can also be obtained from the characters in Theorem 5.8 by setting $j = (q - 1)/2$. The character η annihilates the subgroup of $\mathbb{F}_q^*$ consisting of the squares of elements of $\mathbb{F}_q^*$, and by Theorem 5.6 it is the only nontrivial character of $\mathbb{F}_q^*$ with this property. This uniquely determined character η is called the *quadratic character* of $\mathbb{F}_q$. If q is an odd prime, then for $c \in \mathbb{F}_q^*$ we have $\eta(c) = \left(\dfrac{c}{q}\right)$, the Legendre symbol from elementary number theory.   □

The orthogonality relations (5.3) and (5.4), when applied to additive or multiplicative characters of $\mathbb{F}_q$, yield several fundamental identities. We consider first the case of additive characters, in which we use the notation from Theorem 5.7. Then, for additive characters χ_a and χ_b we have

$$\sum_{c \in \mathbf{F}_q} \chi_a(c)\overline{\chi_b(c)} = \begin{cases} 0 & \text{for } a \neq b, \\ q & \text{for } a = b. \end{cases} \tag{5.8}$$

In particular,

$$\sum_{c \in \mathbf{F}_q} \chi_a(c) = 0 \quad \text{for } a \neq 0. \tag{5.9}$$

Furthermore, for elements $c, d \in \mathbf{F}_q$ we obtain

$$\sum_{b \in \mathbf{F}_q} \chi_b(c)\overline{\chi_b(d)} = \begin{cases} 0 & \text{for } c \neq d, \\ q & \text{for } c = d. \end{cases} \tag{5.10}$$

For multiplicative characters ψ and τ of $\mathbf{F}_q$ we have

$$\sum_{c \in \mathbf{F}_q^*} \psi(c)\overline{\tau(c)} = \begin{cases} 0 & \text{for } \psi \neq \tau, \\ q-1 & \text{for } \psi = \tau. \end{cases} \tag{5.11}$$

In particular,

$$\sum_{c \in \mathbf{F}_q^*} \psi(c) = 0 \text{ for } \psi \neq \psi_0. \tag{5.12}$$

If $c, d \in \mathbf{F}_q^*$, then

$$\sum_{\psi} \psi(c)\overline{\psi(d)} = \begin{cases} 0 & \text{for } c \neq d, \\ q-1 & \text{for } c = d, \end{cases} \tag{5.13}$$

where the sum is extended over all multiplicative characters ψ of $\mathbf{F}_q$.

2. GAUSSIAN SUMS

Let ψ be a multiplicative and χ an additive character of $\mathbf{F}_q$. Then the *Gaussian sum* $G(\psi, \chi)$ is defined by

$$G(\psi, \chi) = \sum_{c \in \mathbf{F}_q^*} \psi(c)\chi(c).$$

The absolute value of $G(\psi, \chi)$ can obviously be at most $q - 1$, but is in general much smaller, as the following theorem shows. We recall that ψ_0 denotes the trivial multiplicative character and χ_0 the trivial additive character of $\mathbf{F}_q$.

    5.11.   Theorem.   *Let ψ be a multiplicative and χ an additive character of $\mathbf{F}_q$. Then the Gaussian sum $G(\psi, \chi)$ satisfies*

$$G(\psi, \chi) = \begin{cases} q-1 & \text{for } \psi = \psi_0, \chi = \chi_0, \\ -1 & \text{for } \psi = \psi_0, \chi \neq \chi_0, \\ 0 & \text{for } \psi \neq \psi_0, \chi = \chi_0. \end{cases} \tag{5.14}$$

If $\psi \neq \psi_0$ and $\chi \neq \chi_0$, then

$$|G(\psi, \chi)| = q^{1/2}. \qquad (5.15)$$

Proof. The first case in (5.14) is trivial, the third case follows from (5.12), and in the second case we have

$$G(\psi_0, \chi) = \sum_{c \in \mathbb{F}_q^*} \chi(c) = \sum_{c \in \mathbb{F}_q} \chi(c) - \chi(0) = -1$$

by (5.9). For $\psi \neq \psi_0$ and $\chi \neq \chi_0$ we get

$$|G(\psi, \chi)|^2 = \overline{G(\psi, \chi)}\, G(\psi, \chi)$$

$$= \sum_{c \in \mathbb{F}_q^*} \sum_{c_1 \in \mathbb{F}_q^*} \overline{\psi(c)}\ \overline{\chi(c)}\, \psi(c_1)\chi(c_1)$$

$$= \sum_{c \in \mathbb{F}_q^*} \sum_{c_1 \in \mathbb{F}_q^*} \psi(c^{-1}c_1)\chi(c_1 - c).$$

In the inner sum we substitute $c^{-1}c_1 = d$. Then,

$$|G(\psi, \chi)|^2 = \sum_{c \in \mathbb{F}_q^*} \sum_{d \in \mathbb{F}_q^*} \psi(d)\chi(c(d-1))$$

$$= \sum_{d \in \mathbb{F}_q^*} \psi(d)\left(\sum_{c \in \mathbb{F}_q} \chi(c(d-1)) - \chi(0) \right)$$

$$= \sum_{d \in \mathbb{F}_q^*} \psi(d) \sum_{c \in \mathbb{F}_q} \chi(c(d-1))$$

by (5.12). The inner sum has the value q if $d = 1$ and the value 0 if $d \neq 1$, according to (5.9). Therefore, $|G(\psi, \chi)|^2 = \psi(1)q = q$, and (5.15) is established. $\qquad \square$

The study of the behavior of Gaussian sums under various transformations of the additive or multiplicative character leads to a number of useful identities.

5.12. Theorem. *Gaussian sums for the finite field $\mathbb{F}_q$ satisfy the following properties*:

   (i) $G(\psi, \chi_{ab}) = \overline{\psi(a)}\, G(\psi, \chi_b)$ *for* $a \in \mathbb{F}_q^*, b \in \mathbb{F}_q$;

   (ii) $G(\psi, \bar{\chi}) = \psi(-1)\overline{G(\psi, \chi)}$;

  (iii) $G(\bar{\psi}, \chi) = \psi(-1)\overline{G(\psi, \chi)}$;

  (iv) $G(\psi, \chi)G(\bar{\psi}, \chi) = \psi(-1)q$ *for* $\psi \neq \psi_0, \chi \neq \chi_0$;

   (v) $G(\psi^p, \chi_b) = G(\psi, \chi_{\sigma(b)})$ *for* $b \in \mathbb{F}_q$, *where p is the characteristic of $\mathbb{F}_q$ and $\sigma(b) = b^p$.*

Proof. (i) For $c \in \mathbb{F}_q$ we have $\chi_{ab}(c) = \chi_1(abc) = \chi_b(ac)$ by the definition in Theorem 5.7. Therefore,

$$G(\psi, \chi_{ab}) = \sum_{c \in \mathbb{F}_q^*} \psi(c)\chi_{ab}(c) = \sum_{c \in \mathbb{F}_q^*} \psi(c)\chi_b(ac).$$

Now set $ac = d$. Then

$$G(\psi, \chi_{ab}) = \sum_{d \in \mathbb{F}_q^*} \psi(a^{-1}d)\chi_b(d)$$

$$= \psi(a^{-1}) \sum_{d \in \mathbb{F}_q^*} \psi(d)\chi_b(d)$$

$$= \overline{\psi(a)} \; G(\psi, \chi_b).$$

(ii) We have $\chi = \chi_b$ for a suitable $b \in \mathbb{F}_q$ and $\bar{\chi}(c) = \chi_b(-c) = \chi_{-b}(c)$ for $c \in \mathbb{F}_q$. Therefore, by using (i) with $a = -1$ and noting that $\psi(-1) = \pm 1$, we get

$$G(\psi, \bar{\chi}) = G(\psi, \chi_{-b}) = \overline{\psi(-1)} \; G(\psi, \chi_b) = \psi(-1)G(\psi, \chi).$$

(iii) It follows from (ii) that $G(\bar{\psi}, \chi) = \bar{\psi}(-1)G(\bar{\psi}, \bar{\chi}) = \psi(-1)G(\psi, \chi)$.

(iv) By combining (iii) and (5.15), we obtain $G(\psi, \chi)G(\bar{\psi}, \chi) = \psi(-1)G(\psi, \chi)\overline{G(\psi, \chi)} = \psi(-1)|G(\psi, \chi)|^2 = \psi(-1)q.$

(v) Since $\mathrm{Tr}(a) = \mathrm{Tr}(a^p)$ for $a \in \mathbb{F}_q$ by Theorem 2.23(v), we have $\chi_1(a) = \chi_1(a^p)$ according to (5.6). Thus, for $c \in \mathbb{F}_q$ we get $\chi_b(c) = \chi_1(bc) = \chi_1(b^p c^p) = \chi_{\sigma(b)}(c^p)$, and so

$$G(\psi^p, \chi_b) = \sum_{c \in \mathbb{F}_q^*} \psi^p(c)\chi_b(c) = \sum_{c \in \mathbb{F}_q^*} \psi(c^p)\chi_{\sigma(b)}(c^p).$$

But c^p runs through $\mathbb{F}_q^*$ as c runs through $\mathbb{F}_q^*$, and the desired result follows.
$\square$

5.13. Remark. In connection with the properties above, the value $\psi(-1)$ is of interest. We obviously have $\psi(-1) = \pm 1$. Let m be the *order* of ψ; that is, m is the least positive integer such that $\psi^m = \psi_0$. Then m divides $q-1$ since $\psi^{q-1} = \psi_0$. The values of ψ are mth roots of unity; in particular, -1 can only appear as a value of ψ if m is even. If g is a primitive element of $\mathbb{F}_q$, then $\psi(g) = \zeta$, a primitive mth root of unity. If m is even (and so q odd), then $\psi(-1) = \psi(g^{(q-1)/2}) = \zeta^{(q-1)/2}$, which is -1 precisely if $(q-1)/2 \equiv m/2 \bmod m$, or, equivalently, $(q-1)/m \equiv 1 \bmod 2$. Therefore, $\psi(-1) = -1$ if and only if m is even and $(q-1)/m$ is odd. In all other cases we have $\psi(-1) = 1$.
$\square$

Gaussian sums occur in a variety of contexts, for example in the following. Let ψ be a multiplicative character of $\mathbb{F}_q$; then, using (5.10), we may write

$$\psi(c) = \frac{1}{q} \sum_{d \in \mathbb{F}_q^*} \psi(d) \sum_{b \in \mathbb{F}_q} \chi_b(c)\overline{\chi_b(d)}$$

$$= \frac{1}{q} \sum_{b \in F_q} \chi_b(c) \sum_{d \in F_q^*} \psi(d) \bar{\chi}_b(d)$$

for any $c \in F_q^*$. Therefore,

$$\psi(c) = \frac{1}{q} \sum_{\chi} G(\psi, \bar{\chi}) \chi(c) \quad \text{for } c \in F_q^*, \tag{5.16}$$

where the sum is extended over all additive characters χ of F_q. This may be thought of as the Fourier expansion of ψ in terms of the additive characters of F_q, with Gaussian sums appearing as Fourier coefficients.

Similarly, if χ is an additive character of F_q, then, using (5.13), we may write

$$\chi(c) = \frac{1}{q-1} \sum_{d \in F_q^*} \chi(d) \sum_{\psi} \psi(c) \overline{\psi(d)}$$

$$= \frac{1}{q-1} \sum_{\psi} \psi(c) \sum_{d \in F_q^*} \bar{\psi}(d) \chi(d) \quad \text{for } c \in F_q^*.$$

Thus we obtain

$$\chi(c) = \frac{1}{q-1} \sum_{\psi} G(\bar{\psi}, \chi) \psi(c) \quad \text{for } c \in F_q^*, \tag{5.17}$$

where the sum is extended over all multiplicative characters ψ of F_q. This can be interpreted as the Fourier expansion of the restriction of χ to F_q^* in terms of the multiplicative characters of F_q, again with Gaussian sums as Fourier coefficients. Therefore, Gaussian sums are instrumental in the transition from the additive to the multiplicative structure (or vice versa) of a finite field.

Before we establish further properties of Gaussian sums, we develop a useful general principle. Let Φ be the set of monic polynomials over F_q, and let λ be a complex-valued function on Φ which is multiplicative in the sense that

$$\lambda(gh) = \lambda(g)\lambda(h) \quad \text{for all } g, h \in \Phi, \tag{5.18}$$

and which satisfies $|\lambda(g)| \leq 1$ for all $g \in \Phi$ and $\lambda(1) = 1$. With Φ_k denoting the subset of Φ containing the polynomials of degree k, consider the power series

$$L(z) = \sum_{k=0}^{\infty} \left(\sum_{g \in \Phi_k} \lambda(g) \right) z^k \tag{5.19}$$

Since there are q^k polynomials in Φ_k, the coefficient of z^k is in absolute value $\leq q^k$, and so the power series converges absolutely for $|z| < q^{-1}$. Because of (5.18) and unique factorization in $F_q[x]$, we may write

$$L(z) = \sum_{g \in \Phi} \lambda(g) z^{\deg(g)}$$

$$= \prod_f \left(1 + \lambda(f) z^{\deg(f)} + \lambda(f^2) z^{\deg(f^2)} + \cdots\right)$$

$$= \prod_f \left(1 + \lambda(f) z^{\deg(f)} + \lambda(f)^2 z^{2\deg(f)} + \cdots\right),$$

where the product is taken over all monic irreducible polynomials f in $\mathbb{F}_q[x]$. It follows that

$$L(z) = \prod_f \left(1 - \lambda(f) z^{\deg(f)}\right)^{-1}$$

Now apply logarithmic differentiation and multiply the result by z to get

$$z \frac{d \log L(z)}{dz} = \sum_f \frac{\lambda(f) \deg(f) z^{\deg(f)}}{1 - \lambda(f) z^{\deg(f)}}.$$

Expansion of $(1 - \lambda(f) z^{\deg(f)})^{-1}$ into a geometric series leads to

$$z \frac{d \log L(z)}{dz} = \sum_f \lambda(f) \deg(f) z^{\deg(f)}$$

$$\cdot \left(1 + \lambda(f) z^{\deg(f)} + \lambda(f)^2 z^{2\deg(f)} + \cdots\right)$$

$$= \sum_f \deg(f) \left(\lambda(f) z^{\deg(f)} + \lambda(f)^2 z^{2\deg(f)}\right.$$

$$\left. + \lambda(f)^3 z^{3\deg(f)} + \cdots\right),$$

and collecting equal powers of z we obtain

$$z \frac{d \log L(z)}{dz} = \sum_{s=1}^{\infty} L_s z^s \tag{5.20}$$

with

$$L_s = \sum_f \deg(f) \lambda(f)^{s/\deg(f)}, \tag{5.21}$$

where the sum is extended over all monic irreducible polynomials f in $\mathbb{F}_q[x]$ with $\deg(f)$ dividing s.

Now suppose there exists a positive integer t such that

$$\sum_{g \in \Phi_k} \lambda(g) = 0 \quad \text{for all } k > t. \tag{5.22}$$

Then $L(z)$ is a complex polynomial of degree $\leqslant t$ with constant term 1, so that we can write

$$L(z) = (1 - \omega_1 z)(1 - \omega_2 z) \cdots (1 - \omega_t z) \tag{5.23}$$

with complex numbers $\omega_1, \omega_2, \ldots, \omega_t$. It follows that

$$
z\frac{d\log L(z)}{dz} = -\sum_{m=1}^{t} \frac{\omega_m z}{1 - \omega_m z}
$$

$$
= -\sum_{m=1}^{t} \omega_m z \sum_{j=0}^{\infty} \omega_m^j z^j
$$

$$
= -\sum_{j=0}^{\infty} \left(\sum_{m=1}^{t} \omega_m^{j+1} \right) z^{j+1} = -\sum_{s=1}^{\infty} \left(\sum_{m=1}^{t} \omega_m^s \right) z^s,
$$

and comparison with (5.20) yields

$$
L_s = -\omega_1^s - \omega_2^s - \cdots - \omega_t^s \quad \text{for all } s \geqslant 1. \tag{5.24}
$$

As an application of the principle expressed in (5.24), we consider the following situation. Let χ be an additive and ψ a multiplicative character of $\mathbb{F}_q$, and let E be a finite extension field of $\mathbb{F}_q$. Then χ and ψ can be "lifted" to E by setting $\chi'(\beta) = \chi(\mathrm{Tr}_{E/\mathbb{F}_q}(\beta))$ for $\beta \in E$ and $\psi'(\beta) = \psi(\mathrm{N}_{E/\mathbb{F}_q}(\beta))$ for $\beta \in E^*$. From the additivity of the trace and the multiplicativity of the norm it follows that χ' is an additive and ψ' a multiplicative character of E. The following theorem establishes an important relationship between the Gaussian sum $G(\psi, \chi)$ in $\mathbb{F}_q$ and the Gaussian sum $G(\psi', \chi')$ in E.

5.14. Theorem (Davenport-Hasse Theorem). *Let χ be an additive and ψ a multiplicative character of $\mathbb{F}_q$, not both of them trivial. Suppose χ and ψ are lifted to characters χ' and ψ', respectively, of the finite extension field E of $\mathbb{F}_q$ with $[E:\mathbb{F}_q] = s$. Then*

$$
G(\psi', \chi') = (-1)^{s-1} G(\psi, \chi)^s
$$

Proof. It is convenient to extend the definition of ψ by setting $\psi(0) = 0$. We use the notation of the discussion leading to (5.24); in particular, Φ denotes again the set of monic polynomials over $\mathbb{F}_q$. We define λ by setting $\lambda(1) = 1$ as required, and for $g \in \Phi$ of positive degree, say $g(x) = x^k - c_1 x^{k-1} + \cdots + (-1)^k c_k$, we set $\lambda(g) = \psi(c_k)\chi(c_1)$. The multiplicative property (5.18) is then easily checked. For $k > 1$ we split up Φ_k according to the values of c_1 and c_k. Each given pair (c_1, c_k) occurs q^{k-2} times in Φ_k, and so

$$
\sum_{g \in \Phi_k} \lambda(g) = q^{k-2} \sum_{c_1, c_k \in \mathbb{F}_q} \psi(c_k)\chi(c_1)
$$

$$
= q^{k-2} \left(\sum_{c \in \mathbb{F}_q^*} \psi(c) \right) \left(\sum_{c \in \mathbb{F}_q} \chi(c) \right).
$$

Since one of χ and ψ is nontrivial, it follows from either (5.9) or (5.12) that

$$\sum_{g \in \Phi_k} \lambda(g) = 0 \text{ for } k > 1.$$

Therefore, (5.22) is satisfied with $t = 1$. Furthermore, Φ_1 comprises the linear polynomials $x - c$ with $c \in \mathbb{F}_q$, and so

$$\sum_{g \in \Phi_1} \lambda(g) = \sum_{c \in \mathbb{F}_q} \psi(c)\chi(c) = \sum_{c \in \mathbb{F}_q^*} \psi(c)\chi(c) = G(\psi, \chi).$$

Thus, $L(z) = 1 + G(\psi, \chi)z$ from (5.19), hence $\omega_1 = -G(\psi, \chi)$ by (5.23). Now we consider L_s, which, by (5.21) and the multiplicativity of λ, is given by

$$L_s = \sum_f \deg(f)\lambda(f)^{s/\deg(f)}$$

$$= \sum_f {}^* \deg(f)\lambda(f^{s/\deg(f)}),$$

where the sum is extended over all monic irreducible polynomials f in $\mathbb{F}_q[x]$ with $\deg(f)$ dividing s, and where the asterisk indicates that $f(x) = x$ is excluded. Each such f has $\deg(f)$ distinct nonzero roots in E, and each root β of f has as its characteristic polynomial over $\mathbb{F}_q$ the polynomial

$$f(x)^{s/\deg(f)} = x^s - c_1 x^{s-1} + \cdots + (-1)^s c_s,$$

say, where $c_1 = \text{Tr}_{E/\mathbb{F}_q}(\beta)$ and $c_s = \text{N}_{E/\mathbb{F}_q}(\beta)$ by (2.2) and (2.3). Therefore,

$$\lambda(f^{s/\deg(f)}) = \psi(c_s)\chi(c_1) = \psi\big(\text{N}_{E/\mathbb{F}_q}(\beta)\big)\chi\big(\text{Tr}_{E/\mathbb{F}_q}(\beta)\big)$$

$$= \psi'(\beta)\chi'(\beta),$$

and so

$$L_s = \sum_f {}^* \deg(f)\lambda(f^{s/\deg(f)}) = \sum_f {}^* \sum_{\substack{\beta \in E \\ f(\beta) = 0}} \psi'(\beta)\chi'(\beta).$$

If f runs through the range of summation above, then β runs exactly through all elements of E^*. Consequently,

$$L_s = \sum_{\beta \in E^*} \psi'(\beta)\chi'(\beta) = G(\psi', \chi'),$$

and an application of (5.24) yields

$$G(\psi', \chi') = -(-G(\psi, \chi))^s,$$

which completes the proof. $\square$

For certain special characters, the associated Gaussian sums can be evaluated explicitly. We thereby obtain formulas that go beyond the trivial cases listed in (5.14). A celebrated formula of this kind holds for the quadratic character η considered in Example 5.10.

5.15. Theorem. *Let $\mathbb{F}_q$ be a finite field with $q = p^s$, where p is an odd prime and $s \in \mathbb{N}$. Let η be the quadratic character of $\mathbb{F}_q$ and let χ_1 be the canonical additive character of $\mathbb{F}_q$. Then*

$$G(\eta, \chi_1) = \begin{cases} (-1)^{s-1}q^{1/2} & \text{if } p \equiv 1 \bmod 4, \\ (-1)^{s-1}i^s q^{1/2} & \text{if } p \equiv 3 \bmod 4. \end{cases}$$

Proof. Using Theorem 5.12(iv) and $\bar{\eta} = \eta$, we obtain $G(\eta, \chi_1)^2 = \eta(-1)q$, and since $\eta(-1) = 1$ for $q \equiv 1 \bmod 4$ and $\eta(-1) = -1$ for $q \equiv 3 \bmod 4$ by Remark 5.13, it follows that

$$G(\eta, \chi_1) = \begin{cases} \pm q^{1/2} & \text{if } q \equiv 1 \bmod 4, \\ \pm i q^{1/2} & \text{if } q \equiv 3 \bmod 4. \end{cases} \tag{5.25}$$

The difficulty of the proof lies in the determination of the correct signs.

We first consider the case $s = 1$. Let V be the set of all complex-valued functions on $\mathbb{F}_p^*$; it is a $(p-1)$-dimensional vector space over the complex numbers. A basis for V is formed by the characteristic functions $f_1, f_2, \ldots, f_{p-1}$ of elements of $\mathbb{F}_p^*$; that is, $f_j(c) = 1$ if $c = j$ and 0 otherwise, where $j = 1, 2, \ldots, p-1$. From the orthogonality relation (5.11) it follows easily that the multiplicative characters $\psi_0, \psi_1, \ldots, \psi_{p-2}$ of $\mathbb{F}_p$ described in Theorem 5.8 also form a basis for V. Let $\zeta = e^{2\pi i/p}$, and define a linear operator T on V by letting Th for $h \in V$ be given by

$$(Th)(c) = \sum_{k=1}^{p-1} \zeta^{ck} h(k) \quad \text{for } c = 1, 2, \ldots, p-1. \tag{5.26}$$

Then Theorem 5.12(i) implies that $T\psi = G(\psi, \chi_1)\bar{\psi}$ for every multiplicative character ψ of $\mathbb{F}_p$. Since $\psi = \bar{\psi}$ precisely for the trivial character and the quadratic character, the matrix T in the basis $\psi_0, \psi_1, \ldots, \psi_{p-2}$ contains two diagonal entries—namely, $G(\psi_0, \chi_1) = -1$ and $G(\eta, \chi_1)$—and a collection of blocks

$$\begin{pmatrix} 0 & G(\bar{\psi}, \chi_1) \\ G(\psi, \chi_1) & 0 \end{pmatrix}$$

corresponding to pairs $\psi, \bar{\psi}$ of conjugate characters that are nontrivial and nonquadratic. If we compute the determinant of T, then each block contributes

$$-G(\psi, \chi_1)G(\bar{\psi}, \chi_1) = -\psi(-1)p$$

by Theorem 5.12(iv). Thus we obtain

$$\det(T) = -G(\eta, \chi_1)(-p)^{(p-3)/2} \prod_{j=1}^{(p-3)/2} \psi_j(-1). \qquad (5.27)$$

Now $\psi_j(-1) = \psi_j'(-1) = (-1)^j$, and so

$$\prod_{j=1}^{(p-3)/2} \psi_j(-1) = (-1)^{1+2+\cdots+(p-3)/2} = (-1)^{(p-1)(p-3)/8} \qquad (5.28)$$

Furthermore, since

$$i^{(p-1)^2/4} = \begin{cases} 1 & \text{if } p \equiv 1 \bmod 4, \\ i & \text{if } p \equiv 3 \bmod 4, \end{cases}$$

it follows from (5.25) that

$$G(\eta, \chi_1) = \pm i^{(p-1)^2/4} p^{1/2}. \qquad (5.29)$$

Combining (5.27), (5.28), and (5.29), we get

$$\det(T) = \pm (-1)^{(p-1)/2} i^{(p-1)^2/4} (-1)^{(p-1)(p-3)/8} p^{(p-2)/2}$$

$$= \pm (-1)^{(p-1)/2} i^{(p-1)^2/4 + (p-1)(p-3)/4} p^{(p-2)/2},$$

hence

$$\det(T) = \pm (-1)^{(p-1)/2} i^{(p-1)(p-2)/2} p^{(p-2)/2}. \qquad (5.30)$$

Now we compute $\det(T)$ utilizing the matrix of T in the basis $f_1, f_2, \ldots, f_{p-1}$. From (5.26) we find

$$\det(T) = \det\left((\zeta^{jk})_{1 \le j, k \le p-1}\right) = \det\left((\zeta^j \zeta^{j(k-1)})_{1 \le j, k \le p-1}\right)$$

$$= \zeta^{1+2+\cdots+(p-1)} \det\left((\zeta^{j(k-1)})_{1 \le j, k \le p-1}\right)$$

$$= \det\left((\zeta^{j(k-1)})_{1 \le j, k \le p-1}\right),$$

which is a Vandermonde determinant. Therefore,

$$\det(T) = \prod_{1 \le m < n \le p-1} (\zeta^n - \zeta^m).$$

With $\delta = e^{\pi i/p}$ we get

$$\det(T) = \prod_{1 \le m < n \le p-1} (\delta^{2n} - \delta^{2m})$$

$$= \prod_{1 \leqslant m < n \leqslant p-1} \delta^{n+m} \left(\delta^{n-m} - \delta^{-(n-m)} \right)$$

$$= \prod_{1 \leqslant m < n \leqslant p-1} \delta^{n+m} \prod_{1 \leqslant m < n \leqslant p-1} \left(2i \sin \frac{\pi(n-m)}{p} \right).$$

Since

$$\sum_{1 \leqslant m < n \leqslant p-1} (n+m) = \sum_{n=2}^{p-1} \sum_{m=1}^{n-1} (n+m)$$

$$= \frac{3}{2} \sum_{n=2}^{p-1} n(n-1) = \frac{3}{2} \sum_{n=1}^{p-2} (n^2+n)$$

$$= \frac{3}{2} \left(\frac{(p-2)(p-1)(2p-3)}{6} + \frac{(p-2)(p-1)}{2} \right)$$

$$= \frac{p(p-1)(p-2)}{2},$$

the first product is equal to

$$\delta^{p(p-1)(p-2)/2} = (-1)^{(p-1)(p-2)/2} = \left((-1)^{p-2} \right)^{(p-1)/2} = (-1)^{(p-1)/2}.$$

Furthermore,

$$A = \prod_{1 \leqslant m < n \leqslant p-1} \left(2 \sin \frac{\pi(n-m)}{p} \right) > 0,$$

and so

$$\det(T) = (-1)^{(p-1)/2} i^{(p-1)(p-2)/2} A \quad \text{with } A > 0.$$

Comparison with (5.30) shows that the plus sign always applies in (5.29), and the theorem is established for $s = 1$.

The general case follows from Theorem 5.14 since the canonical additive character of $\mathbb{F}_p$ is lifted to the canonical additive character of $\mathbb{F}_q$ by (5.7) and the quadratic character of $\mathbb{F}_p$ is lifted to the quadratic character of $\mathbb{F}_q$. $\qquad\square$

Because of (5.14) and Theorem 5.12(i), a formula for $G(\eta, \chi)$ can also be established for any additive character χ of $\mathbb{F}_q$.

We turn to another special formula for Gaussian sums which applies to a wider range of multiplicative characters but needs a restriction on the underlying field. We shall have to use the notion of order of a multiplicative character as introduced in Remark 5.13.

5.16. Theorem (Stickelberger's Theorem). *Let q be a prime power, let ψ be a nontrivial multiplicative character of $\mathbb{F}_{q^2}$ of order m dividing $q+1$, and let χ_1 be the canonical additive character of $\mathbb{F}_{q^2}$. Then,*

$$G(\psi, \chi_1) = \begin{cases} q & \text{if } m \text{ odd or } \dfrac{q+1}{m} \text{ even,} \\ -q & \text{if } m \text{ even and } \dfrac{q+1}{m} \text{ odd.} \end{cases}$$

Proof. We write $E = \mathbb{F}_{q^2}$ and $F = \mathbb{F}_q$. Let γ be a primitive element of E and set $g = \gamma^{q+1}$. Then $g^{q-1} = 1$, so that $g \in F$; furthermore, g is a primitive element of F. Every $\alpha \in E^*$ can be written in the form $\alpha = g^j \gamma^k$ with $0 \leqslant j < q-1$ and $0 \leqslant k < q+1$. Since $\psi(g) = \psi^{q+1}(\gamma) = 1$, we have

$$G(\psi, \chi_1) = \sum_{j=0}^{q-2} \sum_{k=0}^{q} \psi(g^j \gamma^k) \chi_1(g^j \gamma^k)$$

$$= \sum_{k=0}^{q} \psi^k(\gamma) \sum_{j=0}^{q-2} \chi_1(g^j \gamma^k)$$

$$= \sum_{k=0}^{q} \psi^k(\gamma) \sum_{b \in F^*} \chi_1(b\gamma^k). \tag{5.31}$$

If τ_1 is the canonical additive character of F, then $\chi_1(b\gamma^k) = \tau_1(\mathrm{Tr}_{E/F}(b\gamma^k))$ by (5.7). Therefore,

$$\sum_{b \in F^*} \chi_1(b\gamma^k) = \sum_{b \in F^*} \tau_1(b\mathrm{Tr}_{E/F}(\gamma^k))$$

$$= \begin{cases} -1 & \text{for } \mathrm{Tr}_{E/F}(\gamma^k) \neq 0, \\ q-1 & \text{for } \mathrm{Tr}_{E/F}(\gamma^k) = 0, \end{cases} \tag{5.32}$$

because of (5.9). Now $\mathrm{Tr}_{E/F}(\gamma^k) = \gamma^k + \gamma^{kq}$, and so

$$\mathrm{Tr}_{E/F}(\gamma^k) = 0 \quad \text{if and only if} \quad \gamma^{k(q-1)} = -1. \tag{5.33}$$

If q is odd, the last condition is equivalent to $k = (q+1)/2$, and then by (5.32),

$$\sum_{b \in F^*} \chi_1(b\gamma^k) = \begin{cases} -1 & \text{for } 0 \leqslant k < q+1, k \neq \dfrac{q+1}{2}, \\ q-1 & \text{for } k = \dfrac{q+1}{2}. \end{cases}$$

Together with (5.31) we get

$$G(\psi, \chi_1) = - \sum_{\substack{k=0 \\ k \neq (q+1)/2}}^{q} \psi^k(\gamma) + (q-1)\psi^{(q+1)/2}(\gamma)$$

$$= - \sum_{k=0}^{q} \psi^k(\gamma) + q\psi^{(q+1)/2}(\gamma)$$

$$= q\psi^{(q+1)/2}(\gamma)$$

since $\psi(\gamma) \neq 1$ and $\psi^{q+1}(\gamma) = 1$. Now $\psi^{(q+1)/2}(\gamma) = 1$ if $(q+1)/m$ is even and -1 if $(q+1)/m$ is odd, and thus for q odd we have

$$G(\psi, \chi_1) = \begin{cases} q & \text{if } \dfrac{q+1}{m} \text{ even,} \\[2mm] -q & \text{if } \dfrac{q+1}{m} \text{ odd.} \end{cases} \tag{5.34}$$

If q is even, then the condition in (5.33) is equivalent to $\gamma^{k(q-1)} = 1$, and the only k with $0 \leqslant k < q+1$ satisfying this property is $k = 0$. Then by (5.32),

$$\sum_{b \in F^*} \chi_1(b\gamma^k) = \begin{cases} -1 & \text{for } 1 \leqslant k \leqslant q, \\ q-1 & \text{for } k = 0, \end{cases}$$

and (5.31) yields

$$G(\psi, \chi_1) = - \sum_{k=1}^{q} \psi^k(\gamma) + q - 1 = - \sum_{k=0}^{q} \psi^k(\gamma) + q = q.$$

Combined with (5.34), this implies the theorem. □

We show how to use Gaussian sums to establish a classical result of number theory, namely the law of quadratic reciprocity. We recall from Example 5.10 that if p is an odd prime and η is the quadratic character of $\mathbb{F}_p$, then for $c \not\equiv 0 \bmod p$ the Legendre symbol $\left(\dfrac{c}{p}\right)$ is defined by $\left(\dfrac{c}{p}\right) = \eta(c)$.

5.17. Theorem (Law of Quadratic Reciprocity). *For any distinct odd primes p and r we have*

$$\left(\frac{p}{r}\right)\left(\frac{r}{p}\right) = (-1)^{(p-1)(r-1)/4}$$

Proof. Let η be the quadratic character of $\mathbb{F}_p$, let χ_1 be the canonical additive character of $\mathbb{F}_p$, and put $G = G(\eta, \chi_1)$. Then it follows from (5.25) that $G^2 = (-1)^{(p-1)/2}p = \tilde{p}$, and so

$$G^r = (G^2)^{(r-1)/2}G = \tilde{p}^{(r-1)/2}G. \tag{5.35}$$

Let R be the ring of algebraic integers; that is, R consists of all complex numbers that are roots of monic polynomials with integer coefficients. Since the values of (additive and multiplicative) characters of finite fields are complex roots of unity, and since every complex root of unity is an algebraic integer, the values of Gaussian sums are algebraic integers. In particular, $G \in R$. Let (r) be the principal ideal of R generated by r. Then the residue class ring $R/(r)$ has characteristic r, and thus an application of Theorem 1.46 yields

$$G^r = \left(\sum_{c \in \mathbb{F}_p^*} \eta(c) \chi_1(c) \right)^r \equiv \sum_{c \in \mathbb{F}_p^*} \eta^r(c) \chi_1^r(c) \bmod(r).$$

Now

$$\sum_{c \in \mathbb{F}_p^*} \eta^r(c) \chi_1^r(c) = \sum_{c \in \mathbb{F}_p^*} \eta(c) \chi_r(c) = G(\eta, \chi_r) = \eta(r) G$$

by Theorem 5.12(i), and so

$$G^r \equiv \eta(r) G \bmod(r).$$

Together with (5.35) we get

$$\tilde{p}^{(r-1)/2} G \equiv \eta(r) G \bmod(r),$$

and multiplication by G leads to

$$\tilde{p}^{(r-1)/2} \tilde{p} \equiv \eta(r) \tilde{p} \bmod(r)$$

because of $G^2 = \tilde{p}$. Since the numbers on both sides of the congruence above are, in fact, elements of $\mathbb{Z}$, it follows that

$$\tilde{p}^{(r-1)/2} \tilde{p} \equiv \eta(r) \tilde{p} \bmod r$$

as a congruence in $\mathbb{Z}$. But $\tilde{p}$ and r are relatively prime, hence

$$\tilde{p}^{(r-1)/2} \equiv \eta(r) \bmod r.$$

Now $\tilde{p} = (-1)^{(p-1)/2} p$ and $p^{r-1} \equiv 1 \bmod r$, thus multiplication by $p^{(r-1)/2}$ yields

$$(-1)^{(p-1)(r-1)/4} \equiv p^{(r-1)/2} \eta(r) \bmod r. \tag{5.36}$$

We have $p^{(r-1)/2} \equiv \pm 1 \bmod r$, and the plus sign applies if and only if p is congruent to a square mod r. Thus,

$$p^{(r-1)/2} \equiv \left(\frac{p}{r} \right) \bmod r.$$

Since $\eta(r) = \left(\frac{r}{p} \right)$, we get from (5.36)

$$(-1)^{(p-1)(r-1)/4} \equiv \left(\frac{p}{r} \right) \left(\frac{r}{p} \right) \bmod r.$$

But the integers on both sides of this congruence can only be ± 1, and since $r \geqslant 3$, the congruence holds only if the two sides are identical.   □

We consider now character sums involving the quadratic character η of $\mathbb{F}_q$, q odd, and having a quadratic polynomial in the argument. The following explicit formula will be needed in Chapter 7, Section 2.

5.18. Theorem. *Let* $f(x) = a_2 x^2 + a_1 x + a_0 \in \mathbb{F}_q[x]$ *with* q *odd and* $a_2 \neq 0$. *Put* $d = a_1^2 - 4a_0 a_2$ *and let* η *be the quadratic character of* $\mathbb{F}_q$. *Then*

$$\sum_{c \in \mathbf{F}_q} \eta(f(c)) = \begin{cases} -\eta(a_2) & \text{if } d \neq 0, \\ (q-1)\eta(a_2) & \text{if } d = 0. \end{cases}$$

Proof. Multiplying the sum by $\eta(4a_2^2) = 1$, we get

$$\sum_{c \in \mathbf{F}_q} \eta(f(c)) = \eta(a_2) \sum_{c \in \mathbf{F}_q} \eta(4a_2^2 c^2 + 4a_1 a_2 c + 4a_0 a_2)$$

$$= \eta(a_2) \sum_{c \in \mathbf{F}_q} \eta((2a_2 c + a_1)^2 - d) = \eta(a_2) \sum_{b \in \mathbf{F}_q} \eta(b^2 - d).$$

$$(5.37)$$

The result for the case $d = 0$ follows now immediately. For $d \neq 0$ we write

$$\sum_{b \in \mathbf{F}_q} \eta(b^2 - d) = -q + \sum_{b \in \mathbf{F}_q} (1 + \eta(b^2 - d)),$$

and since $1 + \eta(b^2 - d)$ is the number of $c \in \mathbf{F}_q$ with $c^2 = b^2 - d$, we obtain

$$\sum_{b \in \mathbf{F}_q} \eta(b^2 - d) = -q + S(d),\qquad (5.38)$$

where $S(d)$ is the number of ordered pairs (b, c) with $b, c \in \mathbf{F}_q$ and $b^2 - c^2 = d$. To solve this equation, we put $b + c = u$, $b - c = v$ and note that the ordered pairs (b, c) and (u, v) are in one-to-one correspondence since q is odd. Thus $S(d)$ is equal to the number of ordered pairs (u, v) with $u, v \in \mathbf{F}_q$ and $uv = d$, hence $S(d) = q - 1$. Together with (5.37) and (5.38), this implies the desired formula. $\qquad\square$

EXERCISES

5.1. Let G be a finite abelian group, H a proper subgroup of G, and $g \in G$, $g \notin H$. Prove that there exists a character χ of G that annihilates H, but for which $\chi(g) \neq 1$.

5.2. Let H be a subgroup of the finite abelian group G. Prove that the annihilator A of H in $G^\wedge$ is isomorphic to G/H and that $G^\wedge/A$ is isomorphic to H.

5.3. Let G be a finite abelian group and $m \in \mathbb{N}$. Prove that $g \in G$ is an mth power of an element of G if and only if $\chi(g) = 1$ for all characters χ of G for which χ^m is trivial.

5.4. Let $G_1, \ldots, G_k$ be finite abelian groups. Define multiplication of k-tuples $(g_1, \ldots, g_k)$, $(h_1, \ldots, h_k)$ with $g_i, h_i \in G_i$ for $1 \leqslant i \leqslant k$ by

$$(g_1, \ldots, g_k)(h_1, \ldots, h_k) = (g_1 h_1, \ldots, g_k h_k).$$

Show that with this operation the set of all such k-tuples forms again a finite abelian group, the so-called *direct product* $G_1 \otimes \cdots \otimes G_k$.

Then prove that $(G_1 \otimes \cdots \otimes G_k)^\wedge$ is isomorphic to $G_1^\wedge \otimes \cdots \otimes G_k^\wedge$.

5.5. Use the structure theorem for finite abelian groups, which says in its simplest form that every such group is isomorphic to a direct product of finite cyclic groups, to prove that $G^\wedge$ is isomorphic to G whenever G is a finite abelian group.

5.6. For additive characters of $\mathbb{F}_q$ in the notation of Theorem 5.7, show that $\chi_a \chi_b = \chi_{a+b}$ for all $a, b \in \mathbb{F}_q$. Thus prove without reference to Exercise 5.5 that the group of additive characters of $\mathbb{F}_q$ is isomorphic to the additive group of $\mathbb{F}_q$.

5.7. If χ_1 is the canonical additive character of the finite field $\mathbb{F}_q$ of characteristic p, prove that $\chi_1(c^{p^j}) = \chi_1(c)$ for all $c \in \mathbb{F}_q$ and $j \in \mathbb{N}$.

5.8. If ψ is a multiplicative character of $\mathbb{F}_{q^s}$ of order m, prove that the restriction of ψ to $\mathbb{F}_q$ is a multiplicative character of order $m/\gcd(m, (q^s - 1)/(q - 1))$.

5.9. With the notation of Exercise 5.8, prove that the restriction of ψ to $\mathbb{F}_q$ is the trivial character if and only if m divides $(q^s - 1)/(q - 1)$.

5.10. Let ψ be a multiplicative character of $\mathbb{F}_q$ and let ψ' be the lifted character of the extension field $\mathbb{F}_{q^s}$. Prove that $\psi'(c) = \psi^s(c)$ for $c \in \mathbb{F}_q^*$.

5.11. Prove that a multiplicative character τ of $\mathbb{F}_{q^s}$ is equal to a character ψ' lifted from $\mathbb{F}_q$ if and only if τ^{q-1} is trivial.

5.12. If $q \equiv 1 \bmod m$ and ψ varies over all multiplicative characters of $\mathbb{F}_q$ of order dividing m, prove that the lifted character ψ' of $\mathbb{F}_{q^s}$ varies over all multiplicative characters of $\mathbb{F}_{q^s}$ of order dividing m.

5.13. Prove that an additive character χ of the finite extension field E of $\mathbb{F}_q$ is equal to a character lifted from $\mathbb{F}_q$ if and only if $\chi = \mu_b$ with $b \in \mathbb{F}_q$, where μ_1 is the canonical additive character of E.

5.14. Prove for $c \in \mathbb{F}_q^*$ that

$$\sum_{d \mid (q-1)} \frac{\mu(d)}{\phi(d)} \sum_{\psi^{(d)}} \psi^{(d)}(c)$$

$$= \begin{cases} \dfrac{q-1}{\phi(q-1)} & \text{if } c \text{ is a primitive element of } \mathbb{F}_q, \\ 0 & \text{otherwise,} \end{cases}$$

where in the outer sum d runs through all positive divisors of $q - 1$ and in the inner sum $\psi^{(d)}$ runs through the $\phi(d)$ multiplicative characters of $\mathbb{F}_q$ of order d. Here μ denotes the Moebius function (see Definition 3.22) and ϕ Euler's function (see Theorem 1.15 (iv)).

5.15. Show that $\eta(2) = (-1)^{(q^2 - 1)/8}$, where η is the quadratic character of $\mathbb{F}_q$, q odd.

5.16. For $r \in \mathbb{N}$ prove $G(\psi^{p^r}, \chi_b) = G(\psi, \chi_{\rho(b)})$, where $\rho(b) = b^{p^r}$ for $b \in \mathbb{F}_q$ and p is the characteristic of $\mathbb{F}_q$.

5.17. Prove $\sum_{\chi} G(\psi, \chi) = 0$ for all multiplicative characters ψ of $\mathbb{F}_q$, where the sum is extended over all additive characters χ of $\mathbb{F}_q$.

5.18. Prove $\sum_{\psi} G(\psi, \chi) = (q-1)\chi(1)$ for all additive characters χ of $\mathbb{F}_q$, where the sum is extended over all multiplicative characters ψ of $\mathbb{F}_q$.

5.19. For the quadratic character η of $\mathbb{F}_q$, $q = p^s$, p an odd prime, $s \in \mathbb{N}$, and an additive character χ_b, $b \in \mathbb{F}_q$, in the notation of Theorem 5.7, prove that

$$G(\eta, \chi_b) = \eta(b)(-1)^{(q+1)/2} i^{s(p^2 + 2p + 5)/4} q^{1/2}.$$

5.20. If q is odd and η is the quadratic character of $\mathbb{F}_q$, prove that $G(\eta, \chi_a) G(\eta, \chi_b) = \eta(-ab)q$ for $a, b \in \mathbb{F}_q^*$.

5.21. Use the law of quadratic reciprocity to evaluate the Legendre symbols $(\frac{13}{59})$ and $(\frac{7}{61})$.

5.22. Determine all primes p such that $\left(\dfrac{-3}{p}\right) = 1$.

5.23. Determine all odd prime powers q such that the quadratic character η of $\mathbb{F}_q$ satisfies $\eta(3) = 1$.

5.24. Prove that the polynomial $x^2 + ax + b \in \mathbb{F}_q[x]$, q odd, is irreducible in $\mathbb{F}_q[x]$ if and only if $\eta(a^2 - 4b) = -1$.

5.25. Determine whether the polynomial $x^2 + 12x + 41$ is irreducible in $\mathbb{F}_{227}[x]$.

5.26. Let p and r be distinct odd primes, let $s \in \mathbb{N}$ be such that $r^s \equiv 1 \bmod p$, and let ζ be an element of order p in $\mathbb{F}_{r^s}^*$. For $k \in \mathbb{Z}$ define

$$G_k = \sum_{c=1}^{p-1} \left(\frac{c}{p}\right) \zeta^{kc} \in \mathbb{F}_{r^s}.$$

Prove the following properties: (i) $G_k = \left(\dfrac{k}{p}\right) G_1$; (ii) $G_1^2 = (-1)^{(p-1)/2} p$, where the last expression is viewed as an element of $\mathbb{F}_r$.

5.27. Use the results of Exercise 5.26 to prove the law of quadratic reciprocity.

5.28. Prove that

$$\sum_{c \in \mathbb{F}_q} \psi(c+a)\bar{\psi}(c+b) = -1$$

for $a, b \in \mathbb{F}_q$ with $a \neq b$, where ψ is a nontrivial multiplicative character of $\mathbb{F}_q$.

5.29. Let ψ be a nontrivial multiplicative character of $\mathbb{F}_q$ and let S be a subset of $\mathbb{F}_q$ with h elements. Prove that

$$\sum_{c \in \mathbb{F}_q} \left| \sum_{a \in S} \psi(c+a) \right|^2 = h(q-h).$$

5.30. Let $\lambda_1, \lambda_2, \lambda_3$ be nontrivial multiplicative characters of $\mathbb{F}_q$ and let

$a_1, a_2 \in \mathbb{F}_q$ with $a_1 \neq a_2$. Prove that

$$\sum_{b \in \mathbb{F}_q} \left| \sum_{c \in \mathbb{F}_q} \lambda_1(c + a_1)\lambda_2(c + a_2)\lambda_3(c + b) \right|^2$$

$$= \begin{cases} q^2 - 3q & \text{if } \lambda_1\lambda_2 \text{ nontrivial,} \\ q^2 - 2q - 1 & \text{if } \lambda_1\lambda_2 \text{ trivial.} \end{cases}$$

5.31.   Let ψ be a multiplicative character of $\mathbb{F}_q$ of order $m > 1$. For $a \in \mathbb{F}_q$ prove

$$\sum_{c \in \mathbb{F}_q} \psi(ac^n) = \begin{cases} (q - 1)\psi(a) & \text{if } m \text{ divides } n, \\ 0 & \text{otherwise.} \end{cases}$$

5.32.   Prove that $\sum_{c \in \mathbb{F}_q} \eta(f(c)) = 0$ if $q \equiv 3 \bmod 4$, η is the quadratic character of $\mathbb{F}_q$, and $f \in \mathbb{F}_q[x]$ is an odd polynomial—that is, a polynomial with $f(-x) = -f(x)$.

Chapter 6

Linear Recurring Sequences

Sequences in finite fields whose terms depend in a simple manner on their predecessors are of importance for a variety of applications. Such sequences are easy to generate by recursive procedures, which is certainly an advantageous feature from the computational viewpoint, and they also tend to have useful structural properties. Of particular interest is the case where the terms depend linearly on a fixed number of predecessors, resulting in a so-called linear recurring sequence. These sequences are employed, for instance, in coding theory (see Chapter 8, Section 2), in cryptography (see Chapter 9, Section 2), and in several branches of electrical engineering. In these applications, the underlying field is often taken to be $\mathbb{F}_2$, but the theory can be developed quite generally for any finite field.

In Section 1 we show how to implement the generation of linear recurring sequences on special switching circuits called feedback shift registers. We discuss also some basic periodicity properties of such sequences. Section 2 introduces the concept of an impulse response sequence, which is of both practical and theoretical interest. Further relations to periodicity properties are found in this way, and also through the use of the so-called characteristic polynomial of a linear recurring sequence. Another application of the characteristic polynomial yields explicit formulas for the terms of a linear recurring sequence. Maximal period sequences are also defined in this section. These sequences will appear in various applications in later chapters.

The theory of linear recurring sequences can be approached via linear algebra, ideal theory, or formal power series. An approach based on the latter is presented in Section 3. This leads to a computation-oriented way of introducing the minimal polynomial of a linear recurring sequence in the next section. The minimal polynomial is of crucial importance for the linear recurring sequence, since the order of the minimal polynomial gives the least period of the sequence.

In Section 5 we study the collection of all sequences satisfying a given linear recurrence relation. This information is useful in the discussion of operations with linear recurring sequences, such as termwise addition and multiplication for sequences in general finite fields and binary complementation for sequences in $\mathbb{F}_2$. We consider also the problem of determining the various least periods of the sequences generated by a fixed linear recurrence relation. Section 6 presents some determinantal criteria characterizing linear recurring sequences as well as the Berlekamp-Massey algorithm for the calculation of minimal polynomials.

Section 7 is devoted to distribution properties of linear recurring sequences. Exponential sums with linear recurring sequences are the main tools for studying such properties.

1. FEEDBACK SHIFT REGISTERS, PERIODICITY PROPERTIES

Let k be a positive integer, and let $a, a_0, \ldots, a_{k-1}$ be given elements of a finite field $\mathbb{F}_q$. A sequence $s_0, s_1, \ldots$ of elements of $\mathbb{F}_q$ satisfying the relation

$$s_{n+k} = a_{k-1}s_{n+k-1} + a_{k-2}s_{n+k-2} + \cdots + a_0 s_n + a \quad \text{for } n = 0, 1, \ldots$$

$$(6.1)$$

is called a (*kth-order*) *linear recurring sequence* in $\mathbb{F}_q$. The terms s_0, $s_1, \ldots, s_{k-1}$, which determine the rest of the sequence uniquely, are referred to as the *initial values*. A relation of the form (6.1) is called a (*kth-order*) *linear recurrence relation*. In the older literature one may also find the term "difference equation." We speak of a *homogeneous* linear recurrence relation if $a = 0$; otherwise the linear recurrence relation is *inhomogeneous*. The sequence $s_0, s_1, \ldots$ itself is called a *homogeneous*, or *inhomogeneous, linear recurring sequence* in $\mathbb{F}_q$, respectively.

The generation of linear recurring sequences can be implemented on a *feedback shift register*. This is a special kind of electronic switching circuit handling information in the form of elements of $\mathbb{F}_q$, which are represented suitably. Four types of devices are used. The first is an *adder*, which has two inputs and one output, the output being the sum in $\mathbb{F}_q$ of the two inputs. The second is a *constant multiplier*, which has one input and yields as the output the product of the input with a constant element of $\mathbb{F}_q$. The third is a *constant adder*, which is analogous to a constant multiplier, but adds a constant element of $\mathbb{F}_q$ to the input. The fourth type of device is a *delay element* ("flip-flop"), which has one input and one output and is regulated by an external synchronous clock so that its input at a particular time appears as its output one unit of time later. We shall not be concerned here with the physical realization of these devices. The representation of the components in circuit diagrams is shown in Figure 6.1.

A feedback shift register is built by interconnecting a finite number of adders, constant multipliers, constant adders, and delay elements along a closed loop in such a way that two outputs are never connected together. Actually, for the purpose of generating linear recurring sequences, it suffices to connect the components in a rather special manner. A feedback shift register that generates a linear recurring sequence satisfying (6.1) is shown in Figure 6.2.

At the outset, each delay element D_j, $j = 0, 1, \ldots, k - 1$, contains the initial value s_j. If we think of the arithmetic operations and the transfer along the wires to be performed instantaneously, then after one time unit each D_j will contain s_{j+1}. Continuing in this manner, we see that the output of the feedback shift register is the string of element $s_0, s_1, s_2, \ldots$, received in intervals of one time unit. In most of the applications the desired linear recurring sequence is homogeneous, in which case the constant adder is not needed.

6.1. Example. In order to generate a linear recurring sequence in $\mathbb{F}_5$ satisfying the homogeneous linear recurrence relation

$$s_{n+6} = s_{n+5} + 2s_{n+4} + s_{n+1} + 3s_n \quad \text{for } n = 0, 1, \ldots,$$

(a) Adder   (b) Constant multiplier for multiplying by a (c) Constant adder for adding a (d) Delay element

FIGURE 6.1 The building blocks of feedback shift registers. (a) Adder. (b) Constant multiplier for multiplying by a. (c) Constant adder for adding a. (d) Delay element.

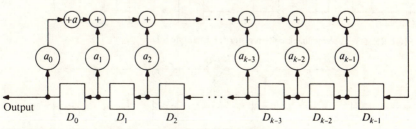

FIGURE 6.2 The general form of a feedback shift register.

one may use the feedback shift register shown in Figure 6.3. Since $a_2 = a_3 = 0$, no connections are necessary at these points.   □

6.2.   Example.   Consider the homogeneous linear recurrence relation

$$s_{n+7} = s_{n+4} + s_{n+3} + s_{n+2} + s_n, \quad n = 0, 1, \ldots, \text{ in } \mathbb{F}_2.$$

A feedback shift register corresponding to this linear recurrence relation is shown in Figure 6.4. Since multiplication by a constant in $\mathbb{F}_2$ either preserves or annihilates elements, the effect of a constant multiplier can be simulated by a wire connection or a disconnection. Therefore, a feedback shift register for the generation of binary homogeneous linear recurring sequences requires only delay elements, adders, and wire connections.   □

Let $s_0, s_1, \ldots$ be a kth-order linear recurring sequence in $\mathbb{F}_q$ satisfying (6.1). As we have noted, this sequence can be generated by the feedback shift register in Figure 6.2. If n is a nonnegative integer, then after n time units the delay element $D_j, j = 0, 1, \ldots, k - 1$, will contain s_{n+j}. It is therefore natural to call the row vector $\mathbf{s}_n = (s_n, s_{n+1}, \ldots, s_{n+k-1})$ the *nth state vector* of the linear recurring sequence (or of the feedback shift register). The state vector $\mathbf{s}_0 = (s_0, s_1, \ldots, s_{k-1})$ is also referred to as the *initial state vector*.

It is a characteristic feature of linear recurring sequences in finite fields that, after a possibly irregular behavior in the beginning, such sequences are eventually of a periodic nature (or ultimately periodic in the sense of Definition 6.3 below). Before studying this property in detail, we introduce some terminology and mention a few general facts about ultimately periodic sequences.

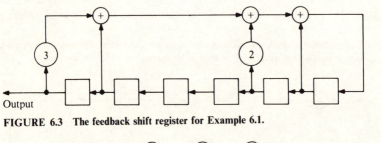

FIGURE 6.3   **The feedback shift register for Example 6.1.**

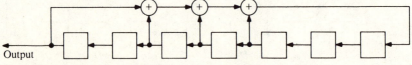

FIGURE 6.4   **The feedback shift register for Example 6.2.**

6.3.   Definition.   Let S be an arbitrary nonempty set, and let $s_0, s_1, \ldots$ be a sequence of elements of S. If there exist integers $r > 0$ and $n_0 \geq 0$ such that $s_{n+r} = s_n$ for all $n \geq n_0$, then the sequence is called *ultimately periodic* and r is called a *period* of the sequence. The smallest number among all the possible periods of an ultimately periodic sequence is called the *least period* of the sequence.

     6.4.   Lemma.   *Every period of an ultimately periodic sequence is divisible by the least period.*

     Proof.   Let r be an arbitrary period of the ultimately periodic sequence $s_0, s_1, \ldots$ and let r_1 be its least period, so that we have $s_{n+r} = s_n$ for all $n \geq n_0$ and $s_{n+r_1} = s_n$ for all $n \geq n_1$ with suitable nonnegative integers n_0 and n_1. If r were not divisible by r_1, we could use the division algorithm for integers to write $r = mr_1 + t$ with integers $m \geq 1$ and $0 < t < r_1$. Then, for all $n \geq \max(n_0, n_1)$ we get

$$s_n = s_{n+r} = s_{n+mr_1+t} = s_{n+(m-1)r_1+t} = \cdots = s_{n+t},$$

and so t is a period of the sequence, which contradicts the definition of the least period.   □

6.5.   Definition.   An ultimately periodic sequence $s_0, s_1, \ldots$ with least period r is called *periodic* if $s_{n+r} = s_n$ holds for all $n = 0, 1, \ldots$.

     The following condition, which is sometimes found in the literature, is equivalent to the definition of a periodic sequence.

     6.6.   Lemma.   *The sequence $s_0, s_1, \ldots$ is periodic if and only if there exists an integer $r > 0$ such that $s_{n+r} = s_n$ for all $n = 0, 1, \ldots$.*

     Proof.   The necessity of the condition is obvious. Conversely, if the condition is satisfied, then the sequence is ultimately periodic and has a least period r_1. Therefore, with a suitable n_0 we have $s_{n+r_1} = s_n$ for all $n \geq n_0$. Now let n be an arbitrary nonnegative integer, and choose an integer $m \geq n_0$ with $m \equiv n \bmod r$. Then $s_{n+r_1} = s_{m+r_1} = s_m = s_n$, which shows that the sequence is periodic in the sense of Definition 6.5.   □

     If $s_0, s_1, \ldots$ is ultimately periodic with least period r, then the least nonnegative integer n_0 such that $s_{n+r} = s_n$ for all $n \geq n_0$ is called the *preperiod*. The sequence is periodic precisely if the preperiod is 0.

     We return now to linear recurring sequences in finite fields and establish the basic results concerning the periodicity behavior of such sequences.

     6.7.   Theorem.   *Let $\mathbb{F}_q$ be any finite field and k any positive integer. Then every kth-order linear recurring sequence in $\mathbb{F}_q$ is ultimately periodic with least period r satisfying $r \leq q^k$, and $r \leq q^k - 1$ if the sequence is homogeneous.*

Proof. We note that there are exactly q^k distinct k-tuples of elements of $\mathbb{F}_q$. Therefore, by considering the state vectors $\mathbf{s}_m$, $0 \leqslant m \leqslant q^k$, of a given kth-order linear recurring sequence in $\mathbb{F}_q$, it follows that $\mathbf{s}_j = \mathbf{s}_i$ for some i and j with $0 \leqslant i < j \leqslant q^k$. Using the linear recurrence relation and induction, we arrive at $\mathbf{s}_{n+j-i} = \mathbf{s}_n$ for all $n \geqslant i$, which shows that the linear recurring sequence itself is ultimately periodic with least period $r \leqslant j - i \leqslant q^k$ In case the linear recurring sequence is homogeneous and no state vector is the zero vector, one can go through the same argument, but with q^k replaced by $q^k - 1$, to obtain $r \leqslant q^k - 1$. If, however, one of the state vectors of a homogeneous linear recurring sequence is the zero vector, then all subsequent state vectors are zero vectors, and so the sequence has least period $r = 1 \leqslant q^k - 1$.   □

6.8.   Example.   The first-order linear recurring sequence $s_0, s_1, \ldots$ in $\mathbb{F}_p$, p prime, with $s_{n+1} = s_n + 1$ for $n = 0, 1, \ldots$ and arbitrary $s_0 \in \mathbb{F}_p$ shows that the upper bound for r in Theorem 6.7 may be attained. If $\mathbb{F}_q$ is any finite field and g is a primitive element of $\mathbb{F}_q$ (see Definition 2.9), then the first-order homogeneous linear recurring sequence $s_0, s_1, \ldots$ in $\mathbb{F}_q$ with $s_{n+1} = g s_n$ for $n = 0, 1, \ldots$ and $s_0 \neq 0$ has least period $r = q - 1$. Therefore, the upper bound for r in the homogeneous case may also be attained. Later on, we shall show that in any $\mathbb{F}_q$ and for any $k \geqslant 1$ there exist kth-order homogeneous linear recurring sequences with least period $r = q^k - 1$ (see Theorem 6.33).   □

6.9.   Example.   For a first-order homogeneous linear recurring sequence in $\mathbb{F}_q$, it is easily seen that the least period divides $q - 1$. However, if $k \geqslant 2$, then the least period of a kth-order homogeneous linear recurring sequence need not divide $q^k - 1$. Consider, for instance, the sequence $s_0, s_1, \ldots$ in $\mathbb{F}_5$ with $s_0 = 0$, $s_1 = 1$, and $s_{n+2} = s_{n+1} + s_n$ for $n = 0, 1, \ldots$, which has least period 20, as is shown by inspection.   □

6.10.   Example.   A linear recurring sequence in a finite field is ultimately periodic, but it need not be periodic, as is illustrated by a second-order linear recurring sequence $s_0, s_1, \ldots$ in $\mathbb{F}_q$ with $s_0 \neq s_1$ and $s_{n+2} = s_{n+1}$ for $n = 0, 1, \ldots$.   □

An important sufficient condition for the periodicity of a linear recurring sequence is provided by the following result.

6.11.   Theorem.   *If $s_0, s_1, \ldots$ is a linear recurring sequence in a finite field satisfying the linear recurrence relation (6.1), and if the coefficient a_0 in (6.1) is nonzero, then the sequence $s_0, s_1, \ldots$ is periodic.*

Proof. According to Theorem 6.7, the given linear recurring sequence is ultimately periodic. If r is its least period and n_0 its preperiod, then $s_{n+r} = s_n$ for all $n \geqslant n_0$. Suppose we had $n_0 \geqslant 1$. From (6.1) with

$n = n_0 + r - 1$ and the fact that $a_0 \neq 0$, we obtain

$$s_{n_0-1+r} = a_0^{-1}\left(s_{n_0+k-1+r} - a_{k-1}s_{n_0+k-2+r} - \cdots - a_1 s_{n_0+r} - a\right)$$

$$= a_0^{-1}\left(s_{n_0+k-1} - a_{k-1}s_{n_0+k-2} - \cdots - a_1 s_{n_0} - a\right).$$

Using (6.1) with $n = n_0 - 1$, we find the same expression for s_{n_0-1}, and so $s_{n_0-1+r} = s_{n_0-1}$. This is a contradiction to the definition of the preperiod. $\square$

Let $s_0, s_1, \ldots$ be a kth-order homogeneous linear recurring sequence in $\mathbb{F}_q$ satisfying the linear recurrence relation

$$s_{n+k} = a_{k-1}s_{n+k-1} + a_{k-2}s_{n+k-2} + \cdots + a_0 s_n \quad \text{for } n = 0, 1, \ldots, \quad (6.2)$$

where $a_j \in \mathbb{F}_q$ for $0 \leqslant j \leqslant k-1$. With this linear recurring sequence we associate the $k \times k$ matrix A over $\mathbb{F}_q$ defined by

$$A = \begin{pmatrix} 0 & 0 & 0 & \cdots & 0 & a_0 \\ 1 & 0 & 0 & \cdots & 0 & a_1 \\ 0 & 1 & 0 & \cdots & 0 & a_2 \\ \vdots & \vdots & \vdots & & \vdots & \vdots \\ 0 & 0 & 0 & \cdots & 1 & a_{k-1} \end{pmatrix}. \quad (6.3)$$

If $k = 1$, then A is understood to be the 1×1 matrix (a_0). We note that the matrix A depends only on the linear recurrence relation satisfied by the given sequence.

6.12. Lemma. *If $s_0, s_1, \ldots$ is a homogeneous linear recurring sequence in $\mathbb{F}_q$ satisfying (6.2) and A is the matrix in (6.3) associated with it, then for the state vectors of the sequence we have*

$$\mathbf{s}_n = \mathbf{s}_0 A^n \quad \text{for } n = 0, 1, \ldots. \quad (6.4)$$

Proof. Since $\mathbf{s}_n = (s_n, s_{n+1}, \ldots, s_{n+k-1})$, one checks easily that $\mathbf{s}_{n+1} = \mathbf{s}_n A$ for all $n \geqslant 0$, so that (6.4) follows by induction. $\square$

We note that the set of all nonsingular $k \times k$ matrices over $\mathbb{F}_q$ forms a finite group under matrix multiplication, called the *general linear group* $GL(k, \mathbb{F}_q)$.

6.13. Theorem. *If $s_0, s_1, \ldots$ is a kth-order homogeneous linear recurring sequence in $\mathbb{F}_q$ satisfying (6.2) with $a_0 \neq 0$, then the least period of the sequence divides the order of the associated matrix A from (6.3) in the general linear group $GL(k, \mathbb{F}_q)$.*

Proof. We have $\det A = (-1)^{k-1} a_0 \neq 0$, so that A is indeed an element of $GL(k, \mathbb{F}_q)$. If m is the order of A in $GL(k, \mathbb{F}_q)$, then from Lemma 6.12 we obtain $\mathbf{s}_{n+m} = \mathbf{s}_0 A^{n+m} = \mathbf{s}_0 A^n = \mathbf{s}_n$ for all $n \geq 0$, and so m is a period of the linear recurring sequence. The rest follows from Lemma 6.4. $\square$

We remark that the above argument, together with Lemma 6.6, yields an alternative proof for Theorem 6.11 in the homogeneous case. From Theorem 6.13 it follows, in particular, that the least period of the sequence $s_0, s_1, \ldots$ divides the order of $GL(k, \mathbb{F}_q)$, which is known to be $q^{(k^2 - k)/2} (q-1)(q^2 - 1) \cdots (q^k - 1)$.

Let now $s_0, s_1, \ldots$ be a kth-order *inhomogeneous* linear recurring sequence in $\mathbb{F}_q$ satisfying (6.1). By using (6.1) with n replaced by $n+1$ and subtracting from the resulting identity the original form of (6.1) we obtain

$$s_{n+k+1} = b_k s_{n+k} + b_{k-1} s_{n+k-1} + \cdots + b_0 s_n \quad \text{for } n = 0, 1, \ldots, \quad (6.5)$$

where $b_0 = -a_0$, $b_j = a_{j-1} - a_j$ for $j = 1, 2, \ldots, k-1$, and $b_k = a_{k-1} + 1$. Therefore, the sequence $s_0, s_1, \ldots$ can be interpreted as a $(k+1)$st-order homogeneous linear recurring sequence in $\mathbb{F}_q$. Consequently, results on homogeneous linear recurring sequences yield information for the inhomogeneous case as well.

An alternative approach to the *inhomogeneous case* proceeds as follows. Let $s_0, s_1, \ldots$ be a kth-order inhomogeneous linear recurring sequence in $\mathbb{F}_q$ satisfying (6.1), and consider the $(k+1) \times (k+1)$ matrix C over $\mathbb{F}_q$ defined by

$$C = \begin{pmatrix} 1 & 0 & 0 & \cdots & 0 & a \\ 0 & 0 & 0 & \cdots & 0 & a_0 \\ 0 & 1 & 0 & \cdots & 0 & a_1 \\ 0 & 0 & 1 & \cdots & 0 & a_2 \\ \vdots & \vdots & \vdots & & \vdots & \vdots \\ 0 & 0 & 0 & \cdots & 1 & a_{k-1} \end{pmatrix}$$

If $k = 1$, take

$$C = \begin{pmatrix} 1 & a \\ 0 & a_0 \end{pmatrix}.$$

We introduce modified state vectors by setting

$$\mathbf{s}'_n = (1, s_n, s_{n+1}, \ldots, s_{n+k-1}) \quad \text{for } n = 0, 1, \ldots .$$

Then it is easily seen that $\mathbf{s}'_{n+1} = \mathbf{s}'_n C$ for all $n \geq 0$, and so $\mathbf{s}'_n = \mathbf{s}'_0 C^n$ for all $n \geq 0$ by induction. If $a_0 \neq 0$ in (6.1), then $\det C = (-1)^{k-1} a_0 \neq 0$, so that the matrix C is an element of $GL(k+1, \mathbb{F}_q)$. One shows then as in the proof of Theorem 6.13 that the least period of $s_0, s_1, \ldots$ divides the order of C in $GL(k+1, \mathbb{F}_q)$.

2. IMPULSE RESPONSE SEQUENCES, CHARACTERISTIC POLYNOMIAL

Among all the homogeneous linear recurring sequences in $\mathbb{F}_q$ satisfying a given kth-order linear recurrence relation such as (6.2), we can single out one that yields the maximal value for the least period in this class of sequences. This is the *impulse response sequence* $d_0, d_1, \ldots$ determined uniquely by its initial values $d_0 = \cdots = d_{k-2} = 0$, $d_{k-1} = 1$ ($d_0 = 1$ if $k = 1$) and the linear recurrence relation

$$d_{n+k} = a_{k-1}d_{n+k-1} + a_{k-2}d_{n+k-2} + \cdots + a_0 d_n \quad \text{for } n = 0, 1, \ldots. \quad (6.6)$$

6.14. Example. Consider the linear recurrence relation

$$s_{n+5} = s_{n+1} + s_n, \quad n = 0, 1, \ldots, \text{ in } \mathbb{F}_2.$$

The impulse response sequence $d_0, d_1, \ldots$ corresponding to it is given by the string of binary digits

$$0\,0\,0\,0\,1\,0\,0\,0\,1\,1\,0\,0\,1\,0\,1\,0\,1\,1\,1\,1\,1\,0\,0\,0\,0\,1\cdots$$

of least period 21. A feedback shift register generating this sequence is shown in Figure 6.5. We can think of this sequence as being obtained by starting with the state in which each delay element is "empty" (i.e., contains 0) and then sending the "impulse" 1 into the rightmost delay element. This explains the term "impulse response sequence." □

6.15. Lemma. *Let* $d_0, d_1, \ldots$ *be the impulse response sequence in* $\mathbb{F}_q$ *satisfying (6.6), and let A be the matrix in (6.3). Then two state vectors* $\mathbf{d}_m$ *and* $\mathbf{d}_n$ *are identical if and only if* $A^m = A^n$.

Proof. The sufficiency follows from Lemma 6.12. Conversely, suppose that $\mathbf{d}_m = \mathbf{d}_n$. From the linear recurrence relation (6.6) we obtain then $\mathbf{d}_{m+t} = \mathbf{d}_{n+t}$ for all $t \geq 0$. By Lemma 6.12 we get $\mathbf{d}_t A^m = \mathbf{d}_t A^n$ for all $t \geq 0$. But since the vectors $\mathbf{d}_0, \mathbf{d}_1, \ldots, \mathbf{d}_{k-1}$ obviously form a basis for the k-dimensional vector space $\mathbb{F}_q^k$ over $\mathbb{F}_q$, we conclude that $A^m = A^n$. □

6.16. Theorem. *The least period of a homogeneous linear recurring sequence in* $\mathbb{F}_q$ *divides the least period of the corresponding impulse response sequence.*

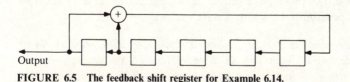

FIGURE 6.5 The feedback shift register for Example 6.14.

Proof. Let $s_0, s_1, \ldots$ be a homogeneous linear recurring sequence in $\mathbb{F}_q$ satisfying (6.2), let $d_0, d_1, \ldots$ be the corresponding impulse response sequence, and let A be the matrix in (6.3). If r is the least period of $d_0, d_1, \ldots$ and n_0 the preperiod, then $\mathbf{d}_{n+r} = \mathbf{d}_n$ for all $n \geqslant n_0$. It follows from Lemma 6.15 that $A^{n+r} = A^n$ for all $n \geqslant n_0$, and so $\mathbf{s}_{n+r} = \mathbf{s}_n$ for all $n \geqslant n_0$ by Lemma 6.12. Therefore, r is a period of $s_0, s_1, \ldots$, and an application of Lemma 6.4 completes the proof. $\square$

6.17.   Theorem.   *If $d_0, d_1, \ldots$ is a kth-order impulse response sequence in $\mathbb{F}_q$ satisfying (6.6) with $a_0 \neq 0$ and A is the matrix in (6.3) associated with it, then the least period of the sequence is equal to the order of A in the general linear group $GL(k, \mathbb{F}_q)$.*

Proof. If r is the least period of $d_0, d_1, \ldots$, then r divides the order of A according to the Theorem 6.13. On the other hand, we have $\mathbf{d}_r = \mathbf{d}_0$ by Theorem 6.11, and so Lemma 6.15 yields $A^r = A^0$, which implies already the desired result. $\square$

6.18.   Example.   For the linear recurrence relation $s_{n+5} = s_{n+1} + s_n$, $n = 0, 1, \ldots$, in $\mathbb{F}_2$ considered in Example 6.14 we have seen that the least period of the corresponding impulse response sequence is equal to 21, which is the same as the order of the matrix

$$A = \begin{pmatrix} 0 & 0 & 0 & 0 & 1 \\ 1 & 0 & 0 & 0 & 1 \\ 0 & 1 & 0 & 0 & 0 \\ 0 & 0 & 1 & 0 & 0 \\ 0 & 0 & 0 & 1 & 0 \end{pmatrix}$$

in $GL(5, \mathbb{F}_2)$. If the initial state vector of a linear recurring sequence in $\mathbb{F}_2$ satisfying the given linear recurrence relation is equal to one of the 21 different state vectors appearing in the impulse response sequence, then the least period is again 21 (since such a sequence is just a shifted impulse response sequence). If we choose the initial state vector $(1, 1, 1, 0, 1)$, we get the string of binary digits $1\,1\,1\,0\,1\,0\,0\,1\,1\,1\,0\,1 \cdots$ of least period 7, and the same least period results from any one of the 7 different state vectors of this sequence in the role of the initial state vector. If the initial state vector is $(1, 1, 0, 1, 1)$, then we obtain the string of binary digits $1\,1\,0\,1\,1\,0\,1\,1 \cdots$ of least period 3, and the same least period results if any one of the 3 different state vectors of this sequence is taken as the initial state vector. The initial state vector $(0, 0, 0, 0, 0)$ produces a sequence of least period 1. We have now exhausted all 32 possibilities for initial state vectors. $\square$

6.19.   Theorem.   *Let $s_0, s_1, \ldots$ be a kth-order homogeneous linear recurring sequence in $\mathbb{F}_q$ with preperiod n_0. If there exist k state vectors $\mathbf{s}_{m_1}, \mathbf{s}_{m_2}, \ldots, \mathbf{s}_{m_k}$ with $m_j \geqslant n_0 \ (1 \leqslant j \leqslant k)$ that are linearly independent over $\mathbb{F}_q$,* *then both* $s_0, s_1, \ldots$ *and its corresponding impulse response sequence are periodic and they have the same least period.*

Proof. Let r be the least period of $s_0, s_1, \ldots$. For $1 \leqslant j \leqslant k$ we have $\mathbf{s}_{m_j} A^r = \mathbf{s}_{m_j + r} = \mathbf{s}_{m_j}$ by using Lemma 6.12, and so A^r is the $k \times k$ identity matrix over $\mathbb{F}_q$. Thus we get $\mathbf{s}_r = \mathbf{s}_0 A^r = \mathbf{s}_0$, which shows that $s_0, s_1, \ldots$ is periodic. Similarly, if $\mathbf{d}_n$ denotes the nth state vector of the impulse response sequence, then $\mathbf{d}_r = \mathbf{d}_0 A^r = \mathbf{d}_0$, and an application of Theorem 6.16 completes the proof. $\qquad \square$

6.20. Example. The condition $m_j \geqslant n_0$ in Theorem 6.19 is needed since there are kth-order homogeneous linear recurring sequences that are not periodic but contain k linearly independent state vectors. Let $d_0, d_1, \ldots$ be the second-order impulse response sequence in $\mathbb{F}_q$ with $d_{n+2} = d_{n+1}$ for $n = 0, 1, \ldots$. The terms of this sequence are $0, 1, 1, 1, \ldots$. Clearly, the state vectors $\mathbf{d}_0$ and $\mathbf{d}_1$ are linearly independent over $\mathbb{F}_q$, but the sequence is not periodic (note that $n_0 = 1$ in this case). The converse of Theorem 6.19 is not true. Consider the third-order linear recurring sequence $s_0, s_1, \ldots$ in $\mathbb{F}_2$ with $s_{n+3} = s_n$ for $n = 0, 1, \ldots$ and $\mathbf{s}_0 = (1, 1, 0)$. Then both $s_0, s_1, \ldots$ and its corresponding impulse response sequence are periodic with least period 3, but any three state vectors of $s_0, s_1, \ldots$ are linearly dependent over $\mathbb{F}_2$. $\qquad \square$

Let $s_0, s_1, \ldots$ be a kth-order homogeneous linear recurring sequence in $\mathbb{F}_q$ satisfying the linear recurrence relation

$$s_{n+k} = a_{k-1} s_{n+k-1} + a_{k-2} s_{n+k-2} + \cdots + a_0 s_n \quad \text{for } n = 0, 1, \ldots, \quad (6.7)$$

where $a_j \in \mathbb{F}_q$ for $0 \leqslant j \leqslant k - 1$. The polynomial

$$f(x) = x^k - a_{k-1} x^{k-1} - a_{k-2} x^{k-2} - \cdots - a_0 \in \mathbb{F}_q[x]$$

is called the *characteristic polynomial* of the linear recurring sequence. It depends, of course, only on the linear recurrence relation (6.7). If A is the matrix in (6.3), then it is easily seen that $f(x)$ is identical with the characteristic polynomial of A in the sense of linear algebra—that is, $f(x) = \det(xI - A)$ with I being the $k \times k$ identity matrix over $\mathbb{F}_q$. On the other hand, the matrix A may be thought of as the companion matrix of the monic polynomial $f(x)$.

As a first application of the characteristic polynomial, we show how the terms of a linear recurring sequence may be represented explicitly in an important special case.

6.21. Theorem. *Let $s_0, s_1, \ldots$ be a kth-order homogeneous linear recurring sequence in $\mathbb{F}_q$ with characteristic polynomial $f(x)$. If the roots $\alpha_1, \ldots, \alpha_k$ of $f(x)$ are all distinct, then*

$$s_n = \sum_{j=1}^{k} \beta_j \alpha_j^n \quad \text{for } n = 0, 1, \ldots, \quad (6.8)$$

where $\beta_1,\dots,\beta_k$ are elements that are uniquely determined by the initial values of the sequence and belong to the splitting field of $f(x)$ over $\mathbb{F}_q$.

Proof. The constants $\beta_1,\dots,\beta_k$ can be determined from the system of linear equations

$$\sum_{j=1}^{k} \alpha_j^n \beta_j = s_n, \quad n = 0,1,\dots,k-1.$$

Since the determinant of this system is a Vandermonde determinant, which is nonzero by the condition on $\alpha_1,\dots,\alpha_k$, the elements $\beta_1,\dots,\beta_k$ are uniquely determined and belong to the splitting field $\mathbb{F}_q(\alpha_1,\dots,\alpha_k)$ of $f(x)$ over $\mathbb{F}_q$, as is seen from Cramer's rule. To prove the identity (6.8) for all $n \geq 0$, it suffices now to check whether the elements on the right-hand side of (6.8), with these specific values for $\beta_1,\dots,\beta_k$, satisfy the linear recurrence relation (6.7). But

$$\sum_{j=1}^{k} \beta_j \alpha_j^{n+k} - a_{k-1} \sum_{j=1}^{k} \beta_j \alpha_j^{n+k-1} - a_{k-2} \sum_{j=1}^{k} \beta_j \alpha_j^{n+k-2} - \cdots - a_0 \sum_{j=1}^{k} \beta_j \alpha_j^{n}$$

$$= \sum_{j=1}^{k} \beta_j f(\alpha_j) \alpha_j^n = 0$$

for all $n \geq 0$, and the proof is complete. $\qquad\square$

6.22. Example. Consider the linear recurring sequence $s_0, s_1,\dots$ in $\mathbb{F}_2$ with $s_0 = s_1 = 1$ and $s_{n+2} = s_{n+1} + s_n$ for $n = 0,1,\dots$. The characteristic polynomial is $f(x) = x^2 - x - 1 \in \mathbb{F}_2[x]$. If $\mathbb{F}_4 = \mathbb{F}_2(\alpha)$, then the roots of $f(x)$ are $\alpha_1 = \alpha$ and $\alpha_2 = 1 + \alpha$. Using the given initial values, we obtain $\beta_1 + \beta_2 = 1$ and $\beta_1 \alpha + \beta_2(1 + \alpha) = 1$, hence $\beta_1 = \alpha$ and $\beta_2 = 1 + \alpha$. By Theorem 6.21 it follows that $s_n = \alpha^{n+1} + (1 + \alpha)^{n+1}$ for all $n \geq 0$. Since $\beta^3 = 1$ for every nonzero $\beta \in \mathbb{F}_4$, we deduce that $s_{n+3} = s_n$ for all $n \geq 0$, which is in accordance with the fact that the least period of the sequence is 3. $\qquad\square$

6.23. Remark. A formula similar to (6.8) is valid if the multiplicity of each root of $f(x)$ is at most the characteristic p of $\mathbb{F}_q$. In detail, let $\alpha_1,\dots,\alpha_m$ be the distinct roots of $f(x)$, and suppose that each α_i, $i = 1,2,\dots,m$, has multiplicity $e_i \leq p$ and that $e_i = 1$ if $\alpha_i = 0$. Then we have

$$s_n = \sum_{i=1}^{m} P_i(n) \alpha_i^n \quad \text{for } n = 0,1,\dots,$$

where each P_i, $i = 1,2,\dots,m$, is a polynomial of degree less than e_i whose coefficients are uniquely determined by the initial values of the sequence and belong to the splitting field of $f(x)$ over $\mathbb{F}_q$. The integer n is of course identified in the usual way with an element of $\mathbb{F}_q$. The reader familiar with differential equations will observe a certain analogy with the general solution of a homogeneous linear differential equation with constant coefficients. □

In case the characteristic polynomial is irreducible, the elements of the linear recurring sequence can be represented in terms of a suitable trace function (see Definition 2.22 and Theorem 2.23 for the definition and basic properties of trace functions).

6.24. Theorem. *Let* $s_0, s_1, \ldots$ *be a kth-order homogeneous linear recurring sequence in* $K = \mathbb{F}_q$ *whose characteristic polynomial* $f(x)$ *is irreducible over* K. *Let* α *be a root of* $f(x)$ *in the extension field* $F = \mathbb{F}_{q^k}$. *Then there exists a uniquely determined* $\theta \in F$ *such that*

$$s_n = \mathrm{Tr}_{F/K}(\theta \alpha^n) \quad for\ n = 0, 1, \ldots .$$

Proof. Since $\{1, \alpha, \ldots, \alpha^{k-1}\}$ constitutes a basis of F over K, we can define a uniquely determined linear mapping L from F into K by setting $L(\alpha^n) = s_n$ for $n = 0, 1, \ldots, k-1$. By Theorem 2.24 there exists a uniquely determined $\theta \in F$ such that $L(\gamma) = \mathrm{Tr}_{F/K}(\theta \gamma)$ for all $\gamma \in F$. In particular, we have

$$s_n = \mathrm{Tr}_{F/K}(\theta \alpha^n) \quad for\ n = 0, 1, \ldots, k-1.$$

It remains to show that the elements $\mathrm{Tr}_{F/K}(\theta \alpha^n)$, $n = 0, 1, \ldots$, form a homogeneous linear recurring sequence with characteristic polynomial $f(x)$. But if $f(x) = x^k - a_{k-1} x^{k-1} - \cdots - a_0 \in K[x]$, then using properties of the trace function we get

$$\mathrm{Tr}_{F/K}(\theta \alpha^{n+k}) - a_{k-1} \mathrm{Tr}_{F/K}(\theta \alpha^{n+k-1}) - \cdots - a_0 \mathrm{Tr}_{F/K}(\theta \alpha^n)$$

$$= \mathrm{Tr}_{F/K}(\theta \alpha^{n+k} - a_{k-1} \theta \alpha^{n+k-1} - \cdots - a_0 \theta \alpha^n)$$

$$= \mathrm{Tr}_{F/K}(\theta \alpha^n f(\alpha)) = 0$$

for all $n \geqslant 0$. □

Further relations between linear recurring sequences and their characteristic polynomials can be found on the basis of the following polynomial identity.

6.25. Theorem. *Let* $s_0, s_1, \ldots$ *be a kth-order homogeneous linear recurring sequence in* $\mathbb{F}_q$ *that satisfies the linear recurrence relation* (6.7) *and is periodic with period* r. *Let* $f(x)$ *be the characteristic polynomial of the sequence. Then the identity*

$$f(x) s(x) = (1 - x^r) h(x) \tag{6.9}$$

holds with

$$s(x) = s_0 x^{r-1} + s_1 x^{r-2} + \cdots + s_{r-2} x + s_{r-1} \in \mathbb{F}_q[x]$$

and

$$h(x) = \sum_{j=0}^{k-1} \sum_{i=0}^{k-1-j} a_{i+j+1} s_i x^j \in \mathbb{F}_q[x], \qquad (6.10)$$

where we set $a_k = -1$.

Proof. We compare the coefficients on both sides of (6.9). For $0 \leq t \leq k + r - 1$, let c_t (resp. d_t) be the coefficient of x^t on the left-hand side (resp. right-hand side) of (6.9). Since $f(x) = -\sum_{i=0}^k a_i x^i$, we have

$$c_t = - \sum_{\substack{0 \leq i \leq k, 0 \leq j \leq r-1 \\ i+j=t}} a_i s_{r-1-j} \quad \text{for } 0 \leq t \leq k+r-1. \qquad (6.11)$$

We note also that the linear recurrence relation (6.7) may be written in the form

$$\sum_{i=0}^k a_i s_{n+i} = 0 \quad \text{for all } n \geq 0. \qquad (6.12)$$

We distinguish now four cases. If $k \leq t \leq r - 1$, then by (6.11) and (6.12),

$$c_t = - \sum_{i=0}^k a_i s_{r-1-t+i} = 0 = d_t.$$

If $t \leq r - 1$ and $t < k$, then by (6.11), (6.12), and the periodicity of the given sequence,

$$c_t = - \sum_{i=0}^t a_i s_{r-1-t+i} = \sum_{i=t+1}^k a_i s_{r-1-t+i}$$

$$= \sum_{i=t+1}^k a_i s_{i-t-1} = \sum_{i=0}^{k-1-t} a_{i+t+1} s_i = d_t.$$

If $t \geq r$ and $t \geq k$, then by (6.11),

$$c_t = - \sum_{i=t-r+1}^k a_i s_{r-1-t+i} = - \sum_{i=0}^{k-1-t+r} a_{i+t-r+1} s_i = d_t.$$

If $r \leq t < k$, then by (6.11) and the periodicity of the given sequence,

$$c_t = - \sum_{i=t-r+1}^t a_i s_{r-1-t+i} = - \sum_{i=0}^{r-1} a_{i+t-r+1} s_i$$

$$= \sum_{i=r}^{k-1-t+r} a_{i+t-r+1} s_i - \sum_{i=0}^{k-1-t+r} a_{i+t-r+1} s_i$$

$$= \sum_{i=0}^{k-1-t} a_{i+t+1} s_{i+r} - \sum_{i=0}^{k-1-t+r} a_{i+t-r+1} s_i$$

$$= \sum_{i=0}^{k-1-t} a_{i+t+1} s_i - \sum_{i=0}^{k-1-t+r} a_{i+t-r+1} s_i = d_t. \qquad \square$$

In Lemma 3.1 we have seen that for any polynomial $f(x) \in \mathbb{F}_q[x]$ with $f(0) \neq 0$ there exists a positive integer e such that $f(x)$ divides $x^e - 1$. This gave rise to the definition of the order of f (see Definition 3.2). We give the following interpretation of $\text{ord}(f)$.

6.26. Lemma. *Let*

$$f(x) = x^k - a_{k-1}x^{k-1} - a_{k-2}x^{k-2} - \cdots - a_0 \in \mathbb{F}_q[x]$$

with $k \geq 1$ and $a_0 \neq 0$. Then $\text{ord}(f(x))$ is equal to the order of the matrix A from (6.3) in the general linear group $GL(k, \mathbb{F}_q)$.

Proof. Since A is the companion matrix of $f(x)$, the polynomial $f(x)$ is, in turn, the minimal polynomial of A. Consequently, if I is the $k \times k$ identity matrix over $\mathbb{F}_q$, then we have $A^e = I$ for some positive integer e if and only if $f(x)$ divides $x^e - 1$. The result follows now from the definitions of the order of $f(x)$ and the order of A. $\square$

6.27. Theorem. *Let $s_0, s_1, \ldots$ be a homogeneous linear recurring sequence in $\mathbb{F}_q$ with characteristic polynomial $f(x) \in \mathbb{F}_q[x]$. Then the least period of the sequence divides $\text{ord}(f(x))$, and the least period of the corresponding impulse response sequence is equal to $\text{ord}(f(x))$. If $f(0) \neq 0$, then both sequences are periodic.*

Proof. If $f(0) \neq 0$, then in the light of Lemma 6.26 the result is essentially a restatement of Theorems 6.13 and 6.17. In this case, the periodicity property follows from Theorem 6.11. If $f(0) = 0$, then we write $f(x) = x^h g(x)$ as in Definition 3.2 and set $t_n = s_{n+h}$ for $n = 0, 1, \ldots$. Then $t_0, t_1, \ldots$ is a homogeneous linear recurring sequence with characteristic polynomial $g(x)$, provided that $\deg(g(x)) > 0$. Its least period is the same as that of the sequence $s_0, s_1, \ldots$. Therefore, by what we have already shown, the least period of $s_0, s_1, \ldots$ divides $\text{ord}(g(x)) = \text{ord}(f(x))$. The desired result concerning the impulse response sequence follows in a similar way. If $g(x)$ is constant, the theorem is trivial. $\square$

We remark that for $f(0) \neq 0$ the least period of the impulse response sequence may also be obtained from the identity (6.9) in the following way. For the impulse response sequence with characteristic polynomial $f(x)$, the polynomial $h(x)$ in (6.10) is given by $h(x) = -1$. Therefore, if r is the least period of the impulse response sequence, then $f(x)$ divides $x^r - 1$ by (6.9) and so $r \geq \text{ord}(f(x))$. On the other hand, r must divide $\text{ord}(f(x))$ by the first part of Theorem 6.27, and so $r = \text{ord}(f(x))$.

6.28. Theorem. *Let $s_0, s_1, \ldots$ be a homogeneous linear recurring sequence in $\mathbb{F}_q$ with nonzero initial state vector, and suppose the characteristic polynomial $f(x) \in \mathbb{F}_q[x]$ is irreducible over $\mathbb{F}_q$ and satisfies $f(0) \neq 0$. Then the sequence is periodic with least period equal to $\text{ord}(f(x))$.*

Proof. The sequence is periodic and its least period r divides ord($f(x)$) by Theorem 6.27. On the other hand, it follows from (6.9) that $f(x)$ divides $(x^r - 1)h(x)$. Since $s(x)$, and therefore $h(x)$, is a nonzero polynomial and since deg($h(x)$) < deg($f(x)$), the irreducibility of $f(x)$ implies that $f(x)$ divides $x^r - 1$, and so $r \geqslant$ ord($f(x)$).   □

Now we present a different proof of Corollary 3.4, which we restate for convenience.

6.29.   Theorem.   *Let $f(x) \in \mathbb{F}_q[x]$ be irreducible over $\mathbb{F}_q$ with deg($f(x)$) = k. Then* ord($f(x)$) *divides $q^k - 1$.*

Proof. We may assume without loss of generality that $f(0) \neq 0$ and that $f(x)$ is monic. We take a homogeneous linear recurring sequence in $\mathbb{F}_q$ that has $f(x)$ as its characteristic polynomial and has a nonzero initial state vector. According to Theorem 6.28, this sequence is periodic with least period ord($f(x)$), so that altogether ord($f(x)$) different state vectors appear in it. If ord($f(x)$) is less than $q^k - 1$, the total number of nonzero k-tuples of elements of $\mathbb{F}_q$, we can choose such a k-tuple that does not appear as a state vector in the sequence above and use it as an initial state vector for another homogeneous linear recurring sequence in $\mathbb{F}_q$ with characteristic polynomial $f(x)$. None of the ord($f(x)$) different state vectors of the second sequence is equal to a state vector of the first sequence, for otherwise the two sequences would be identical from some points onwards and the initial state vector of the second sequence would eventually appear as a state vector in the first sequence—a contradiction. By continuing to generate linear recurring sequences of the type above, we arrive at a partition of the set of $q^k - 1$ nonzero k-tuples of elements of $\mathbb{F}_q$ into subsets of cardinality ord($f(x)$), and the conclusion of the theorem follows.   □

6.30.   Example.   Consider the linear recurrence relation $s_{n+6} = s_{n+4} + s_{n+2} + s_{n+1} + s_n$, $n = 0, 1, \ldots$, in $\mathbb{F}_2$. The corresponding characteristic polynomial is $f(x) = x^6 - x^4 - x^2 - x - 1 \in \mathbb{F}_2[x]$. The polynomial $f(x)$ is irreducible over $\mathbb{F}_2$. Furthermore, $f(x)$ divides $x^{21} - 1$ and no polynomial $x^e - 1$ with $0 < e < 21$, so that ord($f(x)$) = 21. The impulse response sequence corresponding to the linear recurrence relation is given by the string of binary digits

$$00000101001001100101100001 \cdots$$

of least period 21, as it should be. If $(0, 0, 0, 0, 1, 1)$ is taken as the initial state vector, we arrive at the string of binary digits

$$00001111011010101110100011 \cdots$$

of least period 21, and if $(0, 0, 0, 1, 0, 0)$ is taken as the initial state vector, we obtain the string of binary digits

$$.000100011011111100111000100 \cdots$$

of least period 21. Each one of the nonzero sextuples of elements of $\mathbb{F}_2$ appears as a state vector in exactly one of the three sequences. Any other nonzero initial state vector will produce a shifted version of one of the three sequences, which is again a sequence of least period 21.   □

6.31.   Example.   If $f(x) \in \mathbb{F}_q[x]$ with $\deg(f(x)) = k$ is reducible, then $\operatorname{ord}(f(x))$ need not divide $q^k - 1$. Consider $f(x) = x^5 + x + 1 \in \mathbb{F}_2[x]$. Then $f(x)$ is reducible since

$$x^5 + x + 1 = (x^3 + x^2 + 1)(x^2 + x + 1).$$

It follows, for instance, from Theorem 6.27 and Example 6.14 that $\operatorname{ord}(f(x)) = 21$, and this is not a divisor of $2^5 - 1 = 31$.   □

Linear recurring sequences whose least periods are very large are of particular importance in applications. We know from Theorem 6.7 that for a kth-order homogeneous linear recurring sequence in $\mathbb{F}_q$ the least period can be at most $q^k - 1$. In order to generate such sequences for which the least period is actually equal to $q^k - 1$, we have to use the notion of a primitive polynomial (see Definition 3.15).

6.32.   Definition.   A homogeneous linear recurring sequence in $\mathbb{F}_q$ whose characteristic polynomial is a primitive polynomial over $\mathbb{F}_q$ and which has a nonzero initial state vector is called a *maximal period sequence* in $\mathbb{F}_q$.

6.33.   Theorem.   *Every kth-order maximal period sequence in $\mathbb{F}_q$ is periodic and its least period is equal to the largest possible value for the least period of any kth-order homogeneous linear recurring sequence in $\mathbb{F}_q$ — namely, $q^k - 1$.*

Proof.   The fact that the sequence is periodic and that the least period is $q^k - 1$ is a consequence of Theorem 6.28 and Theorem 3.16. The remaining assertion follows from Theorem 6.7.   □

6.34.   Example.   The linear recurrence relation $s_{n+7} = s_{n+4} + s_{n+3} + s_{n+2} + s_n$, $n = 0, 1, \ldots$, in $\mathbb{F}_2$ considered in Example 6.2 has the polynomial $f(x) = x^7 - x^4 - x^3 - x^2 - 1 \in \mathbb{F}_2[x]$ as its characteristic polynomial. Since $f(x)$ is a primitive polynomial over $\mathbb{F}_2$, any sequence with nonzero initial state vector arising from this linear recurrence relation is a maximal period sequence in $\mathbb{F}_2$. If we choose one particular nonzero initial state vector, then the resulting sequence $s_0, s_1, \ldots$ has least period $2^7 - 1 = 127$ according to Theorem 6.33. Therefore, all possible nonzero vectors of $\mathbb{F}_2^7$ appear as state vectors in this sequence. Any other maximal period sequence arising from the given linear recurrence relation is just a shifted version of the sequence $s_0, s_1, \ldots$.   □

3. GENERATING FUNCTIONS

So far, our approach to linear recurring sequences has employed only linear algebra, polynomial algebra, and the theory of finite fields. By using the algebraic apparatus of formal power series, other remarkable facts about linear recurring sequences can be established.

Given an arbitrary sequence $s_0, s_1, \ldots$ of elements of $\mathbb{F}_q$, we associate with it its *generating function*, which is a purely formal expression of the type

$$G(x) = s_0 + s_1 x + s_2 x^2 + \cdots + s_n x^n + \cdots = \sum_{n=0}^{\infty} s_n x^n \qquad (6.13)$$

with an indeterminate x. The underlying idea is that in $G(x)$ we have "stored" all the terms of the sequence in the correct order, so that $G(x)$ should somehow reflect the properties of the sequence. The name "generating function" is, strictly speaking, a misnomer since we do not consider $G(x)$ in any way as a function, but just as a formal object (in an obvious analogy, polynomials are essentially formal objects not to be confused with functions). The term is carried over from the case of real or complex sequences, where it may often turn out that the series analogous to the one in (6.13) is convergent after substitution of a real or complex number x_0 for x, thus enabling us to attach a meaning to $G(x_0)$. In our present situation, the question of the convergence or divergence of the expression in (6.13) is moot, since we think of $G(x)$ as being nothing but a hieroglyph for the sequence $s_0, s_1, \ldots$.

In general, an object of the type

$$B(x) = b_0 + b_1 x + b_2 x^2 + \cdots + b_n x^n + \cdots = \sum_{n=0}^{\infty} b_n x^n,$$

with $b_0, b_1, \ldots$ being a sequence of elements of $\mathbb{F}_q$, is called a *formal power series* (over $\mathbb{F}_q$). In this context, the terms $b_0, b_1, \ldots$ of the sequence are also called the *coefficients* of the formal power series. The adjective "formal" refers again to the idea that the convergence or divergence (whatever that may mean) of these expressions is irrelevant for their study. Two such formal power series

$$B(x) = \sum_{n=0}^{\infty} b_n x^n \quad \text{and} \quad C(x) = \sum_{n=0}^{\infty} c_n x^n$$

over $\mathbb{F}_q$ are considered identical if $b_n = c_n$ for all $n = 0, 1, \ldots$. The set of all formal power series over $\mathbb{F}_q$ is then in an obvious one-to-one correspondence with the set of all sequences of elements of $\mathbb{F}_q$. Thus, it seems as if we have not gained anything from the transition to formal power series (save a conceptual complication). The *raison d'être* of these objects is the fact that we can endow the set of all formal power series over $\mathbb{F}_q$ with a rich and interesting algebraic structure in a fairly natural way. This will be discussed in the sequel.

We note first that we may think of a polynomial

$$p(x) = p_0 + p_1 x + \cdots + p_k x^k \in \mathbb{F}_q[x]$$

as a formal power series over $\mathbb{F}_q$ by identifying it with

$$P(x) = p_0 + p_1 x + \cdots + p_k x^k + 0 \cdot x^{k+1} + 0 \cdot x^{k+2} + \cdots.$$

We introduce now the algebraic operations of addition and multiplication for formal power series in such a way that they extend the corresponding operations for polynomials. In detail, if

$$B(x) = \sum_{n=0}^{\infty} b_n x^n \quad \text{and} \quad C(x) = \sum_{n=0}^{\infty} c_n x^n$$

are two formal power series over $\mathbb{F}_q$, we define their *sum* to be the formal power series

$$B(x) + C(x) = \sum_{n=0}^{\infty} (b_n + c_n) x^n$$

and their *product* to be the formal power series

$$B(x)C(x) = \sum_{n=0}^{\infty} d_n x^n, \quad \text{where } d_n = \sum_{k=0}^{n} b_k c_{n-k} \quad \text{for } n = 0, 1, \dots.$$

If $B(x)$ and $C(x)$ are both polynomials over $\mathbb{F}_q$, then the operations above obviously coincide with polynomial addition and multiplication, respectively. It should be observed at this point that the substitution principle, which is so useful in polynomial algebra, is not valid for formal power series, the simple reason being that the expression $B(a)$ with $a \in \mathbb{F}_q$ and $B(x)$ a formal power series over $\mathbb{F}_q$ may be meaningless. This is, of course, the price we have to pay for disregarding convergence questions.

6.35. Example. Let

$$B(x) = 2 + x^2$$

and

$$C(x) = 1 + x + x^2 + \cdots + x^n + \cdots = \sum_{n=0}^{\infty} 1 \cdot x^n$$

be formal power series over $\mathbb{F}_3$. Then

$$B(x) + C(x) = x + 2x^2 + x^3 + \cdots + x^n + \cdots = \sum_{n=0}^{\infty} d_n x^n$$

with $d_0 = 0$, $d_1 = 1$, $d_2 = 2$, and $d_n = 1$ for $n \geqslant 3$, and

$$B(x)C(x) = 2 + 2x + 0 \cdot x^2 + 0 \cdot x^3 + \cdots = 2 + 2x. \qquad \square$$

Addition of formal power series over $\mathbb{F}_q$ is clearly associative and commutative. The formal power series $0 = \sum_{n=0}^{\infty} 0 \cdot x^n$ serves as an identity element for addition, and if $B(x) = \sum_{n=0}^{\infty} b_n x^n$ is an arbitrary formal power series over $\mathbb{F}_q$, then it has the additive inverse $\sum_{n=0}^{\infty} (-b_n) x^n$, denoted by $-B(x)$. As usual, we shall write $B(x) - C(x)$ instead of $B(x) + (-C(x))$.

Evidently, multiplication of formal power series over $\mathbb{F}_q$ is commutative, and the formal power series $1 = 1 + 0 \cdot x + 0 \cdot x^2 + \cdots + 0 \cdot x^n + \cdots$ acts as a multiplicative identity. Multiplication is associative, for if

$$B(x) = \sum_{n=0}^{\infty} b_n x^n, \quad C(x) = \sum_{n=0}^{\infty} c_n x^n, \quad \text{and} \quad D(x) = \sum_{n=0}^{\infty} d_n x^n,$$

then $(B(x)C(x))D(x)$ and $B(x)(C(x)D(x))$ are both identical with

$$\sum_{n=0}^{\infty} \left(\sum_{(i,j,k) \in L(n)} b_i c_j d_k \right) x^n,$$

where $L(n)$ is the set of all ordered triples (i, j, k) of nonnegative integers with $i + j + k = n$. Furthermore, the distributive law is satisfied since

$$B(x)(C(x) + D(x)) = \sum_{n=0}^{\infty} \left(\sum_{k=0}^{n} b_k (c_{n-k} + d_{n-k}) \right) x^n$$

$$= \sum_{n=0}^{\infty} \left(\sum_{k=0}^{n} b_k c_{n-k} + \sum_{k=0}^{n} b_k d_{n-k} \right) x^n$$

$$= \sum_{n=0}^{\infty} \left(\sum_{k=0}^{n} b_k c_{n-k} \right) x^n + \sum_{n=0}^{\infty} \left(\sum_{k=0}^{n} b_k d_{n-k} \right) x^n$$

$$= B(x)C(x) + B(x)D(x).$$

Altogether, we have shown that the set of all formal power series over $\mathbb{F}_q$, furnished with this addition and multiplication, is a commutative ring with identity, called the *ring of formal power series* over $\mathbb{F}_q$ and denoted by $\mathbb{F}_q[[x]]$. The polynomial ring $\mathbb{F}_q[x]$ is contained as a subring in $\mathbb{F}_q[[x]]$. We collect and extend the information on $\mathbb{F}_q[[x]]$ in the following theorem.

6.36. Theorem. *The ring $\mathbb{F}_q[[x]]$ of formal power series over $\mathbb{F}_q$ is an integral domain containing $\mathbb{F}_q[x]$ as a subring.*

Proof. It remains to verify that $\mathbb{F}_q[[x]]$ has no zero divisors—that is, that a product in $\mathbb{F}_q[[x]]$ can only be zero if one of the factors is zero. Suppose, on the contrary, that we have $B(x)C(x) = 0$ with

$$B(x) = \sum_{n=0}^{\infty} b_n x^n \neq 0 \quad \text{and} \quad C(x) = \sum_{n=0}^{\infty} c_n x^n \neq 0 \quad \text{in } \mathbb{F}_q[[x]].$$

Let k be the least nonnegative integer for which $b_k \neq 0$, and let m be the least nonnegative integer for which $c_m \neq 0$. Then the coefficient of x^{k+m} in $B(x)C(x)$ is $b_k c_m \neq 0$, which contradicts $B(x)C(x) = 0$. $\qquad\qquad\qquad\square$

It will be important for the applications to linear recurring sequences to find those $B(x) \in \mathbb{F}_q[[x]]$ that possess a multiplicative inverse—that is, for which there exists a $C(x) \in \mathbb{F}_q[[x]]$ with $B(x)C(x) = 1$. These formal power series can, in fact, be characterized easily.

6.37. Theorem. *The formal power series*

$$B(x) = \sum_{n=0}^{\infty} b_n x^n \in \mathbb{F}_q[[x]]$$

has a multiplicative inverse if and only if $b_0 \neq 0$.

Proof. If

$$C(x) = \sum_{n=0}^{\infty} c_n x^n \in \mathbb{F}_q[[x]]$$

is such that $B(x)C(x) = 1$, then the following infinite system of equations must be satisfied:

$$b_0 c_0 = 1$$
$$b_0 c_1 + b_1 c_0 = 0$$
$$b_0 c_2 + b_1 c_1 + b_2 c_0 = 0$$
$$\vdots \qquad \vdots$$
$$b_0 c_n + b_1 c_{n-1} + \cdots + b_n c_0 = 0$$
$$\vdots \qquad \vdots$$

From the first equation we conclude that necessarily $b_0 \neq 0$. However, if this condition is satisfied, then c_0 is uniquely determined by the first equation. Passing to the second equation, we see that c_1 is then uniquely determined. In general, the coefficients $c_0, c_1, \ldots$ can be computed recursively from the first equation and the recurrence relation

$$c_n = -b_0^{-1} \sum_{k=1}^{n} b_k c_{n-k} \quad \text{for } n = 1, 2, \ldots .$$

The resulting formal power series $C(x)$ is then a multiplicative inverse of $B(x)$. $\qquad\qquad\qquad\square$

If a multiplicative inverse of $B(x) \in \mathbb{F}_q[[x]]$ exists, then it is, of course, uniquely determined. We use the notation $1/B(x)$ for it. A product $A(x)(1/B(x))$ with $A(x) \in \mathbb{F}_q[[x]]$ will usually be written in the form $A(x)/B(x)$. Since $\mathbb{F}_q[[x]]$ is an integral domain, the familiar rules for operating with fractions hold. The multiplicative inverse of $B(x)$ or an expression $A(x)/B(x)$ can be computed by the algorithm in the proof of

Theorem 6.37. Long division also provides an effective means for accomplishing such computations.

6.38.   Example.   Let $B(x) = 3 + x + x^2$, considered as a formal power series over $\mathbb{F}_5$. Then $B(x)$ has a multiplicative inverse by Theorem 6.37. We compute $1/B(x)$ by long division:

$$
\begin{array}{r}
2 + x \quad +4x^2 \quad +2x^4 \quad + \cdots \\
3 + x + x^2 \overline{\big)\, 1 + 0\cdot x \; + \; 0\cdot x^2 \quad + \quad 0\cdot x^3 + 0\cdot x^4 + 0\cdot x^5 + 0\cdot x^6 + \cdots} \\
-1 - 2x - \; 2x^2 \\
\hline
3x + \; 3x^2 \; + \; 0\cdot x^3 \\
-3x - \quad x^2 \; - \quad x^3 \\
\hline
2x^2 \; + \quad 4x^3 + 0\cdot x^4 \\
- \; 2x^2 \; - \quad 4x^3 - \; 4x^4 \\
\hline
x^4 + 0\cdot x^5 + 0\cdot x^6
\end{array}
$$

Thus we get

$$
\frac{1}{3 + x + x^2} = 2 + x + 4x^2 + 2x^4 + \cdots. \qquad \square
$$

6.39.   Example.   We compute $A(x)/B(x)$ in $\mathbb{F}_2[[x]]$, where

$$
A(x) = 1 + x + x^2 + x^3 + \cdots = \sum_{n=0}^{\infty} 1 \cdot x^n
$$

and $B(x) = 1 + x + x^3$. Using long division, dropping the terms with zero coefficients, and recalling that $1 = -1$ in $\mathbb{F}_2$, we get:

$$
\begin{array}{r}
1 + x^2 + x^3 + x^7 + \cdots \\
1 + x + x^3 \overline{\big)\, 1 + x \; + x^2 + x^3 + x^4 \; + x^5 + x^6 + x^7 + x^8 + x^9 + x^{10} + \cdots} \\
1 + x \qquad + x^3 \\
\hline
x^2 \qquad + x^4 \; + x^5 \\
x^2 + x^3 \qquad + x^5 \\
\hline
x^3 + x^4 \qquad + x^6 \\
x^3 + x^4 \qquad + x^6 \\
\hline
x^7 + x^8 + x^9 + x^{10} \\
\vdots
\end{array}
$$

Therefore,

$$
\frac{1 + x + x^2 + x^3 + \cdots}{1 + x + x^3} = 1 + x^2 + x^3 + x^7 + \cdots. \qquad \square
$$

In order to apply the theory of formal power series, we consider now a kth-order homogeneous linear recurring sequence $s_0, s_1, \ldots$ in $\mathbb{F}_q$ satisfying the linear recurrence relation (6.7) and define its *reciprocal characteristic polynomial* to be

$$f^*(x) = 1 - a_{k-1}x - a_{k-2}x^2 - \cdots - a_0 x^k \in \mathbb{F}_q[x]. \qquad (6.14)$$

The characteristic polynomial $f(x)$ and the reciprocal characteristic polynomial are related by $f^*(x) = x^k f(1/x)$. The following basic identity can then be shown for the generating function of the given sequence.

6.40. Theorem. *Let $s_0, s_1, \ldots$ be a kth-order homogeneous linear recurring sequence in $\mathbb{F}_q$ satisfying the linear recurrence relation (6.7), let $f^*(x) \in \mathbb{F}_q[x]$ be its reciprocal characteristic polynomial, and let $G(x) \in \mathbb{F}_q[[x]]$ be its generating function in (6.13). Then the identity*

$$G(x) = \frac{g(x)}{f^*(x)} \qquad (6.15)$$

holds with

$$g(x) = - \sum_{j=0}^{k-1} \sum_{i=0}^{j} a_{i+k-j} s_i x^j \in \mathbb{F}_q[x], \qquad (6.16)$$

where we set $a_k = -1$. Conversely, if $g(x)$ is any polynomial over $\mathbb{F}_q$ with $\deg(g(x)) < k$ and if $f^*(x) \in \mathbb{F}_q[x]$ is given by (6.14), then the formal power series $G(x) \in \mathbb{F}_q[[x]]$ defined by (6.15) is the generating function of a kth-order homogeneous linear recurring sequence in $\mathbb{F}_q$ satisfying the linear recurrence relation (6.7).

Proof. We have

$$f^*(x)G(x) = -\left(\sum_{n=0}^{k} a_{k-n}x^n \right)\left(\sum_{n=0}^{\infty} s_n x^n \right)$$

$$= - \sum_{j=0}^{k-1} \left(\sum_{i=0}^{j} a_{i+k-j} s_i \right) x^j - \sum_{j=k}^{\infty} \left(\sum_{i=j-k}^{j} a_{i+k-j} s_i \right) x^j$$

$$= g(x) - \sum_{j=k}^{\infty} \left(\sum_{i=0}^{k} a_i s_{j-k+i} \right) x^j. \qquad (6.17)$$

Thus, if the sequence $s_0, s_1, \ldots$ satisfies (6.7), then $f^*(x)G(x) = g(x)$ because of (6.12). Since $f^*(x)$ has a multiplicative inverse in $\mathbb{F}_q[[x]]$ by Theorem 6.37, the identity (6.15) follows. Conversely, we infer from (6.17) that $f^*(x)G(x)$ is equal to a polynomial of degree less than k only if

$$\sum_{i=0}^{k} a_i s_{j-k+i} = 0 \quad \text{for all } j \geqslant k.$$

But these identities just express the fact that the sequence $s_0, s_1, \ldots$ of coefficients of $G(x)$ satisfies the linear recurrence relation (6.7). $\quad\square$

One may summarize the theorem above by saying that the kth-order homogeneous linear recurring sequences with reciprocal characteristic polynomial $f^*(x)$ are in one-to-one correspondence with the fractions $g(x)/f^*(x)$ with $\deg(g(x)) < k$. The identity (6.15) can be used to compute the terms of a linear recurring sequence by long division.

6.41.   Example.   Consider the linear recurrence relation

$$s_{n+4} = s_{n+3} + s_{n+1} + s_n, \quad n = 0, 1, \ldots, \text{ in } \mathbb{F}_2.$$

Its reciprocal characteristic polynomial is

$$f^*(x) = 1 - x - x^3 - x^4 = 1 + x + x^3 + x^4 \in \mathbb{F}_2[x].$$

If the initial state vector is $(1,1,0,1)$, then the polynomial $g(x)$ in (6.16) turns out to be $g(x) = 1 + x^2$. Therefore, the generating function $G(x)$ of the sequence can be obtained from the following long division:

$$
\begin{array}{l}
1\ \ +x\ \ +x^3+x^4+x^6+\cdots \\
1+x+x^3+x^4\big|\overline{1+x^2}\\
\underline{1\ \ +x+x^3+x^4}\\
x\ +x^2+x^3+x^4\\
\underline{x\ +x^2+x^4+\ x^5}\\
x^3+\ x^5\\
\underline{x^3+x^4+x^6+x^7}\\
x^4+\ \ x^5+x^6+x^7\\
\underline{x^4+\ \ x^5+x^7+x^8}\\
x^6+x^8
\end{array}
$$

The result is

$$G(x) = \frac{1 + x^2}{1 + x + x^3 + x^4} = 1 + x + x^3 + x^4 + x^6 + \cdots,$$

which corresponds to the string of binary digits $1101101\cdots$ of least period 3. The impulse response sequence associated with the given linear recurrence relation can be obtained by observing that $g(x) = x^3$ in this case, so that an appropriate long division yields

$$G(x) = \frac{x^3}{1 + x + x^3 + x^4} = x^3 + x^4 + x^5 + x^9 + x^{10} + x^{11} + \cdots,$$

which corresponds to the string of binary digits $000111000111\cdots$ of least period 6. $\quad\square$

On the basis of the identity (6.15), we present now an *alternative proof* of Theorem 6.25. Since the sequence $s_0, s_1, \ldots$ is periodic with period r, its generating function $G(x)$ can be written in the form

$$G(x) = \left(s_0 + s_1 x + \cdots + s_{r-1} x^{r-1}\right)\left(1 + x^r + x^{2r} + \cdots\right) = \frac{s^*(x)}{1 - x^r}$$

with $s^*(x) = s_0 + s_1 x + \cdots + s_{r-1} x^{r-1}$. On the other hand, using the notation of Theorem 6.40 we have $G(x) = g(x)/f^*(x)$ by (6.15). By equating these expressions for $G(x)$, we arrive at the polynomial identity $f^*(x) s^*(x) = (1 - x^r) g(x)$. If $f(x)$ and $s(x)$ are as in (6.9), then

$$f(x) s(x) = x^k f^*\left(\frac{1}{x}\right) x^{r-1} s^*\left(\frac{1}{x}\right) = (x^r - 1) x^{k-1} g\left(\frac{1}{x}\right),$$

and a comparison of (6.10) and (6.16) shows that

$$x^{k-1} g\left(\frac{1}{x}\right) = -h(x), \tag{6.18}$$

which implies already (6.9).

As another application of (6.15) we derive a *general formula for the terms of a linear recurring sequence*. Let $s_0, s_1, \ldots$ be a kth-order homogeneous linear recurring sequence in $\mathbb{F}_q$ with characteristic polynomial $f(x) \in \mathbb{F}_q[x]$. Let e_0 be the multiplicity of 0 as a root of $f(x)$, where we can have $e_0 = 0$, and let $\alpha_1, \ldots, \alpha_m$ be the distinct nonzero roots of $f(x)$ with multiplicities $e_1, \ldots, e_m$, respectively. For the reciprocal characteristic polynomial we obtain then

$$f^*(x) = x^k f\left(\frac{1}{x}\right) = \prod_{i=1}^{m} (1 - \alpha_i x)^{e_i}.$$

Since $\deg(f^*(x)) = k - e_0$, we get from (6.15)

$$G(x) = \frac{g(x)}{f^*(x)} = \sum_{n=0}^{e_0 - 1} t_n x^n + \frac{b(x)}{f^*(x)} \quad .$$

with $t_n \in \mathbb{F}_q$ and $\deg(b(x)) < k - e_0$. Partial fraction decomposition yields

$$\frac{b(x)}{f^*(x)} = \sum_{i=1}^{m} \sum_{j=0}^{e_i - 1} \frac{\beta_{ij}}{(1 - \alpha_i x)^{j+1}},$$

where the β_{ij} belong to the splitting field of $f(x)$ over $\mathbb{F}_q$. Now

$$\frac{1}{(1 - \alpha_i x)^{j+1}} = \sum_{n=0}^{\infty} \binom{n+j}{j} \alpha_i^n x^n,$$

and so

$$G(x) = \sum_{n=0}^{\infty} s_n x^n = \sum_{n=0}^{e_0 - 1} t_n x^n + \sum_{n=0}^{\infty} \left(\sum_{i=1}^{m} \sum_{j=0}^{e_i - 1} \binom{n+j}{j} \beta_{ij} \alpha_i^n \right) x^n.$$

Comparison of coefficients yields

$$s_n = t_n + \sum_{i=1}^{m} \sum_{j=0}^{e_i-1} \binom{n+j}{j} \beta_{ij} \alpha_i^n \quad \text{for} \quad n = 0, 1, \ldots,$$

where $t_n = 0$ for $n \geqslant e_0$. This is the desired formula. If $e_0 \leqslant 1$ and $e_i \leqslant p$ for $1 \leqslant i \leqslant m$, where p is the characteristic of $\mathbb{F}_q$, then it is easily seen that this formula is equivalent to the one given in Remark 6.23.

4. THE MINIMAL POLYNOMIAL

Although we have not yet pointed it out, it is evident that a linear recurring sequence satisfies many other linear recurrence relations apart from the one by which it is defined. For instance, if the sequence $s_0, s_1, \ldots$ is periodic with period r, it satisfies the linear recurrence relations $s_{n+r} = s_n$ ($n = 0, 1, \ldots$), $s_{n+2r} = s_n$ ($n = 0, 1, \ldots$), and so on. The most extreme case is represented by the sequence $0, 0, 0, \ldots$, which satisfies any homogeneous linear recurrence relation. The following theorem describes the relationship between the various linear recurrence relations valid for a given homogeneous linear recurring sequence.

6.42. Theorem. *Let $s_0, s_1, \ldots$ be a homogeneous linear recurring sequence in $\mathbb{F}_q$. Then there exists a uniquely determined monic polynomial $m(x) \in \mathbb{F}_q[x]$ having the following property: a monic polynomial $f(x) \in \mathbb{F}_q[x]$ of positive degree is a characteristic polynomial of $s_0, s_1, \ldots$ if and only if $m(x)$ divides $f(x)$.*

Proof. Let $f_0(x) \in \mathbb{F}_q[x]$ be the characteristic polynomial of a homogeneous linear recurrence relation satisfied by the sequence, and let $h_0(x) \in \mathbb{F}_q[x]$ be the polynomial in (6.10) determined by $f_0(x)$ and the sequence. If $d(x)$ is the (monic) greatest common divisor of $f_0(x)$ and $h_0(x)$, then we can write $f_0(x) = m(x)d(x)$ and $h_0(x) = b(x)d(x)$ with $m(x), b(x) \in \mathbb{F}_q[x]$. We shall prove that $m(x)$ is the desired polynomial. Clearly, $m(x)$ is monic. Now let $f(x) \in \mathbb{F}_q[x]$ be an arbitrary characteristic polynomial of the given sequence, and let $h(x) \in \mathbb{F}_q[x]$ be the polynomial in (6.10) determined by $f(x)$ and the sequence. By applying Theorem 6.40, we obtain that the generating function $G(x)$ of the sequence satisfies

$$G(x) = \frac{g_0(x)}{f_0^*(x)} = \frac{g(x)}{f^*(x)}$$

with $g_0(x)$ and $g(x)$ determined by (6.16). Therefore $g(x)f_0^*(x) = g_0(x)f^*(x)$, and using (6.18) we arrive at

$$h(x)f_0(x) = -x^{\deg(f(x))-1} g\left(\frac{1}{x}\right) x^{\deg(f_0(x))} f_0^*\left(\frac{1}{x}\right)$$

$$= - x^{\deg(f_0(x)) - 1} g_0 \left(\frac{1}{x} \right) x^{\deg(f(x))} f^* \left(\frac{1}{x} \right) = h_0(x) f(x).$$

After division by $d(x)$ we have $h(x)m(x) = b(x)f(x)$, and since $m(x)$ and $b(x)$ are relatively prime, it follows that $m(x)$ divides $f(x)$.

Now suppose that $f(x) \in F_q[x]$ is a monic polynomial of positive degree that is divisible by $m(x)$, say $f(x) = m(x)c(x)$ with $c(x) \in F_q[x]$. Passing to reciprocal polynomials, we get $f^*(x) = m^*(x)c^*(x)$ in an obvious notation. We also have $h_0(x)m(x) = b(x)f_0(x)$, so that, using the relation (6.18), we obtain

$$g_0(x)m^*(x) = - x^{\deg(f_0(x)) - 1} h_0 \left(\frac{1}{x} \right) x^{\deg(m(x))} m \left(\frac{1}{x} \right)$$

$$= - x^{\deg(m(x)) - 1} b \left(\frac{1}{x} \right) x^{\deg(f_0(x))} f_0 \left(\frac{1}{x} \right).$$

Since $\deg(b(x)) < \deg(m(x))$, the product of the first two factors on the right-hand side (negative sign included) is a polynomial $a(x) \in F_q[x]$. Therefore, we have $g_0(x)m^*(x) = a(x)f_0^*(x)$. It follows then from Theorem 6.40 that the generating function $G(x)$ of the sequence satisfies

$$G(x) = \frac{g_0(x)}{f_0^*(x)} = \frac{a(x)}{m^*(x)} = \frac{a(x)c^*(x)}{m^*(x)c^*(x)} = \frac{a(x)c^*(x)}{f^*(x)}.$$

Since

$$\deg(a(x)c^*(x)) = \deg(a(x)) + \deg(c^*(x))$$

$$< \deg(m(x)) + \deg(c(x)) = \deg(f(x)),$$

the second part of Theorem 6.40 shows that $f(x)$ is a characteristic polynomial of the sequence. It is clear that there can only be one polynomial $m(x)$ with the indicated properties.   □

The uniquely determined polynomial $m(x)$ over F_q associated with the sequence $s_0, s_1, \ldots$ according to Theorem 6.42 is called the *minimal polynomial* of the sequence. If $s_n = 0$ for all $n \geq 0$, the minimal polynomial is equal to the constant polynomial 1. For all other homogeneous linear recurring sequences, $m(x)$ is a monic polynomial with $\deg(m(x)) > 0$ that is, in fact, the characteristic polynomial of the linear recurrence relation of least possible order satisfied by the sequence. Another method of calculating the minimal polynomial will be introduced in Section 6.

6.43. Example. Let $s_0, s_1, \ldots$ be the linear recurring sequence in F_2 with

$$s_{n+4} = s_{n+3} + s_{n+1} + s_n, \quad n = 0, 1, \ldots,$$

and initial state vector $(1, 1, 0, 1)$. To find the minimal polynomial, we proceed as in the proof of Theorem 6.42. We may take $f_0(x) = x^4 - x^3 - x - 1 = x^4 + x^3 + x + 1 \in F_2[x]$. Then by (6.10) the polynomial $h_0(x)$ is given by $h_0(x) = x^3 + x$. The greatest common divisor of $f_0(x)$ and $h_0(x)$ is $d(x) = x^2 + 1$, and so the minimal polynomial of the sequence is $m(x) = f_0(x)/d(x) = x^2 + x + 1$. One checks easily that the sequence satisfies the linear recurrence relation

$$s_{n+2} = s_{n+1} + s_n, \quad n = 0, 1, \ldots,$$

as it should according to the general theory. We note that $\operatorname{ord}(m(x)) = 3$, which is identical with the least period of the sequence (compare with Example 6.41). We shall see in Theorem 6.44 below that this is true in general.   □

The minimal polynomial plays a decisive role in the determination of the least period of a linear recurring sequence. This is shown by the following result.

6.44. Theorem. Let $s_0, s_1, \ldots$ be a homogeneous linear recurring sequence in $\mathbb{F}_q$ with minimal polynomial $m(x) \in \mathbb{F}_q[x]$. Then the least period of the sequence is equal to $\operatorname{ord}(m(x))$.

Proof. If r is the least period of the sequence and n_0 its preperiod, then we have $s_{n+r} = s_n$ for all $n \geq n_0$. Therefore, the sequence satisfies the homogeneous linear recurrence relation

$$s_{n+n_0+r} = s_{n+n_0} \quad \text{for } n = 0, 1, \ldots.$$

Then, according to Theorem 6.42, $m(x)$ divides $x^{n_0+r} - x^{n_0} = x^{n_0}(x^r - 1)$, so that $m(x)$ is of the form $m(x) = x^h g(x)$ with $h \leq n_0$ and $g(x) \in \mathbb{F}_q[x]$, where $g(0) \neq 0$ and $g(x)$ divides $x^r - 1$. It follows from the definition of the order of a polynomial that $\operatorname{ord}(m(x)) = \operatorname{ord}(g(x)) \leq r$. On the other hand, r divides $\operatorname{ord}(m(x))$ by Theorem 6.27, and so $r = \operatorname{ord}(m(x))$.   □

6.45. Example. Let $s_0, s_1, \ldots$ be the linear recurring sequence in $\mathbb{F}_2$ with $s_{n+5} = s_{n+1} + s_n$, $n = 0, 1, \ldots$, and initial state vector $(1, 1, 1, 0, 1)$. Following the method in the proof of Theorem 6.42, we take $f_0(x) = x^5 - x - 1 = x^5 + x + 1 \in \mathbb{F}_2[x]$ and get $h_0(x) = x^4 + x^3 + x^2$ from (6.10). Then $d(x) = x^2 + x + 1$, and so the minimal polynomial $m(x)$ of the sequence is given by $m(x) = f_0(x)/d(x) = x^3 + x^2 + 1$. We have $\operatorname{ord}(m(x)) = 7$, and so Theorem 6.44 implies that the least period of the sequence is 7 (compare with Example 6.18).   □

The argument in the example above shows how to find the least period of a linear recurring sequence without evaluating its terms. The method is particularly effective if a table of orders of polynomials is available. Since such tables usually incorporate only irreducible polynomials (see Chapter 10, Section 2), the results in Theorems 3.8 and 3.9 may have to be used to find the order of a given polynomial (compare with Example 3.10).

6.46.   Example. The method in Example 6.45 can also be applied to inhomogeneous linear recurring sequences. Let $s_0, s_1, \ldots$ be such a sequence in $\mathbb{F}_2$ with

$$s_{n+4} = s_{n+3} + s_{n+1} + s_n + 1 \quad \text{for } n = 0, 1, \ldots$$

and initial state vector $(1, 1, 0, 1)$. According to (6.5), the sequence is also given by the homogeneous linear recurrence relation $s_{n+5} = s_{n+3} + s_{n+2} + s_n$, $n = 0, 1, \ldots$, with initial state vector $(1, 1, 0, 1, 0)$. Proceeding as in Example 6.45, we find that the characteristic polynomial

$$f(x) = x^5 + x^3 + x^2 + 1 = (x + 1)^3 (x^2 + x + 1) \in \mathbb{F}_2[x]$$

is in the present case identical with the minimal polynomial $m(x)$ of the sequence. Since $\mathrm{ord}((x + 1)^3) = 4$ by Theorem 3.8 and $\mathrm{ord}(x^2 + x + 1) = 3$, it follows from Theorem 3.9 that $\mathrm{ord}(m(x)) = 12$. Therefore, the sequence $s_0, s_1, \ldots$ is periodic with least period 12.   □

6.47.   Example. Consider the linear recurring sequence $s_0, s_1, \ldots$ in $\mathbb{F}_2$ with

$$s_{n+4} = s_{n+2} + s_{n+1} \quad \text{for } n = 0, 1, \ldots$$

and initial state vector $(1, 0, 1, 0)$. Then

$$f(x) = x^4 + x^2 + x = x(x^3 + x + 1) \in \mathbb{F}_2[x]$$

is a characteristic polynomial of the sequence, and since neither x nor $x^3 + x + 1$ is a characteristic polynomial, we have $m(x) = x^4 + x^2 + x$. The sequence is not periodic, but ultimately periodic with least period $\mathrm{ord}(m(x)) = 7$.   □

6.48.   Theorem. *Let $s_0, s_1, \ldots$ be a homogeneous linear recurring sequence in $\mathbb{F}_q$ and let b be a positive integer. Then the minimal polynomial $m_1(x)$ of the shifted sequence $s_b, s_{b+1}, \ldots$ divides the minimal polynomial $m(x)$ of the original sequence. If $s_0, s_1, \ldots$ is periodic, then $m_1(x) = m(x)$.*

Proof. To prove the first assertion, it suffices to show because of Theorem 6.42 that every homogeneous linear recurrence relation satisfied by the original sequence is also satisfied by the shifted sequence. But this is immediately evident. For the second part, let

$$s_{n+b+k} = a_{k-1} s_{n+b+k-1} + \cdots + a_0 s_{n+b}, \quad n = 0, 1, \ldots,$$

be a homogeneous linear recurrence relation satisfied by the shifted sequence. Let r be a period of $s_0, s_1, \ldots$, so that $s_{n+r} = s_n$ for all $n \geqslant 0$, and choose an integer c with $cr \geqslant b$. Then, by using the linear recurrence relation with n replaced by $n + cr - b$ and invoking the periodicity property, we find that

$$s_{n+k} = a_{k-1} s_{n+k-1} + \cdots + a_0 s_n \quad \text{for all } n \geqslant 0,$$

that is, that the sequence $s_0, s_1, \ldots$ satisfies the same linear recurrence relation as the shifted sequence. By applying again Theorem 6.42, we conclude that $m_1(x) = m(x)$.   □

6.49.   Example.   Let $s_0, s_1, \ldots$ be the linear recurring sequence in $\mathbb{F}_2$ considered in Example 6.47. Its minimal polynomial is $x^4 + x^2 + x$, whereas the minimal polynomial of the shifted sequence $s_1, s_2, \ldots$ is $x^3 + x + 1$, which is a proper divisor of $x^4 + x^2 + x$. This example shows that the second assertion in Theorem 6.48 need not hold if $s_0, s_1, \ldots$ is only ultimately periodic, but not periodic.   □

6.50.   Theorem.   *Let $f(x) \in \mathbb{F}_q[x]$ be monic and irreducible over $\mathbb{F}_q$, and let $s_0, s_1, \ldots$ be a homogeneous linear recurring sequence in $\mathbb{F}_q$ not all of whose terms are 0. If the sequence has $f(x)$ as a characteristic polynomial, then the minimal polynomial of the sequence is equal to $f(x)$.*

Proof.   Since the minimal polynomial $m(x)$ of the sequence divides $f(x)$ according to Theorem 6.42, the irreducibility of $f(x)$ implies that either $m(x) = 1$ or $m(x) = f(x)$. But $m(x) = 1$ holds only for the sequence all of whose terms are 0, and so the result follows.   □

There is a general criterion for deciding whether the characteristic polynomial of the linear recurrence relation defining a given linear recurring sequence is already the minimal polynomial of the sequence.

6.51.   Theorem.   *Let $s_0, s_1, \ldots$ be a sequence in $\mathbb{F}_q$ satisfying a kth-order homogeneous linear recurrence relation with characteristic polynomial $f(x) \in \mathbb{F}_q[x]$. Then $f(x)$ is the minimal polynomial of the sequence if and only if the state vectors $\mathbf{s}_0, \mathbf{s}_1, \ldots, \mathbf{s}_{k-1}$ are linearly independent over $\mathbb{F}_q$.*

Proof.   Suppose $f(x)$ is the minimal polynomial of the sequence. If $\mathbf{s}_0, \mathbf{s}_1, \ldots, \mathbf{s}_{k-1}$ were linearly dependent over $\mathbb{F}_q$, we would have $b_0 \mathbf{s}_0 + b_1 \mathbf{s}_1 + \cdots + b_{k-1} \mathbf{s}_{k-1} = \mathbf{0}$ with coefficients $b_0, b_1, \ldots, b_{k-1} \in \mathbb{F}_q$ not all of which are zero. Multiplying from the right by powers of the matrix A in (6.3) associated with the given linear recurrence relation yields

$$b_0 \mathbf{s}_n + b_1 \mathbf{s}_{n+1} + \cdots + b_{k-1} \mathbf{s}_{n+k-1} = \mathbf{0} \quad \text{for } n = 0, 1, \ldots,$$

because of (6.4). In particular, we obtain

$$b_0 s_n + b_1 s_{n+1} + \cdots + b_{k-1} s_{n+k-1} = 0 \text{ for } n = 0, 1, \ldots.$$

If $b_j = 0$ for $1 \leqslant j \leqslant k-1$, it follows that $s_n = 0$ for all $n \geqslant 0$, a contradiction to the fact that the minimal polynomial $f(x)$ of the sequence has positive degree. In the remaining case, let $j \geqslant 1$ be the largest index with $b_j \neq 0$. Then it follows that the sequence $s_0, s_1, \ldots$ satisfies a jth-order homogeneous linear recurrence relation with $j < k$, which again contradicts the assumption that $f(x)$ is the minimal polynomial. Therefore we have shown that $\mathbf{s}_0, \mathbf{s}_1, \ldots, \mathbf{s}_{k-1}$ are linearly independent over $\mathbb{F}_q$.

Conversely, suppose that $\mathbf{s}_0, \mathbf{s}_1, \ldots, \mathbf{s}_{k-1}$ are linearly independent over $\mathbb{F}_q$. Since $\mathbf{s}_0 \neq \mathbf{0}$, the minimal polynomial has positive degree. If $f(x)$ were not the minimal polynomial, the sequence $s_0, s_1, \ldots$ would satisfy an

mth-order homogeneous linear recurrence relation with $1 \leqslant m < k$, say

$$s_{n+m} = a_{m-1}s_{n+m-1} + \cdots + a_0 s_n \quad \text{for } n = 0, 1, \dots$$

with coefficients from $\mathbb{F}_q$. But this would imply $s_m = a_{m-1}s_{m-1} + \cdots + a_0 s_0$, a contradiction to the given linear independence property.   □

6.52. Corollary. *If $s_0, s_1, \dots$ is an impulse response sequence for some homogeneous linear recurrence relation in $\mathbb{F}_q$, then its minimal polynomial is equal to the characteristic polynomial of that linear recurrence relation.*

Proof. This follows from Theorem 6.51 since the required linear independence property is obviously satisfied for an impulse response sequence.   □

5.   FAMILIES OF LINEAR RECURRING SEQUENCES

Let $f(x) \in \mathbb{F}_q[x]$ be a monic polynomial of positive degree. We denote the set of all homogeneous linear recurring sequences in $\mathbb{F}_q$ with characteristic polynomial $f(x)$ by $S(f(x))$. In other words, $S(f(x))$ consists of all sequences in $\mathbb{F}_q$ satisfying the homogeneous linear recurrence relation determined by $f(x)$. If $\deg(f(x)) = k$, then $S(f(x))$ contains exactly q^k sequences, corresponding to the q^k different choices for initial state vectors. The set $S(f(x))$ may be considered as a vector space over $\mathbb{F}_q$ if operations for sequences are defined termwise. In detail, if σ is the sequence $s_0, s_1, \dots$ and τ the sequence $t_0, t_1, \dots$ in $\mathbb{F}_q$, then the sum $\sigma + \tau$ is taken to be the sequence $s_0 + t_0, s_1 + t_1, \dots$. Furthermore, if $c \in \mathbb{F}_q$, then $c\sigma$ is defined as the sequence $cs_0, cs_1, \dots$. It is seen immediately from the recurrence relation that $S(f(x))$ is closed under this addition and scalar multiplication. The required axioms are easily checked, and so $S(f(x))$ is indeed a vector space over $\mathbb{F}_q$. The role of the zero vector is played by the *zero sequence*, all of whose terms are 0. Since $S(f(x))$ has q^k elements, the dimension of the vector space is k. We obtain k linearly independent elements of $S(f(x))$ by choosing k linearly independent k-tuples $\mathbf{y}_1, \dots, \mathbf{y}_k$ of elements of $\mathbb{F}_q$ and considering the sequences $\sigma_1, \dots, \sigma_k$ belonging to $S(f(x))$, where each σ_j, $1 \leqslant j \leqslant k$, has $\mathbf{y}_j$ as its initial state vector. A natural choice for $\mathbf{y}_1, \dots, \mathbf{y}_k$ is to take the standard basis vectors

$$\mathbf{e}_1 = (1, 0, \dots, 0), \mathbf{e}_2 = (0, 1, \dots, 0), \dots, \mathbf{e}_k = (0, \dots, 0, 1).$$

Another possibility that is often advantageous is to consider the impulse response sequence $d_0, d_1, \dots$ belonging to $S(f(x))$ and to choose for $\mathbf{y}_1, \dots, \mathbf{y}_k$ the state vectors $\mathbf{d}_0, \dots, \mathbf{d}_{k-1}$ of this impulse response sequence.

In the following discussion, we shall explore the relationship between the various sets $S(f(x))$.

6.53. Theorem. *Let $f(x)$ and $g(x)$ be two nonconstant monic polynomials over $\mathbb{F}_q$. Then $S(f(x))$ is a subset of $S(g(x))$ if and only if $f(x)$ divides $g(x)$.*

Proof. Suppose $S(f(x))$ is contained in $S(g(x))$. Consider the impulse response sequence belonging to $S(f(x))$. This sequence has $f(x)$ as its minimal polynomial because of Corollary 6.52. By hypothesis, the sequence belongs also to $S(g(x))$. Therefore, according to Theorem 6.42, its minimal polynomial $f(x)$ divides $g(x)$. Conversely, if $f(x)$ divides $g(x)$ and $s_0, s_1, \ldots$ is any sequence belonging to $S(f(x))$, then the minimal polynomial $m(x)$ of the sequence divides $f(x)$ by Theorem 6.42. Consequently, $m(x)$ divides $g(x)$, and so another application of Theorem 6.42 shows that the sequence $s_0, s_1, \ldots$ belongs to $S(g(x))$. Therefore, $S(f(x))$ is a subset of $S(g(x))$. □

6.54. Theorem. *Let $f_1(x), \ldots, f_h(x)$ be nonconstant monic polynomials over $\mathbb{F}_q$. If $f_1(x), \ldots, f_h(x)$ are relatively prime, then the intersection*

$$S(f_1(x)) \cap \cdots \cap S(f_h(x))$$

consists only of the zero sequence. If $f_1(x), \ldots, f_h(x)$ have a (monic) greatest common divisor $d(x)$ of positive degree, then

$$S(f_1(x)) \cap \cdots \cap S(f_h(x)) = S(d(x)).$$

Proof. The minimal polynomial $m(x)$ of a sequence in the intersection must divide $f_1(x), \ldots, f_h(x)$. In the case of relative primality, $m(x)$ is necessarily the constant polynomial 1; but only the zero sequence has this minimal polynomial. In the second case, we conclude that $m(x)$ divides $d(x)$, and then Theorem 6.42 implies that $S(f_1(x)) \cap \cdots \cap S(f_h(x))$ is contained in $S(d(x))$. The fact that $S(d(x))$ is a subset of $S(f_1(x)) \cap \cdots \cap S(f_h(x))$ follows immediately from Theorem 6.53. □

We define $S(f(x)) + S(g(x))$ to be the set of all sequences $\sigma + \tau$ with $\sigma \in S(f(x))$ and $\tau \in S(g(x))$. This definition can, of course, be extended to any finite number of such sets.

6.55. Theorem. *Let $f_1(x), \ldots, f_h(x)$ be nonconstant monic polynomials over $\mathbb{F}_q$. Then*

$$S(f_1(x)) + \cdots + S(f_h(x)) = S(c(x)),$$

where $c(x)$ is the (monic) least common multiple of $f_1(x), \ldots, f_h(x)$.

Proof. It suffices to consider the case $h = 2$ since the general case follows easily by induction. We note first that, according to Theorem 6.53, each sequence belonging to $S(f_1(x))$ or to $S(f_2(x))$ belongs to $S(c(x))$, and since the latter is a vector space, it follows that $S(f_1(x)) + S(f_2(x))$ is contained in $S(c(x))$. We compare now the dimensions of these vector spaces over $\mathbb{F}_q$. Writing $V_1 = S(f_1(x))$ and $V_2 = S(f_2(x))$ and letting $d(x)$ be the (monic) greatest common divisor of $f_1(x)$ and $f_2(x)$, we get

$$\dim(V_1 + V_2) = \dim(V_1) + \dim(V_2) - \dim(V_1 \cap V_2)$$
$$= \deg(f_1(x)) + \deg(f_2(x)) - \deg(d(x)),$$

where we have applied Theorem 6.54. But $c(x) = f_1(x)f_2(x)/d(x)$, and so

$$\dim(V_1 + V_2) = \deg(c(x)) = \dim(S(c(x))).$$

Therefore, the linear subspace $S(f_1(x)) + S(f_2(x))$ has the same dimension as the vector space $S(c(x))$, and so $S(f_1(x)) + S(f_2(x)) = S(c(x))$. $\qquad\square$

In the special case where $f(x)$ and $g(x)$ are relatively prime nonconstant monic polynomials over $\mathbb{F}_q$, we will have

$$S(f(x)g(x)) = S(f(x)) + S(g(x)).$$

Since, in this case, Theorem 6.54 shows that $S(f(x)) \cap S(g(x))$ consists only of the zero sequence, $S(f(x)g(x))$ is (in the language of linear algebra) the direct sum of the linear subspaces $S(f(x))$ and $S(g(x))$. In other words, every sequence $\sigma \in S(f(x)g(x))$ can be expressed uniquely in the form $\sigma = \sigma_1 + \sigma_2$ with $\sigma_1 \in S(f(x))$ and $\sigma_2 \in S(g(x))$.

Let us recall that $S(f(x))$ is a vector space over $\mathbb{F}_q$ whose dimension is equal to the degree of $f(x)$. This vector space has an interesting additional property: if the sequence $s_0, s_1, \ldots$ belongs to $S(f(x))$, then for every integer $b \geqslant 0$ the shifted sequence $s_b, s_{b+1}, \ldots$ again belongs to $S(f(x))$. This follows, of course, immediately from the linear recurrence relation. We express this property by saying that $S(f(x))$ is *closed under shifts of sequences*. Taken together, the properties listed here characterize the sets $S(f(x))$ completely.

6.56. Theorem. *Let E be a set of sequences in $\mathbb{F}_q$. Then $E = S(f(x))$ for some monic polynomial $f(x) \in \mathbb{F}_q[x]$ of positive degree if and only if E is a vector space over $\mathbb{F}_q$ of positive finite dimension (under the usual addition and scalar multiplication of sequences) which is closed under shifts of sequences.*

Proof. We have already noted above that these conditions are necessary. To establish the converse, consider an arbitrary sequence $\sigma \in E$ that is not the zero sequence. If $s_0, s_1, \ldots$ are the terms of σ and $b \geqslant 0$ is an integer, we denote by $\sigma^{(b)}$ the shifted sequence $s_b, s_{b+1}, \ldots$. By hypothesis, the sequences $\sigma^{(0)}, \sigma^{(1)}, \sigma^{(2)}, \ldots$ all belong to E. But E is a finite set, and so there exist nonnegative integers $i < j$ with $\sigma^{(i)} = \sigma^{(j)}$. It follows that the original sequence σ satisfies the homogeneous linear recurrence relation $s_{n+j} = s_{n+i}$, $n = 0, 1, \ldots$. According to Theorem 6.42, the sequence σ has then a minimal polynomial $m_\sigma(x) \in \mathbb{F}_q[x]$ of positive degree k, say. The state vectors $\mathbf{s}_0, \mathbf{s}_1, \ldots, \mathbf{s}_{k-1}$ of the sequence σ are thus linearly independent over $\mathbb{F}_q$ by virtue of Theorem 6.51. Consequently, the sequences

$\sigma^{(0)}, \sigma^{(1)}, \ldots, \sigma^{(k-1)}$ are linearly independent elements of $S(m_\sigma(x))$ and hence form a basis for $S(m_\sigma(x))$. Since $\sigma^{(0)}, \sigma^{(1)}, \ldots, \sigma^{(k-1)}$ belong to the vector space E, it follows that $S(m_\sigma(x))$ is a linear subspace of E. Letting E^* denote the set E with the zero sequence deleted and carrying out the argument above for every $\sigma \in E^*$, we arrive at the statement that the finite sum $\sum_{\sigma \in E^*} S(m_\sigma(x))$ of vector spaces is a linear subspace of E. On the other hand, it is trivial that E is contained in $\sum_{\sigma \in E^*} S(m_\sigma(x))$, and so $E = \sum_{\sigma \in E^*} S(m_\sigma(x))$. By invoking Theorem 6.55, we get

$$E = \sum_{\sigma \in E^*} S(m_\sigma(x)) = S(f(x)),$$

where $f(x)$ is the least common multiple of all the polynomials $m_\sigma(x)$ with σ running through E^*.   □

It follows from Theorem 6.55 that the sum of two or more homogeneous linear recurring sequences in $\mathbb{F}_q$ is again a homogeneous linear recurring sequence. A characteristic polynomial of the sum sequence is also obtained from this theorem. In important special cases, the minimal polynomial and the least period of the sum sequence can be determined directly on the basis of the corresponding information for the original sequences.

6.57. Theorem. *For each $i = 1, 2, \ldots, h$, let σ_i be a homogeneous linear recurring sequence in $\mathbb{F}_q$ with minimal polynomial $m_i(x) \in \mathbb{F}_q[x]$. If the polynomials $m_1(x), \ldots, m_h(x)$ are pairwise relatively prime, then the minimal polynomial of the sum $\sigma_1 + \cdots + \sigma_h$ is equal to the product $m_1(x) \cdots m_h(x)$.*

Proof. It suffices to consider the case $h = 2$ since the general case follows then by induction. If $m_1(x)$ or $m_2(x)$ is the constant polynomial 1, the result is trivial. Similarly, if the minimal polynomial $m(x) \in \mathbb{F}_q[x]$ of $\sigma_1 + \sigma_2$ is the constant polynomial 1, we obtain a trivial case. Therefore, we assume that the polynomials $m_1(x)$, $m_2(x)$, and $m(x)$ have positive degrees. Since

$$\sigma_1 + \sigma_2 \in S(m_1(x)) + S(m_2(x)) = S(m_1(x) m_2(x))$$

on account of Theorem 6.55, it follows that $m(x)$ divides $m_1(x) m_2(x)$. Now suppose that the terms of σ_1 are $s_0, s_1, \ldots$, that those of σ_2 are $t_0, t_1, \ldots$, and that

$$m(x) = x^k - a_{k-1} x^{k-1} - \cdots - a_0.$$

Then

$$s_{n+k} + t_{n+k} = a_{k-1}(s_{n+k-1} + t_{n+k-1}) + \cdots + a_0(s_n + t_n) \quad \text{for } n = 0, 1, \ldots.$$

If we set

$$u_n = s_{n+k} - a_{k-1} s_{n+k-1} - \cdots - a_0 s_n$$
$$= -t_{n+k} + a_{k-1} t_{n+k-1} + \cdots + a_0 t_n \quad \text{for } n = 0, 1, \ldots$$

and recall that $S(m_1(x))$ and $S(m_2(x))$ are vector spaces over $\mathbb{F}_q$ closed under shifts of sequences (see Theorem 6.56), then we can conclude that the sequence $u_0, u_1, \ldots$ belongs to both $S(m_1(x))$ and $S(m_2(x))$ and is thus the zero sequence, according to Theorem 6.54. But this shows that both $m_1(x)$ and $m_2(x)$ divide $m(x)$, hence $m_1(x)m_2(x)$ divides $m(x)$, and so $m(x) = m_1(x)m_2(x)$. □

If the minimal polynomials $m_1(x), \ldots, m_h(x)$ of the individual sequences $\sigma_1, \ldots, \sigma_h$ are not pairwise relatively prime, then the special nature of the sequences $\sigma_1, \ldots, \sigma_h$ has to be taken into account in order to determine the minimal polynomial of the sum sequence $\sigma = \sigma_1 + \cdots + \sigma_h$. The most feasible method is based on the use of generating functions. Suppose that for $i = 1, 2, \ldots, h$ the generating function of σ_i is $G_i(x) \in \mathbb{F}_q[[x]]$. Then the generating function of σ is given by $G(x) = G_1(x) + \cdots + G_h(x)$. By Theorem 6.40, each $G_i(x)$ can be written as a fraction with, for instance, the reciprocal polynomial of $m_i(x)$ as denominator. We add these fractions, reduce the resulting fraction to lowest terms, and combine the second part of Theorem 6.40 and the method in the proof of Theorem 6.42 to find the minimal polynomial of σ. This technique yields also an alternative proof for Theorem 6.57.

6.58. Example. Let σ_1 be the impulse response sequence in $\mathbb{F}_2$ belonging to $S(x^4 + x^3 + x + 1)$ and σ_2 the impulse response sequence in $\mathbb{F}_2$ belonging to $S(x^5 + x^4 + 1)$. Then, according to Corollary 6.52, the corresponding minimal polynomials are

$$m_1(x) = x^4 + x^3 + x + 1 = (x^2 + x + 1)(x + 1)^2 \in \mathbb{F}_2[x]$$

and

$$m_2(x) = x^5 + x^4 + 1 = (x^2 + x + 1)(x^3 + x + 1) \in \mathbb{F}_2[x].$$

Using Theorem 6.40, the generating function $G(x)$ of the sum sequence $\sigma = \sigma_1 + \sigma_2$ turns out to be

$$G(x) = \frac{x^3}{(x^2 + x + 1)(x + 1)^2} + \frac{x^4}{(x^2 + x + 1)(x^3 + x^2 + 1)}$$

$$= \frac{x^3}{(x^3 + x^2 + 1)(x + 1)^2}.$$

By the second part of Theorem 6.40, the reciprocal polynomial $f_0(x) = (x^3 + x + 1)(x + 1)^2$ of the denominator is a characteristic polynomial of σ. According to (6.18), the associated polynomial $h_0(x)$ is given by $h_0(x) = -x^4(1/x)^3 = -x$. Since $f_0(x)$ and $h_0(x)$ are relatively prime, the method in the proof of Theorem 6.42 yields the minimal polynomial

$$m(x) = (x^3 + x + 1)(x + 1)^2$$

for σ. We note that $m(x)$ is a proper divisor of the least common multiple of $m_1(x)$ and $m_2(x)$, which is

$$(x^2 + x + 1)(x + 1)^2(x^3 + x + 1). \qquad \square$$

From the information about the minimal polynomial contained in Theorem 6.57, one can immediately deduce a useful result concerning the least period of a sum sequence.

6.59. Theorem. *For each $i = 1,2,\ldots,h$, let σ_i be a homogeneous linear recurring sequence in $\mathbb{F}_q$ with minimal polynomial $m_i(x) \in \mathbb{F}_q[x]$ and least period r_i. If the polynomials $m_1(x),\ldots,m_h(x)$ are pairwise relatively prime, then the least period of the sum $\sigma_1 + \cdots + \sigma_h$ is equal to the least common multiple of $r_1,\ldots,r_h$.*

Proof. We consider only the case $h = 2$, the general result following by induction. If r is the least period of $\sigma_1 + \sigma_2$, then $r = \mathrm{ord}(m_1(x)m_2(x))$ by Theorems 6.44 and 6.57. An application of Theorem 3.9 shows that r is the least common multiple of $\mathrm{ord}(m_1(x))$ and $\mathrm{ord}(m_2(x))$, and so of r_1 and r_2. $\qquad \square$

6.60. Example. Let the sequences σ_1 and σ_2 be as in Example 6.58. Then the least periods of σ_1 and σ_2 are $r_1 = \mathrm{ord}(m_1(x)) = 6$ and $r_2 = \mathrm{ord}(m_2(x)) = 21$, respectively. The least period r of $\sigma_1 + \sigma_2$ is $r = \mathrm{ord}(m(x)) = 14$. In these computations of orders we use, of course, Theorem 3.9. The arguments above have been carried out without having evaluated the terms of the sequences involved. In this special case we may, of course, compare the results with explicit computations of the least periods:

σ_1:	000111000111000111000111100$\cdots$	least period $r_1 = 6$
σ_2:	000011111010100110001000001$\cdots$	least period $r_2 = 21$
$\sigma_1 + \sigma_2$:	000100111101100001001111101$\cdots$	least period $r = 14$

Notice that r is a proper divisor of the least common multiple of r_1 and r_2. $\square$

6.61. Theorem. *For each $i = 1,2,\ldots,h$, let σ_i be an ultimately periodic sequence in $\mathbb{F}_q$ with least period r_i. If $r_1,\ldots,r_h$ are pairwise relatively prime, then the least period of the sum $\sigma_1 + \cdots + \sigma_h$ is equal to the product $r_1 \cdots r_h$.*

Proof. It suffices to consider the case $h = 2$ since the general case follows then by induction. It is obvious that $r_1 r_2$ is a period of $\sigma_1 + \sigma_2$, so that the least period r of $\sigma_1 + \sigma_2$ divides $r_1 r_2$. Therefore, r is of the form

$r = d_1 d_2$ with d_1 and d_2 being positive divisors of r_1 and r_2, respectively. In particular, $d_1 r_2$ is a period of $\sigma_1 + \sigma_2$. Consequently, if the terms of σ_1 are $s_0, s_1, \ldots$ and those of σ_2 are $t_0, t_1, \ldots$, then we have

$$s_{n + d_1 r_2} + t_{n + d_1 r_2} = s_n + t_n$$

for all sufficiently large n. But $t_{n + d_1 r_2} = t_n$ for all sufficiently large n, and so $s_{n + d_1 r_2} = s_n$ for all sufficiently large n. Therefore, r_1 divides $d_1 r_2$, and since r_1 and r_2 are relatively prime, r_1 divides d_1, which implies $d_1 = r_1$. Similarly, one shows that $d_2 = r_2$. $\qquad \square$

In the finite field $\mathbb{F}_2$, there is an interesting operation on sequences called binary complementation. If σ is a sequence in $\mathbb{F}_2$, then its *binary complement*, denoted by $\bar{\sigma}$, is obtained by replacing each digit 0 in σ by 1 and each digit 1 in σ by 0. Binary complementation is, in fact, a special case of addition of sequences since the binary complement $\bar{\sigma}$ of σ arises by adding to σ the sequence all of whose terms are 1. Therefore, if σ is a homogeneous linear recurring sequence, then $\bar{\sigma}$ is one as well. Clearly, the least period of $\bar{\sigma}$ is the same as that of σ. The minimal polynomial of $\bar{\sigma}$ can be obtained from that of σ in an easy manner.

6.62. Theorem. *Let σ be a homogeneous linear recurring sequence in $\mathbb{F}_2$ with binary complement $\bar{\sigma}$. Write the minimal polynomial $m(x) \in \mathbb{F}_2[x]$ of σ in the form $m(x) = (x + 1)^h m_1(x)$ with an integer $h \geq 0$ and $m_1(x) \in \mathbb{F}_2[x]$ satisfying $m_1(1) = 1$. Then the minimal polynomial $\bar{m}(x)$ of $\bar{\sigma}$ is given by $\bar{m}(x) = (x + 1)m(x)$ if $h = 0$, $\bar{m}(x) = m_1(x)$ if $h = 1$, and $\bar{m}(x) = m(x)$ if $h > 1$.*

Proof. Let ε be the sequence in $\mathbb{F}_2$ all of whose terms are 1. Since $\bar{\sigma} = \sigma + \varepsilon$ and the minimal polynomial of ε is $x + 1$, the case $h = 0$ is settled by invoking Theorem 6.57. If $h \geq 1$, then $\bar{\sigma} = \sigma + \varepsilon \in S(m(x))$ because of Theorem 6.55, and so $\bar{m}(x)$ divides $m(x)$. If $\bar{m}(x)$ is the constant polynomial 1, then $\bar{\sigma}$ is necessarily the zero sequence and $\sigma = \varepsilon$, and the theorem holds. Therefore, we assume from now on that $\bar{m}(x)$ is of positive degree. We get $\sigma = \bar{\sigma} + \varepsilon \in S(\bar{m}(x)(x + 1))$ because of Theorems 6.53 and 6.55, thus $m(x)$ divides $\bar{m}(x)(x + 1)$, and so for $h \geq 1$ we have either $\bar{m}(x) = m(x)$ or $\bar{m}(x) = (x + 1)^{h-1} m_1(x)$. If $h > 1$, it follows that $\sigma = \bar{\sigma} + \varepsilon \in S(\bar{m}(x))$, which yields $\bar{m}(x) = m(x)$. If $h = 1$, let the terms of σ be $s_0, s_1, \ldots$ and let

$$m_1(x) = x^k + a_{k-1} x^{k-1} + \cdots + a_0$$

be of positive degree, the excluded case being trivial. We set

$$u_n = s_{n+k} + a_{k-1} s_{n+k-1} + \cdots + a_0 s_n \quad \text{for } n = 0, 1, \ldots.$$

Since the sequence $s_0, s_1, \ldots$ has $m(x) = (x + 1)m_1(x)$ as a characteristic polynomial, it follows easily that $u_{n+1} = u_n$ for all $n \geq 0$. Therefore, $u_n = u_0$ for all $n \geq 0$, and we must have $u_0 = 1$, for otherwise $m_1(x)$ would be a characteristic polynomial of σ. Consequently,

$$s_{n+k} + 1 = a_{k-1} s_{n+k-1} + \cdots + a_0 s_n \quad \text{for all } n \geq 0.$$

Since $m_1(1) = 1 + a_{k-1} + \cdots + a_0 = 1$, we obtain

$$s_{n+k} + 1 = a_{k-1}(s_{n+k-1} + 1) + \cdots + a_0(s_n + 1) \quad \text{for all } n \geq 0,$$

and this means that $m_1(x)$ is a characteristic polynomial of $\bar{\sigma}$. Thus, $\overline{m}(x) = m_1(x)$ in the case where $h = 1$. $\qquad\square$

We recall that $S(f(x))$ denotes the set of all homogeneous linear recurring sequences in $\mathbb{F}_q$ with characteristic polynomial $f(x)$, where $f(x) \in \mathbb{F}_q[x]$ is a monic polynomial of positive degree. We want to determine the *positive integers that appear as least periods* of sequences from $S(f(x))$, and also, *for how many sequences* from $S(f(x))$ such a positive integer is attained as a least period.

The polynomial $f(x)$ can be written in the form $f(x) = x^h g(x)$, where $h \geq 0$ is an integer and $g(x) \in \mathbb{F}_q[x]$ with $g(0) \neq 0$. The case in which $g(x)$ is a constant polynomial can be dealt with immediately, since then every sequence from $S(f(x))$ has least period 1. If $h \geq 1$ and $g(x)$ is of positive degree, then, by the discussion following Theorem 6.55, every sequence $\sigma \in S(f(x))$ can be expressed uniquely in the form $\sigma = \sigma_1 + \sigma_2$ with $\sigma_1 \in S(x^h)$ and $\sigma_2 \in S(g(x))$. Apart from finitely many initial terms, all terms of σ_1 are zero, so that the least period of σ is equal to the least period of σ_2. Furthermore, a given sequence $\sigma_2 \in S(g(x))$ leads to q^h different sequences from $S(f(x))$ by adding to it all the q^h sequences from $S(x^h)$. Consequently, if $r_1, \ldots, r_t$ are the least periods of sequences from $S(g(x))$ and $N_1, \ldots, N_t$ are the corresponding numbers of sequences from $S(g(x))$ having these least periods, then, for $1 \leq i \leq t$, there are exactly $q^h N_i$ sequences belonging to $S(f(x))$ with least period r_i, and no other least periods occur among the sequences from $S(f(x))$.

We may assume from now on that $h = 0$—that is, that $f(0) \neq 0$. Suppose first that $f(x)$ is irreducible over $\mathbb{F}_q$. Then, according to Theorems 6.44 and 6.50, every sequence from $S(f(x))$ with nonzero initial state vector has least period $\text{ord}(f(x))$. Therefore, one sequence from $S(f(x))$ has least period 1 and $q^{\deg(f(x))} - 1$ sequences from $S(f(x))$ have least period $\text{ord}(f(x))$.

Next, we consider the case that $f(x)$ is a power of an irreducible polynomial. Thus, let $f(x) = g(x)^b$ with $g(x) \in \mathbb{F}_q[x]$ monic and irreducible over $\mathbb{F}_q$ and $b \geq 2$ an integer. The minimal polynomial of any sequence from $S(f(x))$ with nonzero initial state vector is then of the form $g(x)^c$ with $1 \leq c \leq b$. According to Theorem 6.53, we have

$$S(g(x)) \subseteq S(g(x)^2) \subseteq \cdots \subseteq S(f(x)).$$

Therefore, if $\deg(g(x)) = k$, then there are $q^k - 1$ sequences from $S(f(x))$

with minimal polynomial $g(x)$, $q^{2k} - q^k$ sequences from $S(f(x))$ with minimal polynomial $g(x)^2$, and, in general, for $c = 1, 2, \ldots, b$ there are $q^{ck} - q^{(c-1)k}$ sequences from $S(f(x))$ with minimal polynomial $g(x)^c$. By combining this information with Theorems 3.8 and 6.44, we arrive at the following result.

6.63. Theorem. *Let $f(x) = g(x)^b$ with $g(x) \in \mathbb{F}_q[x]$ monic and irreducible over $\mathbb{F}_q$, $g(0) \neq 0$, $\deg(g(x)) = k$, $\mathrm{ord}(g(x)) = e$, and b a positive integer. Let t be the smallest integer with $p^t \geq b$, where p is the characteristic of $\mathbb{F}_q$. Then $S(f(x))$ contains the following numbers of sequences with the following least periods: one sequence with least period 1, $q^k - 1$ sequences with least period e, and for $b \geq 2$, $q^{kp^j} - q^{kp^{j-1}}$ sequences with least period ep^j $(j = 1, 2, \ldots, t-1)$ and $q^{kb} - q^{kp^{t-1}}$ sequences with least period ep^t.*

In the case of an arbitrary monic polynomial $f(x) \in \mathbb{F}_q[x]$ of positive degree with $f(0) \neq 0$, we start from the canonical factorization

$$f(x) = \prod_{i=1}^{h} g_i(x)^{b_i},$$

where the $g_i(x)$ are distinct monic irreducible polynomials over $\mathbb{F}_q$ and the b_i are positive integers. It follows then from Theorem 6.55 that

$$S(f(x)) = S\left(g_1(x)^{b_1}\right) + \cdots + S\left(g_h(x)^{b_h}\right).$$

In fact, every sequence from $S(f(x))$ is obtained exactly once by forming all possible sums $\sigma_1 + \cdots + \sigma_h$ with $\sigma_i \in S(g_i(x)^{b_i})$ for $1 \leq i \leq h$. Since the least periods attained by sequences from $S(g_i(x)^{b_i})$ are known from Theorem 6.63, the analogous information about $S(f(x))$ can thus be deduced from Theorem 6.59.

6.64. Example. Let

$$f(x) = (x^2 + x + 1)^2(x^4 + x^3 + 1) \in \mathbb{F}_2[x].$$

According to Theorem 6.63, $S((x^2 + x + 1)^2)$ contains one sequence with least period 1, 3 sequences with least period 3, and 12 sequences with least period 6, whereas $S(x^4 + x^3 + 1)$ contains one sequence with least period 1 and 15 sequences with least period 15. Therefore, by forming all possible sums of sequences from $S((x^2 + x + 1)^2)$ and $S(x^4 + x^3 + 1)$ and using Theorem 6.59, we conclude that $S(f(x))$ contains one sequence with least period 1, 3 sequences with least period 3, 12 sequences with least period 6, 60 sequences with least period 15, and 180 sequences with least period 30. □

We have already investigated the behavior of linear recurring sequences under termwise addition. A similar theory can be developed for the operation of termwise multiplication, although it presents greater difficulties. If σ is the sequence of elements $s_0, s_1, \ldots$ of $\mathbb{F}_q$ and τ is the sequence of elements $t_0, t_1, \ldots$ of $\mathbb{F}_q$, then the product sequence $\sigma\tau$ has terms $s_0 t_0, s_1 t_1, \ldots$. Analogously, one defines the product of any finite number of sequences. Let S be the vector space over $\mathbb{F}_q$ consisting of all sequences of elements of $\mathbb{F}_q$, under the usual addition and scalar multiplication of sequences. For nonconstant monic polynomials $f_1(x), \ldots, f_h(x)$ over $\mathbb{F}_q$, let $S(f_1(x)) \cdots S(f_h(x))$ be the subspace of S spanned by all products $\sigma_1 \cdots \sigma_h$ with $\sigma_i \in S(f_i(x))$, $1 \leqslant i \leqslant h$. The following result is basic.

6.65. Theorem. *If $f_1(x), \ldots, f_h(x)$ are nonconstant monic polynomials over $\mathbb{F}_q$, then there exists a nonconstant monic polynomial $g(x) \in \mathbb{F}_q[x]$ such that*

$$S(f_1(x)) \cdots S(f_h(x)) = S(g(x)).$$

Proof. Set $E = S(f_1(x)) \cdots S(f_h(x))$. Since each $S(f_i(x))$, $1 \leqslant i \leqslant h$, contains a sequence with initial term 1, the vector space E contains a nonzero sequence. Furthermore, E is spanned by finitely many sequences and thus finite-dimensional. From the fact that each $S(f_i(x))$, $1 \leqslant i \leqslant h$, is closed under shifts of sequences it follows that E has the same property, and then the argument is complete by Theorem 6.56. $\square$

6.66. Corollary. *The product of finitely many linear recurring sequences in $\mathbb{F}_q$ is again a linear recurring sequence in $\mathbb{F}_q$.*

Proof. By the remarks following (6.5), the given linear recurring sequences can be taken to be homogeneous. The result is then implicit in Theorem 6.65. $\square$

The explicit determination of the polynomial $g(x)$ in Theorem 6.65 is, in general, not easy. There is, however, a special case that allows a simpler treatment of the problem.

For nonconstant polynomials $f_1(x), \ldots, f_h(x)$ over $\mathbb{F}_q$, we define $f_1(x) \vee \cdots \vee f_h(x)$ to be the monic polynomial whose roots are the distinct elements of the form $\alpha_1 \cdots \alpha_h$, where each α_i is a root of $f_i(x)$ in the splitting field of $f_1(x) \cdots f_h(x)$ over $\mathbb{F}_q$. Since the conjugates (over $\mathbb{F}_q$) of such a product $\alpha_1 \cdots \alpha_h$ are again elements of this form, it follows that $f_1(x) \vee \cdots \vee f_h(x)$ is a polynomial over $\mathbb{F}_q$.

6.67. Theorem. *For each $i = 1, 2, \ldots, h$, let $f_i(x)$ be a nonconstant monic polynomial over $\mathbb{F}_q$ without multiple roots. Then we have*

$$S(f_1(x)) \cdots S(f_h(x)) = S(f_1(x) \vee \cdots \vee f_h(x)).$$

We need a preparatory lemma and some notation for the proof of this result. For a finite extension field F of $\mathbb{F}_q$, let S_F be the vector space over F consisting of all sequences of elements of F, under termwise addition and scalar multiplication of sequences. Thus, in particular, $S_{\mathsf{F}_q} = S$. By the product $V_1 \cdots V_h$ of h subspaces $V_1, \ldots, V_h$ of S_F we mean the subspace of S_F spanned by all products $\sigma_1 \cdots \sigma_h$ with $\sigma_i \in V_i$, $1 \leq i \leq h$. For a nonconstant monic polynomial $f(x) \in F[x]$, let $S_F(f(x))$ be the vector space over F consisting of all homogeneous linear recurring sequences in F with characteristic polynomial $f(x)$.

6.68. Lemma. *Let F be a finite extension field of F_q, and let $f_1(x), \ldots, f_h(x)$ be nonconstant monic polynomials over F_q. Then,*

$$S(f_1(x)) \cdots S(f_h(x)) = S \cap (S_F(f_1(x)) \cdots S_F(f_h(x))).$$

Proof. Clearly, the vector space on the left-hand side is contained in the vector space on the right-hand side. To show the converse, we note first that each $S(f_i(x))$, $1 \leq i \leq h$, spans $S_F(f_i(x))$ over F. Therefore, $S(f_1(x)) \cdots S(f_h(x))$ spans $S_F(f_1(x)) \cdots S_F(f_h(x))$ over F. Let $\rho_1, \ldots, \rho_m$ be a basis of $S(f_1(x)) \cdots S(f_h(x))$ over F_q, and let $\omega_1, \ldots, \omega_k$ be a basis of F over F_q with $\omega_1 \in \mathsf{F}_q$. Then any $\sigma \in S_F(f_1(x)) \cdots S_F(f_h(x))$ can be written in the form

$$\sigma = \sum_{i=1}^{k} \sum_{j=1}^{m} c_{ij} \omega_i \rho_j,$$

where the coefficients c_{ij} are in F_q. Let the terms of the sequence ρ_j, $1 \leq j \leq m$, be the elements $r_{j0}, r_{j1}, \ldots$ of F_q. If now $\sigma \in S$, then for the terms s_n, $n = 0, 1, \ldots$, of σ we get

$$s_n = \sum_{i=1}^{k} \left(\sum_{j=1}^{m} c_{ij} r_{jn} \right) \omega_i \in \mathsf{F}_q \quad \text{for } n = 0, 1, \ldots.$$

Since the coefficient of each ω_i is in F_q, it follows from the definition of $\omega_1, \ldots, \omega_k$ that $\sum_{j=1}^{m} c_{ij} r_{jn} = 0$ for $2 \leq i \leq k$ and all n. Consequently,

$$\sigma = \sum_{j=1}^{m} c_{1j} \omega_1 \rho_j \in S(f_1(x)) \cdots S(f_h(x))$$

and the proof is complete.   □

Proof of Theorem 6.67. Let F be the splitting field of $f_1(x) \cdots f_h(x)$ over F_q. For $1 \leq i \leq h$, let α_i run through the roots of $f_i(x)$. Then by Theorem 6.55,

$$S_F(f_i(x)) = \sum_{\alpha_i} S_F(x - \alpha_i) \quad \text{for } 1 \leq i \leq h.$$

We note that we have the distributive law $V_1(V_2 + V_3) = V_1 V_2 + V_1 V_3$ for subspaces V_1, V_2, V_3 of S_F, which is shown by observing that the left-hand vector space is contained in the right-hand vector space (by the distributive law for sequences) and that $V_1 V_2 \subseteq V_1(V_2 + V_3)$ and $V_1 V_3 \subseteq V_1(V_2 + V_3)$ imply $V_1 V_2 + V_1 V_3 \subseteq V_1(V_2 + V_3)$. On the basis of the distributive law, it follows that

$$S_F(f_1(x)) \cdots S_F(f_h(x)) = \sum_{\alpha_1, \ldots, \alpha_h} S_F(x - \alpha_1) \cdots S_F(x - \alpha_h).$$

It is easy to check directly that

$$S_F(x - \alpha_1) \cdots S_F(x - \alpha_h) = S_F(x - \alpha_1 \cdots \alpha_h),$$

and so

$$S_F(f_1(x)) \cdots S_F(f_h(x)) = \sum_{\alpha_1, \ldots, \alpha_h} S_F(x - \alpha_1 \cdots \alpha_h)$$

$$= S_F(f_1(x) \vee \cdots \vee f_h(x))$$

by Theorem 6.55. The result of Theorem 6.67 follows now from Lemma 6.68. □

Theorem 6.67 shows, in particular, how to find a characteristic polynomial for the product of homogeneous linear recurring sequences, at least in the special case considered there. For this purpose, an *alternative argument* may be based on Theorem 6.21. It suffices to carry out the details for the product of two homogeneous linear recurring sequences. Let the sequence $s_0, s_1, \ldots$ belong to $S(f(x))$ and let $t_0, t_1, \ldots$ belong to $S(g(x))$. If $f(x)$ has only the simple roots $\alpha_1, \ldots, \alpha_k$ and $g(x)$ has only the simple roots $\beta_1, \ldots, \beta_m$, then by (6.8),

$$s_n = \sum_{i=1}^{k} b_i \alpha_i^n \quad \text{and} \quad t_n = \sum_{j=1}^{m} c_j \beta_j^n \quad \text{for } n = 0, 1, \ldots,$$

where the coefficients b_i and c_j belong to a finite extension field of $\mathbb{F}_q$. If $\gamma_1, \ldots, \gamma_r$ are the distinct values of the products $\alpha_i \beta_j$, $1 \le i \le k$, $1 \le j \le m$, then

$$u_n = s_n t_n = \sum_{i=1}^{k} \sum_{j=1}^{m} b_i c_j (\alpha_i \beta_j)^n = \sum_{i=1}^{r} d_i \gamma_i^n \quad \text{for } n = 0, 1, \ldots,$$

with suitable coefficients $d_1, \ldots, d_r$ in a finite extension field of $\mathbb{F}_q$. Now let

$$h(x) = f(x) \vee g(x) = x^r - a_{r-1} x^{r-1} - \cdots - a_0 \in \mathbb{F}_q[x].$$

Then for $n = 0, 1, \ldots$ we have

$$u_{n+r} - a_{r-1} u_{n+r-1} - \cdots - a_0 u_n = \sum_{i=1}^{r} d_i \gamma_i^n h(\gamma_i) = 0,$$

and so the product sequence $u_0, u_1, \ldots$ has $h(x)$ as a characteristic polynomial.

6.69. Example. Consider the sequence $0, 1, 0, 1, \ldots$ in $\mathbb{F}_2$ with the least period 2 and minimal polynomial $(x - 1)^2$. If we multiply this sequence with itself, we get back the same sequence. On the other hand, $(x - 1)^2 \vee (x - 1)^2 = x - 1$, which is not a characteristic polynomial of the product sequence. Therefore, the identity in Theorem 6.67 may cease to hold if some of the polynomials $f_i(x)$ are allowed to have multiple roots.   □

There is an analog of Theorem 6.61 for multiplication of sequences. For obvious reasons, sequences for which all but finitely many terms are zero have to be excluded from consideration.

6.70. Theorem. *For each $i = 1, 2, \ldots, h$, let σ_i be an ultimately periodic sequence in $\mathbb{F}_q$ with infinitely many nonzero terms and with least period r_i. If $r_1, \ldots, r_h$ are pairwise relatively prime, then the least period of the product $\sigma_1 \cdots \sigma_h$ is equal to $r_1 \cdots r_h$.*

Proof. We consider only the case $h = 2$ since the general case follows then by induction. As in the proof of Theorem 6.61 one shows that the least period r of $\sigma_1 \sigma_2$ must be of the form $r = d_1 d_2$ with d_1 and d_2 being positive divisors of r_1 and r_2, respectively. In particular, $d_1 r_2$ is a period of $\sigma_1 \sigma_2$. Thus, if the terms of σ_1 are $s_0, s_1, \ldots$ and those of σ_2 are $t_0, t_1, \ldots$, then we have

$$s_{n + d_1 r_2} t_n = s_{n + d_1 r_2} t_{n + d_1 r_2} = s_n t_n$$

for all sufficiently large n. Since there exists an integer b with $t_n \neq 0$ for all sufficiently large $n \equiv b \bmod r_2$, it follows that $s_{n + d_1 r_2} = s_n$ for all such n. Now fix a sufficiently large n; by the Chinese remainder theorem, we can choose an integer $m \geq n$ with $m \equiv n \bmod r_1$ and $m \equiv b \bmod r_2$. Then

$$s_n = s_m = s_{m + d_1 r_2} = s_{n + d_1 r_2},$$

and so $d_1 r_2$ is a period of σ_1. Therefore, r_1 divides $d_1 r_2$, and since r_1 and r_2 are relatively prime, r_1 divides d_1, which implies $d_1 = r_1$. Similarly, one shows that $d_2 = r_2$.   □

Multiplication of sequences can be used to describe the relation between homogeneous linear recurring sequences belonging to characteristic polynomials that are powers of each other. The case in which one of the characteristic polynomials is linear has to be considered first.

6.71. Lemma. *If c is a nonzero element of $\mathbb{F}_q$ and k is a positive integer, then*

$$S\big((x - c)^k\big) = S(x - c)S\big((x - 1)^k\big).$$

Proof. Let the sequence $s_0, s_1, \ldots$ belong to $S(x - c)$, and let $t_0, t_1, \ldots$

belong to $S((x-1)^k)$. Then $s_n = c^n s_0$ for $n = 0, 1, \dots$ and

$$\sum_{i=0}^{k} \binom{k}{i}(-1)^{k-i} t_{n+i} = 0 \quad \text{for } n = 0, 1, \dots .$$

It follows that

$$\sum_{i=0}^{k} \binom{k}{i}(-c)^{k-i} s_{n+i} t_{n+i} = c^{n+k} s_0 \sum_{i=0}^{k} \binom{k}{i}(-1)^{k-i} t_{n+i} = 0$$

for $n = 0, 1, \dots$, and so

$$\sum_{i=0}^{k} \binom{k}{i}(-c)^{k-i} x^i = (x-c)^k$$

is a characteristic polynomial of the product sequence $s_0 t_0, s_1 t_1, \dots$. Consequently, the vector space $S(x-c)S((x-1)^k)$ is a subspace of $S((x-c)^k)$. Since $c \neq 0$, the first vector space has dimension k over $\mathbb{F}_q$ and is thus equal to $S((x-c)^k)$, which has the same dimension over $\mathbb{F}_q$.   □

6.72. Theorem. *Let $f(x) \in \mathbb{F}_q[x]$ be a nonconstant monic polynomial with $f(0) \neq 0$ and without multiple roots, and let k be a positive integer. Then,*

$$S(f(x)^k) = S(f(x))S((x-1)^k).$$

Proof. Let F be the splitting field of $f(x)$ over $\mathbb{F}_q$. Then, with α running through the roots of $f(x)$, we get

$$S_F(f(x)^k) = \sum_{\alpha} S_F((x-\alpha)^k)$$

by Theorem 6.55. Using Lemma 6.71 and the distributive law shown in the proof of Theorem 6.67, we obtain

$$S_F(f(x)^k) = \sum_{\alpha} S_F((x-1)^k) S_F(x-\alpha) = S_F((x-1)^k) \sum_{\alpha} S_F(x-\alpha)$$

$$= S_F((x-1)^k) S_F(f(x)),$$

where we applied Theorem 6.55 in the last step. The desired result follows now from Lemma 6.68.   □

6. CHARACTERIZATION OF LINEAR RECURRING SEQUENCES

It is an important problem to decide whether a given sequence of elements of $\mathbb{F}_q$ is a linear recurring sequence or not. From the theoretical point of view, the question can be settled immediately since *the linear recurring sequences in $\mathbb{F}_q$ are precisely the ultimately periodic sequences*. However, the periods of a linear recurring sequence (even of one of moderately low order) can be extremely long, so that in practice it may not be feasible to determine the nature of the sequence on the basis of this criterion. Alternative ways of characterizing linear recurring sequences employ techniques from linear algebra.

Let $s_0, s_1, \ldots$ be an arbitrary sequence of elements of $\mathbb{F}_q$. For integers $n \geqslant 0$ and $r \geqslant 1$, we introduce the *Hankel determinants*

$$
D_n^{(r)} = \begin{vmatrix} s_n & s_{n+1} & \cdots & s_{n+r-1} \\ s_{n+1} & s_{n+2} & \cdots & s_{n+r} \\ \vdots & \vdots & & \vdots \\ s_{n+r-1} & s_{n+r} & \cdots & s_{n+2r-2} \end{vmatrix}.
$$

It will transpire that linear recurring sequences can be characterized in terms of the vanishing of sufficiently many of these Hankel determinants.

6.73. Lemma. *Let $s_0, s_1, \ldots$ be an arbitrary sequence in $\mathbb{F}_q$, and let $n \geqslant 0$ and $r \geqslant 1$ be integers. Then $D_n^{(r)} = D_n^{(r+1)} = 0$ implies $D_{n+1}^{(r)} = 0$.*

Proof. For $m \geqslant 0$ define the vector $\mathbf{s}_m = (s_m, s_{m+1}, \ldots, s_{m+r-1})$. From $D_n^{(r)} = 0$ it follows that the vectors $\mathbf{s}_n, \mathbf{s}_{n+1}, \ldots, \mathbf{s}_{n+r-1}$ are linearly dependent over $\mathbb{F}_q$. If $\mathbf{s}_{n+1}, \ldots, \mathbf{s}_{n+r-1}$ are already linearly dependent over $\mathbb{F}_q$, we immediately get $D_{n+1}^{(r)} = 0$. Otherwise, $\mathbf{s}_n$ is a linear combination of $\mathbf{s}_{n+1}, \ldots, \mathbf{s}_{n+r-1}$. Set $\mathbf{s}'_m = (s_m, s_{m+1}, \ldots, s_{m+r})$ for $m \geqslant 0$. Then the vectors $\mathbf{s}'_n, \mathbf{s}'_{n+1}, \ldots, \mathbf{s}'_{n+r}$, being the row vectors of the vanishing determinant $D_n^{(r+1)}$, are linearly dependent over $\mathbb{F}_q$. If $\mathbf{s}'_n, \mathbf{s}'_{n+1}, \ldots, \mathbf{s}'_{n+r-1}$ are already linearly dependent over $\mathbb{F}_q$, then an application of the linear transformation

$$
L_1: (a_0, a_1, \ldots, a_r) \in \mathbb{F}_q^{r+1} \mapsto (a_1, \ldots, a_r) \in \mathbb{F}_q^r
$$

shows that $\mathbf{s}_{n+1}, \mathbf{s}_{n+2}, \ldots, \mathbf{s}_{n+r}$ are linearly dependent over $\mathbb{F}_q$, and so $D_{n+1}^{(r)} = 0$. Otherwise, $\mathbf{s}'_{n+r}$ is a linear combination of $\mathbf{s}'_n, \mathbf{s}'_{n+1}, \ldots, \mathbf{s}'_{n+r-1}$, and by an application of the linear transformation

$$
L_2: (a_0, \ldots, a_{r-1}, a_r) \in \mathbb{F}_q^{r+1} \mapsto (a_0, \ldots, a_{r-1}) \in \mathbb{F}_q^r
$$

we obtain that $\mathbf{s}_{n+r}$ is a linear combination of $\mathbf{s}_n, \mathbf{s}_{n+1}, \ldots, \mathbf{s}_{n+r-1}$. But in the case under consideration $\mathbf{s}_n$ is a linear combination of $\mathbf{s}_{n+1}, \ldots, \mathbf{s}_{n+r-1}$, so that the row vectors $\mathbf{s}_{n+1}, \ldots, \mathbf{s}_{n+r-1}, \mathbf{s}_{n+r}$ of $D_{n+1}^{(r)}$ are linearly dependent over $\mathbb{F}_q$, which implies $D_{n+1}^{(r)} = 0$. □

6.74. Theorem. *The sequence $s_0, s_1, \ldots$ in $\mathbb{F}_q$ is a linear recurring sequence if and only if there exists a positive integer r such that $D_n^{(r)} = 0$ for all but finitely many $n \geqslant 0$.*

Proof. Suppose $s_0, s_1, \ldots$ satisfies a kth-order homogeneous linear recurrence relation. For any fixed $n \geq 0$, consider the determinant $D_n^{(k+1)}$. Because of the linear recurrence relation, the $(k+1)$st row of $D_n^{(k+1)}$ is a linear combination of the first k rows, and so $D_n^{(k+1)} = 0$. The inhomogeneous case reduces to the homogeneous case by (6.5).

To show sufficiency, let $k+1$ be the least positive integer such that $D_n^{(k+1)} = 0$ for all but finitely many $n \geq 0$. If $k+1 = 1$, then we are done, and so we may assume $k \geq 1$. There is an integer $m \geq 0$ with $D_n^{(k+1)} = 0$ for all $n \geq m$. If we had $D_{n_0}^{(k)} = 0$ for some $n_0 \geq m$, then $D_n^{(k)} = 0$ for all $n \geq n_0$ by Lemma 6.73, which contradicts the definition of $k+1$. Therefore, $D_n^{(k)} \neq 0$ for all $n \geq m$. Setting $\mathbf{s}_n = (s_n, s_{n+1}, \ldots, s_{n+k})$, we note that for $n \geq m$ the vectors $\mathbf{s}_n, \mathbf{s}_{n+1}, \ldots, \mathbf{s}_{n+k}$, being the row vectors of $D_n^{(k+1)}$, are linearly dependent over $\mathbb{F}_q$. Since $D_n^{(k)} \neq 0$, the vectors $\mathbf{s}_n, \mathbf{s}_{n+1}, \ldots, \mathbf{s}_{n+k-1}$ are linearly independent over $\mathbb{F}_q$, and so $\mathbf{s}_{n+k}$ is a linear combination of $\mathbf{s}_n, \mathbf{s}_{n+1}, \ldots, \mathbf{s}_{n+k-1}$. It follows then by induction that each $\mathbf{s}_n$ with $n \geq m$ is a linear combination of $\mathbf{s}_m, \mathbf{s}_{m+1}, \ldots, \mathbf{s}_{m+k-1}$. The latter are k vectors in $\mathbb{F}_q^{k+1}$, therefore there exists a nonzero vector $(a_0, a_1, \ldots, a_k) \in \mathbb{F}_q^{k+1}$ with

$$a_0 s_n + a_1 s_{n+1} + \cdots + a_k s_{n+k} = 0 \quad \text{for } m \leq n \leq m+k-1.$$

This implies

$$a_0 s_n + a_1 s_{n+1} + \cdots + a_k s_{n+k} = 0 \quad \text{for all } n \geq m,$$

or

$$a_0 s_{n+m} + a_1 s_{n+m+1} + \cdots + a_k s_{n+m+k} = 0 \text{ for all } n \geq 0.$$

Thus, the sequence $s_0, s_1, \ldots$ satisfies a homogeneous linear recurrence relation of order at most $m+k$. $\qquad\square$

6.75. Theorem. *The sequence $s_0, s_1, \ldots$ in $\mathbb{F}_q$ is a homogeneous linear recurring sequence with minimal polynomial of degree k if and only if $D_0^{(r)} = 0$ for all $r \geq k+1$ and $k+1$ is the least positive integer for which this holds.*

Proof. If a given linear recurring sequence is the zero sequence, the necessity of the condition is clear. Otherwise, we have $k \geq 1$, and $D_0^{(r)} = 0$ for all $r \geq k+1$ follows since the $(k+1)$st row of $D_0^{(r)}$ is a linear combination of the first k rows. Moreover, we get $D_0^{(k)} \neq 0$ from Theorem 6.51, and so the necessity of the condition is shown in all cases.

Conversely, suppose the condition on the Hankel determinants is satisfied. By using Lemma 6.73 and induction on n, one establishes that $D_n^{(r)} = 0$ for all $r \geq k+1$ and all $n \geq 0$. In particular, $D_n^{(k+1)} = 0$ for all $n \geq 0$, and so $s_0, s_1, \ldots$ is a linear recurring sequence by Theorem 6.74. If its minimal polynomial has degree d, then, by what we have already shown in the first part, we know that $D_0^{(r)} = 0$ for all $r \geqslant d + 1$ and that $d + 1$ is the least positive integer for which this holds. It follows that $d = k$. $\square$

We note that if a homogeneous linear recurring sequence is known to have a minimal polynomial of degree $k \geqslant 1$, then the minimal polynomial is determined by the first $2k$ terms of the sequence. To see this, write down the equations (6.2) for $n = 0, 1, \ldots, k - 1$, thereby obtaining a system of k linear equations for the unknown coefficients $a_0, a_1, \ldots, a_{k-1}$ of the minimal polynomial. The determinant of this system is $D_0^{(k)}$, which is $\neq 0$ by Theorem 6.51. Therefore, the system can be solved uniquely.

An important question is that of the actual *computation of the minimal polynomial* of a given homogeneous linear recurring sequence. To be sure, a method of finding the minimal polynomial was already presented in the course of the proof of Theorem 6.42. This method depends on the prior knowledge of a characteristic polynomial of the sequence and on the determination of a greatest common divisor in $\mathbb{F}_q[x]$. We shall now discuss a recursive algorithm (called *Berlekamp-Massey algorithm*) which produces the minimal polynomial after finitely many steps, provided we know an upper bound for the degree of the minimal polynomial.

Let $s_0, s_1, \ldots$ be a sequence of elements of $\mathbb{F}_q$ with generating function $G(x) = \sum_{n=0}^{\infty} s_n x^n$. For $j = 0, 1, \ldots$ we define polynomials $g_j(x)$ and $h_j(x)$ over $\mathbb{F}_q$, integers m_j, and elements b_j of $\mathbb{F}_q$ as follows. Initially, we set

$$g_0(x) = 1, \quad h_0(x) = x, \quad \text{and} \quad m_0 = 0. \qquad (6.19)$$

Then we proceed recursively by letting b_j be the coefficient of x^j in $g_j(x)G(x)$ and setting:

$$g_{j+1}(x) = g_j(x) - b_j h_j(x),$$

$$h_{j+1}(x) = \begin{cases} b_j^{-1} x g_j(x) & \text{if } b_j \neq 0 \text{ and } m_j \geqslant 0, \\ x h_j(x) & \text{otherwise,} \end{cases} \qquad (6.20)$$

$$m_{j+1} = \begin{cases} -m_j & \text{if } b_j \neq 0 \text{ and } m_j \geqslant 0, \\ m_j + 1 & \text{otherwise.} \end{cases}$$

If $s_0, s_1, \ldots$ is a homogeneous linear recurring sequence with a minimal polynomial of degree k, then it turns out that $g_{2k}(x)$ is equal to the reciprocal minimal polynomial. Thus, the minimal polynomial $m(x)$ itself is given by $m(x) = x^k g_{2k}(1/x)$. If it is only known that the minimal polynomial is of degree $\leqslant k$, then set $r = \lfloor k + \frac{1}{2} - \frac{1}{2} m_{2k} \rfloor$, where $\lfloor y \rfloor$ denotes the greatest integer $\leqslant y$, and *the minimal polynomial $m(x)$ is given by* $m(x) = x^r g_{2k}(1/x)$. In both cases, it is seen immediately from the algorithm that $m(x)$ depends only on the $2k$ terms $s_0, s_1, \ldots, s_{2k-1}$ of the sequence.

Therefore, one may replace the generating function $G(x)$ in the algorithm by the polynomial

$$G_{2k-1}(x) = \sum_{n=0}^{2k-1} s_n x^n.$$

6.76.   Example.   The first 8 terms of a homogeneous linear recurring sequence in $\mathbb{F}_3$ of order $\leqslant 4$ are given by $0,2,1,0,1,2,1,0$. To find the minimal polynomial, we use the Berlekamp-Massey algorithm with

$$G_7(x) = 2x + x^2 + x^4 + 2x^5 + x^6 \in \mathbb{F}_3[x]$$

in place of $G(x)$. The computation is summarized in the following table.

j	$g_j(x)$	$h_j(x)$	m_j	b_j
0	1	x	0	0
1	1	x^2	1	2
2	$1+x^2$	$2x$	-1	1
3	$1+x+x^2$	$2x^2$	0	0
4	$1+x+x^2$	$2x^3$	1	2
5	$1+x+x^2+2x^3$	$2x+2x^2+2x^3$	-1	2
6	$1+x^3$	$2x^2+2x^3+2x^4$	0	1
7	$1+x^2+2x^3+x^4$	$x+x^4$	0	1
8	$1+2x+x^2+2x^3$		0	

Then, $r = \lfloor 4+\tfrac{1}{2}-\tfrac{1}{2}m_8 \rfloor = 4$, and so $m(x) = x^4 + 2x^3 + x^2 + 2x$. The homogeneous linear recurrence relation of least order satisfied by the sequence is therefore $s_{n+4} = s_{n+3} + 2s_{n+2} + s_{n+1}$ for $n = 0,1,\dots$.   □

6.77.   Example.   Find the homogeneous linear recurring sequence in $\mathbb{F}_2$ of least order whose first 8 terms are $1,1,0,0,1,0,1,1$. We use the Berlekamp-Massey algorithm with $G_7(x) = 1 + x + x^4 + x^6 + x^7 \in \mathbb{F}_2[x]$ in place of $G(x)$. The computation is summarized in the following table.

j	$g_j(x)$	$h_j(x)$	m_j	b_j
0	1	x	0	1
1	$1+x$	x	0	0
2	$1+x$	x^2	1	1
3	$1+x+x^2$	$x+x^2$	-1	1
4	1	x^2+x^3	0	1
5	$1+x^2+x^3$	x	0	0
6	$1+x^2+x^3$	x^2	1	0
7	$1+x^2+x^3$	x^3	2	0
8	$1+x^2+x^3$		3	

Then, $r = \lfloor 4 + \frac{1}{2} - \frac{1}{2}m_8 \rfloor = 3$, and so $m(x) = x^3 + x + 1$. Therefore, the given terms form the initial segment of a homogeneous linear recurring sequence $s_0, s_1, \ldots$ satisfying $s_{n+3} = s_{n+1} + s_n$ for $n = 0, 1, \ldots$, and no such sequence of lower order with these initial terms exists. $\square$

We shall now prove, in general, that *the Berlekamp-Massey algorithm yields the minimal polynomial* after the indicated number of steps. To this end, we define auxiliary polynomials $u_j(x)$ and $v_j(x)$ over $\mathbb{F}_q$ recursively by setting

$$u_0(x) = 0 \quad \text{and} \quad v_0(x) = -1, \tag{6.21}$$

and then for $j = 0, 1, \ldots$,

$$u_{j+1}(x) = u_j(x) - b_j v_j(x),$$

$$v_{j+1}(x) = \begin{cases} b_j^{-1} x u_j(x) & \text{if } b_j \neq 0 \text{ and } m_j \geq 0, \\ x v_j(x) & \text{otherwise.} \end{cases} \tag{6.22}$$

We claim that for each $j \geq 0$ we have

$$\deg(g_j(x)) \leq \tfrac{1}{2}(j + 1 - m_j) \quad \text{and} \quad \deg(h_j(x)) \leq \tfrac{1}{2}(j + 2 + m_j). \tag{6.23}$$

This is obvious for $j = 0$ because of the initial conditions in (6.19), and assuming the inequalities to be shown for some $j \geq 0$, we get from (6.20) in the case where $b_j \neq 0$ and $m_j \geq 0$,

$$\deg(g_{j+1}(x)) \leq \max(\deg(g_j(x)), \deg(h_j(x)))$$
$$\leq \tfrac{1}{2}(j + 2 + m_j) = \tfrac{1}{2}(j + 2 - m_{j+1}).$$

Otherwise,

$$\deg(g_{j+1}(x)) \leq \tfrac{1}{2}(j + 1 - m_j) = \tfrac{1}{2}(j + 2 - m_{j+1}).$$

The same distinction of cases proves the second inequality in (6.23). A similar inductive argument shows that for each $j \geq 0$ we have

$$\deg(u_j(x)) \leq \tfrac{1}{2}(j - 1 - m_j) \quad \text{and} \quad \deg(v_j(x)) \leq \tfrac{1}{2}(j + m_j). \tag{6.24}$$

The auxiliary polynomials $u_j(x)$ and $v_j(x)$ are related to the polynomials $g_j(x)$ and $h_j(x)$ occurring in the algorithm by means of the following congruences, valid for each $j \geq 0$:

$$g_j(x)G(x) \equiv u_j(x) + b_j x^j \bmod x^{j+1}, \tag{6.25}$$

$$h_j(x)G(x) \equiv v_j(x) + x^j \bmod x^{j+1}. \tag{6.26}$$

Both (6.25) and (6.26) are true for $j = 0$ because of (6.19), (6.21), and the definition of b_0. Assuming that both congruences have been shown for some

$j \geqslant 0$, we get

$$g_{j+1}(x)G(x) = g_j(x)G(x) - b_j h_j(x)G(x)$$

$$\equiv u_j(x) + b_j x^j + c_{j+1} x^{j+1} - b_j\left(v_j(x) + x^j + d_{j+1} x^{j+1}\right)$$

$$\equiv u_{j+1}(x) + e_{j+1} x^{j+1} \bmod x^{j+2}$$

with suitable coefficients $c_{j+1}, d_{j+1}, e_{j+1} \in \mathbb{F}_q$. Since $|m_j| \leqslant j$, as is seen easily by induction, we have $\deg(u_{j+1}(x)) \leqslant j$ from (6.24). Therefore, e_{j+1} is the coefficient of x^{j+1} in $g_{j+1}(x)G(x)$, and so $e_{j+1} = b_{j+1}$. The induction step for (6.26) is carried out similarly.

Next, one establishes by a straightforward induction argument that

$$h_j(x)u_j(x) - g_j(x)v_j(x) = x^j \quad \text{for each } j \geqslant 0. \tag{6.27}$$

Now let $s(x)$ and $u(x)$ be polynomials over $\mathbb{F}_q$ with $s(x)G(x) = u(x)$ and $s(0) = 1$. Then by (6.26),

$$h_j(x)u(x) - s(x)v_j(x) = s(x)\left(h_j(x)G(x) - v_j(x)\right)$$

$$\equiv s(x)x^j \equiv x^j \bmod x^{j+1},$$

and so for some $U_j(x) \in \mathbb{F}_q[x]$ we have

$$h_j(x)u(x) - s(x)v_j(x) = x^j U_j(x) \quad \text{with } U_j(0) = 1. \tag{6.28}$$

Similarly, one uses (6.25) to show that there exists $V_j(x) \in \mathbb{F}_q[x]$ with

$$g_j(x)u(x) - s(x)u_j(x) = x^j V_j(x). \tag{6.29}$$

Now suppose the minimal polynomial $m(x)$ of the given homogeneous linear recurring sequence satisfies $\deg(m(x)) \leqslant k$, and let $s(x)$ be the reciprocal minimal polynomial. Then $s(0) = 1$ and $\deg(s(x)) \leqslant k$, and from (6.15) we know that there exists $u(x) \in \mathbb{F}_q[x]$ with $s(x)G(x) = u(x)$ and $\deg(u(x)) \leqslant \deg(m(x)) - 1 \leqslant k - 1$. Consider (6.28) with $j = 2k$. Using (6.23) and (6.24), we obtain

$$\deg(h_{2k}(x)u(x)) \leqslant \tfrac{1}{2}(2k + 2 + m_{2k}) + k - 1 = 2k + \tfrac{1}{2}m_{2k}$$

and

$$\deg(s(x)v_{2k}(x)) \leqslant k + \tfrac{1}{2}(2k + m_{2k}) = 2k + \tfrac{1}{2}m_{2k},$$

and so

$$\deg(h_{2k}(x)u(x) - s(x)v_{2k}(x)) \leqslant 2k + \tfrac{1}{2}m_{2k}.$$

On the other hand,

$$\deg(h_{2k}(x)u(x) - s(x)v_{2k}(x)) = \deg(x^{2k}U_{2k}(x)) \geqslant 2k,$$

and these inequalities are only compatible if $m_{2k} \geqslant 0$. Using again (6.23) and (6.24), one verifies that $\deg(g_{2k}(x)u(x))$ and $\deg(s(x)u_{2k}(x))$ are both

$\leqslant 2k - \frac{1}{2} - \frac{1}{2}m_{2k}$, hence (6.29) shows that

$$\deg\left(x^{2k}V_{2k}(x)\right) = \deg\left(g_{2k}(x)u(x) - s(x)u_{2k}(x)\right) < 2k.$$

But this is only possible if $V_{2k}(x)$ is the zero polynomial. Consequently, (6.29) yields $g_{2k}(x)u(x) = s(x)u_{2k}(x)$, and multiplying (6.28) for $j = 2k$ by $g_{2k}(x)$ leads to

$$h_{2k}(x)g_{2k}(x)u(x) - s(x)g_{2k}(x)v_{2k}(x)$$
$$= s(x)\left(h_{2k}(x)u_{2k}(x) - g_{2k}(x)v_{2k}(x)\right) = x^{2k}U_{2k}(x)g_{2k}(x).$$

Together with (6.27), we get $s(x) = U_{2k}(x)g_{2k}(x)$, which implies $u(x) = U_{2k}(x)u_{2k}(x)$. Since $s(x)$ is the reciprocal minimal polynomial, it follows from the second part of Theorem 6.40 that $s(x)$ and $u(x)$ are relatively prime. Because of this fact, $U_{2k}(x)$ must be a constant polynomial, and since $U_{2k}(0) = 1$ by (6.28), we actually have $U_{2k}(x) = 1$. Therefore $s(x) = g_{2k}(x)$, and as a by-product we obtain $u(x) = u_{2k}(x)$. If $\deg(m(x)) = k$, then

$$m(x) = x^k s\left(\frac{1}{x}\right) = x^k g_{2k}\left(\frac{1}{x}\right),$$

as we claimed earlier. If $\deg(m(x)) = t \leqslant k$, then we have $s(x) = g_{2t}(x)$, $u(x) = u_{2t}(x)$, and $m_{2t} \geqslant 0$. Clearly, $\max(\deg(s(x)), 1 + \deg(u(x))) \leqslant t$, and the second part of Theorem 6.40 implies that

$$t = \max(\deg(s(x)), 1 + \deg(u(x))).$$

It follows then from (6.23) and (6.24) that

$$t = \max(\deg(g_{2t}(x)), 1 + \deg(u_{2t}(x))) \leqslant t + \tfrac{1}{2} - \tfrac{1}{2}m_{2t},$$

and so $m_{2t} = 0$ or 1. Furthermore, we note that $g_j(x) = s(x)$ and $b_j = 0$ for all $j \geqslant 2t$, so that $m_j = m_{2t} + j - 2t$ for all $j \geqslant 2t$ by the definition of m_j. Setting $j = 2k$, we obtain $t = k + \frac{1}{2}m_{2t} - \frac{1}{2}m_{2k}$, and since $m_{2t} = 0$ or 1, we conclude that

$$t = \lfloor k + \tfrac{1}{2} - \tfrac{1}{2}m_{2k} \rfloor = r.$$

Therefore,

$$m(x) = x^r s\left(\frac{1}{x}\right) = x^r g_{2k}\left(\frac{1}{x}\right),$$

in accordance with our claim.

7. DISTRIBUTION PROPERTIES OF LINEAR RECURRING SEQUENCES

We are interested in the number of occurrences of a given element of $\mathbb{F}_q$ in either the full period or parts of the period of a linear recurring sequence in

$\mathbb{F}_q$. In order to provide general information on this question, we first carry out a detailed study of exponential sums that involve linear recurring sequences. It will then become apparent that in the case of linear recurring sequences for which the least period is large, the elements of the underlying finite field appear about equally often in the full period and also in large segments of the full period.

Let $s_0, s_1, \ldots$ be a kth-order linear recurring sequence in $\mathbb{F}_q$ satisfying (6.1), let r be its least period and n_0 its preperiod, so that $s_{n+r} = s_n$ for $n \geqslant n_0$. With this sequence we associate a positive integer R in the following way. Consider the impulse response sequence $d_0, d_1, \ldots$ satisfying (6.6), let r_1 be its least period and n_1 its preperiod; then we set $R = r_1 + n_1$. Of course, R depends only on the linear recurrence relation (6.1) and not on the specific form of the sequence. If $s_0, s_1, \ldots$ is a homogeneous linear recurring sequence with characteristic polynomial $f(x) \in \mathbb{F}_q[x]$, then $r_1 = \mathrm{ord}(f(x))$, and if in addition $f(0) \neq 0$, then $R = \mathrm{ord}(f(x))$, as implied by Theorem 6.27. By the same theorem, r divides r_1 and $r \leqslant R$ in the homogeneous case.

In the exponential sums to be considered, we use additive characters of $\mathbb{F}_q$ as discussed in Chapter 5 and weights defined in terms of the function $e(t) = e^{2\pi i t}$ for real t.

6.78. Theorem. *Let $s_0, s_1, \ldots$ be a kth-order linear recurring sequence in $\mathbb{F}_q$ with least period r and preperiod n_0, and let R be the positive integer introduced above. Let χ be a nontrivial additive character of $\mathbb{F}_q$. Then for every integer h we have*

$$\left| \sum_{n=u}^{u+r-1} \chi(s_n) e\left(\frac{hn}{r}\right) \right| \leqslant \left(\frac{r}{R}\right)^{1/2} q^{k/2} \quad \text{for all } u \geqslant n_0. \tag{6.30}$$

In particular, we have

$$\left| \sum_{n=u}^{u+r-1} \chi(s_n) \right| \leqslant \left(\frac{r}{R}\right)^{1/2} q^{k/2} \quad \text{for all } u \geqslant n_0. \tag{6.31}$$

Proof. By changing the initial state vector from $\mathbf{s}_0$ to $\mathbf{s}_u$, which does not affect the upper bound in (6.30), we may assume, without loss of generality, that the sequence $s_0, s_1, \ldots$ is periodic and that $u = 0$. For a column vector $\mathbf{b} = (b_0, b_1, \ldots, b_{k-1})^{\mathrm{T}}$ in $\mathbb{F}_q^k$ and an integer h, we set

$$\sigma(\mathbf{b}; h) = \sigma(b_0, b_1, \ldots, b_{k-1}; h)$$
$$= \sum_{n=0}^{r-1} \chi(b_0 s_n + b_1 s_{n+1} + \cdots + b_{k-1} s_{n+k-1}) e\left(\frac{hn}{r}\right).$$

Since the general term of this sum has period r as a function of n, we can write

$$\sigma(\mathbf{b}; h) = \sum_{n=0}^{r-1} \chi(b_0 s_{n+1} + b_1 s_{n+2} + \cdots + b_{k-1} s_{n+k}) e\left(\frac{h(n+1)}{r}\right).$$

Using the linear recurrence relation (6.1), we get

$$|\sigma(\mathbf{b};h)| = \left| \sum_{n=0}^{r-1} \chi \left(b_0 s_{n+1} + b_1 s_{n+2} + \cdots + b_{k-2} s_{n+k-1} + b_{k-1} a_0 s_n \right. \right.$$

$$\left. \left. + b_{k-1} a_1 s_{n+1} + \cdots + b_{k-1} a_{k-1} s_{n+k-1} + b_{k-1} a \right) e\left(\frac{hn}{r} \right) \right|$$

$$= \left| \sum_{n=0}^{r-1} \chi \left(b_{k-1} a_0 s_n + (b_0 + b_{k-1} a_1) s_{n+1} + \cdots \right. \right.$$

$$\left. \left. + (b_{k-2} + b_{k-1} a_{k-1}) s_{n+k-1} \right) e\left(\frac{hn}{r} \right) \right|$$

$$= \left| \sigma(b_{k-1} a_0, b_0 + b_{k-1} a_1, \ldots, b_{k-2} + b_{k-1} a_{k-1}; h) \right|.$$

This identity can be written in the form

$$|\sigma(\mathbf{b};h)| = |\sigma(A\mathbf{b};h)|,$$

where A is the matrix in (6.3). It follows by induction that

$$|\sigma(\mathbf{b};h)| = |\sigma(A^j \mathbf{b};h)| \quad \text{for all } j \geqslant 0. \tag{6.32}$$

Let $\mathbf{d}$ be the column vector $\mathbf{d} = (1, 0, \ldots, 0)^T$ in $\mathbb{F}_q^k$, and let $\mathbf{d}_0, \mathbf{d}_1, \ldots$ be the state vectors of the impulse response sequence $d_0, d_1, \ldots$ satisfying (6.6). Then we claim that two state vectors $\mathbf{d}_m$ and $\mathbf{d}_n$ are identical if and only if $A^m \mathbf{d} = A^n \mathbf{d}$. For if $\mathbf{d}_m = \mathbf{d}_n$, then $A^m \mathbf{d} = A^n \mathbf{d}$ follows from Lemma 6.15. On the other hand, if $A^m \mathbf{d} = A^n \mathbf{d}$, then $A^{m+j} \mathbf{d} = A^{n+j} \mathbf{d}$, and so $A^m(A^j \mathbf{d}) = A^n(A^j \mathbf{d})$, for all $j \geqslant 0$. But since the vectors $\mathbf{d}, A\mathbf{d}, A^2\mathbf{d}, \ldots, A^{k-1}\mathbf{d}$ form a basis for the vector space $\mathbb{F}_q^k$ over $\mathbb{F}_q$, we get $A^m = A^n$, which implies $\mathbf{d}_m = \mathbf{d}_n$ by Lemma 6.15.

The distinct vectors in the sequence $\mathbf{d}_0, \mathbf{d}_1, \ldots$ are exactly given by $\mathbf{d}_0, \mathbf{d}_1, \ldots, \mathbf{d}_{R-1}$. Therefore, by what we have just shown, the distinct vectors among $\mathbf{d}, A\mathbf{d}, A^2\mathbf{d}, \ldots$ are exactly given by $\mathbf{d}, A\mathbf{d}, \ldots, A^{R-1}\mathbf{d}$. Using (6.32), we get

$$R|\sigma(\mathbf{d};h)|^2 = \sum_{j=0}^{R-1} \left| \sigma(A^j \mathbf{d};h) \right|^2 \leqslant \sum_{\mathbf{b}} |\sigma(\mathbf{b};h)|^2, \tag{6.33}$$

where the last sum is taken over all column vectors $\mathbf{b}$ in $\mathbb{F}_q^k$. Now

$$\sum_{\mathbf{b}} |\sigma(\mathbf{b};h)|^2 = \sum_{\mathbf{b}} \sigma(\mathbf{b};h) \overline{\sigma(\mathbf{b};h)}$$

$$= \sum_{b_0, b_1, \ldots, b_{k-1} \in \mathbb{F}_q} \sum_{m,n=0}^{r-1} \chi \left(b_0(s_m - s_n) + b_1(s_{m+1} - s_{n+1}) \right.$$

$$\left. + \cdots + b_{k-1}(s_{m+k-1} - s_{n+k-1}) \right) e\left(\frac{h(m-n)}{r} \right)$$

$$= \sum_{m,n=0}^{r-1} e\left(\frac{h(m-n)}{r} \right) \tag{6.34}$$

$$\sum_{b_0, b_1, \ldots, b_{k-1} \in \mathbb{F}_q} \chi(b_0(s_m - s_n)) \chi(b_1(s_{m+1} - s_{n+1})) \cdots$$

$$\cdot \chi(b_{k-1}(s_{m+k-1} - s_{n+k-1}))$$

$$= \sum_{m,n=0}^{r-1} e\left(\frac{h(m-n)}{r}\right) \left(\sum_{b_0 \in \mathbb{F}_q} \chi(b_0(s_m - s_n))\right) \cdots$$

$$\cdot \left(\sum_{b_{k-1} \in \mathbb{F}_q} \chi(b_{k-1}(s_{m+k-1} - s_{n+k-1}))\right).$$

We note that for $c \in \mathbb{F}_q$ we have

$$\sum_{b \in \mathbb{F}_q} \chi(bc) = \begin{cases} 0 & \text{if } c \neq 0, \\ q & \text{if } c = 0, \end{cases}$$

according to (5.9). Therefore, in the last expression in (6.34) one only gets a contribution from those ordered pairs (m, n) for which simultaneously $s_m = s_n, \ldots, s_{m+k-1} = s_{n+k-1}$. But since $0 \leqslant m, n \leqslant r - 1$, this is only possible for $m = n$. It follows that

$$\sum_{\mathbf{b}} |\sigma(\mathbf{b}; h)|^2 = rq^k.$$

By combining this with (6.33), we arrive at

$$|\sigma(\mathbf{d}; h)| \leqslant \left(\frac{r}{R}\right)^{1/2} q^{k/2},$$

which proves (6.30). The inequality (6.31) results from (6.30) by setting $h = 0$. $\qquad \square$

6.79. Remark. Let χ be a nontrivial additive character of $\mathbb{F}_q$ and let ψ be an arbitrary multiplicative character of $\mathbb{F}_q$. Then the Gaussian sum

$$G(\psi, \chi) = \sum_{c \in \mathbb{F}_q^*} \psi(c) \chi(c)$$

can be considered as a special case of the sum in (6.30). To see this, let g be a primitive element of $\mathbb{F}_q$ and introduce the first-order linear recurring sequence $s_0, s_1, \ldots$ in $\mathbb{F}_q$ with $s_0 = 1$ and $s_{n+1} = gs_n$ for $n = 0, 1, \ldots$. Then $r = R = q - 1$ and $n_0 = 0$. We note that $\psi(g) = e(h/r)$ for some integer h. Thus we can write

$$G(\psi, \chi) = \sum_{n=0}^{r-1} \chi(g^n) \psi(g^n) = \sum_{n=0}^{r-1} \chi(s_n) e\left(\frac{hn}{r}\right).$$

If ψ is nontrivial, then in this special case both sides of (6.30) are identical according to (5.15). $\qquad \square$

The sums in Theorem 6.78 are extended over a full period of the given linear recurring sequence. An estimate for character sums over segments of the period can be deduced from this result. We need the following auxiliary inequality.

6.80. Lemma. *For any positive integers r and N we have*

$$\sum_{h=0}^{r-1}\left|\sum_{j=0}^{N-1} e\left(\frac{hj}{r}\right)\right| < \frac{2}{\pi}r\log r + \frac{2}{5}r + N. \tag{6.35}$$

Proof. The inequality is trivial for $r=1$. For $r \geqslant 2$ we have

$$\left|\sum_{j=0}^{N-1} e\left(\frac{hj}{r}\right)\right| = \frac{|e(hN/r)-1|}{|e(h/r)-1|} \leqslant \frac{1}{\sin\pi\|h/r\|}$$

$$= \csc\pi\left\|\frac{h}{r}\right\| \quad \text{for } 1 \leqslant h \leqslant r-1,$$

where $\|t\|$ denotes the absolute distance from the real number t to the nearest integer. It follows that

$$\sum_{h=0}^{r-1}\left|\sum_{j=0}^{N-1} e\left(\frac{hj}{r}\right)\right| \leqslant \sum_{h=1}^{r-1} \csc\pi\left\|\frac{h}{r}\right\| + N \leqslant 2\sum_{h=1}^{\lfloor r/2\rfloor} \csc\frac{\pi h}{r} + N. \tag{6.36}$$

By comparing sums with integrals, we obtain

$$\sum_{h=1}^{\lfloor r/2\rfloor} \csc\frac{\pi h}{r} = \csc\frac{\pi}{r} + \sum_{h=2}^{\lfloor r/2\rfloor} \csc\frac{\pi h}{r} \leqslant \csc\frac{\pi}{r} + \int_1^{\lfloor r/2\rfloor} \csc\frac{\pi x}{r}\,dx$$

$$\leqslant \csc\frac{\pi}{r} + \frac{r}{\pi}\int_{\pi/r}^{\pi/2} \csc t\,dt$$

$$= \csc\frac{\pi}{r} + \frac{r}{\pi}\log\cot\frac{\pi}{2r} \leqslant \csc\frac{\pi}{r} + \frac{r}{\pi}\log\frac{2r}{\pi}.$$

For $r \geqslant 6$ we have $(\pi/r)^{-1}\sin(\pi/r) \geqslant (\pi/6)^{-1}\sin(\pi/6)$, hence $\sin(\pi/r) \geqslant 3/r$. This implies

$$\sum_{h=1}^{\lfloor r/2\rfloor} \csc\frac{\pi h}{r} \leqslant \frac{1}{\pi}r\log r + \left(\frac{1}{3} - \frac{1}{\pi}\log\frac{\pi}{2}\right)r \quad \text{for } r \geqslant 6,$$

and so

$$\sum_{h=1}^{\lfloor r/2\rfloor} \csc\frac{\pi h}{r} < \frac{1}{\pi}r\log r + \frac{1}{5}r \quad \text{for } r \geqslant 6.$$

This inequality is easily checked for $r = 3$, 4, and 5, so that (6.35) holds for $r \geqslant 3$ in view of (6.36). For $r = 2$ the inequality (6.35) is shown by inspection. $\square$

6.81. Theorem. *Let* $s_0, s_1, \ldots$ *be a kth-order linear recurring sequence in* $\mathbb{F}_q$, *and let* r, n_0, *and* R *be as in Theorem 6.78. Then, for any nontrivial additive character* χ *of* $\mathbb{F}_q$ *we have*

$$\left| \sum_{n=u}^{u+N-1} \chi(s_n) \right| < \left(\frac{r}{R} \right)^{1/2} q^{k/2} \left(\frac{2}{\pi} \log r + \frac{2}{5} + \frac{N}{r} \right) \quad \text{for } u \geq n_0 \text{ and } 1 \leq N \leq r.$$

Proof. We start from the identity

$$\sum_{n=u}^{u+N-1} \chi(s_n) = \sum_{n=u}^{u+r-1} \chi(s_n) \sum_{j=0}^{N-1} \frac{1}{r} \sum_{h=0}^{r-1} e\left(\frac{h(n-u-j)}{r} \right) \quad \text{for } 1 \leq N \leq r,$$

which is valid since the sum over j is 1 for $u \leq n \leq u+N-1$ and 0 for $u + N \leq n \leq u + r - 1$. Rearranging terms, we get

$$\sum_{n=u}^{u+N-1} \chi(s_n) = \frac{1}{r} \sum_{h=0}^{r-1} \left(\sum_{j=0}^{N-1} e\left(\frac{-h(u+j)}{r} \right) \right) \left(\sum_{n=u}^{u+r-1} \chi(s_n) e\left(\frac{hn}{r} \right) \right),$$

and so by (6.30),

$$\left| \sum_{n=u}^{u+N-1} \chi(s_n) \right| \leq \frac{1}{r} \sum_{h=0}^{r-1} \left| \sum_{j=0}^{N-1} e\left(\frac{-h(u+j)}{r} \right) \right| \left| \sum_{n=u}^{u+r-1} \chi(s_n) e\left(\frac{hn}{r} \right) \right|$$

$$\leq \frac{1}{r} \left(\frac{r}{R} \right)^{1/2} q^{k/2} \sum_{h=0}^{r-1} \left| \sum_{j=0}^{N-1} e\left(\frac{hj}{r} \right) \right|.$$

An application of Lemma 6.80 yields the desired inequality. □

It should be noted that the inequalities in Theorems 6.78 and 6.81 are only of interest if the least period r of $s_0, s_1, \ldots$ is sufficiently large. For small r, these results are actually weaker than the trivial estimate

$$\left| \sum_{n=u}^{u+N-1} \chi(s_n) \right| \leq N \quad \text{for } 1 \leq N \leq r.$$

In order to obtain nontrivial statements, r should be somewhat larger than $q^{k/2}$.

Let $s_0, s_1, \ldots$ be a linear recurring sequence in $\mathbb{F}_q$ with least period r and preperiod n_0. For $b \in \mathbb{F}_q$ we denote by $Z(b)$ the number of n, $n_0 \leq n \leq n_0 + r - 1$, with $s_n = b$. Therefore $Z(b)$ is the number of occurrences of b in a full period of the linear recurring sequence.

If $s_0, s_1, \ldots$ is a kth-order maximal period sequence, then $Z(b)$ can be determined explicitly. We have $r = q^k - 1$ and $n_0 = 0$ according to Theorem 6.33, and so the state vectors $\mathbf{s}_0, \mathbf{s}_1, \ldots, \mathbf{s}_{r-1}$ of the sequence run exactly through all nonzero vectors in $\mathbb{F}_q^k$. Consequently, $Z(b)$ is equal to the number of nonzero vectors in $\mathbb{F}_q^k$ that have b as a first coordinate. Elementary counting arguments show then that $Z(b) = q^{k-1}$ for $b \neq 0$ and

$Z(0) = q^{k-1} - 1$. Therefore, up to a slight aberration for the zero element, the elements of $\mathbb{F}_q$ occur equally often in a full period of a maximal period sequence.

In the general case, one cannot expect such an equitable distribution of elements. One may, however, estimate the deviation between the actual number of occurrences and the ideal number r/q. If r is sufficiently large, then this deviation is comparatively small.

6.82. Theorem. *Let $s_0, s_1, \ldots$ be a kth-order linear recurring sequence in $\mathbb{F}_q$ with least period r, and let R be as in Theorem 6.78. Then, for any $b \in \mathbb{F}_q$ we have*

$$\left| Z(b) - \frac{r}{q} \right| \leqslant \left(1 - \frac{1}{q} \right) \left(\frac{r}{R} \right)^{1/2} q^{k/2}.$$

Proof. For given $b \in \mathbb{F}_q$, let the real-valued function δ_b on $\mathbb{F}_q$ be defined by $\delta_b(b) = 1$ and $\delta_b(c) = 0$ for $c \neq b$. Because of (5.10), the function δ_b can be represented in the form

$$\delta_b(c) = \frac{1}{q} \sum_\chi \chi(c - b) \quad \text{for all } c \in \mathbb{F}_q,$$

where the sum is extended over all additive characters χ of $\mathbb{F}_q$. It follows that

$$Z(b) = \sum_{n=n_0}^{n_0+r-1} \delta_b(s_n) = \sum_{n=n_0}^{n_0+r-1} \frac{1}{q} \sum_\chi \chi(s_n - b)$$

$$= \frac{1}{q} \sum_\chi \bar{\chi}(b) \sum_{n=n_0}^{n_0+r-1} \chi(s_n).$$

By separating the contribution from the trivial additive character of $\mathbb{F}_q$ and using an asterisk to indicate the deletion of this character from the range of summation, we get

$$Z(b) - \frac{r}{q} = \frac{1}{q} \sum_\chi {}^* \bar{\chi}(b) \sum_{n=n_0}^{n_0+r-1} \chi(s_n).$$

Thus, by using (6.31), we obtain

$$\left| Z(b) - \frac{r}{q} \right| \leqslant \frac{1}{q} \sum_\chi {}^* \left| \sum_{n=n_0}^{n_0+r-1} \chi(s_n) \right| \leqslant \left(1 - \frac{1}{q} \right) \left(\frac{r}{R} \right)^{1/2} q^{k/2},$$

since there are $q - 1$ nontrivial additive characters of $\mathbb{F}_q$.   □

6.83. Corollary. *Let $s_0, s_1, \ldots$ be a homogeneous linear recurring* *sequence in* $\mathbb{F}_q$ *with least period r whose minimal polynomial* $m(x) \in \mathbb{F}_q[x]$ *has degree* $k \geqslant 1$ *and satisfies* $m(0) \neq 0$. *Then, for every* $b \in \mathbb{F}_q$ *we have*

$$\left| Z(b) - \frac{r}{q} \right| \leqslant \left(1 - \frac{1}{q} \right) q^{k/2}.$$

Proof. We have $r = \mathrm{ord}(m(x))$ according to Theorem 6.44. Furthermore, $R = \mathrm{ord}(m(x))$ by a remark preceding Theorem 6.78, and Theorem 6.82 yields the desired result.   □

If the linear recurring sequence has an irreducible minimal polynomial, then an alternative method based on Gaussian sums leads to somewhat better estimates. In the subsequent proof, we shall use the formulas for Gaussian sums in Theorem 5.11.

6.84.   Theorem. *Let* $s_0, s_1, \ldots$ *be a homogeneous linear recurring sequence in* $\mathbb{F}_q$ *with least period r. Suppose the minimal polynomial* $m(x)$ *of the sequence is irreducible over* $\mathbb{F}_q$, *has degree k, and satisfies* $m(0) \neq 0$. *Let h be the least common multiple of r and* $q - 1$. *Then,*

$$\left| Z(0) - \frac{(q^{k-1} - 1)r}{q^k - 1} \right| \leqslant \left(1 - \frac{1}{q} \right) \left(\frac{r}{h} - \frac{r}{q^k - 1} \right) q^{k/2} \qquad (6.37)$$

and

$$\left| Z(b) - \frac{q^{k-1}r}{q^k - 1} \right| \leqslant \left(\frac{r}{h} - \frac{r}{q^k - 1} + \frac{h-r}{h} q^{1/2} \right) q^{(k/2)-1} \quad \text{for } b \neq 0.$$

$$(6.38)$$

Proof. Set $K = \mathbb{F}_q$, and let F be the splitting field of $m(x)$ over K. Let α be a fixed root of $m(x)$ in F; then $\alpha \neq 0$ because of $m(0) \neq 0$. By Theorem 6.24, there exists $\theta \in F$ such that

$$s_n = \mathrm{Tr}_{F/K}(\theta \alpha^n) \quad \text{for } n = 0, 1, \ldots. \qquad (6.39)$$

We clearly have $\theta \neq 0$. Let λ' be the canonical additive character of K. Then, for any given $b \in K$, the character relation (5.9) yields

$$\frac{1}{q} \sum_{c \in K} \lambda'(c(b - s_n)) = \begin{cases} 1 & \text{if } s_n = b, \\ 0 & \text{if } s_n \neq b, \end{cases}$$

and so, together with (6.39),

$$Z(b) = \frac{1}{q} \sum_{n=0}^{r-1} \sum_{c \in K} \lambda'(bc) \lambda'\left(\mathrm{Tr}_{F/K}(-c\theta \alpha^n) \right).$$

If λ denotes the canonical additive character of F, then λ' and λ are related by $\lambda'(\mathrm{Tr}_{F/K}(\beta)) = \lambda(\beta)$ for all $\beta \in F$ (see (5.7)). Therefore,

$$Z(b) = \frac{1}{q} \sum_{c \in K} \lambda'(bc) \sum_{n=0}^{r-1} \bar{\lambda}(c\theta\alpha^n)$$

$$= \frac{r}{q} + \frac{1}{q} \sum_{c \in K^*} \lambda'(bc) \sum_{n=0}^{r-1} \bar{\lambda}(c\theta\alpha^n). \tag{6.40}$$

Now by (5.17),

$$\bar{\lambda}(\beta) = \frac{1}{q^k - 1} \sum_{\psi} G(\bar{\psi}, \bar{\lambda}) \psi(\beta) \quad \text{for } \beta \in F^*,$$

where the sum is extended over all multiplicative characters ψ of F. For $c \in K^*$ it follows that

$$\sum_{n=0}^{r-1} \bar{\lambda}(c\theta\alpha^n) = \frac{1}{q^k - 1} \sum_{n=0}^{r-1} \sum_{\psi} G(\bar{\psi}, \bar{\lambda}) \psi(c\theta\alpha^n)$$

$$= \frac{1}{q^k - 1} \sum_{\psi} \psi(c\theta) G(\bar{\psi}, \bar{\lambda}) \sum_{n=0}^{r-1} \psi(\alpha)^n.$$

The inner sum in the last expression is a finite geometric series that vanishes if $\psi(\alpha) \neq 1$, because of $\psi(\alpha)^r = \psi(\alpha^r) = \psi(1) = 1$. Therefore, we only have to sum over the set J of those characters ψ for which $\psi(\alpha) = 1$, and so

$$\sum_{n=0}^{r-1} \bar{\lambda}(c\theta\alpha^n) = \frac{r}{q^k - 1} \sum_{\psi \in J} \psi(c\theta) G(\bar{\psi}, \bar{\lambda}).$$

Substituting this in (6.40), we get

$$Z(b) = \frac{r}{q} + \frac{r}{q(q^k - 1)} \sum_{c \in K^*} \lambda'(bc) \sum_{\psi \in J} \psi(c\theta) G(\bar{\psi}, \bar{\lambda})$$

$$= \frac{r}{q} + \frac{r}{q(q^k - 1)} \sum_{\psi \in J} \psi(\theta) G(\bar{\psi}, \bar{\lambda}) \sum_{c \in K^*} \psi(c) \lambda'(bc).$$

If we consider the restriction ψ' of ψ to K^*, then the inner sum may be viewed as a Gaussian sum in K with an additive character $\lambda'_b(c) = \lambda'(bc)$ for $c \in K$. Thus,

$$Z(b) = \frac{r}{q} + \frac{r}{q(q^k - 1)} \sum_{\psi \in J} \psi(\theta) G(\bar{\psi}, \bar{\lambda}) G(\psi', \lambda'_b). \tag{6.41}$$

Now let $b = 0$. Then λ'_b is the trivial additive character of K, and so the Gaussian sum $G(\psi', \lambda'_b)$ vanishes unless ψ' is trivial, in which case $G(\psi', \lambda'_b) = q - 1$. Consequently, it suffices to extend the sum in (6.41) over the set A of characters ψ for which $\psi(\alpha) = 1$ and ψ' is trivial, so that

$$Z(0) = \frac{r}{q} + \frac{(q-1)r}{q(q^k - 1)} \sum_{\psi \in A} \psi(\theta) G(\bar{\psi}, \bar{\lambda}).$$

The trivial multiplicative character contributes -1 to the sum, hence we get

$$Z(0) - \frac{(q^{k-1}-1)r}{q^k-1} = \frac{(q-1)r}{q(q^k-1)} \sum_{\psi \in A}^* \psi(\theta) G(\bar{\psi}, \bar{\lambda}),$$

where the asterisk indicates that the trivial multiplicative character is deleted from the range of summation. Since λ is nontrivial, we have $|G(\bar{\psi}, \bar{\lambda})| = q^{k/2}$ for every nontrivial ψ, and so

$$\left| Z(0) - \frac{(q^{k-1}-1)r}{q^k-1} \right| \leqslant \frac{(q-1)r}{q(q^k-1)} (|A|-1) q^{k/2}. \tag{6.42}$$

Let H be the smallest subgroup of F^* containing α and K^*. The element α has order r in the cyclic group F^*, therefore $|H| = h$, the least common multiple of r and $q-1$. Furthermore, we have $\psi \in A$ if and only if $\psi(\beta) = 1$ for all $\beta \in H$. In other words, A is the annihilator of H in $(F^*)^{\wedge}$ (see p. 165), and so

$$|A| = \frac{|F^*|}{|H|} = \frac{q^k-1}{h} \tag{6.43}$$

by Theorem 5.6. The inequality (6.37) follows now from (6.42) and (6.43).

For $b \neq 0$, we go back to (6.41) and note first that the additive character λ_b' is then nontrivial. Therefore, the trivial multiplicative character contributes 1 to the sum in (6.41), so that we can write

$$Z(b) - \frac{q^{k-1}r}{q^k-1} = \frac{r}{q(q^k-1)} \sum_{\psi \in J}^* \psi(\theta) G(\bar{\psi}, \bar{\lambda}) G(\psi', \lambda_b').$$

Now $G(\psi', \lambda_b') = -1$ if ψ' is trivial and $|G(\psi', \lambda_b')| = q^{1/2}$ if ψ' is nontrivial, which implies

$$\left| Z(b) - \frac{q^{k-1}r}{q^k-1} \right| \leqslant \frac{r}{q^k-1} (|A|-1+(|J|-|A|)q^{1/2}) q^{(k/2)-1}.$$

Since J is the annihilator in $(F^*)^{\wedge}$ of the subgroup of F^* generated by α, we have $|J| = (q^k-1)/r$ by Theorem 5.6. This is combined with (6.43) to complete the proof of (6.38). $\qquad\qquad\qquad\qquad\qquad\qquad\qquad\qquad\qquad$ $\square$

One can also obtain results about the distribution of elements in parts of the period. Let $s_0, s_1, \ldots$ be an arbitrary linear recurring sequence in $\mathbb{F}_q$ with least period r and preperiod n_0. For $b \in \mathbb{F}_q$, for $N_0 \geqslant n_0$ and $1 \leqslant N \leqslant r$, let $Z(b; N_0, N)$ be the number of n, $N_0 \leqslant n \leqslant N_0 + N - 1$, with $s_n = b$.

6.85. Theorem. *Let $s_0, s_1, \ldots$ be a kth-order linear recurring sequence in $\mathbb{F}_q$ with least period r and preperiod n_0, and let R be as in Theorem 6.78. Then, for any $b \in \mathbb{F}_q$ we have*

$$\left| Z(b; N_0, N) - \frac{N}{q} \right| \leqslant \left(1 - \frac{1}{q}\right)\left(\frac{r}{R}\right)^{1/2} q^{k/2}\left(\frac{2}{\pi}\log r + \frac{2}{5} + \frac{N}{r}\right)$$

for $N_0 \geqslant n_0$ and $1 \leqslant N \leqslant r$.

Proof. Proceeding as in the proof of Theorem 6.82 and using the same notation as there, we arrive at the identity

$$Z(b; N_0, N) - \frac{N}{q} = \frac{1}{q}\sum_{\chi}^{*}\overline{\chi}(b)\sum_{n=N_0}^{N_0+N-1} \chi(s_n).$$

On the basis of Theorem 6.81 we obtain then

$$\left| Z(b; N_0, N) - \frac{N}{q} \right| \leqslant \frac{1}{q}\sum_{\chi}^{*}\left| \sum_{n=N_0}^{N_0+N-1} \chi(s_n)\right|$$

$$\leqslant \left(1 - \frac{1}{q}\right)\left(\frac{r}{R}\right)^{1/2} q^{k/2}\left(\frac{2}{\pi}\log r + \frac{2}{5} + \frac{N}{r}\right),$$

since there are $q - 1$ nontrivial additive characters of $\mathbb{F}_q$. $\square$

The method in the proof of Theorem 6.84 can also be adapted to produce results on the distribution of elements in parts of the period (compare with Exercises 6.69, 6.70, and 6.71).

EXERCISES

6.1. Design a feedback shift register implementing the linear recurrence relation $s_{n+5} = s_{n+4} - s_{n+3} - s_{n+1} + s_n$, $n = 0, 1, \ldots$, in $\mathbb{F}_3$.
6.2. Design a feedback shift register implementing the linear recurrence relation $s_{n+7} = 3s_{n+5} - 2s_{n+4} + s_{n+3} + 2s_n + 1$, $n = 0, 1, \ldots$, in $\mathbb{F}_7$.
6.3. Let r be a period of the ultimately periodic sequence $s_0, s_1, \ldots$ and let n_0 be the least nonnegative integer such that $s_{n+r} = s_n$ for all $n \geqslant n_0$. Prove that n_0 is equal to the preperiod of the sequence.
6.4. Determine the order of the matrix

$$A = \begin{pmatrix} 0 & 0 & 0 & -1 \\ 1 & 0 & 0 & 1 \\ 0 & 1 & 0 & 1 \\ 0 & 0 & 1 & -1 \end{pmatrix}$$

in the general linear group $GL(4, \mathbb{F}_3)$.
6.5. Obtain the results of Example 6.18 by the methods of Section 5.
6.6. Use (6.8) to give an explicit formula for the terms of the linear recurring sequence in $\mathbb{F}_3$ with $s_0 = s_1 = 1$, $s_2 = 0$, and $s_{n+3} = -s_{n+1} + s_n$ for $n = 0, 1, \ldots$.
6.7. Use the result in Remark 6.23 to give an explicit formula for the

terms of the linear recurring sequence in $\mathbb{F}_4$ with $s_0 = s_1 = s_2 = 0$, $s_3 = 1$, and $s_{n+4} = \alpha s_{n+3} + s_{n+1} + \alpha s_n$ for $n = 0, 1, \ldots$, where α is a primitive element of $\mathbb{F}_4$.

6.8. Prove that the terms s_n given by the formula in Remark 6.23 satisfy the homogeneous linear recurrence relation with characteristic polynomial $f(x)$.

6.9. Prove the result in Remark 6.23 for the case where $e_i \leqslant 2$ for $i = 1, 2, \ldots, m$ and $e_i = 1$ if $\alpha_i = 0$.

6.10. Represent the elements of the linear recurring sequence in $\mathbb{F}_2$ with $s_0 = 0$, $s_1 = s_2 = 1$, and $s_{n+3} = s_{n+2} + s_n$ for $n = 0, 1, \ldots$ in terms of a suitable trace function.

6.11. Prove Lemma 6.26 by using linear recurring sequences.

6.12. Determine the least period of the impulse response sequence in $\mathbb{F}_2$ satisfying the linear recurrence relation $s_{n+7} = s_{n+6} + s_{n+5} + s_{n+1} + s_n$ for $n = 0, 1, \ldots$.

6.13. Calculate the least period of the impulse response sequence associated with the linear recurrence relation $s_{n+10} = s_{n+7} + s_{n+2} + s_{n+1} + s_n$ in $\mathbb{F}_2$.

6.14. Prove Theorem 6.27 by using generating functions.

6.15. Find a linear recurring sequence of least order in $\mathbb{F}_2$ whose least period is 21.

6.16. Find a linear recurring sequence of least order in $\mathbb{F}_2$ whose least period is 24.

6.17. Let r be the least period of the *Fibonacci sequence* in $\mathbb{F}_q$—that is, of the sequence with $s_0 = 0$, $s_1 = 1$, and $s_{n+2} = s_{n+1} + s_n$ for $n = 0, 1, \ldots$. Let p be the characteristic of $\mathbb{F}_q$. Prove that $r = 20$ if $p = 5$, that r divides $p - 1$ if $p \equiv \pm 1 \bmod 5$, and that r divides $p^2 - 1$ in all other cases.

6.18. Construct a maximal period sequence in $\mathbb{F}_3$ of least period 80.

6.19. An (m, k) *de Bruijn sequence* is a finite sequence $s_0, s_1, \ldots, s_{N-1}$ with $N = m^k$ terms from a set of m elements such that the k-tuples $(s_n, s_{n+1}, \ldots, s_{n+k-1})$, $n = 0, 1, \ldots, N-1$, with subscripts considered modulo N are all different. Prove that if $d_0, d_1, \ldots$ is a kth-order impulse response sequence and maximal period sequence in $\mathbb{F}_q$, then $s_0 = 0$, $s_n = d_{n-1}$ for $1 \leqslant n \leqslant q^k - 1$ yields a (q, k) de Bruijn sequence.

6.20. Construct a $(2, 5)$ de Bruijn sequence.

6.21. Let $B(x) = 2 - x + x^3 \in \mathbb{F}_7[x]$. Calculate the first six nonzero terms of the formal power series $1/B(x)$.

6.22. Let

$$A(x) = -1 - x + x^2, \qquad B(x) = \sum_{n=0}^{\infty} (-1)^n x^n \in \mathbb{F}_3[[x]].$$

Calculate the first five nonzero terms of the formal power series $A(x)/B(x)$.

6.23. Consider the linear recurring sequence in $\mathbb{F}_3$ with $s_0 = s_1 = s_2 = 1$, $s_3 = s_4 = -1$, and $s_{n+5} = s_{n+4} + s_{n+2} - s_{n+1} + s_n$ for $n = 0, 1, \ldots$. Represent the generating function of the sequence in the form (6.15).

6.24. Calculate the first eight terms of the impulse response sequence associated with the linear recurrence relation $s_{n+5} = s_{n+3} + s_{n+2} + s_n$ in $\mathbb{F}_2$ by long division.

6.25. Let $s_0, s_1, \ldots$ be a homogeneous linear recurring sequence in $\mathbb{F}_q$. Prove that the set of all polynomials $f(x) = a_k x^k + \cdots + a_1 x + a_0 \in \mathbb{F}_q[x]$ such that $a_k s_{n+k} + \cdots + a_1 s_{n+1} + a_0 s_n = 0$ for $n = 0, 1, \ldots$ forms an ideal of $\mathbb{F}_q[x]$. Thus show the existence of a uniquely determined minimal polynomial of the sequence.

6.26. Consider the linear recurring sequence in $\mathbb{F}_2$ with $s_0 = s_3 = s_4 = s_5 = s_6 = 0$, $s_1 = s_2 = s_7 = 1$, and $s_{n+8} = s_{n+7} + s_{n+6} + s_{n+5} + s_n$ for $n = 0, 1, \ldots$. Use the method in the proof of Theorem 6.42 to determine the minimal polynomial of the sequence.

6.27. Consider the linear recurring sequence in $\mathbb{F}_5$ with $s_0 = s_1 = s_2 = 1$, $s_3 = -1$, and $s_{n+4} = 3s_{n+2} - s_{n+1} + s_n$ for $n = 0, 1, \ldots$. Use the method in the proof of Theorem 6.42 to determine the minimal polynomial of the sequence.

6.28. Prove that a homogeneous linear recurring sequence in a finite field is periodic if and only if its minimal polynomial $m(x)$ satisfies $m(0) \neq 0$.

6.29. Given a homogeneous linear recurring sequence in a finite field with minimal polynomial $m(x)$, prove that the preperiod of the sequence is equal to the multiplicity of 0 as a root of $m(x)$.

6.30. Prove Corollary 6.52 by using the construction of the minimal polynomial in the proof of Theorem 6.42.

6.31. Use the criterion in Theorem 6.51 to determine the minimal polynomial of the linear recurring sequence in $\mathbb{F}_2$ with $s_{n+6} = s_{n+3} + s_{n+2} + s_{n+1} + s_n$ for $n = 0, 1, \ldots$ and initial state vector $(1, 1, 1, 0, 0, 1)$.

6.32. Find the least period of the linear recurring sequence in Exercise 6.26.

6.33. Find the least period of the linear recurring sequence in Exercise 6.27.

6.34. Find the least period of the linear recurring sequence in $\mathbb{F}_2$ with $s_0 = s_1 = s_2 = s_6 = s_7 = 0$, $s_3 = s_4 = s_5 = s_8 = 1$, and $s_{n+9} = s_{n+7} + s_{n+4} + s_{n+1} + s_n$ for $n = 0, 1, \ldots$.

6.35. Find the least period of the linear recurring sequence in $\mathbb{F}_3$ with $s_0 = s_1 = 1$, $s_2 = s_3 = 0$, $s_4 = -1$, and $s_{n+5} = s_{n+4} - s_{n+3} + s_{n+2} + s_n$ for $n = 0, 1, \ldots$.

6.36. Find the least period of the linear recurring sequence in $\mathbb{F}_3$ with

$s_{n+4} = s_{n+3} + s_{n+2} - s_n - 1$ for $n = 0, 1, \ldots$ and initial state vector $(0, -1, 1, 0)$.

6.37.   Prove that a kth-order linear recurring sequence $s_0, s_1, \ldots$ in $\mathbb{F}_q$ has least period q^k exactly in the following cases:
    (a) $k = 1$, q prime, $s_{n+1} = s_n + a$ for $n = 0, 1, \ldots$ with $a \in \mathbb{F}_q^*$;
    (b) $k = 2$, $q = 2$, $s_{n+2} = s_n + 1$ for $n = 0, 1, \ldots$.

6.38.   Given a homogeneous linear recurring sequence in $\mathbb{F}_q$ with a nonconstant minimal polynomial $m(x) \in \mathbb{F}_q[x]$ whose roots are nonzero and simple, prove that the least period of the sequence is equal to the least positive integer r such that $\alpha^r = 1$ for all roots α of $m(x)$.

6.39.   Prove: if the homogeneous linear recurring sequence σ in $\mathbb{F}_q$ has minimal polynomial $f(x) \in \mathbb{F}_q[x]$ with $\deg(f(x)) = n \geq 1$, then every sequence in $S(f(x))$ can be expressed uniquely as a linear combination of $\sigma = \sigma^{(0)}$ and the shifted sequences $\sigma^{(1)}, \sigma^{(2)}, \ldots, \sigma^{(n-1)}$ with coefficients in $\mathbb{F}_q$.

6.40.   Let $f_1(x), \ldots, f_k(x)$ be nonconstant monic polynomials over $\mathbb{F}_q$ that are pairwise relatively prime. Prove that $S(f_1(x) \cdots f_k(x))$ is the direct sum of the linear subspaces $S(f_1(x)), \ldots, S(f_k(x))$.

6.41.   Let $s_0, s_1, \ldots$ be a homogeneous linear recurring sequence in $K = \mathbb{F}_q$ with characteristic polynomial $f(x) = f_1(x) \cdots f_r(x)$, where the $f_i(x)$ are distinct monic irreducible polynomials over K. For $i = 1, \ldots, r$, let α_i be a fixed root of $f_i(x)$ in its splitting field F_i over K. Prove that there exist uniquely determined elements $\theta_1 \in F_1, \ldots, \theta_r \in F_r$ such that
$$s_n = \mathrm{Tr}_{F_1/K}(\theta_1 \alpha_1^n) + \cdots + \mathrm{Tr}_{F_r/K}(\theta_r \alpha_r^n) \quad \text{for } n = 0, 1, \ldots .$$

6.42.   With the notation of Exercise 6.41, prove that the sequence $s_0, s_1, \ldots$ has $f(x)$ as its minimal polynomial if and only if $\theta_i \neq 0$ for $1 \leq i \leq r$. Thus show that the number of sequences in $S(f(x))$ that have $f(x)$ as minimal polynomial is given by $(q^{k_1} - 1) \cdots (q^{k_r} - 1)$, where $k_i = \deg(f_i(x))$ for $1 \leq i \leq r$.

6.43.   Let σ_1 and σ_2 be the impulse response sequences in $\mathbb{F}_2$ associated with the linear recurrence relations $s_{n+6} = s_{n+3} + s_n (n = 0, 1, \ldots)$ and $s_{n+3} = s_{n+1} + s_n (n = 0, 1, \ldots)$, respectively. Find the least period of $\sigma_1 + \sigma_2$.

6.44.   Let σ_1 be the linear recurring sequence in $\mathbb{F}_3$ with $s_{n+3} = s_{n+2} - s_{n+1} - s_n$ for $n = 0, 1, \ldots$ and initial state vector $(0, 1, 0)$, and let σ_2 be the linear recurring sequence in $\mathbb{F}_3$ with $s_{n+5} = -s_{n+3} - s_{n+2} + s_n$ for $n = 0, 1, \ldots$ and initial state vector $(1, 1, 1, 0, 1)$. Use the method of Example 6.58 to determine the minimal polynomial of the sum sequence $\sigma_1 + \sigma_2$.

6.45.   Find the least period of the sum sequence in Exercise 6.44.

6.46.   Given a homogeneous linear recurring sequence in $\mathbb{F}_2$ with minimal polynomial $x^6 + x^5 + x^4 + 1 \in \mathbb{F}_2[x]$, determine the minimal polynomial of its binary complement.

6.47.   Let $f(x) = x^9 + x^7 + x^4 + x^3 + x^2 + x + 1 \in \mathbb{F}_2[x]$. Determine the least periods of sequences from $S(f(x))$ and the number of sequences attaining each possible least period.

6.48. Let $f(x) = (x+1)^3(x^3 - x + 1) \in \mathbb{F}_3[x]$. Determine the least periods of sequences from $S(f(x))$ and the number of sequences attaining each possible least period.

6.49. Let $f(x) = x^5 - 2x^4 - x^2 - 1 \in \mathbb{F}_5[x]$. Determine the least periods of sequences from $S(f(x))$ and the number of sequences attaining each possible least period.

6.50. Find a monic polynomial $g(x) \in \mathbb{F}_3[x]$ such that

$$S(x+1)S(x^2 + x - 1)S(x^2 - x - 1) = S(g(x)).$$

6.51. Find a monic polynomial $g(x) \in \mathbb{F}_2[x]$ such that

$$S(x^2 + x + 1)S(x^5 + x^4 + 1) = S(g(x)).$$

6.52. For odd q determine a monic $g(x) \in \mathbb{F}_q[x]$ for which

$$S((x-1)^2)S((x-1)^2) = S(g(x)).$$

What is the situation for even q?

6.53. Prove that $f \vee (gh) = (f \vee g)(f \vee h)$ for nonconstant polynomials $f, g, h \in \mathbb{F}_q[x]$, provided the two factors on the right-hand side are relatively prime.

6.54. Consider the impulse response sequence in $\mathbb{F}_2$ associated with the linear recurrence relation $s_{n+4} = s_{n+2} + s_n$, $n = 0, 1, \ldots$, and the linear recurring sequence in $\mathbb{F}_2$ with $s_{n+4} = s_n$, $n = 0, 1, \ldots$, and initial state vector $(0, 1, 1, 1)$. Use these sequences to show that there is no analog of Theorem 6.59 for multiplication of sequences.

6.55. For $r \in \mathbb{N}$ and $f \in \mathbb{F}_q[x]$ with $\deg(f) > 0$, let $\sigma_r(f)$ be the sum of the rth powers of the distinct roots of f. Prove that $\sigma_r(f \vee g) = \sigma_r(f)\sigma_r(g)$ for nonconstant polynomials $f, g \in \mathbb{F}_q[x]$, provided that the number of distinct roots of $f \vee g$ is equal to the product of the numbers of distinct roots of f and g, respectively.

6.56. Let $s_0, s_1, \ldots$ be an arbitrary sequence in $\mathbb{F}_q$, and let $n \geqslant 0$ and $r \geqslant 1$ be integers. Prove that if both Hankel determinants $D_{n+2}^{(r)}$ and $D_n^{(r+1)}$ are 0, then also $D_{n+1}^{(r)} = 0$.

6.57. Prove that the sequence $s_0, s_1, \ldots$ in $\mathbb{F}_q$ is a homogeneous linear recurring sequence with minimal polynomial of degree k if and only if $D_n^{(k+1)} = 0$ for all $n \geqslant 0$ and $k + 1$ is the least positive integer for which this holds.

6.58. Give a complete proof for the second inequality in (6.23).

6.59. Prove the inequalities in (6.24).

6.60. Give a complete proof for (6.26).

6.61. Prove (6.27).

6.62. The first 10 terms of a homogeneous linear recurring sequence in $\mathbb{F}_2$ of order $\leqslant 5$ are given by $0, 1, 1, 0, 0, 0, 0, 1, 1, 1$. Determine its minimal polynomial by the Berlekamp-Massey algorithm.

6.63. The first 8 terms of a homogeneous linear recurring sequence in $\mathbb{F}_5$ of order $\leqslant 4$ are given by $2, 1, 0, 1, -2, 0, -2, -1$. Determine its minimal polynomial by the Berlekamp-Massey algorithm.

6.64. The first 10 terms of a homogeneous linear recurring sequence in $\mathbb{F}_3$ of order $\leqslant 5$ are given by $1, -1, 0, -1, 0, 0, 0, 0, 1, 0$. Determine its minimal polynomial by the Berlekamp-Massey algorithm.

6.65. Find the homogeneous linear recurring sequence in $\mathbb{F}_5$ of least order whose first 10 terms are $2, 0, -1, -2, 0, 0, -2, 2, -1, -2$.

6.66. Suppose the conditions of Theorem 6.78 hold and assume in addition that the characteristic polynomial $f(x)$ of the sequence $s_0, s_1, \ldots$ satisfies $f(0) \neq 0$. Establish the following improvement of (6.31):

$$\left| \sum_{n=u}^{u+r-1} \chi(s_n) \right| \leqslant \left(\frac{r}{R} \right)^{1/2} (q^k - r)^{1/2} \quad \text{for all } u \geqslant 0.$$

(*Hint*: Note that $\mathbf{b} = \mathbf{0}$ can be excluded in (6.33).)

6.67. Suppose the conditions of Theorem 6.84 hold, let r be a multiple of $(q^k - 1)/(q - 1)$ and let $(q^k - 1)/r$ and k be relatively prime. Prove that $Z(0) = (q^{k-1} - 1)r/(q^k - 1)$.

6.68. Suppose the conditions of Theorem 6.84 hold, let q be odd and $h = (q^k - 1)/2$. Prove that equality holds in (6.37).

6.69. Let $Z(b; N_0, N)$ be as in Theorem 6.85. Under the conditions of Theorem 6.84 and using the notation in the proof of this theorem, show that

$$Z(b; N_0, N)$$

$$= \frac{N}{r} Z(b) + \frac{1}{q(q^k - 1)}$$

$$\sum_{\substack{\psi \\ \psi(\alpha) \neq 1}} \psi(\theta) G(\bar{\psi}, \bar{\lambda}) G(\psi', \lambda'_b) \frac{\psi(\alpha)^{N_0 + N} - \psi(\alpha)^{N_0}}{\psi(\alpha) - 1}.$$

6.70. Deduce from the result of Exercise 6.69 that

$$\left| Z(0; N_0, N) - \frac{(q^{k-1} - 1)N}{q^k - 1} \right| \leqslant \left(1 - \frac{1}{q} \right) \left(\frac{N}{h} - \frac{N}{q^k - 1} \right) q^{k/2}$$

$$+ q^{(k/2)-1} \left(\frac{2}{\pi} \log \frac{h}{q - 1} + \varepsilon_h \right),$$

where $\varepsilon_h = 0$ for $h = q - 1$ and $\varepsilon_h = \frac{2}{5}$ for $h > q - 1$.

6.71. Deduce from the result of Exercise 6.69 that

$$\left| Z(b; N_0, N) - \frac{q^{k-1}N}{q^k - 1} \right| \leqslant \left(\frac{2}{\pi} \log r + \frac{2}{5} + \frac{N(h - r)}{hr} \right) q^{(k-1)/2}$$

$$+ \left(\frac{N}{h} - \frac{N}{q^k - 1} \right) q^{(k/2)-1}$$

for $b \neq 0$.

Chapter 7

Theoretical Applications of Finite Fields

Finite fields play a fundamental role in some of the most fascinating applications of modern algebra to the real world. These applications occur in the general area of data communication, a vital concern in our information society. Technological breakthroughs like space and satellite communications and mundane matters like guarding the privacy of information in data banks all depend in one way or another on the use of finite fields. Because of the importance of these applications to communication and information theory, we will present them in greater detail in the following chapters. Chapter 8 discusses applications of finite fields to coding theory, the science of reliable transmission of messages, and Chapter 9 deals with applications to cryptology, the art of enciphering and deciphering secret messages.

This chapter is devoted to applications of finite fields within mathematics. These applications are indeed numerous, so we can only offer a selection of possible topics. Section 1 contains some results on the use of finite fields in affine and projective geometry and illustrates in particular their role in the construction of projective planes with a finite number of points and lines. Section 2 on combinatorics demonstrates the variety of applications of finite fields to this subject and points out their usefulness in problems of design of statistical experiments.

In Section 3 we give the definition of a linear modular system and show how finite fields are involved in this theory. A system is regarded as a structure into which something (matter, energy, or information) may be put at certain times and that itself puts out something at certain times. For instance, we may visualize a system as an electrical circuit whose input is a voltage signal and whose output is a current reading. Or we may think of a system as a network of switching elements whose input is an on/off setting of a number of input switches and whose output is the on/off pattern of an array of lights.

Some applications of finite fields to the simulation of randomness are discussed in Section 4. In particular, we show how certain linear recurring sequences can be used to simulate random sequences of bits. In numerical analysis one often has to simulate random sequences of real numbers; it is perhaps surprising that linear recurring sequences in finite fields can also be instrumental in this task.

We emphasize that the applications are only described to give examples for the use of various properties of finite fields. Therefore, the examples contain rather the algebraic and combinatorial aspects, without regard to their practical application or indeed other usefulness. For instance, we are not going to discuss the analysis of experimental design or the analysis or synthesis of linear modular systems, nor do we explain geometric properties that are not directly connected with finite fields.

1. FINITE GEOMETRIES

In this section we describe the use of finite fields in geometric problems. A projective plane consists of a set of points and a set of lines together with an incidence relation that allows us to state for every point and for every line either that the point is on the line or is not on the line. In order to have a proper definition, certain axioms have to be satisfied.

7.1. Definition. A *projective plane* is defined as a set of elements, called *points*, together with distinguished sets of points, called *lines*, as well as a relation *I*, called *incidence*, between points and lines subject to the following conditions:

(i) every pair of distinct lines is incident with a unique point (i.e., to every pair of distinct lines there is one point contained in both lines, called their *intersection*);

(ii) every pair of distinct points is incident with a unique line (i.e., to every pair of distinct points there is exactly one line which contains both points);

(iii) there exist four points such that no three of them are incident with a single line (i.e., there exist four points such that no three of them are on the same line).

It follows that each line contains at least three points and that through each point there must be at least three lines. If the set of points is finite, we speak of a *finite projective plane*. From the three axioms above one deduces that (iii) holds also with the concepts of "point" and "line" interchanged. This establishes a *principle of duality* between points and lines, from which one can derive the following result.

7.2. **Theorem.** *Let* Π *be a finite projective plane. Then*:

(i) *there is an integer* $m \geqslant 2$ *such that every point (line) of* Π *is incident with exactly* $m + 1$ *lines (points) of* Π;

(ii) Π *contains exactly* $m^2 + m + 1$ *points (lines)*.

7.3. **Example.** The simplest finite projective plane is that with $m = 2$; there are precisely three lines through each point and three points on each line. Altogether there are 7 points and 7 lines in the plane. This projective plane is called the *Fano plane* and it may be illustrated as shown in Figure 7.1. The points are A, B, C, D, E, F, and G and the lines are ADC, AGE, AFB, CGF, CEB, DGB, and DEF. Since straightness is not a meaningful concept in a finite plane, the subset DEF is a line in the finite projective plane.   □

The integer m in Theorem 7.2 is called the *order* of the finite projective plane. We will see that finite projective planes of order m exist for every integer m of the form $m = p^n$, where p is a prime. It is known that there is no plane for $m = 6$, but it is not known whether a plane exists for $m = 10$. Many planes have been found for $m = 9$, but no plane has yet been found for which m is not a power of a prime.

In ordinary analytic geometry we represent points of the plane as ordered pairs (x, y) of real numbers and lines are sets of points that satisfy real equations of the form $ax + by + c = 0$ with a and b not both 0. Now the field of real numbers can be replaced by any other field, in particular a finite field. This type of geometry is known as affine geometry (or euclidean geometry) and leads to the concept of an affine plane.

7.4. **Definition.** An *affine plane* is a triple $(\mathcal{P}, \mathcal{L}, I)$ consisting of a set $\mathcal{P}$ of points, a set $\mathcal{L}$ of lines, and an incidence relation I such that:

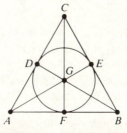

FIGURE 7.1 The Fano plane.

(i)   every pair of distinct points is incident with a unique line;

(ii)   every point $P \in \mathcal{P}$ not on a line $L \in \mathcal{L}$ lies on a unique line $M \in \mathcal{L}$ which does not intersect L;

(iii) there exist four points such that no three of them are incident with a single line.

The proof of the following theorem is straightforward.

7.5. Theorem. *Let K be any field. Let $\mathcal{P}$ denote the set of ordered pairs (x, y) with $x, y \in K$, and let $\mathcal{L}$ consist of those subsets L of $\mathcal{P}$ which satisfy linear equations, i.e., $L \in \mathcal{L}$ if for some $a, b, c \in K$ with $(a, b) \neq (0,0)$ we have $L = \{(x, y): ax + by + c = 0\}$. A point $P \in \mathcal{P}$ is incident with a line $L \in \mathcal{L}$ if and only if $P \in L$. Then $(\mathcal{P}, \mathcal{L}, I)$ is an affine plane, denoted by $AG(2, K)$.*

It can be shown readily that if $|K| = m$, then each line of $AG(2, K)$ contains exactly m points. We can construct a projective plane from $AG(2, K)$ by adding a line to it (and, conversely, we can obtain an affine plane from any projective plane by deleting one line and all the points on it).

We change the notation in $AG(2, K)$ and rename all the points as $(x, y, 1)$, that is, (x, y, z) with $z = 1$, and use the equation $ax + by + cz = 0$ with $(a, b) \neq (0,0)$ as the equation of a line. Now add the set of points

$$L_\infty = \{(1,0,0)\} \cup \{(x, 1, 0) : x \in K\}$$

to $\mathcal{P}$ to form a new set $\mathcal{P}' = \mathcal{P} \cup L_\infty$. The points of L_∞ can be represented by the equation $z = 0$ and so can be interpreted as a line. Let this new line L_∞ be added to $\mathcal{L}$ to form the set $\mathcal{L}' = \mathcal{L} \cup \{L_\infty\}$. With the natural extended notion of incidence, it can be verified that $(\mathcal{P}', \mathcal{L}', I')$ satisfies all the axioms for a projective plane.

7.6. Theorem. *Let $AG(2, K) = (\mathcal{P}, \mathcal{L}, I)$ and let*

$$\mathcal{P}' = \mathcal{P} \cup \{(1,0,0)\} \cup \{(x, 1, 0) : x \in K\} = \mathcal{P} \cup L_\infty,$$

$$\mathcal{L}' = \mathcal{L} \cup \{L_\infty\},$$

and let the extended incidence relation be denoted by I'. Then $(\mathcal{P}', \mathcal{L}', I')$ is a projective plane, denoted by $PG(2, K)$.

7.7. Example. The plane $PG(2, \mathbb{F}_2)$—that is, the projective plane over the field $\mathbb{F}_2$—has seven points: $(0,0,1)$, $(1,0,1)$, $(0,1,1)$, and $(1,1,1)$ with $z \neq 0$ and the three distinct points on the line $z = 0$, namely, $(1,0,0)$, $(0,1,0)$, and $(1,1,0)$. It can be verified that $PG(2, \mathbb{F}_2)$ also contains seven lines and that this projective plane is the Fano plane of Example 7.3.   □

In constructing $PG(2, K)$, every line of $AG(2, K)$ must meet the new line L_∞, so there will be an additional point on each line; also L_∞ contains

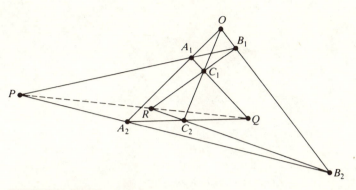

FIGURE 7.2 Desargues's theorem.

$m + 1$ points if K contains m elements. Since for every prime power $m = p^n = q$ there are finite fields $\mathbb{F}_q$, we have the following theorem.

7.8. Theorem. *For every prime power $q = p^n$, p prime, $n \in \mathbb{N}$, there exists a finite projective plane of order q — namely, $PG(2, \mathbb{F}_q)$.*

The additional line L_∞ added to an affine plane to obtain a projective plane is sometimes called the *line at infinity*. If two lines intersect on L_∞, they are called *parallel*.

Next we present without proof two interesting theorems, which hold in all projective planes that can be represented analytically in terms of fields. Two triangles $\triangle A_1 B_1 C_1$ and $\triangle A_2 B_2 C_2$ are said to be *in perspective* from a point O if the lines $A_1 A_2$, $B_1 B_2$, and $C_1 C_2$ pass through O. Points on the same line are said to be *collinear*.

7.9. Theorem (Desargues's Theorem). *If $\triangle A_1 B_1 C_1$ and $\triangle A_2 B_2 C_2$ are in perspective from O, then the intersections of the lines $A_1 B_1$ and $A_2 B_2$, of $A_1 C_1$ and $A_2 C_2$, and of $B_1 C_1$ and $B_2 C_2$, are collinear.*

The theorem is illustrated in Figure 7.2; the intersections of corresponding lines are P, Q, and R and are collinear.

7.10. Theorem (Theorem of Pappus). *If A_1, B_1, C_1 are points of a line and A_2, B_2, C_2 are points of another line in the same plane, and if $A_1 B_2$ and $A_2 B_1$ intersect in P, $A_1 C_2$ and $A_2 C_1$ intersect in Q, and $B_1 C_2$ and $B_2 C_1$ intersect in R, then P, Q, and R are collinear.*

The theorem is illustrated in Figure 7.3. Both theorems play an important role in projective geometry. If Desargues's theorem holds in some projective plane, then coordinates can be defined in terms of elements from a division ring. Here we define a point as an ordered triple (x_0, x_1, x_2) of three *homogeneous coordinates*, where the x_i are elements of a division ring R, not all of them simultaneously 0. The triples (ax_0, ax_1, ax_2), $0 \neq a \in R$,

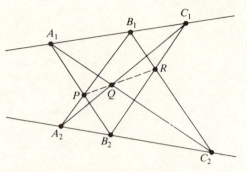

FIGURE 7.3   The theorem of Pappus.

shall denote the same point. Thus each point is represented in $m - 1$ ways if $|R| = m$, and because there are $m^3 - 1$ possible triples of coordinates, the total number of different points is

$$(m^3 - 1)/(m - 1) = m^2 + m + 1.$$

A line is defined as the set of all those points whose coordinates satisfy an equation of the form $x_0 + a_1 x_1 + a_2 x_2 = 0$, or of the form $x_1 + a_2 x_2 = 0$, or of the form $x_2 = 0$, where $a_i \in R$. There are $m^2 + m + 1$ such lines in the plane and it is straightforward to show that the points and lines thus defined satisfy the axioms of a finite projective plane.

From Theorem 2.55—that is, Wedderburn's theorem—we know that any finite division ring is a field, a finite field $\mathbb{F}_q$. In that case the equation of any line can be written as $a_0 x_0 + a_1 x_1 + a_2 x_2 = 0$, where the a_i are not simultaneously 0, and $(aa_0)x_0 + (aa_1)x_1 + (aa_2)x_2 = 0$ with $a \in \mathbb{F}_q^*$ is the same line. The line connecting the points (y_0, y_1, y_2) and (z_0, z_1, z_2) may then also be defined as the set of all points with coordinates

$$(ay_0 + bz_0, ay_1 + bz_1, ay_2 + bz_2),$$

where a and b are in $\mathbb{F}_q$, not both equal to 0. There are $q^2 - 1$ such triples, and since simultaneous multiplication of a and b by the same nonzero element produces the same point, they yield $q + 1$ different points.

In $PG(2, \mathbb{F}_q)$ Desargues's theorem and its converse hold, and the proof relies on commutativity of multiplication in $\mathbb{F}_q$. In general, Desargues's theorem and its converse do not both apply if the coordinatizing ring does not have commutativity of multiplication. Thus Wedderburn's theorem plays an important role in this context.

A projective plane in which Desargues's theorem holds is called *Desarguesian*; otherwise it is called *non-Desarguesian*. Desarguesian planes of order m exist only if m is the power of a prime, and up to isomorphism there exists only one Desarguesian plane for any given prime power $m = p^n$.

A finite Desarguesian plane can always be coordinatized by a finite field. Since such fields exist only when the order is a prime power, a projective plane with exactly $m + 1$ points on each line, m not a prime power, will have to be non-Desarguesian. It is not known whether such planes for m not a prime power exist. If it can be proved that up to isomorphism there exists only one finite projective plane of order m, and if m is a prime power, then this plane must be Desarguesian. This is the case for $m = 2, 3, 4, 5, 7$, and 8. For m prime, only Desarguesian planes are known. But it has been shown that for all prime powers $m = p^n$, $n \geqslant 2$, except for 4 and 8, there exist non-Desarguesian planes of order m.

The theorem of Pappus implies the theorem of Desargues. If the theorem of Pappus holds in some projective plane, then the multiplication in the coordinatizing ring is necessarily commutative. The theorem of Pappus holds in $PG(2, \mathbb{F}_q)$ for any prime power q. A finite Desarguesian plane also satisfies the theorem of Pappus.

A remarkable distinction between the properties of a $PG(2, \mathbb{F}_q)$ with q even and a $PG(2, \mathbb{F}_q)$ with q odd is given in the following theorem.

7.11. Theorem. *The diagonal points of a complete quadrangle in $PG(2, \mathbb{F}_q)$ are collinear if and only if q is even.*

Proof. We assume, without loss of generality, that the vertices of the quadrangle are $(1,0,0)$, $(0,1,0)$, $(0,0,1)$, and $(1,1,1)$. Its six sides are $x_2 = 0$, $x_1 = 0$, $x_1 - x_2 = 0$, $x_0 = 0$, $x_0 - x_2 = 0$, and $x_0 - x_1 = 0$, while the three diagonal points are $(1,1,0)$, $(1,0,1)$, and $(0,1,1)$. The line through the first two points contains all points with coordinates $(a + b, a, b)$, where $(a, b) \neq (0,0)$, and the third point is one of these if and only if $a = b$ and $a + b = 0$. In a finite field $\mathbb{F}_q$ this is only possible if the characteristic is 2. □

The latter case is illustrated in Example 7.3. Let the vertices of the complete quadrangle be C, D, E, G. In this case, the diagonal points are A, F, B, and they are collinear.

We introduce now concepts analogous to those with which we are familiar in analytic geometry, and we restrict ourselves to Desarguesian planes, coordinatized by a finite field $\mathbb{F}_q$.

Let the equations of two distinct lines be

$$a_{01}x_0 + a_{11}x_1 + a_{21}x_2 = 0,$$
$$a_{02}x_0 + a_{12}x_1 + a_{22}x_2 = 0. \tag{7.1}$$

Let the point of intersection of these two lines be P. All lines through P form a pencil and each line in this pencil has an equation of the form

$$(ra_{01} + sa_{02})x_0 + (ra_{11} + sa_{12})x_1 + (ra_{21} + sa_{22})x_2 = 0,$$

where $r, s \in \mathbb{F}_q$ are not both 0. There are $q + 1$ lines in the pencil: the two lines (7.1) given above corresponding to $s = 0$ and $r = 0$, respectively, and those corresponding to $q-1$ different ratios rs^{-1} with $r \neq 0$ and $s \neq 0$. Let another pencil through a point $Q \neq P$ be given by

$$(rb_{01} + sb_{02})x_0 + (rb_{11} + sb_{12})x_1 + (rb_{21} + sb_{22})x_2 = 0.$$

A projective correspondence between the lines of the two pencils is defined by letting a line of the first, given by a pair (r, s), correspond to the line of the second pencil that belongs to the same pair. Two corresponding lines meet in a unique point, except when the line PQ corresponds to itself, and the coordinates of all the points satisfy the equation

$$(a_{01}x_0 + a_{11}x_1 + a_{21}x_2)(b_{02}x_0 + b_{12}x_1 + b_{22}x_2)$$
$$- (a_{02}x_0 + a_{12}x_1 + a_{22}x_2)(b_{01}x_0 + b_{11}x_1 + b_{21}x_2) = 0, \qquad (7.2)$$

obtained by eliminating r and s from the equations of the two pencils.

7.12.   Definition.   The set of points whose coordinates satisfy equation (7.2) is called a *conic*. If the line PQ corresponds to itself under the correspondence above, then the conic is called *degenerate*. It consists then of the $2q+1$ points of two intersecting lines. A *nondegenerate* conic consists of the $q+1$ points of intersection of corresponding lines. A line that has precisely one point in common with a conic is called a *tangent* of it; a line that has two points in common is a *secant*.

The equation of a nondegenerate conic is quadratic, therefore it cannot have more than two points in common with any line. Take one point of a nondegenerate conic and connect it by lines to the other q points. Then the resulting lines are secants and the remaining one of the $q+1$ lines through that point must be a tangent.

The $q+1$ points of a nondegenerate conic thus have the property that no three of them are collinear. It can be shown that any set of $q+1$ points in a $PG(2, \mathbb{F}_q)$, q odd, such that no three of them are collinear is a nondegenerate conic.

The following theorem, which we prove only in part, exhibits a difference between conics in Desarguesian planes of odd and of even order.

7.13.   Theorem.   (i) *In a Desarguesian plane of odd order there pass two or no tangents of a nondegenerate conic through a point not on the conic.*

(ii) *In a Desarguesian plane of even order all the tangents of a nondegenerate conic meet in a single point.*

Proof.   We prove (ii) as an example of how properties of finite fields are used in the theory of finite projective planes. Assume without loss of generality that three points on a nondegenerate conic in a plane of even order are $A(1,0,0)$, $B(0,1,0)$, $C(0,0,1)$ and that the tangents through these three points are, respectively, $x_1 - k_0 x_2 = 0$, $x_2 - k_1 x_0 = 0$, $x_0 - k_2 x_1 = 0$. Let $P(t_0, t_1, t_2)$ be another point of the conic. None of the t_i can be 0, because then P would be on a line through two of the points A, B, and C, contradicting the fact that no three points of the conic are collinear. Therefore we can write $x_1 - t_1 t_2^{-1} x_2 = 0$ for PA, $x_2 - t_2 t_0^{-1} x_0 = 0$ for PB, and $x_0 - t_0 t_1^{-1} x_1 = 0$ for PC.

Consider the equation for the line PA. As we choose for P the various points of the conic, leaving out A, B, and C, the ratio $t_1 t_2^{-1}$ runs through the elements of $\mathbb{F}_q$ apart from 0 and k_0. Since

$$\prod_{c \in \mathbb{F}_q^*} (x - c) = x^{q-1} - 1,$$

the product of all nonzero elements of $\mathbb{F}_q$ is $(-1)^q$. Thus, multiplying the product of the $q - 2$ values $t_1 t_2^{-1}$ assumes by k_0, we obtain $(-1)^q = 1$, since q is even. We have

$$k_0 \prod t_1 t_2^{-1} = 1, \quad k_1 \prod t_2 t_0^{-1} = 1, \quad \text{and} \quad k_2 \prod t_0 t_1^{-1} = 1,$$

where the product extends over all points of the conic except A, B, and C. Multiplying the three products above we get $k_0 k_1 k_2 = 1$. Therefore the points $(1, k_0 k_1, k_1)$, $(k_2, 1, k_1 k_2)$, and $(k_0 k_2, k_0, 1)$ are identical. The three tangents at A, B, and C pass through this point; and because these points were arbitrary, any three tangents meet in the same point. $\qquad\square$

Analogs of the concept of a projective plane can be defined for dimensions higher than 2.

7.14. Definition. A *projective space*, or a *projective geometry*, or an *m-space* is a set of points, together with distinguished sets of points, called lines, subject to the following conditions:

(i)   There is a unique line through any pair of distinct points.
(ii) A line that intersects two lines of a triangle intersects the third line as well.
(iii) Every line contains at least three points.
(iv) Define a k-space as follows. A 0-space is a point. If $A_0, \ldots, A_k$ are points not all in the same $(k-1)$-space, then all points collinear with A_0 and any point in the $(k-1)$-space defined by $A_1, \ldots, A_k$ form a k-space. Thus a line is a 1-space, and all the other spaces are defined recursively. Axiom (iv) demands: If $k < m$, then not all points considered are in the same k-space.
(v)   There exists no $(m+1)$-space in the set of points considered.

We say that an m-space has m dimensions, and if we refer to a k-space as a subspace of a projective space of higher dimension, we call it a k-*flat*. An $(m-1)$-flat in a projective space of m dimensions is called a *hyperplane*. A 2-space is a projective plane in the sense of Definition 7.1. It can be proved that in any 2-flat in a projective space of at least three dimensions the theorem of Desargues (Theorem 7.9) is always valid. Desargues's theorem can only fail to be true in projective planes that cannot be embedded in a projective space of at least three dimensions.

A projective space containing only finitely many points is called a *finite projective space* (or *finite projective geometry*, or *finite m-space*). In analogy with $PG(2, \mathbb{F}_q)$, we can construct the finite *m*-space $PG(m, \mathbb{F}_q)$. Define a point as an ordered $(m + 1)$-tuple $(x_0, x_1, \ldots, x_m)$, where the coordinates $x_i \in \mathbb{F}_q$ are not simultaneously 0. The $(m + 1)$-tuples $(ax_0, ax_1, \ldots, ax_m)$ with $a \in \mathbb{F}_q^*$ define the same point. There are therefore $(q^{m+1} - 1)/(q - 1)$ points in $PG(m, \mathbb{F}_q)$.

A *k*-flat in $PG(m, \mathbb{F}_q)$ is the set of all those points whose coordinates satisfy $m - k$ linearly independent homogeneous linear equations

$$
\begin{aligned}
a_{10}x_0 &+ \cdots + a_{1m}x_m = 0 \\
&\vdots \\
a_{m-k,0}x_0 &+ \cdots + a_{m-k,m}x_m = 0
\end{aligned}
$$

with coefficients $a_{ij} \in \mathbb{F}_q$. Alternatively, a *k*-flat consists of all those points with coordinates

$$
(a_0 x_{00} + \cdots + a_k x_{k0}, \ldots, a_0 x_{0m} + \cdots + a_k x_{km})
$$

with the $a_i \in \mathbb{F}_q$ not simultaneously 0 and the $k + 1$ given points

$$
(x_{00}, \ldots, x_{0m}), \ldots, (x_{k0}, \ldots, x_{km})
$$

being linearly independent; that is, the matrix

$$
\begin{pmatrix}
x_{00} & \cdots & x_{0m} \\
x_{10} & \cdots & x_{1m} \\
\vdots & & \vdots \\
x_{k0} & \cdots & x_{km}
\end{pmatrix}
$$

has rank $k + 1$. The number of points in a *k*-flat is $(q^{k+1} - 1)/(q - 1)$; there are $q + 1$ points on a line and $q^2 + q + 1$ on a plane. That $PG(m, \mathbb{F}_q)$ satisfies the five axioms for an *m*-space is easily verified.

We know that in $\mathbb{F}_{q^{m+1}}$ all powers of a primitive element α can be represented as polynomials in α of degree at most *m* with coefficients in $\mathbb{F}_q$. If

$$
\alpha^i = a_m \alpha^m + \cdots + a_0,
$$

we may consider α^i as representing a point in $PG(m, \mathbb{F}_q)$ with coordinates $(a_0, \ldots, a_m)$. Two powers α^i, α^j represent the same point if and only if $\alpha^i = a\alpha^j$ for some $a \in \mathbb{F}_q^*$—that is, if and only if

$$
i \equiv j \bmod (q^{m+1} - 1)/(q - 1).
$$

A k-flat S through $k+1$ linearly independent points represented by $\alpha^{i_0}, \ldots,$ α^{i_k} will contain all points represented by $\sum_{r=0}^{k} a_r \alpha^{i_r}, a_r \in \mathbb{F}_q$ not simultaneously 0. For each $h = 0, 1, \ldots, v-1$ with $v = (q^{m+1}-1)/(q-1)$, the points $\sum_{r=0}^{k} a_r \alpha^{i_r+h}, a_r \in \mathbb{F}_q$ not simultaneously 0, form k-flats, and we denote the k-flat with given h by S_h. We have $S_v = S_0 = S$ because $\alpha^v \in \mathbb{F}_q$. Let j be the least positive integer for which $S_j = S$. Then from $S_{nj} = S$ for all $n \in \mathbb{N}$ it follows that j divides v, say $v = tj$. We call j the *cycle* of S.

If α^{d_0} is a point of the k-flat S, then so are the points with exponents

$$d_0, d_0 + j, \ldots, d_0 + (t-1)j,$$

because $S_{nj} = S$ for $n = 0, 1, \ldots, t-1$. Further points on S can be written with the following exponents of α:

$$d_1, \quad d_1 + j \quad, \ldots, d_1 + (t-1)j$$
$$\vdots \qquad \vdots \qquad\qquad \vdots$$
$$d_{u-1}, d_{u-1} + j, \ldots, d_{u-1} + (t-1)j,$$

where $d_{r_1} - d_{r_2}$ is not divisible by j for $r_1 \neq r_2$. The number of all these distinct points is $tu = (q^{k+1}-1)/(q-1)$.

If $tj = (q^{m+1}-1)/(q-1)$ and $tu = (q^{k+1}-1)(q-1)$ are relatively prime, then $t = 1$, $j = v$, and all k-flats have cycle v. This is the case for $k = m-1$, and for $k = 1$ when m is even.

7.15. Example. Consider $PG(3, \mathbb{F}_2)$ with 15 points, 35 lines, 15 planes, and $q^{m+1} = 16$. Using a root $\alpha \in \mathbb{F}_{16}$ of the primitive polynomial $x^4 + x + 1$ over $\mathbb{F}_2$, we can establish a correspondence between the powers of α and the points of $PG(3, \mathbb{F}_2)$. We obtain:

$A(0,0,0,1) \cdots \alpha^3$	$F(0,1,1,0) \cdots \alpha^5$	$K(1,0,1,1) \cdots \alpha^{13}$
$B(0,0,1,0) \cdots \alpha^2$	$G(0,1,1,1) \cdots \alpha^{11}$	$L(1,1,0,0) \cdots \alpha^4$
$C(0,0,1,1) \cdots \alpha^6$	$H(1,0,0,0) \cdots \alpha^0$	$M(1,1,0,1) \cdots \alpha^7$
$D(0,1,0,0) \cdots \alpha^1$	$I(1,0,0,1) \cdots \alpha^{14}$	$N(1,1,1,0) \cdots \alpha^{10}$
$E(0,1,0,1) \cdots \alpha^9$	$J(1,0,1,0) \cdots \alpha^8$	$O(1,1,1,1) \cdots \alpha^{12}$

The plane

$$S = S_0 = \{a_0 \alpha^0 + a_1 \alpha^1 + a_2 \alpha^{2} \cdot \quad a_0, a_1, a_2 \in \mathbb{F}_2 \text{ not all } 0\}$$

is the same as the plane $x_3 = 0$. It contains the points B, D, F, H, J, L, and N. It has cycle 15, as has any other hyperplane. The plane

$$S_1 = \{a_0 \alpha^1 + a_1 \alpha^2 + a_2 \alpha^3 : \quad a_0, a_1, a_2 \in \mathbb{F}_2 \text{ not all } 0\}$$

is the same as the plane $x_0 = 0$ and contains the points A, B, C, D, E, F, and G; and so on. The line

$$\{a_0 \alpha^3 + a_1 \alpha^8 : \quad a_0, a_1 \in \mathbb{F}_2 \text{ not both } 0\},$$

that is, the line AJK, has cycle 5, the lines ABC and ADE both have cycle 15, and this accounts for all the $5 + 15 + 15 = 35$ lines.   □

A *finite affine* (or *euclidean*) *geometry*, denoted by $AG(m, \mathbb{F}_q)$, is the set of flats that remain when a hyperplane with all its flats is removed from $PG(m, \mathbb{F}_q)$. Those flats that were removed are called *flats at infinity*. Those remaining flats that intersect in a flat at infinity are called *parallel*. It is convenient to consider the excluded hyperplane as the one whose equation is $x_m = 0$. Then we may fix x_m for all points in $AG(m, \mathbb{F}_q)$ at 1, and consider only the remaining coordinates as those of a point in $AG(m, \mathbb{F}_q)$. Since there are $q^m + \cdots + q + 1$ points in $PG(m, \mathbb{F}_q)$, and the $q^{m-1} + \cdots + q + 1$ points of a hyperplane were removed, there remain q^m points in $AG(m, \mathbb{F}_q)$.

A k-flat within $AG(m, \mathbb{F}_q)$ contains all those q^k points that satisfy a system of equations of the form

$$a_{i0}x_0 + \cdots + a_{i,m-1}x_{m-1} + a_{im} = 0, \quad i = 1, \ldots, m - k,$$

where the coefficient matrix has rank $m - k$. In particular, a hyperplane is defined by

$$a_0 x_0 + \cdots + a_{m-1}x_{m-1} + a_m = 0,$$

where $a_0, \ldots, a_{m-1}$ are not all 0. If $a_0, \ldots, a_{m-1}$ are kept constant and a_m runs through all elements of $\mathbb{F}_q$, then we obtain a pencil of parallel hyperplanes.

2. COMBINATORICS

In this section we describe some of the useful aspects of finite fields in combinatorics.

There is a close connection between finite geometries and *designs*. The designs we wish to consider consist of two nonempty sets of objects, with an incidence relation between objects of different sets. For instance, the objects may be points and lines, with a given point lying or not lying on a given line. The terminology that is normally used in this area has its origin in the applications in statistics, in connection with the design of experiments. The two types of objects are called *varieties* (in early applications these were plants or fertilizers) and *blocks*. The number of varieties will, as a rule, be denoted by v, and the number of blocks by b.

A design for which every block is incident with the same number k of varieties and every variety is incident with the same number r of blocks is called a *tactical configuration*. Clearly

$$vr = bk. \tag{7.3}$$

If $v = b$, and hence $r = k$, the tactical configuration is called *symmetric*. For instance, the points and lines of a $PG(2, \mathbb{F}_q)$ form a symmetric tactical configuration with $v = b = q^2 + q + 1$ and $r = k = q + 1$. The property of a finite projective plane that every pair of distinct points is incident with a unique line may serve to motivate the following definition.

7.16. Definition. A tactical configuration is called a *balanced incomplete block design (BIBD)*, or (v, k, λ) *block design*, if $v \geqslant k \geqslant 2$ and every pair of distinct varieties is incident with the same number λ of blocks.

If for a fixed variety a_1 we count in two ways all the ordered pairs (a_2, B) with a variety $a_2 \neq a_1$ and a block B incident with a_1, a_2, we obtain the identity

$$r(k - 1) = \lambda(v - 1) \tag{7.4}$$

for any (v, k, λ) block design. Thus, the parameters b and r of a BIBD are determined by v, k, and λ because of (7.3) and (7.4).

7.17. Example. Let the set of varieties be $\{0, 1, 2, 3, 4, 5, 6\}$ and let the blocks be the subsets $\{0, 1, 3\}$, $\{1, 2, 4\}$, $\{2, 3, 5\}$, $\{3, 4, 6\}$, $\{4, 5, 0\}$, $\{5, 6, 1\}$, and $\{6, 0, 2\}$, with the obvious incidence relation between varieties and blocks. This is a symmetric BIBD with $v = b = 7$, $r = k = 3$, and $\lambda = 1$. It is equivalent to the Fano plane in Example 7.3. A BIBD with $k = 3$ and $\lambda = 1$ is called a *Steiner triple system*. $\square$

7.18. Example. More generally, a BIBD is obtained by taking the points of a projective geometry $PG(m, \mathbb{F}_q)$ or of an affine geometry $AG(m, \mathbb{F}_q)$ as varieties and its t-flats for some fixed t, $1 \leqslant t < m$, as blocks. In the projective case, the parameters of the resulting BIBD are as follows:

$$v = \frac{q^{m+1} - 1}{q - 1}, \quad b = \prod_{i=1}^{t+1} \frac{q^{m-t+i} - 1}{q^i - 1}, \quad r = \prod_{i=1}^{t} \frac{q^{m-t+i} - 1}{q^i - 1},$$

$$k = \frac{q^{t+1} - 1}{q - 1}, \quad \lambda = \prod_{i=1}^{t-1} \frac{q^{m-t+i} - 1}{q^i - 1},$$

where the last product is interpreted to be 1 if $t = 1$. The BIBD is symmetric in case $t = m - 1$—that is, if the blocks are the hyperplanes of $PG(m, \mathbb{F}_q)$. In the affine case, the parameters of the resulting BIBD are as follows:

$$v = q^m, \quad b = q^{m-t} \prod_{i=1}^{t} \frac{q^{m-t+i} - 1}{q^i - 1}, \quad r = \prod_{i=1}^{t} \frac{q^{m-t+i} - 1}{q^i - 1},$$

$$k = q^t, \quad \lambda = \prod_{i=1}^{t-1} \frac{q^{m-t+i} - 1}{q^i - 1},$$

with the same convention for $t = 1$ as above. Such a BIBD is never symmetric. $\square$

A tactical configuration can be described by its *incidence matrix*.

This is a matrix A of v rows and b columns, where the rows correspond to the varieties and the columns to the blocks. We number the varieties and blocks, and if the ith variety is incident with the jth block, we define the (i, j) entry of A to be the integer 1, otherwise 0. The sum of entries in any row is r and that in any column is k.

If A is the incidence matrix of a (v, k, λ) block design, then the inner product of two different rows of A is λ. Thus, if A^T denotes the transpose of A, then

$$AA^T = \begin{pmatrix} r & \lambda & \cdots & \lambda \\ \lambda & r & \cdots & \lambda \\ \vdots & \vdots & & \vdots \\ \lambda & \lambda & \cdots & r \end{pmatrix} = (r - \lambda)I + \lambda J,$$

where I is the $v \times v$ identity matrix and J is the $v \times v$ matrix with all entries equal to 1. We compute the determinant of AA^T by subtracting the first column from the others and then adding to the first row the sum of the others. The result is

$$\det(AA^T) = \begin{vmatrix} rk & 0 & 0 & \cdots & 0 \\ \lambda & r - \lambda & 0 & \cdots & 0 \\ \lambda & 0 & r - \lambda & \cdots & 0 \\ \vdots & \vdots & \vdots & & \vdots \\ \lambda & 0 & 0 & \cdots & r - \lambda \end{vmatrix} = rk(r - \lambda)^{v-1},$$

where we have used (7.4). If $v = k$, the design is trivial, since each block is incident with all v varieties. If $v > k$, then $r > \lambda$ by (7.4), and so AA^T is of rank v. The matrix A cannot have smaller rank, hence we obtain

$$b \geqslant v. \tag{7.5}$$

By (7.3), we must also have $r \geqslant k$.

For a *symmetric* (v, k, λ) block design we have $r = k$, hence $AJ = JA$, and so A commutes with $(r - \lambda)I + \lambda J = AA^T$. Since A is nonsingular if $v > k$, we get $A^T A = AA^T = (r - \lambda)I + \lambda J$. It follows that *any two distinct blocks have exactly λ varieties in common*. This holds trivially if $v = k$.

We have seen that the conditions (7.3) and (7.4), and furthermore (7.5) in the nontrivial case, are necessary for the existence of a BIBD with parameters v, b, r, k, λ. These conditions are, however, not sufficient for the existence of such a design. For instance, a BIBD with $v = b = 43$, $r = k = 7$, and $\lambda = 1$ is known to be impossible.

The varieties and blocks of a symmetric (v, k, λ) block design with $k \geqslant 3$ and $\lambda = 1$ satisfy the conditions for points and lines of a finite projective plane. The converse is also true. Thus, *the concepts of a symmetric $(v, k, 1)$ block design with $k \geqslant 3$ and of a finite projective plane are equivalent.*

Consider the BIBD in Example 7.17 and interpret the varieties

$0, 1, 2, 3, 4, 5, 6$ as integers modulo 7. Each block of this design has the property that the differences between its distinct elements yield all nonzero residues modulo 7. This suggests the following definition.

7.19.   Definition.   A set $D = \{d_1, \ldots, d_k\}$ of $k \geqslant 2$ distinct residues modulo v is called a (v, k, λ) *difference set* if for every $d \not\equiv 0 \bmod v$ there are exactly λ ordered pairs (d_i, d_j) with $d_i, d_j \in D$ such that $d_i - d_j \equiv d \bmod v$.

The following results provide a connection between difference sets, designs, and finite projective planes.

7.20.   Theorem.   *Let $\{d_1, \ldots, d_k\}$ be a (v, k, λ) difference set. Then with all residues modulo v as varieties, the blocks*

$$B_t = \{d_1 + t, \ldots, d_k + t\}, \quad t = 0, 1, \ldots, v - 1,$$

form a symmetric (v, k, λ) block design under the obvious incidence relation.

Proof.   A residue a modulo v occurs exactly in the blocks with subscripts $a - d_1, \ldots, a - d_k$ modulo v, thus every variety is incident with the same number k of blocks. For a pair of distinct residues a, c modulo v, we have $a, c \in B_t$ if and only if $a \equiv d_i + t \bmod v$ and $c \equiv d_j + t \bmod v$ for some d_i, d_j. Consequently, $a - c \equiv d_i - d_j \bmod v$, and conversely, for every solution (d_i, d_j) of the last congruence, both a and c occur in the block with subscript $a - d_i$ modulo v. By hypothesis, there are exactly λ solutions (d_i, d_j) of this congruence, and so all the conditions for a symmetric (v, k, λ) block design are satisfied. $\square$

7.21.   Corollary.   *Let $\{d_1, \ldots, d_k\}$ be a $(v, k, 1)$ difference set with $k \geqslant 3$. Then the residues modulo v and the blocks B_t, $t = 0, 1, \ldots, v - 1$, from Theorem 7.20 satisfy the conditions for points and lines of a finite projective plane of order $k - 1$.*

Proof.   This follows from Theorem 7.20 and the observation above that symmetric $(v, k, 1)$ block designs with $k \geqslant 3$ are finite projective planes. $\square$

It follows from Theorem 7.20 and (7.4) that the parameters v, k, λ of a difference set are linked by the identity $k(k - 1) = \lambda(v - 1)$. This can also be seen directly from the definition of a difference set.

7.22.   Example.   The set $\{0, 1, 2, 4, 5, 8, 10\}$ of residues modulo 15 is a $(15, 7, 3)$ difference set. The blocks

$$B_t = \{t, t + 1, t + 2, t + 4, t + 5, t + 8, t + 10\}, \quad t = 0, 1, \ldots, 14,$$

form a symmetric $(15, 7, 3)$ block design according to Theorem 7.20. The blocks of this design can be interpreted as the 15 planes of the projective geometry $PG(3, \mathbb{F}_2)$, with the 15 residues representing the points. Each plane is a Fano plane $PG(2, \mathbb{F}_2)$. The lines of the block B_t can be obtained by cyclically permuting the points of the line

$$L_t = B_t \cap B_{t-4} = \{t, t+1, t+4\}$$

in the plane B_t according to the permutation

$$t \to t+1 \to t+2 \to t+4 \to t+5 \to t+10 \to t+8 \to t.$$

For instance, the lines in the plane $B_0 = \{0,1,2,4,5,10,8\}$ are

$$\{0,1,4\}, \{1,2,5\}, \{2,4,10\}, \{4,5,8\}, \{5,10,0\}, \{10,8,1\}, \{8,0,2\}. \qquad \square$$

Examples of difference sets can be obtained from finite projective geometries. As in the discussion preceding Example 7.15, we identify points of $PG(m,\mathbb{F}_q)$ with powers of α, where α is a primitive element of $\mathbb{F}_{q^{m+1}}$ and the exponents of α are considered modulo $v = (q^{m+1} - 1)/(q - 1)$. Let S be any hyperplane of $PG(m,\mathbb{F}_q)$. Then S has cycle v, and so the hyperplanes $S_h = \alpha^h S$, $h = 0,1,\ldots,v-1$, are distinct. These are already all hyperplanes of $PG(m,\mathbb{F}_q)$, since v is also the total number of hyperplanes. Thus, the following is the complete list of hyperplanes of $PG(m,\mathbb{F}_q)$, with the points contained in them indicated by the corresponding exponents of α:

$$
\begin{array}{llllll}
S_0 & : d_1 & d_2 & \cdots & d_k \\
S_1 & : d_1+1 & d_2+1 & \cdots & d_k+1 \\
& \vdots & \vdots & \vdots & \vdots \\
S_{v-1} & : d_1+v-1 & d_2+v-1 & \cdots & d_k+v-1
\end{array}
$$

Here $k = (q^m - 1)/(q - 1)$, the number of points in a hyperplane. If we look for those rows that contain a particular value, say 0, then we obtain the k hyperplanes through α^0. These k rows are given by:

$$
\begin{array}{llll}
d_1 - d_1 & d_2 - d_1 & \cdots & d_k - d_1 \\
d_1 - d_2 & d_2 - d_2 & \cdots & d_k - d_2 \\
\vdots & \vdots & & \vdots \\
d_1 - d_k & d_2 - d_k & \cdots & d_k - d_k
\end{array}
$$

Any point $\neq \alpha^0$ appears in as many of those k hyperplanes as there are hyperplanes through two distinct points—that is, $\lambda = (q^{m-1} - 1)/(q - 1)$ of them—so that the off-diagonal entries repeat each nonzero residue modulo v precisely λ times. Hence $\{d_1,\ldots,d_k\}$ is a (v,k,λ) difference set. We summarize this result as follows.

7.23. Theorem. *The points in any hyperplane of $PG(m,\mathbb{F}_q)$ determine a (v,k,λ) difference set with parameters*

$$v = \frac{q^{m+1} - 1}{q - 1}, \quad k = \frac{q^m - 1}{q - 1}, \quad \lambda = \frac{q^{m-1} - 1}{q - 1}.$$

7.24. Example. Consider the hyperplane $x_1 = 0$ of $PG(3,\mathbb{F}_2)$ in Example 7.15. It contains the points A, B, C, H, I, J, K, and so the corresponding exponents of α yield the $(15, 7, 3)$ difference set $\{0, 2, 3, 6, 8, 13, 14\}$. $\square$

Another branch of combinatorics in which finite fields are useful is the theory of orthogonal latin squares.

7.25.   Definition.   An array

$$
L = (a_{ij}) = \begin{pmatrix} a_{11} & a_{12} & \cdots & a_{1n} \\ a_{21} & a_{22} & \cdots & a_{2n} \\ \vdots & \vdots & & \vdots \\ a_{n1} & a_{n2} & \cdots & a_{nn} \end{pmatrix}
$$

is called a *latin square* of order n if each row and each column contains every element of a set of n elements exactly once. Two latin squares (a_{ij}) and (b_{ij}) of order n are said to be *orthogonal* if the n^2 ordered pairs (a_{ij}, b_{ij}) are all different.

7.26.   Theorem.   *A latin square of order n exists for every positive integer n.*

Proof.   Consider (a_{ij}) with $a_{ij} \equiv i + j \bmod n$, $1 \leqslant a_{ij} \leqslant n$. Then $a_{ij} = a_{ik}$ implies $i + j \equiv i + k \bmod n$, and so $j \equiv k \bmod n$, which means $j = k$ since $1 \leqslant i, j, k \leqslant n$. Similarly, $a_{ij} = a_{kj}$ implies $i = k$. Thus the elements of each row and each column are distinct. $\square$

Orthogonal latin squares were first studied by Euler. He conjectured that there did not exist pairs of orthogonal latin squares of order n if n is twice an odd integer. This was disproved in 1959 by the construction of a pair of orthogonal latin squares of order 22. It is now known that the values of n for which there exists a pair of orthogonal latin squares of order n are precisely all $n > 2$ with $n \neq 6$.

For some values of n, more than two latin squares of order n exist that are mutually orthogonal (i.e., orthogonal in pairs). We shall show that if $n = q$, a prime power, then there exist $q - 1$ mutually orthogonal latin squares of order q, by using the existence of finite fields of order q.

7.27.   Theorem.   *Let $a_0 = 0, a_1, a_2, \ldots, a_{q-1}$ be the elements of $\mathbb{F}_q$. Then the arrays*

$$
L_k = \begin{pmatrix} a_0 & a_1 & \cdots & a_{q-1} \\ a_k a_1 & a_k a_1 + a_1 & \cdots & a_k a_1 + a_{q-1} \\ a_k a_2 & a_k a_2 + a_1 & \cdots & a_k a_2 + a_{q-1} \\ \vdots & \vdots & & \vdots \\ a_k a_{q-1} & a_k a_{q-1} + a_1 & \cdots & a_k a_{q-1} + a_{q-1} \end{pmatrix}, \quad k = 1, \ldots, q-1,
$$

form a set of $q - 1$ mutually orthogonal latin squares of order q.

Proof. Each L_k is clearly a latin square. Let $a_{ij}^{(k)} = a_k a_{i-1} + a_{j-1}$ be the (i, j) entry of L_k. For $k \neq m$, suppose

$$\left(a_{ij}^{(k)}, a_{ij}^{(m)} \right) = \left(a_{gh}^{(k)}, a_{gh}^{(m)} \right) \text{ for some } 1 \leqslant i, j, g, h \leqslant q.$$

Then

$$\left(a_k a_{i-1} + a_{j-1}, a_m a_{i-1} + a_{j-1} \right) = \left(a_k a_{g-1} + a_{h-1}, a_m a_{g-1} + a_{h-1} \right),$$

and so

$$a_k(a_{i-1} - a_{g-1}) = a_{h-1} - a_{j-1}, \quad a_m(a_{i-1} - a_{g-1}) = a_{h-1} - a_{j-1}.$$

Since $a_k \neq a_m$, it follows that $a_{i-1} = a_{g-1}, a_{h-1} = a_{j-1}$, hence $i = g, j = h$. Thus the ordered pairs of corresponding entries from L_k and L_m are all different, and so L_k and L_m are orthogonal. $\square$

7.28.   Example.   A set of four mutually orthogonal latin squares of order 5 is given below, using the construction in Theorem 7.27:

$$
L_1 \qquad\qquad\qquad L_2
$$

$$
\begin{pmatrix}
0 & 1 & 2 & 3 & 4 \\
1 & 2 & 3 & 4 & 0 \\
2 & 3 & 4 & 0 & 1 \\
3 & 4 & 0 & 1 & 2 \\
4 & 0 & 1 & 2 & 3
\end{pmatrix}
\qquad
\begin{pmatrix}
0 & 1 & 2 & 3 & 4 \\
2 & 3 & 4 & 0 & 1 \\
4 & 0 & 1 & 2 & 3 \\
1 & 2 & 3 & 4 & 0 \\
3 & 4 & 0 & 1 & 2
\end{pmatrix}
$$

$$
L_3 \qquad\qquad\qquad L_4
$$

$$
\begin{pmatrix}
0 & 1 & 2 & 3 & 4 \\
3 & 4 & 0 & 1 & 2 \\
1 & 2 & 3 & 4 & 0 \\
4 & 0 & 1 & 2 & 3 \\
2 & 3 & 4 & 0 & 1
\end{pmatrix}
\begin{pmatrix}
0 & 1 & 2 & 3 & 4 \\
4 & 0 & 1 & 2 & 3 \\
3 & 4 & 0 & 1 & 2 \\
2 & 3 & 4 & 0 & 1 \\
1 & 2 & 3 & 4 & 0
\end{pmatrix} .
$$

$\square$

The following result, which also yields information for the case where the order n of the latin squares is not a prime power, is proved in the same way as Theorem 7.27. Note that Theorem 7.29 shows, in particular, the existence of a pair of orthogonal latin squares of order n for any $n > 1$ with $n \not\equiv 2 \bmod 4$.

7.29.   Theorem.   *Let $q_1, \ldots, q_s$ be prime powers and let*

$$a_0^{(i)} = 0, a_1^{(i)}, a_2^{(i)}, \ldots, a_{q_i-1}^{(i)}$$

be the elements of $\mathbb{F}_{q_i}$. Define the s-tuples

$$b_k = \left(a_k^{(1)}, \ldots, a_k^{(s)} \right) \quad for\ 0 \leqslant k \leqslant r = \min_{1 \leqslant i \leqslant s} (q_i - 1),$$

and let $b_{r+1}, \ldots, b_{n-1}$ with $n = q_1 \cdots q_s$ be the remaining s-tuples that can be formed by taking in the ith coordinate an element of $\mathbb{F}_{q_i}$. These s-tuples are *added and multiplied by adding and multiplying their coordinates. Then the arrays*

$$
L_k = \begin{pmatrix}
b_0 & b_1 & \cdots & b_{n-1} \\
b_k b_1 & b_k b_1 + b_1 & \cdots & b_k b_1 + b_{n-1} \\
b_k b_2 & b_k b_2 + b_1 & \cdots & b_k b_2 + b_{n-1} \\
\vdots & \vdots & & \vdots \\
b_k b_{n-1} & b_k b_{n-1} + b_1 & \cdots & b_k b_{n-1} + b_{n-1}
\end{pmatrix}, \quad k = 1, \ldots, r,
$$

form a set of r mutually orthogonal latin squares of order n.

Tactical configurations and latin squares are of use in the *design of statistical experiments*. For example, suppose that n varieties of wheat are to be compared as to their mean yield on a certain type of soil. At our disposal is a rectangular field subdivided into n^2 plots. However, even if we are careful in the selection of our field, differences in soil fertility will occur on it. Thus, if all the plots of the first row are occupied by the first variety, it may very well be that the first row is of high fertility and we might obtain a high yield for the first variety although it is not superior to the other varieties. We shall be less likely to vitiate our comparisons if we set every variety once in every row and once in every column. In other words, the varieties should be planted on the n^2 plots in such a way that a latin square of order n is formed.

It is often desirable to test at the same time other factors influencing the yield. For instance, we might want to apply n different fertilizers and evaluate their effectiveness. We will then arrange fertilizers and varieties on the n^2 plots in such a way that both the arrangement of fertilizers and the arrangement of varieties form a latin square of order n, and such that every fertilizer is applied exactly once to every variety. Thus, in the language of combinatorics, the latin squares of fertilizer and variety arrangements should be orthogonal. Similar applications exist for balanced incomplete block designs.

As another example for a combinatorial concept allowing applications of finite fields, we introduce so-called Hadamard matrices. These matrices are useful in coding theory, in communication theory, and physics because of Hadamard transforms, and also in problems of determination of weights, resistances, voltages, and so on.

7.30.   Definition.   A *Hadamard matrix* H_n is an $n \times n$ matrix with integer entries ± 1 that satisfies

$$
H_n H_n^{\mathrm{T}} = nI.
$$

Since $H_n^{-1} = (1/n) H_n^{\mathrm{T}}$, we also have $H_n^{\mathrm{T}} H_n = nI$. Thus, any two distinct rows and any two distinct columns of H_n are orthogonal. The determinant of a Hadamard matrix attains a bound due to Hadamard. We have $\det(H_n H_n^T) = n^n$, and so $|\det(H_n)| = n^{n/2}$, while Hadamard's result states that $|\det(M)| \leq n^{n/2}$ for any real $n \times n$ matrix M with entries of absolute value ≤ 1.

Changing the signs of rows or columns leaves the defining property unaltered, so we may assume that H_n is *normalized* — that is, that all entries in the first row and first column are $+1$. It is easily seen that the order n of a Hadamard matrix (a_{ij}) can only be 1, 2, or a multiple of 4. For we have

$$\sum_{j=1}^{n} (a_{1j} + a_{2j})(a_{1j} + a_{3j}) = \sum_{j=1}^{n} a_{1j}^2 = n$$

for $n \geq 3$ and every term in the first sum is either 0 or 4, hence the result follows. It is conjectured that a Hadamard matrix H_n exists for all those n.

7.31. Example. Hadamard matrices of the lowest orders are:

$$H_1 = (1), \quad H_2 = \begin{pmatrix} 1 & 1 \\ 1 & -1 \end{pmatrix}, \quad H_4 = \begin{pmatrix} 1 & 1 & 1 & 1 \\ 1 & -1 & 1 & -1 \\ 1 & 1 & -1 & -1 \\ 1 & -1 & -1 & 1 \end{pmatrix}. \qquad \square$$

We describe now a construction method for Hadamard matrices using finite fields.

7.32. Theorem. *Let $a_1, \ldots, a_q$ be the elements of $\mathbb{F}_q$, $q \equiv 3 \bmod 4$, and let η be the quadratic character of $\mathbb{F}_q$. Then the matrix*

$$H = \begin{pmatrix} 1 & 1 & 1 & 1 & \cdots & 1 \\ 1 & -1 & b_{12} & b_{13} & \cdots & b_{1q} \\ 1 & b_{21} & -1 & b_{23} & \cdots & b_{2q} \\ 1 & b_{31} & b_{32} & -1 & \cdots & b_{3q} \\ \vdots & \vdots & \vdots & \vdots & & \vdots \\ 1 & b_{q1} & b_{q2} & b_{q3} & \cdots & -1 \end{pmatrix}$$

with $b_{ij} = \eta(a_j - a_i)$ for $1 \leq i, j \leq q, i \neq j$, is a Hadamard matrix of order $q + 1$.

Proof. Since all entries are ± 1, it suffices to show that the inner product of any two distinct rows is 0. The inner product of the first row with the $(i+1)$st row, $1 \leq i \leq q$, is

$$1 + (-1) + \sum_{j \neq i} b_{ij} = \sum_{j \neq i} \eta(a_j - a_i) = \sum_{c \in \mathbb{F}_q^*} \eta(c) = 0$$

by (5.12). The inner product of the $(i+1)$st row with the $(k+1)$st row, $1 \leq i < k \leq q$, is

$$1 - b_{ki} - b_{ik} + \sum_{j \neq i, k} b_{ij} b_{kj}$$

$$= 1 - \eta(a_i - a_k) - \eta(a_k - a_i) + \sum_{j \neq i, k} \eta(a_j - a_i)\eta(a_j - a_k)$$

$$= 1 - [1 + \eta(-1)]\eta(a_i - a_k) + \sum_{c \in \mathbb{F}_q} \eta((c - a_i)(c - a_k)) = 0,$$

since $\eta(-1) = -1$ for $q \equiv 3 \bmod 4$ by Remark 5.13 and the last sum is -1 by Theorem 5.18. $\qquad\qquad\qquad\qquad\qquad\qquad\qquad\qquad\qquad\qquad\qquad\square$

If H_n is a Hadamard matrix of order n, then

$$\begin{pmatrix} H_n & H_n \\ H_n & -H_n \end{pmatrix}$$

is one of order $2n$. Therefore, Hadamard matrices of orders $2^h(q + 1)$ with $h \geqslant 0$ and prime powers $q \equiv 3 \bmod 4$ can be obtained in this manner. By starting from the Hadamard matrix H_1 in Example 7.31, one can also obtain Hadamard matrices of orders 2^h, $h \geqslant 0$.

3. LINEAR MODULAR SYSTEMS

System theory is a discipline that aims at providing a common abstract basis and unified conceptual framework for studying the behavior of various types and forms of systems. It is a collection of methods as well as special techniques and algorithms for dealing with problems in system analysis, synthesis, identification, optimization, and other areas. It is mainly the mathematical structure of a system that is of interest to a system theorist, and not its physical form or area of applications, or whether a system is electrical, mechanical, economic, biological, chemical, and so on. What matters to the theorist is whether it is linear or nonlinear, discrete-time or continuous-time, deterministic or stochastic, discrete-state or continuous-state, and so on.

In the introduction to this chapter we gave an informal description of systems. We present now a rigorous definition of finite-state systems, which provide an idealized model for a large number of physical devices and phenomena. Ideas and techniques developed for finite-state systems have also been found useful in such diverse problems as the investigation of human nervous activity, the analysis of English syntax, and the design of digital computers.

7.33. Definition. A (complete, deterministic) *finite-state system* $\mathfrak{M}$ is defined by the following:

(1) A finite, nonempty set $U = \{\alpha_1, \alpha_2, \ldots, \alpha_h\}$, called the *input*

alphabet of $\mathfrak{M}$. An element of U is called an *input symbol*.

(2) A finite, nonempty set $Y = \{\beta_1, \beta_2, \ldots, \beta_s\}$, called the *output alphabet* of $\mathfrak{M}$. An element of Y is called an *output symbol*.

(3) A finite, nonempty set $S = \{\sigma_1, \sigma_2, \ldots, \sigma_r\}$, called the *state set* of $\mathfrak{M}$. An element of S is called a *state*.

(4) A *next-state function* f that maps the set of all ordered pairs (σ_i, α_j) into S.

(5) An *output function* g that maps the set of all ordered pairs (σ_i, α_j) into Y.

A finite-state system $\mathfrak{M}$ can be interpreted as a device whose input, output, and state at time t are denoted by $u(t)$, $y(t)$, and $s(t)$, respectively, where these variables are defined for integers t only and assume values taken from U, Y, and S, respectively. Given the state and input of $\mathfrak{M}$ at time t, f specifies the state at time $t+1$ and g the output at time t:

$$s(t+1) = f(s(t), u(t)),$$

$$y(t) = g(s(t), u(t)).$$

Linear modular systems constitute a special class of finite-state systems, where the input and output alphabets and the state set carry the structure of a vector space over a finite field $\mathbb{F}_q$ and the next-state and output functions are linear. Linear modular systems have found wide applications in computer control circuitry, implementation of error-correcting codes, random number generation, and other digital tasks.

7.34. Definition. A *linear modular system* (*LMS*) $\mathfrak{M}$ of order n over $\mathbb{F}_q$ is defined by the following:

(1) A k-dimensional vector space U over $\mathbb{F}_q$, called *input space* of $\mathfrak{M}$, the elements of which are called *inputs* and are written as column vectors.

(2) An m-dimensional vector space Y over $\mathbb{F}_q$, called *output space* of $\mathfrak{M}$, the elements of which are called *outputs* and are written as column vectors.

(3) An n-dimensional vector space S over $\mathbb{F}_q$, called *state space* of $\mathfrak{M}$, the elements of which are called *states* and are written as column vectors.

(4) Four *characterizing matrices* over $\mathbb{F}_q$:

$$A = (a_{ij})_{n \times n}, \quad B = (b_{ij})_{n \times k},$$

$$C = (c_{ij})_{m \times n}, \quad D = (d_{ij})_{m \times k}.$$

The matrix A is called the *characteristic matrix* of $\mathfrak{M}$.

(5)   A rule relating the state at time $t + 1$ and output at time t to the state and input at time t:

$$s(t+1) = As(t) + Bu(t),$$

$$y(t) = Cs(t) + Du(t).$$

An LMS over $\mathbb{F}_q$ can be simulated by a switching circuit incorporating adders, constant multipliers, and delay elements (compare with Chapter 6, Section 1). It is convenient here to use adders summing also more than two field elements. Thus, an *adder* has two or more inputs

$$u_1(t), u_2(t), \ldots, u_r(t) \in \mathbb{F}_q$$

and a single output

$$y_1(t) = u_1(t) + u_2(t) + \cdots + u_r(t).$$

A *constant multiplier* with a constant $a \in \mathbb{F}_q$ has a single input $u_1(t) \in \mathbb{F}_q$ and a single output $y_1(t) = au_1(t)$. A *delay element* has a single input $u_1(t) \in \mathbb{F}_q$ and a single output $y_1(t) = u_1(t-1)$. Symbolically, these components are represented as shown in Figure 7.4.

We describe now how we can obtain a realization of an LMS $\mathfrak{M}$ as a circuit simulating the operations of $\mathfrak{M}$:

1.   Draw k input terminals labelled $u_1, \ldots, u_k$, m output terminals labelled $y_1, \ldots, y_m$, and n delay elements, where the output of the ith delay element is $s_i = s_i(t)$ and its input is $s_i' = s_i(t+1)$.
2.   Insert an adder in front of each output terminal y_i and each delay element.
3.   The inputs to the adder associated with the ith delay element are the s_j, each applied via a constant multiplier with constant a_{ij}, $1 \leq i, j \leq n$, and the u_j, each applied via a constant multiplier with constant b_{ij}, $1 \leq j \leq k$.

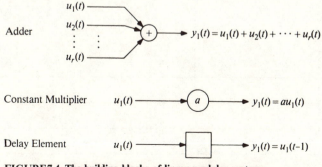

FIGURE 7.4 The building blocks of linear modular systems.

4. The inputs to the adder associated with the output terminal y_i, $1 \leqslant i \leqslant m$, are the s_j, each applied via a constant multiplier with constant c_{ij}, $1 \leqslant j \leqslant n$, and the u_j, each applied via a constant multiplier with constant d_{ij}, $1 \leqslant j \leqslant k$.

If we define

$$\mathbf{u}(t) = \begin{pmatrix} u_1 \\ \vdots \\ u_k \end{pmatrix}, \quad \mathbf{y}(t) = \begin{pmatrix} y_1 \\ \vdots \\ y_m \end{pmatrix}, \quad \mathbf{s}(t) = \begin{pmatrix} s_1 \\ \vdots \\ s_n \end{pmatrix}, \quad \mathbf{s}(t+1) = \begin{pmatrix} s_1' \\ \vdots \\ s_n' \end{pmatrix},$$

then the operation of the circuit represented in Figure 7.5 is precisely that described in Definition 7.34(5).

7.35. Example. Let the characterizing matrices of a fourth-order LMS over $\mathbb{F}_3$ be:

$$A = \begin{pmatrix} 0 & 2 & 0 & 0 \\ 1 & 0 & 2 & 1 \\ 0 & 1 & 1 & 0 \\ 2 & 0 & 1 & 1 \end{pmatrix}, \quad B = \begin{pmatrix} 1 \\ 0 \\ 0 \\ 0 \end{pmatrix}, \quad C = \begin{pmatrix} 0 & 0 & 2 & 1 \\ 0 & 2 & 0 & 0 \end{pmatrix}, \quad D = \begin{pmatrix} 0 \\ 1 \end{pmatrix}.$$

Then its realization as a circuit is shown in Figure 7.6.   □

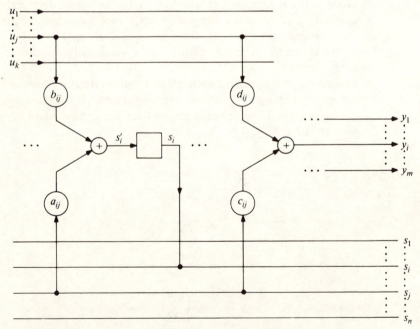

FIGURE 7.5 The realization of an LMS as a switching circuit.

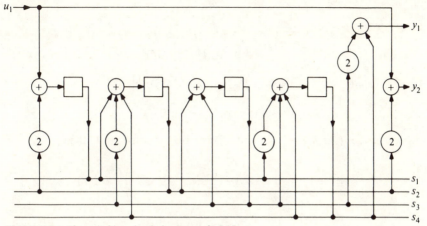

FIGURE 7.6 The switching circuit for Example 7.35.

Conversely, we can describe an arbitrary switching circuit with a finite number of adders, constant multipliers, and delay elements over $\mathbb{F}_q$ as an LMS over $\mathbb{F}_q$ as follows (provided every closed loop contains at least one delay element):

1. Locate in the given circuit all delay elements and all external input and output terminals, and label them as in Figure 7.5.
2. Trace the paths from s_j to s_i' and compute the product of the multiplier constants encountered along each path and add the products. Let a_{ij} denote this sum.
3. Let b_{ij} denote the corresponding sum for the paths from u_j to s_i', c_{ij} for the paths from s_j to y_i, d_{ij} for the paths from u_j to y_i.

Then the circuit is the realization of an LMS over $\mathbb{F}_q$ with characterizing matrices A, B, C, D.

The states and the outputs of an LMS depend on the initial state $s(0)$ and the sequence of inputs $u(t)$, $t = 0, 1, \ldots$. The dependence on these data can be expressed explicitly.

7.36. *Theorem* (General Response Formula). *For an LMS with characterizing matrices A, B, C, D we have*:

(i) $\displaystyle s(t) = A^t s(0) + \sum_{i=0}^{t-1} A^{t-i-1} B u(i)$ *for* $t = 1, 2, \ldots,$

(ii) $\displaystyle y(t) = CA^t s(0) + \sum_{i=0}^{t} H(t-i) u(i)$ *for* $t = 0, 1, \ldots,$

where

$$H(t) = \begin{cases} D & \text{if } t = 0, \\ CA^{t-1}B & \text{if } t \geqslant 1. \end{cases}$$

Proof. (i) Let $t = 0$ in Definition 7.34(5), then

$$\mathbf{s}(1) = A\mathbf{s}(0) + B\mathbf{u}(0),$$

which proves (i) for $t = 1$. Assume (i) is true for some $t \geqslant 1$, then

$$\mathbf{s}(t+1) = A\left(A^t\mathbf{s}(0) + \sum_{i=0}^{t-1} A^{t-i-1}B\mathbf{u}(i) \right) + B\mathbf{u}(t)$$

$$= A^{t+1}\mathbf{s}(0) + \sum_{i=0}^{t} A^{t-i}B\mathbf{u}(i)$$

proves (i) for $t + 1$.

(ii) By (i) and Definition 7.34(5) we have

$$\mathbf{y}(t) = C\left(A^t\mathbf{s}(0) + \sum_{i=0}^{t-1} A^{t-i-1}B\mathbf{u}(i) \right) + D\mathbf{u}(t)$$

$$= CA^t\mathbf{s}(0) + \sum_{i=0}^{t} H(t-i)\mathbf{u}(i),$$

where $H(t-i) = CA^{t-i-1}B$ when $t - i \geqslant 1$ and $H(t-i) = D$ when $t - i = 0$. □

By Theorem 7.36(ii) we can decompose the output of an LMS into two components, the *free component*

$$\mathbf{y}(t)_{\text{free}} = CA^t\mathbf{s}(0)$$

obtained in case $\mathbf{u}(t) = \mathbf{0}$ for all $t \geqslant 0$, and the *forced component*

$$\mathbf{y}(t)_{\text{forced}} = \sum_{i=0}^{t} H(t-i)\mathbf{u}(i)$$

obtained by setting $\mathbf{s}(0) = \mathbf{0}$. Given any input sequence $\mathbf{u}(t), t = 0, 1, \ldots$, and an initial state $\mathbf{s}(0)$, these two components can be found separately and then added up.

In the remainder of this section we study the states of an LMS in the *input-free case* —that is, when $\mathbf{u}(t) = \mathbf{0}$ for all $t \geqslant 0$. Some simple graph-theoretic language will be useful. Given an LMS $\mathfrak{M}$ of order n over $\mathbb{F}_q$ with characteristic matrix A, the *state graph* of $\mathfrak{M}$, or of A, is an oriented graph with q^n vertices, one for each possible state of $\mathfrak{M}$. An arrow points from state $\mathbf{s}_1$ to state $\mathbf{s}_2$ if and only if $\mathbf{s}_2 = A\mathbf{s}_1$. In this case we say that $\mathbf{s}_1$ *leads to* $\mathbf{s}_2$. A *path* of length r in a state graph is a sequence of r arrows $b_1, b_2, \ldots, b_r$ and $r + 1$ vertices $v_1, v_2, \ldots, v_{r+1}$ such that b_i points from v_i to v_{i+1}, $i = 1, 2, \ldots, r$. If the v_i are distinct except $v_{r+1} = v_1$, the path is called a *cycle* of length r. If v_i is the only vertex leading to v_{i+1}, $i = 1, 2, \ldots, r-1$, and the only vertex leading to v_1 is v_r, then the cycle is called a *pure cycle*. For example, a pure cycle of length 8 is given as shown in Figure 7.7.

The *order* of a given state $\mathbf{s}$ is the least positive integer t such that $A^t\mathbf{s} = \mathbf{s}$. Thus, the order of $\mathbf{s}$ is the length of the cycle which includes $\mathbf{s}$. In the following, let A be nonsingular—that is, $\det(A) \ne 0$. It is clear that in this case the corresponding state graph consists of pure cycles only. The order of the characteristic matrix A is the least positive integer t such that $A^t = I$, the $n \times n$ identity matrix.

7.37. Lemma. *If $t_1, \ldots, t_K$ are the orders of the possible states of an LMS with nonsingular characteristic matrix A, then the order of A is $\operatorname{lcm}(t_1, \ldots, t_K)$.*

Proof. Let t be the order of A and $t' = \operatorname{lcm}(t_1, \ldots, t_K)$. Since $A^t\mathbf{s} = \mathbf{s}$ for every $\mathbf{s}$, t must be a multiple of t'. Also, $(A^{t'} - I)\mathbf{s} = \mathbf{0}$ for all $\mathbf{s}$, hence $A^{t'} = I$. Thus $t' \geqslant t$, and therefore $t = t'$. $\qquad\square$

7.38. Lemma. *If A has the form*

$$A = \begin{pmatrix} A_1 & 0 \\ 0 & A_2 \end{pmatrix}$$

with square matrices A_1 and A_2, and $\begin{pmatrix} \mathbf{s}_1 \\ \mathbf{0} \end{pmatrix}$ and $\begin{pmatrix} \mathbf{0} \\ \mathbf{s}_2 \end{pmatrix}$ are two states, partitioned according to the partition of A, with orders t_1 and t_2, respectively, then the order of $\mathbf{s} = \begin{pmatrix} \mathbf{s}_1 \\ \mathbf{s}_2 \end{pmatrix}$ is $\operatorname{lcm}(t_1, t_2)$.

Proof. This follows immediately from the fact that $A^t\begin{pmatrix} \mathbf{s}_1 \\ \mathbf{s}_2 \end{pmatrix} = \begin{pmatrix} \mathbf{s}_1 \\ \mathbf{s}_2 \end{pmatrix}$ if and only if $A_1^t\mathbf{s}_1 = \mathbf{s}_1$ and $A_2^t\mathbf{s}_2 = \mathbf{s}_2$. $\qquad\square$

FIGURE 7.7   A pure cycle of length 8.

Let $\mathfrak{M}$ be an LMS with nonsingular characteristic matrix A. Up to isomorphisms (i.e., one-to-one and onto mappings τ such that $\tau(\mathbf{s}_1)$ leads to $\tau(\mathbf{s}_2)$ whenever $\mathbf{s}_1$ leads to $\mathbf{s}_2$) the state graph of $\mathfrak{M}$ is characterized by the formal sum

$$\Sigma = (n_1, t_1) + (n_2, t_2) + \cdots + (n_R, t_R),$$

which indicates that n_i is the number of cycles of length t_i. Σ is called the *cycle sum* of $\mathfrak{M}$, or of A, and each ordered pair (n_i, t_i) is called a *cycle term*. Cycle terms are assumed to commute with respect to $+$, and we observe the convention $(n', t) + (n'', t) = (n' + n'', t)$.

Consider a matrix A of the form

$$A = \begin{pmatrix} A_1 & 0 \\ 0 & A_2 \end{pmatrix}$$

with square matrices A_1 and A_2, and suppose the state graph of A_i has n_i cycles of length t_i, $i = 1, 2$. Hence there are $n_1 t_1$ states of the form $\begin{pmatrix} \mathbf{s}_1 \\ \mathbf{0} \end{pmatrix}$ of order t_1, and $n_2 t_2$ states of the form $\begin{pmatrix} \mathbf{0} \\ \mathbf{s}_2 \end{pmatrix}$ of order t_2. By Lemma 7.38 the state graph of A must contain $n_1 n_2 t_1 t_2$ states of order $\operatorname{lcm}(t_1, t_2)$ and hence

$$n_1 n_2 t_1 t_2 / \operatorname{lcm}(t_1, t_2) = n_1 n_2 \gcd(t_1, t_2)$$

cycles of length $\operatorname{lcm}(t_1, t_2)$.

The product of two cycle terms is the cycle term defined by

$$(n_1, t_1) \cdot (n_2, t_2) = (n_1 n_2 \gcd(t_1, t_2), \operatorname{lcm}(t_1, t_2)).$$

The product of two cycle sums is defined as the formal sum of all possible products of cycle terms from the two given cycle sums. In other words, the product is calculated by the distributive law.

7.39. Theorem. *If*

$$A = \begin{pmatrix} A_1 & 0 \\ 0 & A_2 \end{pmatrix}$$

and the cycle sums of A_1 and A_2 are Σ_1 and Σ_2, respectively, then the cycle sum of A is $\Sigma_1 \Sigma_2$.

Our aim is to give a procedure for computing the cycle sum of an LMS over $\mathbb{F}_q$ with nonsingular characteristic matrix A. We need some basic facts about matrices. The *characteristic polynomial* of a square matrix M over $\mathbb{F}_q$ is defined by $\det(xI - M)$. The *minimal polynomial* $m(x)$ of M is the monic polynomial over $\mathbb{F}_q$ of least degree such that $m(M) = 0$, the zero matrix. For a monic polynomial

$$g(x) = x^k + a_{k-1} x^{k-1} + \cdots + a_1 x + a_0$$

over $\mathbb{F}_q$, its *companion matrix* is given by

$$M(g(x)) = \begin{pmatrix} 0 & 0 & 0 & \cdots & 0 & -a_0 \\ 1 & 0 & 0 & \cdots & 0 & -a_1 \\ 0 & 1 & 0 & \cdots & 0 & -a_2 \\ \vdots & \vdots & \vdots & & \vdots & \vdots \\ 0 & 0 & 0 & \cdots & 1 & -a_{k-1} \end{pmatrix}.$$

Then $g(x)$ is the characteristic polynomial and the minimal polynomial of $M(g(x))$.

Let M be a square matrix over $\mathbb{F}_q$ with the monic elementary divisors $g_1(x), \ldots, g_w(x)$. Then the product $g_1(x) \cdots g_w(x)$ is equal to the characteristic polynomial of M, and M is similar to

$$M^* = \begin{pmatrix} M(g_1(x)) & 0 & \cdots & 0 \\ 0 & M(g_2(x)) & \cdots & 0 \\ \vdots & \vdots & & \vdots \\ 0 & 0 & \cdots & M(g_w(x)) \end{pmatrix},$$

that is, $M = P^{-1} M^* P$ for some nonsingular matrix P over $\mathbb{F}_q$. The matrix M^* is called the *rational canonical form* of M and the submatrices $M(g_i(x))$ are called the *elementary blocks* of M^*.

Now let the nonsingular matrix A be the characteristic matrix of an LMS over $\mathbb{F}_q$. For the purpose of computing its cycle sum, A can be replaced by a similar matrix. Thus, we consider the rational canonical form A^* of A. Extending Theorem 7.39 by induction, we obtain the following. Let $g_1(x), \ldots, g_w(x)$ be the monic elementary divisors of A and let Σ_i be the cycle sum of the companion matrix $M(g_i(x))$; then the cycle sum Σ of A^*, and so of A, is given by

$$\Sigma = \Sigma_1 \Sigma_2 \cdots \Sigma_w.$$

Let the characteristic polynomial $f(x)$ of A have the canonical factorization

$$f(x) = \prod_{j=1}^{r} p_j(x)^{e_j},$$

where the $p_j(x)$ are distinct monic irreducible polynomials over $\mathbb{F}_q$. Then the elementary divisors of A are of the form

$$p_j(x)^{e_{j1}}, p_j(x)^{e_{j2}}, \ldots, p_j(x)^{e_{jk_j}}, \quad j = 1, 2, \ldots, r,$$

where

$$e_{j1} \geqslant e_{j2} \geqslant \cdots \geqslant e_{jk_j} > 0, \quad e_{j1} + e_{j2} + \cdots + e_{jk_j} = e_j.$$

The minimal polynomial of A is equal to

$$m(x) = \prod_{j=1}^{r} p_j(x)^{e_{j1}}.$$

It remains to consider the question of determining the cycle sum of a typical elementary block $M(g_i(x))$ of A^*, where $g_i(x)$ is of the form $p(x)^e$ for some monic irreducible factor $p(x)$ of $f(x)$. The following result provides the required information.

7.40.   Theorem.   *Let $p(x)$ be a monic irreducible polynomial over $\mathbb{F}_q$ of degree d and let $t_h = \mathrm{ord}(p(x)^h)$. Then the cycle sum of $M(p(x)^e)$ is given by*

$$(1,1) + \left(\frac{q^d - 1}{t_1}, t_1\right) + \left(\frac{q^{2d} - q^d}{t_2}, t_2\right) + \cdots + \left(\frac{q^{ed} - q^{(e-1)d}}{t_e}, t_e\right).$$

In summary, we obtain the following *procedure for determining the cycle sum* of an LMS $\mathfrak{M}$ over $\mathbb{F}_q$ with nonsingular characteristic matrix A:

C1.   Find the elementary divisors of A, say $g_1(x), \ldots, g_w(x)$.
C2.   Let $g_i(x) = f_i(x)^{m_i}$, where $f_i(x)$ is monic and irreducible over $\mathbb{F}_q$. Find the orders $t_1^{(i)} = \mathrm{ord}(f_i(x))$.
C3.   Evaluate the orders $t_h^{(i)} = \mathrm{ord}(f_i(x)^h)$ for $i = 1, 2, \ldots, w$ and $h = 1, 2, \ldots, m_i$ by the formula $t_h^{(i)} = t_1^{(i)} p^{c_h}$, where p is the characteristic of $\mathbb{F}_q$ and c_h is the least integer such that $p^{c_h} \geqslant h$ (see Theorem 3.8).
C4.   Determine the cycle sum $\sum_i$ of $M(g_i(x))$ for $i = 1, 2, \ldots, w$ according to Theorem 7.40.
C5.   The cycle sum $\sum$ of $\mathfrak{M}$ is given by $\sum = \sum_1 \sum_2 \cdots \sum_w$.

7.41.   Example.   Let the characteristic matrix of an LMS $\mathfrak{M}$ over $\mathbb{F}_2$ be given as

$$A = \begin{pmatrix} 0 & 0 & 1 & 0 & 0 \\ 1 & 0 & 1 & 0 & 0 \\ 0 & 1 & 1 & 0 & 0 \\ 0 & 0 & 0 & 0 & 1 \\ 0 & 0 & 0 & 1 & 1 \end{pmatrix}.$$

Here

$$g_1(x) = x^3 + x^2 + x + 1 = (x+1)^3, \quad f_1(x) = x + 1, \quad m_1 = 3,$$

$$g_2(x) = x^2 + x + 1, \quad f_2(x) = x^2 + x + 1, \quad m_2 = 1.$$

Steps C2 and C3 yield $t_1^{(1)} = 1$, $t_2^{(1)} = 2$, $t_3^{(1)} = 4$, $t_1^{(2)} = 3$. Hence by Theorem

7.40,

$$\sum_1 = (1,1) + (1,1) + (1,2) + (1,4) = (2,1) + (1,2) + (1,4),$$
$$\sum_2 = (1,1) + (1,3),$$

and so

$$\sum = \sum_1 \sum_2 = [(2,1) + (1,2) + (1,4)][(1,1) + (1,3)]$$
$$= (2,1) + (1,2) + (2,3) + (1,4) + (1,6) + (1,12).$$

Thus the state graph of $\mathfrak{M}$ consists of two cycles of length 1, one cycle of length 2, two cycles of length 3, and one cycle each of length 4, 6, and 12. □

From C5 it follows that the state orders realizable by $\mathfrak{M}$ are given by

$$\mathrm{lcm}\left(t_{h_1}^{(1)}, t_{h_2}^{(2)}, \ldots, t_{h_w}^{(w)} \right)$$

for every combination of integers $h_1, \ldots, h_w$, $0 \leqslant h_i \leqslant m_i$. If one wishes to compute all possible state orders realizable by $\mathfrak{M}$, without computing its cycle sum, one uses the following theorem.

7.42. Theorem. *Let $\mathfrak{M}$ be an LMS with nonsingular characteristic matrix A. Let the canonical factorization of the minimal polynomial of A be*

$$m(x) = p_1(x)^{b_1} \cdots p_r(x)^{b_r}$$

and let $t_h^{(j)} = \mathrm{ord}(\, p_j(x)^h)$. Then the state orders realizable by $\mathfrak{M}$ are given by all the integers of the form

$$\mathrm{lcm}\left(t_{h_1}^{(1)}, t_{h_2}^{(2)}, \ldots, t_{h_r}^{(r)} \right) \quad \text{with } 0 \leqslant h_j \leqslant b_j \ \text{ for } 1 \leqslant j \leqslant r.$$

4. PSEUDORANDOM SEQUENCES

The notion of a random sequence of events is basic in probability theory and statistics. Let us take a standard model for the description of this notion. Consider an experiment in which an unbiased coin is flipped repeatedly. Mark down 0 for heads and 1 for tails. The result of this experiment is then a sequence of binary digits (or *bits* in the parlance of computer science) which will display typical features of randomness. For instance, the relative frequency of each bit will approach $\frac{1}{2}$ in the long run, and the relative frequency of two successive 0's (or of two successive 1's) will approach $\frac{1}{4}$ in the long run. More generally, for any given block of m bits the relative frequency of this block among all the blocks of m successive bits in the sequence will approach

2^{-m} in the long run. In short, the sequence can be expected to have all the statistical properties satisfied by a sequence of independent random variables which attain each value 0 and 1 with probability $\frac{1}{2}$.

Flipping coins is thus not just an idle pastime, but can serve as a method for generating random sequences of bits. Since there is no guarantee that our coin is truly unbiased, the generated sequence should be subjected to tests for randomness. For instance, we may check the statistical quantities mentioned above—namely, the relative frequency of each bit (*distribution test*) and the relative frequency of blocks of bits (*serial test*). Another popular test for randomness is the *correlation test*, which is based on the calculation of the *correlation coefficients*

$$C_N(h) = \sum_{n=0}^{N-1} (-1)^{s_n - s_{n+h}} \tag{7.6}$$

of the given sequence $s_0, s_1, \ldots$ of bits for positive integers N and h. The correlation coefficient $C_N(h)$ can be interpreted as follows: write the shifted sequence $s_h, s_{h+1}, \ldots$ underneath the original sequence and count the agreements and disagreements among the first N corresponding terms; then $C_N(h)$ is equal to the number of agreements minus the number of disagreements. For a random sequence of bits $C_N(h)$ should be relatively small compared to N.

Random sequences of bits are used frequently for simulation purposes, for various applications in electrical engineering, and also in cryptography (see Chapter 9, Section 2). In practice, the generation of such sequences by coin flipping or similar physical means is problematic. First of all, the practical applications require long strings of bits, and the physical generation of all those bits may simply take too long. Furthermore, it is an established principle that scientific calculations have to be reproducible and verifiable, and this means that all the bits used in a calculation must be stored for later recall. This may tie up a lot of the computer's memory capacity. In many applications it is therefore preferable to work with sequences of bits that can be generated directly in the computer. Since the computer only responds to deterministic programs, the resulting sequences will not be random. However, we can try to generate deterministic sequences of bits that pass various tests for randomness. Such deterministic sequences are called *pseudorandom* sequences of bits.

A commonly employed method of generating pseudorandom sequences of bits is based on the use of suitable linear recurrence relations in the finite field $\mathbb{F}_2$. The sequences that one generates are the maximal period sequences introduced in Chapter 6. We will show that—with certain qualifications—maximal period sequences in $\mathbb{F}_2$ pass the tests for randomness described above—namely, the distribution test, the serial test, and the correlation test. Since there is no extra effort involved, we will establish the relevant facts for maximal period sequences in an arbitrary finite field $\mathbb{F}_q$. We are thus dealing with pseudorandom sequences of elements of $\mathbb{F}_q$.

We recall from Chapter 6 that a kth-order *maximal period sequence* in $\mathbb{F}_q$ is a sequence $s_0, s_1, \ldots$ of elements of $\mathbb{F}_q$ generated by a linear recurrence relation

$$s_{n+k} = a_{k-1}s_{n+k-1} + \cdots + a_0 s_n \quad \text{for} \quad n = 0, 1, \ldots, \tag{7.7}$$

for which the characteristic polynomial $x^k - a_{k-1}x^{k-1} - \cdots - a_0$ is a primitive polynomial over $\mathbb{F}_q$ and not all initial values $s_0, \ldots, s_{k-1}$ are 0. A kth-order maximal period sequence is periodic with least period $r = q^k - 1$ (see Theorem 6.33). A requirement we have to impose is that r be very large, say at least as large as the total number of pseudorandom elements of $\mathbb{F}_q$ to be used in the specific application. In this way the periodicity of the sequence—which is a distinctly nonrandom feature—will not come into play. With this proviso we will now investigate the performance of maximal period sequences under tests for randomness. The distribution test and the serial test can be treated simultaneously. For $\mathbf{b} = (b_1, \ldots, b_m) \in \mathbb{F}_q^m$ let $Z(\mathbf{b})$ be the number of n, $0 \leq n \leq r - 1$, such that $s_{n+i-1} = b_i$ for $1 \leq i \leq m$. The case $m = 1$ corresponds to the distribution test and was already dealt with on p. 240. The case of an $m \geq 2$ corresponds to the serial test for blocks of length m. The following result shows that $Z(\mathbf{b})$ is close to the ideal number rq^{-m} provided that m is not too large.

7.43.   Theorem.   *If $1 \leq m \leq k$ and $\mathbf{b} \in \mathbb{F}_q^m$, then for any kth-order maximal period sequence in $\mathbb{F}_q$ we have*

$$Z(\mathbf{b}) = \begin{cases} q^{k-m} - 1 & \text{for } \mathbf{b} = \mathbf{0}, \\ q^{k-m} & \text{for } \mathbf{b} \neq \mathbf{0}. \end{cases}$$

Proof.   Since $r = q^k - 1$, the state vectors $\mathbf{s}_0, \mathbf{s}_1, \ldots, \mathbf{s}_{r-1}$ of the sequence run exactly through all nonzero vectors in $\mathbb{F}_q^k$. Therefore $Z(\mathbf{b})$ is equal to the number of nonzero vectors $\mathbf{s} \in \mathbb{F}_q^k$ that have $\mathbf{b}$ as the m-tuple of their first m coordinates. For $\mathbf{b} \neq \mathbf{0}$ we can have all possible combinations of elements of $\mathbb{F}_q$ in the remaining $k - m$ coordinates of $\mathbf{s}$, so that $Z(\mathbf{b}) = q^{k-m}$. For $\mathbf{b} = \mathbf{0}$ we have to exclude the possibility that all the remaining $k - m$ coordinates of $\mathbf{s}$ are 0, hence $Z(\mathbf{b}) = q^{k-m} - 1$. $\square$

Theorem 6.85 shows that parts of the period of a maximal period sequence also perform well under the distribution test. We now turn to the correlation test for a maximal period sequence $s_0, s_1, \ldots$ in $\mathbb{F}_q$. We first extend the definition of correlation coefficients in (7.6) to the general case. Let χ be a fixed nontrivial additive character of $\mathbb{F}_q$ (compare with Chapter 5, Section 1) and set

$$C_N(h) = \sum_{n=0}^{N-1} \chi(s_n - s_{n+h}) \tag{7.8}$$

for positive integers N and h. For $q = 2$ this definition reduces to (7.6) since there is only one nontrivial additive character of $\mathbb{F}_2$ and it is given by $\chi(0) = 1, \chi(1) = -1$. In the case $N = r$ we can give explicit formulas for the correlation coefficients.

7.44. Theorem. *For any maximal period sequence in $\mathbb{F}_q$ with least period r we have*

$$C_r(h) = \begin{cases} r & \text{if } h \equiv 0 \mod r, \\ -1 & \text{if } h \not\equiv 0 \mod r. \end{cases}$$

Proof. If $h \equiv 0 \mod r$, then $s_n = s_{n+h}$ for all $n \geq 0$ and the result follows immediately from (7.8). If $h \not\equiv 0 \mod r$, then $u_n - s_{n+h}$, $n = 0, 1, \ldots$, defines a sequence satisfying the same linear recurrence relation as $s_0, s_1, \ldots$. By Lemma 6.4, $u_0, u_1, \ldots$ cannot be the zero sequence, and so it is again a maximal period sequence in $\mathbb{F}_q$. Applying Theorem 7.43 with $m = 1$ to this sequence, we get

$$C_r(h) = \sum_{n=0}^{r-1} \chi(s_n - s_{n+h}) = \sum_{n=0}^{r-1} \chi(u_n) = (q^{k-1} - 1)\chi(0) + q^{k-1} \sum_{b \in \mathbb{F}_q^*} \chi(b)$$

$$= -1 + q^{k-1} \sum_{b \in \mathbb{F}_q} \chi(b) = -1,$$

where we used (5.9) in the last step.   □

7.45. Example. Consider the linear recurring sequence $s_0, s_1, \ldots$ in $\mathbb{F}_2$ with $s_{n+5} = s_{n+2} + s_n$ for $n = 0, 1, \ldots$ and initial values $s_0 = s_2 = s_4 = 1$, $s_1 = s_3 = 0$. Since $x^5 - x^2 - 1$ is a primitive polynomial over $\mathbb{F}_2$, this sequence is a maximal period sequence in $\mathbb{F}_2$ with least period $r = 2^5 - 1 = 31$. Write down the 31 bits making up the period of the sequence and underneath the first 31 terms of the sequence shifted by $h = 3$ terms to the left:

1 0 1 0 1 0 0 0 0 1 0 0 1 0 1 1 0 0 1 1 1 1 1 0 0 0 1 1 0 1 1

0 1 0 0 0 0 1 0 0 1 0 1 1 0 0 1 1 1 1 1 0 0 0 1 1 0 1 1 1 0 1

The number of agreements of corresponding terms is 15, the number of disagreements is 16, hence $C_{31}(3) = 15 - 16 = -1$, in accordance with Theorem 7.44. If we consider the pairs (s_n, s_{n+1}), $n = 0, 1, \ldots, 30$, then there are 7 of type $(0,0)$ and 8 each of type $(0,1)$, $(1,0)$, and $(1,1)$, in accordance with Theorem 7.43.   □

For $N < r$ we can give bounds for the correlation coefficients $C_N(h)$, and in the trivial case $h \equiv 0 \mod r$ we have an explicit formula.

7.46. Theorem. *For any kth-order maximal period sequence in $\mathbb{F}_q$ and $1 \leq N < r = q^k - 1$ we have*

$$C_N(h) = N \quad \text{if } h \equiv 0 \mod r$$

and

$$|C_N(h)| < q^{k/2}\left(\frac{2}{\pi}\log r + \frac{2}{5} + \frac{N}{r}\right) \quad \text{if} \quad h \not\equiv 0 \bmod r.$$

Proof. We proceed as in the proof of Theorem 7.44. If $h \equiv 0 \bmod r$, then $u_n = s_n - s_{n+h} = 0$ for all $n \geq 0$ and so

$$C_N(h) = \sum_{n=0}^{N-1} \chi(u_n) = N.$$

If $h \not\equiv 0 \bmod r$, then $u_0, u_1, \ldots$ is a kth-order maximal period sequence in $\mathbb{F}_q$ and so

$$|C_N(h)| = \left|\sum_{n=0}^{N-1} \chi(u_n)\right| < q^{k/2}\left(\frac{2}{\pi}\log r + \frac{2}{5} + \frac{N}{r}\right)$$

by Theorem 6.81, since $n_0 = 0$ and $R = r$ in this case. $\qquad\qquad\square$

If we take every second term of a random sequence of elements of $\mathbb{F}_q$, we would expect that the resulting subsequence has again randomness properties. More generally, the property of being a random sequence should be invariant under the operations of *decimation* defined as follows. If σ is a given sequence $s_0, s_1, s_2, \ldots$ of elements of $\mathbb{F}_q$ and $d \geq 1$ and $h \geq 0$ are integers, then the decimated sequence $\sigma_d^{(h)}$ has the terms $s_h, s_{h+d}, s_{h+2d}, \ldots$. In other words, $\sigma_d^{(h)}$ is obtained by taking every dth term of σ, starting from s_h. The following result shows that the property of being a maximal period sequence in $\mathbb{F}_q$ is invariant under many decimations. This can be viewed as further evidence that maximal period sequences are good candidates for pseudorandom sequences.

7.47.   Theorem. *Let σ be a given kth-order maximal period sequence in $\mathbb{F}_q$. Then $\sigma_d^{(h)}$ is a kth-order maximal period sequence in $\mathbb{F}_q$ if and only if $\gcd(d, q^k - 1) = 1$, and $\sigma_d^{(h)}$ is a maximal period sequence in $\mathbb{F}_q$ satisfying the same linear recurrence relation as σ (or, equivalently, $\sigma_d^{(h)}$ is a shifted version of σ) if and only if $d \equiv q^j \bmod (q^k - 1)$ for some j with $0 \leq j \leq k - 1$.*

Proof. Denote the terms of σ and $\sigma_d^{(h)}$ by s_n and u_n, respectively. The minimal polynomial of σ is a primitive polynomial $f(x)$ over $K = \mathbb{F}_q$ of degree k. If α is a fixed root of $f(x)$ in $F = \mathbb{F}_{q^k}$, then α is a primitive element of F (see Definition 3.15). By Theorem 6.24 there is a unique $\theta \in F^*$ such that

$$s_n = \mathrm{Tr}_{F/K}(\theta \alpha^n) \quad \text{for all} \quad n \geq 0.$$

It follows that

$$u_n = s_{h+nd} = \mathrm{Tr}_{F/K}(\beta(\alpha^d)^n) \quad \text{for all} \quad n \geq 0,$$

where $\beta = \theta \alpha^h \in F^*$. Let $f_d(x)$ be the minimal polynomial of α^d over K. Then the calculation in the proof of Theorem 6.24 shows that $\sigma_d^{(h)}$ is a linear recurring sequence with characteristic polynomial $f_d(x)$. If $\gcd(d, q^k - 1) = 1$, then α^d is a primitive element of F, and so $f_d(x)$ is a primitive polynomial over K of degree k. Since $\beta \neq 0$, not all u_n are 0, thus $\sigma_d^{(h)}$ is a kth-order maximal period sequence in $\mathbb{F}_q$. If $\gcd(d, q^k - 1) > 1$, then α^d is not a primitive element of F, and so $\sigma_d^{(h)}$ cannot be a kth-order maximal period sequence in $\mathbb{F}_q$. The first part of the theorem is thus shown.

Furthermore, $\sigma_d^{(h)}$ is a maximal period sequence in $\mathbb{F}_q$ satisfying the same linear recurrence relation as σ if and only if $f_d(x) = f(x)$. By Theorem 2.14, this identity holds if and only if $\alpha^d = \alpha^{q^j}$, hence $d \equiv q^j \bmod(q^k - 1)$, for some j with $0 \leqslant j \leqslant k - 1$. Since the state vectors of σ run through all nonzero vectors in $\mathbb{F}_q^k$, the maximal period sequences in $\mathbb{F}_q$ satisfying the same linear recurrence relation as σ are exactly the shifted versions of σ. $\square$

Maximal period sequences possess a universality property, in the sense that a much larger class of linear recurring sequences can be derived from them by applying decimations.

7.48.   Theorem.   *Let σ be a given kth-order maximal period sequence in $\mathbb{F}_q$. Then every linear recurring sequence in $\mathbb{F}_q$ having an irreducible minimal polynomial $g(x)$ with $g(0) \neq 0$ and $\deg(g(x))$ dividing k can be obtained from σ by applying a suitable decimation.*

Proof.   If the terms of σ are denoted by s_n, then as in the proof of Theorem 7.47 we have

$$s_n = \mathrm{Tr}_{F/K}(\theta \alpha^n) \quad \text{for all} \quad n \geqslant 0,$$

where α is a primitive element of $F = \mathbb{F}_{q^k}$, $\theta \in F^*$, and $K = \mathbb{F}_q$. Let $u_0, u_1, \ldots$ be a linear recurring sequence in $\mathbb{F}_q$ with irreducible minimal polynomial $g(x)$, where $g(0) \neq 0$ and $m = \deg(g(x))$ divides k. Then $g(x)$ has a root $\gamma \in E = \mathbb{F}_{q^m}$, and $\gamma \neq 0$ since $g(0) \neq 0$. Furthermore, E is a subfield of F by Theorem 2.6. It follows that there exists an integer $d \geqslant 1$ such that $\gamma = \alpha^d$. By Theorem 6.24 we have

$$u_n = \mathrm{Tr}_{E/K}(\beta \gamma^n) \quad \text{for all} \quad n \geqslant 0,$$

where $\beta \in E^*$. Let $\delta \in F^*$ be such that $\mathrm{Tr}_{F/E}(\delta) = \beta$, and choose an integer $h \geqslant 0$ with $\delta \theta^{-1} = \alpha^h$. Then by the transitivity of the trace (see Theorem 2.26) we have

$$s_{h+nd} = \mathrm{Tr}_{F/K}(\theta \alpha^{h+nd}) = \mathrm{Tr}_{F/K}(\delta \gamma^n) = \mathrm{Tr}_{E/K}(\mathrm{Tr}_{F/E}(\delta \gamma^n))$$

$$= \mathrm{Tr}_{E/K}(\beta \gamma^n) = u_n$$

for all $n \geqslant 0$, and so the sequence $u_0, u_1, \ldots$ is equal to the decimated sequence $\sigma_d^{(h)}$. $\square$

The condition $g(0) \neq 0$ in Theorem 7.48 rules out the case $g(x) = x$ in which the sequence has the form $c, 0, 0, \ldots$ with $c \in \mathbb{F}_q^*$. Such a sequence has preperiod 1, and thus it cannot be derived from σ by a decimation since every decimated sequence $\sigma_d^{(h)}$ is periodic.

In the special case $d = 1$ we write $\sigma_1^{(h)} = \sigma^{(h)}$, which is the sequence obtained by shifting σ by h terms. Maximal period sequences can be characterized in terms of a structural property of the set of all shifted sequences. We use again the termwise operations for sequences introduced in Chapter 6, Section 5.

7.49. Theorem. *If σ is a nonzero periodic sequence of elements of $\mathbb{F}_q$, then the shifted sequences $\sigma^{(h)}, h = 0, 1, \ldots$, together with the zero sequence form a vector space over $\mathbb{F}_q$ under termwise operations for sequences if and only if σ is a maximal period sequence in $\mathbb{F}_q$.*

Proof. If σ is a kth-order maximal period sequence in $\mathbb{F}_q$, then the initial state vectors of the sequences $\sigma^{(h)}$, $h = 0, 1, \ldots, q^k - 2$, and of the zero sequence run exactly through all vectors in $\mathbb{F}_q^k$. From this it follows easily that these sequences form a vector space over $\mathbb{F}_q$. Note also that any shifted sequence $\sigma^{(h)}$, $h \geq 0$, is identical to one with $0 \leq h \leq q^k - 2$.

Conversely, if σ is a nonzero periodic sequence of elements of $\mathbb{F}_q$ with least period r, then the distinct shifted sequences are $\sigma = \sigma^{(0)}, \sigma^{(1)}, \ldots, \sigma^{(r-1)}$. If these together with the zero sequence form a vector space V over $\mathbb{F}_q$, then V is closed under shifts of sequences, and so Theorem 6.56 shows that $V = S(f(x))$ for some monic polynomial $f(x) \in \mathbb{F}_q[x]$ of degree $k \geq 1$. Counting the number of elements of V in two different ways we get $r + 1 = q^k$, hence σ is a kth-order linear recurring sequence in $\mathbb{F}_q$ with least period $r = q^k - 1$. In particular, the state vectors of σ run through all nonzero vectors in $\mathbb{F}_q^k$, and so some $\sigma^{(h)}$ is the impulse response sequence with characteristic polynomial $f(x)$. Theorem 6.27 implies that $\mathrm{ord}\,(f(x)) = r = q^k - 1$. If we had $f(0) = 0$, then $a_0 = 0$ in (7.7) and the sequence in $S(f(x))$ with initial values $1, 0, \ldots, 0$ has all subsequent terms equal to 0; but this nonperiodic sequence cannot belong to V, a contradiction. Thus $f(0) \neq 0$, and so $f(x)$ is a primitive polynomial over $\mathbb{F}_q$ by Theorem 3.16. Consequently, σ is a maximal period sequence in $\mathbb{F}_q$. $\square$

7.50. Example. Let σ be the linear recurring sequence $s_0, s_1, \ldots$ in $\mathbb{F}_2$ with $s_{n+4} = s_{n+1} + s_n$ for $n = 0, 1, \ldots$ and initial values $s_0 = s_1 = s_2 = 0, s_3 = 1$. Since $x^4 - x - 1$ is a primitive polynomial over $\mathbb{F}_2$, σ is a maximal period sequence in $\mathbb{F}_2$ with least period $r = 2^4 - 1 = 15$. The 15 bits making up the period of σ are

$$0 \quad 0 \quad 0 \quad 1 \quad 0 \quad 0 \quad 1 \quad 1 \quad 0 \quad 1 \quad 0 \quad 1 \quad 1 \quad 1 \quad 1.$$

As an illustration of Theorem 7.48 we derive all the linear recurring sequences in $\mathbb{F}_2$ having an irreducible minimal polynomial $g(x) \neq x$ with $\deg(g(x)) = 1$ or 2 by applying a suitable decimation to σ. The constant sequence $1, 1, 1, \ldots$ ($= \sigma_{15}^{(3)}$) has minimal polynomial $x - 1$, and the periodic sequences $0, 1, 1, \ldots$ ($= \sigma_5^{(1)}$), $1, 0, 1, \ldots$ ($= \sigma_5^{(3)}$), and $1, 1, 0, \ldots$ ($= \sigma_5^{(6)}$) with least period 3 represent all the linear recurring sequences in $\mathbb{F}_2$ with minimal polynomial $x^2 - x - 1$. As an illustration of Theorem 7.49 we note that $\sigma + \sigma^{(3)}$ must be either a shifted version of σ or the zero sequence, and in fact $\sigma + \sigma^{(3)} = \sigma^{(14)}$. On the other hand, if τ is the linear recurring sequence $t_0, t_1, \ldots$ in $\mathbb{F}_2$ with $t_{n+4} = t_{n+3} + t_{n+2} + t_{n+1} + t_n$ for $n = 0, 1, \ldots$ and initial values $t_0 = t_1 = t_2 = 0, t_3 = 1$, then τ is the periodic sequence $0, 0, 0, 1, 1, \ldots$ with least period 5 and $\tau + \tau^{(3)}$ is neither a shifted version of τ nor the zero sequence. This is again in accordance with Theorem 7.49 since τ is not a maximal period sequence in $\mathbb{F}_2$. $\square$

For many simulation purposes, and especially for applications in numerical analysis, one needs random sequences of real numbers. These numbers should all belong to a given interval on the real line, which for simplicity we may take to be the interval $[0, 1]$. The generation of a random sequence of numbers in $[0, 1]$ can again be described by a statistical experiment. Pick a number from $[0, 1]$ at random, where the probability that the number belongs to a specific subinterval of $[0, 1]$ should be equal to the length of the subinterval. Repeat this procedure indefinitely, with each selection being statistically independent of all the previous ones. Since we are using here a special probability distribution giving equal likelihood to subintervals of the same length, one often speaks of the resulting sequence as a sequence of *uniform random numbers*. The notion of a sequence of uniform random numbers is an idealized concept, and in practice one works with a deterministic analog called a sequence of *uniform pseudorandom numbers*. Such a sequence is generated by a deterministic method and should pass various tests for randomness. The advantages of such a sequence are similar to those of a pseudorandom sequence of bits described earlier.

Maximal period sequences in finite fields can be used to generate sequences of uniform pseudorandom numbers. Let $\mathbb{F}_p$ be a finite prime field— that is, p is prime—and let $s_0, s_1, \ldots$ be a kth-order maximal period sequence in $\mathbb{F}_p$. In the following we view the terms s_n of the sequence as integers with $0 \leqslant s_n < p$. The integers s_n have to be transformed into numbers in $[0, 1]$. One method of doing this is the *normalization method*, in which one chooses p to be a large prime and normalizes s_n by setting

$$w_n = \frac{s_n}{p} \in [0, 1] \quad \text{for all} \quad n \geqslant 0.$$

Then $w_0, w_1, \ldots$ is taken as a sequence of uniform pseudorandom numbers. Clearly, this sequence is periodic with least period $r = p^k - 1$. Since the sequence $w_0, w_1, \ldots$ differs from the sequence $s_0, s_1, \ldots$ only by a constant factor, the statistical properties of the two sequences will essentially be the same. Thus it suffices to refer to our earlier discussion of statistical properties of maximal period sequences.

A second method of transforming the integers s_n into numbers in $[0, 1]$ is the *digital* (or *Tausworthe*) *method*. Here we let p be a small prime and we choose an integer $m \geqslant 1$. Then we set

$$w_n = \sum_{i=1}^{m} s_{mn+i-1} p^{-i} \quad \text{for all} \quad n \geqslant 0, \tag{7.9}$$

and we use $w_0, w_1, \ldots$ as a sequence of uniform pseudorandom numbers. The formula (7.9) means that the sequence $s_0, s_1, \ldots$ is split up into blocks of length m, and each block is interpreted as the digital representation in the base p of a number in $[0, 1]$. In practice one usually works with the prime $p = 2$, since this facilitates the calculation of the terms s_n by the relation (7.7) and since in this case we get the numbers w_n in their binary representation which is well suited for computer calculations.

   7.51. Lemma. *The sequence $w_0, w_1, \ldots$ of numbers defined by (7.9) is periodic with least period*

$$r = \frac{p^k - 1}{\gcd(m, p^k - 1)}.$$

   Proof. Since mr is a multiple of $p^k - 1$ and thus a period of the maximal period sequence $s_0, s_1, \ldots$, we have

$$w_{n+r} = \sum_{i=1}^{m} s_{mn+i-1+mr} p^{-i} = \sum_{i=1}^{m} s_{mn+i-1} p^{-i} = w_n \quad \text{for all} \quad n \geqslant 0.$$

Therefore $w_0, w_1, \ldots$ is periodic with period r. Now let u be an arbitrary period of this sequence. Then $w_{n+u} = w_n$ for all $n \geqslant 0$, hence

$$\sum_{i=1}^{m} s_{mn+i-1+mu} p^{-i} = \sum_{i=1}^{m} s_{mn+i-1} p^{-i} \quad \text{for all} \quad n \geqslant 0.$$

The uniqueness of digital representations implies that

$$s_{mn+i-1+mu} = s_{mn+i-1} \quad \text{for} \quad 1 \leqslant i \leqslant m \quad \text{and all } n \geqslant 0.$$

Now $mn + i - 1$ runs through all nonnegative integers if i and n run through all integers with $1 \leqslant i \leqslant m$ and $n \geqslant 0$, thus

$$s_{n+mu} = s_n \quad \text{for all} \quad n \geqslant 0.$$

This means that mu is a period of the sequence $s_0, s_1, \ldots$, and so $p^k - 1$ divides mu. It follows that r divides u, and therefore r is the least period of the sequence $w_0, w_1, \ldots$. $\square$

   In order to make the least period of the sequence $w_0, w_1, \ldots$ as large as possible, we will choose the block length m in such a way that $\gcd(m, p^k - 1) = 1$. The least period of the sequence is then equal to $p^k - 1$ by Lemma 7.51. If we want to make this least period large for $p = 2$, then k should not be chosen too small. We note also that if $m > k$, then the last $m - k$ digits of any w_n depend on the first k digits on account of the relation (7.7). Therefore we impose the condition $m \leqslant k$ in order to prevent such obvious dependencies.

   An important test for randomness for a sequence $w_0, w_1, \ldots$ of uniform pseudorandom numbers is the *uniformity test*. We start from the observation that in the ideal case of a sequence $x_0, x_1, \ldots$ of uniform random numbers the probability that the inequality $x_n \leqslant t$ is satisfied for a given $t \in [0, 1]$ is equal to t. We compare this with the elementary probability $P_N(t)$ that the inequality $w_n \leqslant t$ is satisfied among the first N terms of the given sequence $w_0, w_1, \ldots$ — that is, $P_N(t)$ is N^{-1} times the number of $n, 0 \leqslant n < N$, with $w_n \leqslant t$. The largest deviation

$$D_N = \sup_{0 \leqslant t \leqslant 1} |P_N(t) - t| \tag{7.10}$$

between these two probabilities provides a way of measuring the extent to which $w_0, w_1, \ldots$ differs from a sequence of uniform random numbers. For a "good" sequence of uniform pseudorandom numbers the value of D_N should be small for large N. When applying the uniformity test to a periodic sequence $w_0, w_1, \ldots$ with least period r, it suffices to consider the case $1 \leqslant N \leqslant r$ since the behavior of the sequence repeats itself beyond the period.

7.52.   Theorem.   *If $m \leqslant k$ and $\gcd(m, p^k - 1) = 1$, then the sequence $w_0, w_1, \ldots$ of uniform pseudorandom numbers generated by (7.9) satisfies $D_r = p^{-m}$ with $r = p^k - 1$.*

Proof.   The least period of $w_0, w_1, \ldots$ is $r = p^k - 1$ by Lemma 7.51. The sequence of m-tuples

$$\mathbf{s}_n = (s_n, s_{n+1}, \ldots, s_{n+m-1}), \quad n = 0, 1, \ldots,$$

also has least period r; in other words, $\mathbf{s}_n$ just depends on the residue class of n modulo r. From $\gcd(m, r) = 1$ it follows then that the finite sequence $\mathbf{s}_{mn}, n = 0, 1, \ldots, r - 1$, is a rearrangement of the finite sequence $\mathbf{s}_n, n = 0, 1, \ldots, r - 1$. In particular, for any $\mathbf{b} \in \{0, 1, \ldots, p-1\}^m$ the number of n, $0 \leqslant n \leqslant r - 1$, with $\mathbf{s}_{mn} = \mathbf{b}$ is the same as the number of $n, 0 \leqslant n \leqslant r - 1$, with $\mathbf{s}_n = \mathbf{b}$. The latter number is given by Theorem 7.43. Together with (7.9) this yields the following information: the number of n, $0 \leqslant n \leqslant r - 1$, with $w_n = 0$ is equal to $p^{k-m} - 1$, and for any rational number cp^{-m} with $c \in \mathbb{Z}, 1 \leqslant c < p^m$, the number of $n, 0 \leqslant n \leqslant r - 1$, with $w_n = cp^{-m}$ is equal to p^{k-m}; this exhausts all possible values of w_n. For $a \in \mathbb{Z}, 0 \leqslant a < p^m$, consider a real t with $ap^{-m} \leqslant t < (a+1)p^{-m}$. Then

$$P_r(t) = \frac{1}{r}(p^{k-m} - 1 + ap^{k-m}),$$

and so

$$|P_r(t) - t| = \left| \frac{(a+1)p^{k-m} - 1}{p^k - 1} - t \right| = \left| \left(\frac{a+1}{p^m} - t \right) - \frac{1 - (a+1)p^{-m}}{p^k - 1} \right|.$$

Now

$$0 < \frac{a+1}{p^m} - t \leqslant \frac{1}{p^m}$$

and

$$0 \leqslant \frac{1-(a+1)p^{-m}}{p^k-1} \leqslant \frac{1-p^{-m}}{p^k-1} = \frac{1}{p^m} \cdot \frac{p^m-1}{p^k-1} \leqslant \frac{1}{p^m},$$

hence

$$|P_r(t)-t| \leqslant \frac{1}{p^m}.$$

Since

$$\left|P_r\left(1-\frac{1}{p^m}\right)-\left(1-\frac{1}{p^m}\right)\right| = \frac{1}{p^m}$$

and $P_r(1)=1$, it follows from (7.10) that $D_r = p^{-m}$. $\qquad\square$

Theorem 7.52 shows that if m is chosen sufficiently large, then the sequence $w_0, w_1, \ldots$ passes the uniformity test when considered over the full period. For parts of the period—that is, for $1 \leqslant N < r$—we can establish an upper bound for the quantity D_N in (7.10). Let $w_0, w_1, \ldots$ be a sequence of elements of $[0, 1]$ whose terms are given by finite digital representations

$$w_n = \sum_{i=1}^{m} w_p^{(i)} p^{-i}, \quad n = 0, 1, \ldots, \tag{7.11}$$

where the digits $w_n^{(i)}$ belong to the set $\{0, 1, \ldots, p-1\}$ and m is independent of n. For $h \in \mathbb{Z}$ we define $e_p(h) = e(h/p)$, where $e(t)$ is the complex exponential function used in Chapter 6, Section 7.

7.53. Lemma. *Let $w_0, w_1, \ldots$ be a sequence of elements of $[0, 1]$ given by (7.11) and let N be a positive integer. Let B be a constant such that for any $h_1, \ldots, h_m \in \{0, 1, \ldots, p-1\}$ that are not all 0 we have*

$$\left| \frac{1}{N} \sum_{n=0}^{N-1} e_p(h_1 w_n^{(1)} + \cdots + h_m w_n^{(m)}) \right| \leqslant B \tag{7.12}$$

Then the quantity D_N in (7.10) satisfies

$$D_N \leqslant \frac{1}{p^m} + Bm\left(\frac{2}{\pi}\log p + \frac{7}{5}\right).$$

Proof. For $0 \leqslant t < 1$ let

$$t = \sum_{i=1}^{\infty} t_i p^{-i}$$

be the digital representation of t in the base p, with $t_i \in \{0, 1, \ldots, p-1\}$ for all $i \geqslant 1$ and the usual condition that $t_i < p-1$ for infinitely many i. Then we have $w_n \leqslant t$ if and only if $w_n^{(1)} = t_1, \ldots, w_n^{(i-1)} = t_{i-1}, w_n^{(i)} < t_i$ for some i with $1 \leqslant i \leqslant m-1$ (for $i=1$ the condition reduces to $w_n^{(1)} < t_1$) or $w_n^{(1)} = t_1, \ldots, w_n^{(m-1)} = t_{m-1}, w_n^{(m)} \leqslant t_m$. Thus, if we put $u_i = t_i - 1$ for $1 \leqslant i \leqslant m-1$

and $u_m = t_m$ and interpret empty sums to be equal to 0, then

$$P_N(t) = \frac{1}{N} \sum_{i=1}^{m} \sum_{j=0}^{u_i} \sum_{n=0}^{N-1} d_{t_1}(w_n^{(1)}) \cdots (d_{t_{i-1}}(w_n^{(i-1)}) d_j(w_n^{(i)}),$$

where $d_j(w) = 1$ for $j = w$ and $d_j(w) = 0$ for $j \ne w$ with j, $w \in \{0, 1, \ldots, p-1\}$.
Now

$$d_j(w) = \frac{1}{p} \sum_{h=0}^{p-1} e_p(h(w-j)),$$

and so

$$P_N(t) = \frac{1}{N} \sum_{i=1}^{m} \sum_{j=0}^{u_i} \sum_{n=0}^{N-1} \frac{1}{p^i} \sum_{h_1,\ldots,h_i=0}^{p-1} e_p(h_1 w_n^{(1)} + \cdots + h_i w_n^{(i)})$$

$$\cdot e_p(-h_1 t_1 - \cdots - h_{i-1} t_{i-1} - h_i j).$$

Separating the contribution from the choice $h_1 = \cdots = h_i = 0$ in the inner sum
and denoting by an asterisk the deletion of the corresponding term, we get

$$P_N(t) = \sum_{i=1}^{m} \frac{u_i + 1}{p^i} + \sum_{i=1}^{m} \frac{1}{p^i} \sum_{h_1,\ldots,h_i=0}^{p-1} e_p(-h_1 t_1 - \cdots - h_{i-1} t_{i-1})$$

$$\cdot \frac{1}{N} \sum_{n=0}^{N-1} e_p(h_1 w_n^{(1)} + \cdots + h_i w_n^{(i)}) \sum_{j=0}^{u_i} e_p(-h_i j).$$

Using

$$\left| \sum_{i=1}^{m} \frac{u_i + 1}{p^i} - t \right| \le \frac{1}{p^m}$$

and the condition (7.12), we obtain

$$|P_N(t) - t| \le \frac{1}{p^m} + B \sum_{i=1}^{m} \frac{1}{p^i} \sum_{h_1,\ldots,h_i=0}^{p-1}{}^* \left| \sum_{j=0}^{u_i} e_p(h_i j) \right|$$

$$\le \frac{1}{p^m} + B \sum_{i=1}^{m} \frac{1}{p^i} \sum_{h_1,\ldots,h_i=0}^{p-1} \left| \sum_{j=0}^{u_i} e_p(hj) \right|.$$

$$= \frac{1}{p^m} + \frac{B}{p} \sum_{i=1}^{m} \sum_{h=0}^{p-1} \left| \sum_{j=0}^{u_i} e_p(hj) \right|.$$

By Lemma 6.80 we have

$$\sum_{h=0}^{p-1} \left| \sum_{j=0}^{u_i} e_p(hj) \right| < \frac{2}{\pi} p \log p + \frac{2}{5} p + u_i + 1 \le \frac{2}{\pi} p \log p + \frac{7}{5} p,$$

and so

$$|P_N(t) - t| \le \frac{1}{p^m} + Bm \left(\frac{2}{\pi} \log p + \frac{7}{5} \right) \quad \text{for} \quad 0 \le t < 1.$$

Since $P_N(1) = 1$, the desired result follows.   □

7.54. Theorem. *Let $m \leqslant k$ and $\gcd(m, p^k - 1) = 1$, and let $w_0, w_1, \ldots$ be the sequence of uniform pseudorandom numbers generated by (7.9). Then for $1 \leqslant N < r = p^k - 1$ the quantity D_N in (7.10) satisfies*

$$D_N \leqslant \frac{1}{p^m} + \frac{mp^{k/2}}{N}\left(\frac{2}{\pi}\log r + \frac{2}{5} + \frac{N}{r}\right)\left(\frac{2}{\pi}\log p + \frac{7}{5}\right).$$

Proof. The bound for D_N is obtained from Lemma 7.53 by determining a suitable constant B such that (7.12) holds. If the w_n are given by (7.9), then we have $w_n^{(i)} = s_{mn+i-1}$ for $1 \leqslant i \leqslant m$ and all $n \geqslant 0$. Thus

$$\frac{1}{N}\sum_{n=0}^{N-1} e_p(h_1 w_n^{(1)} + \cdots + h_m w_n^{(m)}) = \frac{1}{N}\sum_{n=0}^{N-1} e_p\left(\sum_{i=1}^{m} h_i s_{mn+i-1}\right).$$

We note that for any $h \in \mathbb{Z}$ we have $e_p(h) = \chi_1(h)$, where χ_1 is the canonical additive character of $\mathbb{F}_p$ (see Chapter 5, Section 1) and on the right-hand side we identify h with the corresponding element of $\mathbb{F}_p$—namely, with the residue class of h modulo p. If we now identify all h_i and s_n with the corresponding elements of $\mathbb{F}_p$ and define

$$v_n = \sum_{i=1}^{m} h_i s_{mn+i-1} \in \mathbb{F}_p \quad \text{for all} \quad n \geqslant 0,$$

then we can write

$$\frac{1}{N}\sum_{n=0}^{N-1} e_p(h_1 w_n^{(1)} + \cdots + h_m w_n^{(m)}) = \frac{1}{N}\sum_{n=0}^{N-1} \chi_1(v_n). \tag{7.13}$$

Since $s_0, s_1, \ldots$ is a kth-order maximal period sequence in $K = \mathbb{F}_p$, we get as in the proof of Theorem 7.47

$$s_n = \mathrm{Tr}_{F/K}(\theta\alpha^n) \quad \text{for all} \quad n \geqslant 0,$$

where α is a primitive element of $F = \mathbb{F}_{p^k}$ and $\theta \in F^*$. Suppose $h_1, \ldots, h_m \in \mathbb{F}_p$ are not all 0. Then

$$v_n = \sum_{i=1}^{m} h_i \mathrm{Tr}_{F/K}(\theta\alpha^{mn+i-1}) = \mathrm{Tr}_{F/K}\left(\theta \sum_{i=1}^{m} h_i \alpha^{mn+i-1}\right) = \mathrm{Tr}_{F/K}(\beta\gamma^n)$$

for all $n \geqslant 0$, where

$$\beta = \theta \sum_{i=1}^{m} h_i \alpha^{i-1} \quad \text{and} \quad \gamma = \alpha^m.$$

Since $m \leqslant k$ and $\{1, \alpha, \alpha^2, \ldots, \alpha^{k-1}\}$ is a basis of F over K, we have $\beta \neq 0$. Furthermore, the condition $\gcd(m, p^k - 1) = 1$ implies that γ is a primitive element of F. Theorem 6.24 and its proof show then that $v_0, v_1, \ldots$ is a kth-order maximal period sequence in K. Thus

$$\left|\frac{1}{N}\sum_{n=0}^{N-1} \chi_1(v_n)\right| < \frac{p^{k/2}}{N}\left(\frac{2}{\pi}\log r + \frac{2}{5} + \frac{N}{r}\right) \quad \text{for} \quad 1 \leqslant N < r \tag{7.14}$$

by Theorem 6.81, since $n_0 = 0$ and $R = r$ in this case. Taking into account (7.13), we can therefore use the expression on the right-hand side of (7.14) as a possible value of B in condition (7.12). The rest follows from Lemma 7.53.

$\square$

In the proof of Theorem 7.52 we have obtained the exact distribution of values in the least period of the sequence $w_0, w_1, \ldots$ generated by (7.9), under the conditions $m \leqslant k$ and $\gcd(m, p^k - 1) = 1$. With these hypotheses we can show an analogous result for higher dimensions d as long as $d \leqslant k/m$. For such a d we consider the d-tuples

$$\mathbf{w}_n = (w_n, w_{n+1}, \ldots, w_{n+d-1}) \in [0, 1]^d \quad \text{for} \quad 0 \leqslant n \leqslant r - 1,$$

where $r = p^k - 1$ is the least period of the sequence $w_0, w_1, \ldots$. Each $\mathbf{w}_n$ is a d-tuple of the form

$$\mathbf{c} = (c_1 p^{-m}, \ldots, c_d p^{-m}) \tag{7.15}$$

with $c_j \in \mathbb{Z}$ and $0 \leqslant c_j < p^m$ for $1 \leqslant j \leqslant d$. Consider also the sequence of md-tuples

$$\mathbf{s}_n = (s_n, s_{n+1}, \ldots, s_{n+md-1}), \quad n = 0, 1, \ldots,$$

which has again least period r. The same argument as in the proof of Theorem 7.52 shows that for any $\mathbf{b} \in \{0, 1, \ldots, p-1\}^{md}$ the number of $n, 0 \leqslant n \leqslant r - 1$, with $\mathbf{s}_{mn} = \mathbf{b}$ is the same as the number of $n, 0 \leqslant n \leqslant r - 1$, with $\mathbf{s}_n = \mathbf{b}$. The latter number can be obtained from Theorem 7.43 since the condition on d yields $md \leqslant k$. In this way we arrive at the following result: the number of n, $0 \leqslant n \leqslant r - 1$, with $\mathbf{w}_n = \mathbf{0}$ is equal to $p^{k-md} - 1$, and for any $\mathbf{c} \neq \mathbf{0}$ of the form (7.15) the number of $n, 0 \leqslant n \leqslant r - 1$, with $\mathbf{w}_n = \mathbf{c}$ is equal to p^{k-md}. Thus the d-tuples $\mathbf{w}_n$ show a very regular distribution behavior. The study of the distribution of the $\mathbf{w}_n$ amounts to performing an analog of the serial test for random sequences of bits described earlier in this section. We can therefore say that a sequence $w_0, w_1, \ldots$ of uniform pseudorandom numbers generated by (7.9) passes the serial test for dimensions $d \leqslant k/m$, at least when it is considered over the full period. Results for parts of the period can be obtained by an extension of the method in the proof of Theorem 7.54.

EXERCISES

7.1.   List the points and lines of $PG(2, \mathbb{F}_3)$. Draw a diagram showing all the intersections. Enumerate the points on L_∞ and the families of parallel lines in $AG(2, \mathbb{F}_3)$.

7.2.   In $PG(2, \mathbb{F}_4)$ consider the quadrangle $A(1, 1, 1 + \beta)$, $B(0, 1, \beta)$, $C(1, 1, \beta)$, $D(1, 1 + \beta, \beta)$, where β is a primitive element of $\mathbb{F}_4$. Find its diagonal points and verify that they are collinear.

7.3.   There are six points in $PG(2, \mathbb{F}_4)$, no three of which are collinear.

Four of them are the points A, B, C, D of Exercise 7.2. Find the other two points.

7.4. Find the equation of the conic consisting of the points A, B, C, D of Exercise 7.2 and $E(1, 1 + \beta, 1 + \beta)$, determine all its tangents and the point where they meet.

7.5. Show that for a nondegenerate conic in $PG(2, \mathbb{F}_5)$ the tangents do not all meet in the same point.

7.6. Prove: if L is a set of points of $PG(2, \mathbb{F}_q)$ such that every line of $PG(2, \mathbb{F}_q)$ contains a point of L, then $|L| \geqslant q + 1$ with equality if and only if L is a line.

7.7. Prove that among any $m + 3$ points of a finite projective plane of order m one can find three collinear ones.

7.8. Determine the number of points, lines, planes, and hyperplanes of $PG(4, \mathbb{F}_3)$. How many planes are there through a given line?

7.9. In $PG(4, \mathbb{F}_3)$ determine the 3-flats through the plane given by $(1, 0, 0, 0, 0)$, $(0, 0, 1, 0, 0)$, and $(0, 0, 0, 0, 1)$.

7.10. Prove that the number of k-flats of $PG(m, \mathbb{F}_q)$, $1 \leqslant k < m$, or also within a fixed m-flat of a projective geometry over $\mathbb{F}_q$ of higher dimension, is equal to

$$\frac{(q^{m+1} - 1)(q^m - 1) \cdots (q^{m-k+1} - 1)}{(q^{k+1} - 1)(q^k - 1) \cdots (q - 1)}.$$

7.11. Show that the following system of blocks forms a BIBD and evaluate the parameters v, b, r, k, and λ:

$$\{1, 2, 3\} \quad \{1, 4, 7\} \quad \{1, 5, 9\} \quad \{1, 6, 8\}$$

$$\{4, 5, 6\} \quad \{2, 5, 8\} \quad \{2, 6, 7\} \quad \{2, 4, 9\}$$

$$\{7, 8, 9\} \quad \{3, 6, 9\} \quad \{3, 4, 8\} \quad \{3, 5, 7\}$$

7.12. Solve the following special case of the Kirkman Schoolgirl Problem. A schoolmistress takes 9 girls for a daily walk, the girls arranged in rows of 3 girls. Plan the walk for 4 consecutive days so that no girl walks with any of her classmates in any triplet more than once.

7.13. In a school of b boys, t athletics teams of k boys each are formed in such a way that every boy plays on the same number of teams. Also, the arrangement is such that each pair of boys plays together the same number of times. On how many teams does a boy play and how often do two boys play on the same team?

7.14. Prove: if v is even for a symmetric (v, k, λ) block design, then $k - \lambda$ is a square.

7.15. Verify that $\{0, 1, 2, 3, 5, 7, 12, 13, 16\}$ is a difference set of residues modulo 19. Determine the parameters v, k, and λ.

7.16. Show that $\{0, 4, 5, 7\}$ is a difference set of residues modulo 13 which yields $PG(2, \mathbb{F}_3)$.

7.17. Prove the following generalization of Theorem 7.20. Let

$$\{d_{i1},\ldots,d_{ik}\}, \quad i = 1,\ldots,s,$$

be a system of (v, k, λ) difference sets. Then with all residues modulo v as varieties, the vs blocks

$$\{d_{i1} + t,\ldots,d_{ik} + t\}, \quad t = 0, 1,\ldots,v - 1 \text{ and } i = 1,\ldots,s,$$

form a $(v, k, \lambda s)$ block design.

7.18. Let $L^{(k)} = (a_{ij}^{(k)})$, where $a_{ij}^{(k)} \equiv i + jk \bmod 9$, $0 \leqslant a_{ij}^{(k)} < 9$ for $1 \leqslant i, j \leqslant 9$. Which of the arrays $L^{(k)}$, $k = 1, 2,\ldots,8$, are latin squares? Are $L^{(2)}$ and $L^{(5)}$ orthogonal?

7.19. A latin square of order n is said to be in *normalized* form if the first row and the first column are both the ordered set $\{1, 2,\ldots,n\}$. How many normalized latin squares of each order $n \leqslant 4$ are there?

7.20. Let L be a latin square of order m with entries in $\{1, 2,\ldots,m\}$ and M a latin square of order n with entries in $\{1, 2,\ldots,n\}$. From L and M construct a latin square of order mn with entries in $\{1, 2,\ldots,m\} \times \{1, 2,\ldots,n\}$.

7.21. Construct three mutually orthogonal latin squares of order 4.

7.22. Prove that for $n \geqslant 2$ there can be at most $n - 1$ mutually orthogonal latin squares of order n.

7.23. A *magic square* of order n consists of the integers 1 to n^2 arranged in an $n \times n$ array such that the sums of entries in rows, columns, and diagonals are all the same. Let $A = (a_{ij})$ and $B = (b_{ij})$ be two orthogonal latin squares of order n with entries in $\{0, 1,\ldots,n - 1\}$ such that the sum of entries in each of the diagonals of A and B is $n(n - 1)/2$. Show that $M = (na_{ij} + b_{ij} + 1)$ is a magic square of order n. Construct a magic square of order 4 from two orthogonal latin squares obtained in Exercise 7.21.

7.24. Determine Hadamard matrices of orders 8 and 12.

7.25. If H_m and H_n are Hadamard matrices, show that there exists a Hadamard matrix H_{mn}.

7.26. Show that from a normalized Hadamard matrix of order $4t, t \geqslant 2$, one can construct a symmetric $(4t - 1, 2t - 1, t - 1)$ block design.

7.27. Prove that the state graph of an LMS over $\mathbb{F}_q$ with nonsingular characteristic matrix consists of pure cycles only.

7.28. Prove that the state graphs of similar characteristic matrices over $\mathbb{F}_q$ are isomorphic. (*Note:* Two matrices A, B over $\mathbb{F}_q$ are *similar* if there exists a nonsingular matrix P over $\mathbb{F}_q$ such that $B = PAP^{-1}$.)

7.29. Suppose the characteristic matrix A of an LMS $\mathfrak{M}$ over $\mathbb{F}_2$ has the minimal polynomial $(x + 1)^5 (x^3 + x + 1)^3$. What are the state orders realizable by $\mathfrak{M}$?

7.30. Determine the orders of all states in the LMS $\mathfrak{M}$ of Example 7.41.

7.31. Suppose the characteristic matrix A of an LMS $\mathfrak{M}$ over $\mathbb{F}_q$ is *nonderogatory*; that is, its minimal polynomial is equal to its characteristic polynomial. Let the minimal polynomial of A be of the form $p(x)^e$, where $p(x)$ is a monic irreducible polynomial over $\mathbb{F}_q$ of degree d. Without using Theorem 7.40, prove that the cycle sum of $\mathfrak{M}$ is given by the expression in that theorem.

7.32. Calculate the cycle sum of the LMS $\mathfrak{M}$ over $\mathbb{F}_3$ given in Example 7.35.

7.33. Prove Theorem 7.42.

7.34. Let $s_0, s_1, \ldots$ be a kth-order maximal period sequence in $\mathbb{F}_q$ and let $N = d(q^k - 1)/(q - 1)$ for some positive integer d. If $Z_N(0)$ denotes the number of n, $0 \leqslant n \leqslant N - 1$, such that $s_n = 0$, prove that

$$Z_N(0) = \frac{d(q^{k-1} - 1)}{q - 1}.$$

7.35. Let $s_0, s_1, \ldots$ be a kth-order maximal period sequence in $\mathbb{F}_q$. For $1 \leqslant m \leqslant k$, $1 \leqslant N < r = q^k - 1$, and $\mathbf{b} = (b_1, \ldots, b_m) \in \mathbb{F}_q^m$ let $Z_N(\mathbf{b})$ be the number of n, $0 \leqslant n \leqslant N - 1$, such that $s_{n+i-1} = b_i$ for $1 \leqslant i \leqslant m$. Prove that

$$|Z_N(\mathbf{b}) - Nq^{-m}| \leqslant (1 - q^{-m})q^{k/2}\left(\frac{2}{\pi}\log r + \frac{2}{5} + \frac{N}{r}\right).$$

7.36. Let $s_0, s_1, \ldots$ be a periodic sequence of elements of $\mathbb{F}_q$ with least period r. For fixed $c \in \mathbb{F}_q$ we say that a *run* of c of length $m \geqslant 1$ occurs if $s_n \neq c$, $s_{n+i} = c$ for $1 \leqslant i \leqslant m$, and $s_{n+m+1} \neq c$ for some n with $0 \leqslant n \leqslant r - 1$. Prove that for a kth-order maximal period sequence in $\mathbb{F}_q$ with $r = q^k - 1 \geqslant 2$ exactly the following runs occur. For $1 \leqslant m \leqslant k - 2$ and any $c \in \mathbb{F}_q$ there are $(q - 1)^2 q^{k-m-2}$ runs of c of length m. The number of runs of c of length $k - 1$ is $q - 1$ for $c = 0$ and $q - 2$ for $c \neq 0$. There is no run of 0 of length k, and there is one run of c of length k for every $c \neq 0$. No runs of length $> k$ can occur.

7.37. Prove: If σ is a periodic sequence with period r, then the decimated sequence $\sigma_d^{(h)}$ has period $r/\gcd(d, r)$. Use a suitable decimation of the sequence σ in Example 7.50 to show that this result does not hold in general if "period" is replaced by "least period".

7.38. Prove the following converse of Theorem 7.48: any decimated sequence of a kth-order maximal period sequence in $\mathbb{F}_q$ is either the zero sequence or a linear recurring sequence in $\mathbb{F}_q$ having an irreducible minimal polynomial $g(x)$ with $g(0) \neq 0$ and $\deg(g(x))$ dividing k.

7.39. Let σ be a given kth-order maximal period sequence in $\mathbb{F}_q$. Prove that every kth-order maximal period sequence in $\mathbb{F}_q$ is equal to a shifted version of $\sigma_d^{(0)}$ for some d.

7.40. Let σ be a nonzero periodic sequence of elements of $\mathbb{F}_2$ with least period

r. Prove that if for every h with $1 \leqslant h \leqslant r - 1$ the sequence $\sigma + \sigma^{(h)}$ is a shifted version of σ, then σ is a maximal period sequence in $\mathbb{F}_2$.

7.41. Prove that for any maximal period sequence in $\mathbb{F}_q$ there exists a shifted version σ of the sequence such that $\sigma_q^{(0)} = \sigma$.

7.42. Let $s_0, s_1, \ldots$ be a kth-order maximal period sequence in the finite prime field $\mathbb{F}_p$, and view the terms s_n of the sequence as integers with $0 \leqslant s_n < p$. For positive integers m and d define

$$w_n = \sum_{i=1}^{m} s_{dn+i-1} p^{-i} \quad \text{for} \quad n = 0, 1, \ldots .$$

Prove that if $\gcd(d, p^k - 1) = 1$, then the sequence $w_0, w_1, \ldots$ is periodic with least period $p^k - 1$.

Chapter 8

Algebraic Coding Theory

One of the major applications of finite fields is coding theory. This theory has its origin in a famous theorem of Shannon that guarantees the existence of codes that can transmit information at rates close to the capacity of a communication channel with an arbitrarily small probability of error. One purpose of algebraic coding theory—the theory of error-correcting and error-detecting codes—is to devise methods for the construction of such codes.

During the last two decades more and more abstract algebraic tools such as the theory of finite fields and the theory of polynomials over finite fields have influenced coding. In particular, the description of redundant codes by polynomials over $\mathbb{F}_q$ is a milestone in this development. The fact that one can use shift registers for coding and decoding establishes a connection with linear recurring sequences. In our discussion of algebraic coding theory we do not consider any of the problems of the implementation or technical realization of the codes. We restrict ourselves to the study of basic properties of block codes and the description of some interesting classes of block codes.

Section 1 contains some background on algebraic coding theory and discusses the important class of linear codes in which encoding is performed by a linear transformation. A particularly interesting type of linear code is a cyclic code—that is, a linear code invariant under cyclic shifts. Our study of cyclic codes in Section 2 includes a description of what is possibly the most widely known family of codes, the BCH codes named after Bose, Ray-Chaudhuri, and Hocquenghem. BCH codes can be implemented easily and permit a fast decoding algorithm. The Goppa codes discussed in Section 3 can be viewed as generalized BCH codes. Goppa codes allow a much wider choice of parameters than BCH codes, but can still be decoded efficiently. If the decoding algorithm for Goppa codes is specialized to BCH codes, one obtains a second way of decoding BCH codes.

1. LINEAR CODES

The problem of the communication of information—in particular the coding and decoding of information for the reliable transmission over a "noisy" channel—is of great importance today. Typically, one has to transmit a message which consists of a finite string of symbols that are elements of some finite alphabet. For instance, if this alphabet consists simply of 0 and 1, the message can be described as a binary number. Generally the alphabet is assumed to be a finite field. Now the transmission of finite strings of elements of the alphabet over a communication channel need not be perfect in the sense that each bit of information is transmitted unaltered over this channel. As there is no ideal channel without "noise," the receiver of the transmitted message may obtain distorted information and may make errors in interpreting the transmitted signal.

One of the main problems of coding theory is to make the errors, which occur for instance because of noisy channels, extremely improbable. The methods to improve the reliability of transmission depend on properties of finite fields.

A basic idea in algebraic coding theory is to transmit *redundant* information together with the message one wants to communicate; that is, one extends the string of message symbols to a longer string in a systematic manner.

A simple model of a communication system is shown in Figure 8.1. We assume that the symbols of the message and of the coded message are elements of the same finite field $\mathbb{F}_q$. Coding means to encode a block of k message symbols $a_1 a_2 \cdots a_k$, $a_i \in \mathbb{F}_q$, into a *code word* $c_1 c_2 \cdots c_n$ of n symbols $c_j \in \mathbb{F}_q$, where $n > k$. We regard the code word as an n-dimensional row vector $\mathbf{c}$ in $\mathbb{F}_q^n$. Thus f in Figure 8.1 is a function from $\mathbb{F}_q^k$ into $\mathbb{F}_q^n$, called a *coding scheme*, and $g: \mathbb{F}_q^n \to \mathbb{F}_q^k$ is a *decoding scheme*.

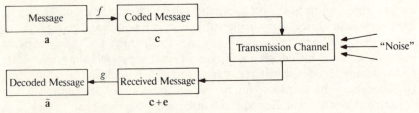

FIGURE 8.1 A communication system.

A simple type of coding scheme arises when each block $a_1 a_2 \cdots a_k$ of message symbols is encoded into a code word of the form

$$a_1 a_2 \cdots a_k c_{k+1} \cdots c_n,$$

where the first k symbols are the original *message symbols* and the additional $n-k$ symbols in $\mathbb{F}_q$ are *control symbols*. Such coding schemes are often presented in the following way. Let H be a given $(n-k) \times n$ matrix with entries in $\mathbb{F}_q$ that is of the special form

$$H = (A, I_{n-k}),$$

where A is an $(n-k) \times k$ matrix and I_{n-k} is the identity matrix of order $n-k$. The control symbols $c_{k+1}, \ldots, c_n$ can then be calculated from the system of equations

$$H\mathbf{c}^{\mathrm{T}} = \mathbf{0}$$

for code words $\mathbf{c}$. The equations of this system are called *parity-check equations*.

8.1. Example. Let H be the following 3×7 matrix over $\mathbb{F}_2$:

$$H = \begin{pmatrix} 1 & 0 & 1 & 1 & 1 & 0 & 0 \\ 1 & 1 & 0 & 1 & 0 & 1 & 0 \\ 1 & 1 & 1 & 0 & 0 & 0 & 1 \end{pmatrix}.$$

Then the control symbols can be calculated by solving $H\mathbf{c}^{\mathrm{T}} = \mathbf{0}$, given c_1, c_2, c_3, c_4:

$$
\begin{aligned}
c_1 \quad\quad + c_3 + c_4 + c_5 \quad\quad\quad &= 0 \\
c_1 + c_2 \quad\quad + c_4 \quad\quad + c_6 \quad &= 0 \\
c_1 + c_2 + c_3 \quad\quad\quad\quad + c_7 &= 0
\end{aligned}
$$

The control symbols c_5, c_6, c_7 can be expressed as

$$
\begin{aligned}
c_5 &= c_1 \quad\quad + c_3 + c_4 \\
c_6 &= c_1 + c_2 \quad\quad + c_4 \\
c_7 &= c_1 + c_2 + c_3
\end{aligned}
$$

Thus the coding scheme in this case is the linear map from $\mathbb{F}_2^4$ into $\mathbb{F}_2^7$ given by

$$(a_1, a_2, a_3, a_4) \mapsto (a_1, a_2, a_3, a_4, a_1 + a_3 + a_4, a_1 + a_2 + a_4, a_1 + a_2 + a_3).$$

$\square$

In general, we use the following terminology in connection with coding schemes that are given by linear maps.

8.2.   Definition.   Let H be an $(n-k)\times n$ matrix of rank $n-k$ with entries in $\mathbb{F}_q$. The set C of all n-dimensional vectors $\mathbf{c} \in \mathbb{F}_q^n$ such that $H\mathbf{c}^T = \mathbf{0}$ is called a *linear (n,k) code* over $\mathbb{F}_q$; n is called the *length* and k the *dimension* of the code. The elements of C are called *code words* (or *code vectors*), the matrix H is a *parity-check matrix* of C. If $q = 2$, C is called a *binary code*. If H is of the form (A, I_{n-k}), then C is called a *systematic code*.

We note that the set C of solutions of the system $H\mathbf{c}^T = \mathbf{0}$ of linear equations is a subspace of dimension k of the vector space $\mathbb{F}_q^n$. Since the code words form an additive group, C is also called a *group code*. Moreover, C can be regarded as the null space of the matrix H.

8.3.   Example (*Parity-Check Code*).   Let $q = 2$ and let the given message be $a_1 \cdots a_k$, then the coding scheme f is defined by

$$f: a_1 \cdots a_k \mapsto b_1 \cdots b_{k+1},$$

where $b_i = a_i$ for $i = 1, \ldots, k$ and

$$b_{k+1} = \sum_{i=1}^{k} a_i.$$

Hence it follows that the sum of digits of any code word $b_1 \cdots b_{k+1}$ is 0. If the sum of digits of the received word is 1, then the receiver knows that a transmission error must have occurred. Let $n = k + 1$, then this code is a binary linear $(n, n-1)$ code with parity-check matrix $H = (11 \cdots 1)$.   □

8.4.   Example (*Repetition Code*).   In a repetition code each code word consists of only one message symbol a_1 and $n-1$ control symbols $c_2 = \cdots = c_n$ all equal to a_1; that is, a_1 is repeated $n-1$ times. This is a linear $(n, 1)$ code with parity-check matrix $H = (-\mathbf{1}, I_{n-1})$.   □

The parity-check equations $H\mathbf{c}^T = \mathbf{0}$ with $H = (A, I_{n-k})$ imply

$$\mathbf{c}^T = \begin{pmatrix} I_k \\ -A \end{pmatrix} \mathbf{a}^T = \left[\mathbf{a}(I_k, -A^T) \right]^T,$$

where $\mathbf{a} = a_1 \cdots a_k$ is the message and $\mathbf{c} = c_1 \cdots c_n$ is the code word. This leads to the following definition.

8.5.   Definition.   The $k \times n$ matrix $G = (I_k, -A^T)$ is called the *canonical generator matrix* of a linear (n, k) code with parity-check matrix $H = (A, I_{n-k})$.

From $H\mathbf{c}^T = \mathbf{0}$ and $\mathbf{c} = \mathbf{a}G$ it follows that H and G are related by

$$GH^T = 0. \tag{8.1}$$

The code C is equal to the row space of the canonical generator matrix G. More generally, any $k \times n$ matrix G whose row space is equal to C is called a *generator matrix* of C. A generator matrix G of C can be used for encoding– namely, a message $\mathbf{a}$ is encoded by $\mathbf{c} = \mathbf{a}G \in C$.

8.6. Example. The canonical generator matrix for the code defined by H in Example 8.1 is given by

$$G = \begin{pmatrix} 1 & 0 & 0 & 0 & 1 & 1 & 1 \\ 0 & 1 & 0 & 0 & 0 & 1 & 1 \\ 0 & 0 & 1 & 0 & 1 & 0 & 1 \\ 0 & 0 & 0 & 1 & 1 & 1 & 0 \end{pmatrix}. \qquad \square$$

8.7. Definition. If $\mathbf{c}$ is a code word and $\mathbf{y}$ is the received word after communication through a "noisy" channel, then $\mathbf{e} = \mathbf{y} - \mathbf{c} = e_1 \cdots e_n$ is called the *error word* or the *error vector*.

8.8. Definition. Let $\mathbf{x}, \mathbf{y}$ be two vectors in $\mathbb{F}_q^n$. Then:

(i) the *Hamming distance* $d(\mathbf{x}, \mathbf{y})$ between $\mathbf{x}$ and $\mathbf{y}$ is the number of coordinates in which $\mathbf{x}$ and $\mathbf{y}$ differ;

(ii) the (*Hamming*) *weight* $w(\mathbf{x})$ of $\mathbf{x}$ is the number of nonzero coordinates of $\mathbf{x}$.

Thus $d(\mathbf{x}, \mathbf{y})$ gives the number of errors if $\mathbf{x}$ is the transmitted code word and $\mathbf{y}$ is the received word. It follows immediately that $w(\mathbf{x}) = d(\mathbf{x}, \mathbf{0})$ and $d(\mathbf{x}, \mathbf{y}) = w(\mathbf{x} - \mathbf{y})$. The proof of the following lemma is left as an exercise.

8.9. Lemma. *The Hamming distance is a metric on $\mathbb{F}_q^n$; that is, for all $\mathbf{x}, \mathbf{y}, \mathbf{z} \in \mathbb{F}_q^n$ we have*:

(i) $d(\mathbf{x}, \mathbf{y}) = 0$ *if and only if* $\mathbf{x} = \mathbf{y}$;

(ii) $d(\mathbf{x}, \mathbf{y}) = d(\mathbf{y}, \mathbf{x})$;

(iii) $d(\mathbf{x}, \mathbf{z}) \leqslant d(\mathbf{x}, \mathbf{y}) + d(\mathbf{y}, \mathbf{z})$.

In decoding received words $\mathbf{y}$, one usually tries to find the code word $\mathbf{c}$ such that $w(\mathbf{y} - \mathbf{c})$ is as small as possible, that is, one assumes that it is more likely that few errors have occurred rather than many. Thus in decoding we are looking for a code word $\mathbf{c}$ that is closest to $\mathbf{y}$ according to the Hamming distance. This rule is called *nearest neighbor decoding*.

8.10. Definition. For $t \in \mathbb{N}$ a code $C \subseteq \mathbb{F}_q^n$ is called *t-error-correcting* if for any $\mathbf{y} \in \mathbb{F}_q^n$ there is at most one $\mathbf{c} \in C$ such that $d(\mathbf{y}, \mathbf{c}) \leqslant t$.

If $\mathbf{c} \in C$ is transmitted and at most t errors occur, then we have $d(\mathbf{y}, \mathbf{c}) \leqslant t$ for the received word $\mathbf{y}$. If C is t-error-correcting, then for all other code words $\mathbf{z} \neq \mathbf{c}$ we have $d(\mathbf{y}, \mathbf{z}) > t$, which means that $\mathbf{c}$ is closest to $\mathbf{y}$ and nearest neighbor decoding gives the correct result. Therefore, one aim in coding theory is to construct codes with code words "far apart." On the other hand, one tries to transmit as much information as possible. To reconcile these two aims is one of the problems of coding.

8.11.   Definition. The number

$$d_C = \min_{\substack{\mathbf{u}, \mathbf{v} \in C \\ \mathbf{u} \neq \mathbf{v}}} d(\mathbf{u}, \mathbf{v}) = \min_{\mathbf{0} \neq \mathbf{c} \in C} w(\mathbf{c})$$

is called the *minimum distance* of the linear code C.

   8.12.   Theorem.   *A code C with minimum distance d_C can correct up to t errors if $d_C \geq 2t + 1$.*

   Proof. A ball $B_t(\mathbf{x})$ of radius t and center $\mathbf{x} \in \mathbb{F}_q^n$ consists of all vectors $\mathbf{y} \in \mathbb{F}_q^n$ such that $d(\mathbf{x}, \mathbf{y}) \leq t$. The nearest neighbor decoding rule ensures that each received word with t or fewer errors must be in a ball of radius t and center the transmitted code word. To correct t errors, the balls with code words $\mathbf{x}$ as centers must not overlap. If $\mathbf{u} \in B_t(\mathbf{x})$ and $\mathbf{u} \in B_t(\mathbf{y})$, $\mathbf{x}, \mathbf{y} \in C$, $\mathbf{x} \neq \mathbf{y}$, then

$$d(\mathbf{x}, \mathbf{y}) \leq d(\mathbf{x}, \mathbf{u}) + d(\mathbf{u}, \mathbf{y}) \leq 2t,$$

a contradiction to $d_C \geq 2t + 1$.   □

8.13.   Example. The code of Example 8.1 has minimum distance $d_C = 3$ and therefore can correct one error.   □

   The following lemma is often useful in determining the minimum distance of a code.

   8.14.   Lemma.   *A linear code C with parity-check matrix H has minimum distance $d_C \geq s + 1$ if and only if any s columns of H are linearly independent.*

   Proof. Assume there are s linearly dependent columns of H, then $H\mathbf{c}^T = \mathbf{0}$ and $w(\mathbf{c}) \leq s$ for suitable $\mathbf{c} \in C$, $\mathbf{c} \neq \mathbf{0}$, hence $d_C \leq s$. Similarly, if any s columns of H are linearly independent, then there is no $\mathbf{c} \in C$, $\mathbf{c} \neq \mathbf{0}$, of weight $\leq s$, hence $d_C \geq s + 1$.   □

   Next we describe a simple decoding algorithm for linear codes. Let C be a linear (n, k) code over $\mathbb{F}_q$. The vector space $\mathbb{F}_q^n / C$ consists of all cosets $\mathbf{a} + C = \{\mathbf{a} + \mathbf{c} : \mathbf{c} \in C\}$ with $\mathbf{a} \in \mathbb{F}_q^n$. Each coset contains q^k vectors and $\mathbb{F}_q^n$ can be regarded as being partitioned into cosets of C—namely,

$$\mathbb{F}_q^n = (\mathbf{a}^{(0)} + C) \cup (\mathbf{a}^{(1)} + C) \cup \cdots \cup (\mathbf{a}^{(s)} + C),$$

where $\mathbf{a}^{(0)} = \mathbf{0}$ and $s = q^{n-k} - 1$. A received vector $\mathbf{y}$ must be in one of the cosets, say in $\mathbf{a}^{(i)} + C$. If the code word $\mathbf{c}$ was transmitted, then the error is given by $\mathbf{e} = \mathbf{y} - \mathbf{c} = \mathbf{a}^{(i)} + \mathbf{z} \in \mathbf{a}^{(i)} + C$ for suitable $\mathbf{z} \in C$. This leads to the following decoding scheme.

8.15.   Decoding of Linear Codes.   All possible error vectors $\mathbf{e}$ of a received vector $\mathbf{y}$ are the vectors in the coset of $\mathbf{y}$. The most likely error vector is the vector $\mathbf{e}$ with minimum weight in the coset of $\mathbf{y}$. Thus we decode $\mathbf{y}$ as $\mathbf{x} = \mathbf{y} - \mathbf{e}$.

The implementation of this procedure can be facilitated by the *coset-leader algorithm* for error correction of linear codes.

8.16.   Definition.   Let $C \subseteq \mathbb{F}_q^n$ be a linear (n, k) code and let $\mathbb{F}_q^n / C$ be the factor space. An element of minimum weight in a coset $\mathbf{a} + C$ is called a *coset leader* of $\mathbf{a} + C$. If several vectors in $\mathbf{a} + C$ have minimum weight, we choose one of them as coset leader.

Let $\mathbf{a}^{(1)}, \ldots, \mathbf{a}^{(s)}$ be the coset leaders of the cosets $\neq C$ and let $\mathbf{c}^{(1)} = \mathbf{0}$, $\mathbf{c}^{(2)}, \ldots, \mathbf{c}^{(q^k)}$ be all code words in C. Consider the following array:

$$
\begin{array}{cccc}
\mathbf{c}^{(1)} & \mathbf{c}^{(2)} & \cdots & \mathbf{c}^{(q^k)} \\
\mathbf{a}^{(1)} + \mathbf{c}^{(1)} & \mathbf{a}^{(1)} + \mathbf{c}^{(2)} & \cdots & \mathbf{a}^{(1)} + \mathbf{c}^{(q^k)} \\
\vdots & \vdots & & \vdots \\
\mathbf{a}^{(s)} + \mathbf{c}^{(1)} & \mathbf{a}^{(s)} + \mathbf{c}^{(2)} & \cdots & \mathbf{a}^{(s)} + \mathbf{c}^{(q^k)}
\end{array}
$$

$\}$ row of code words

$\}$ remaining cosets

column of
coset leaders

If a word $\mathbf{y} = \mathbf{a}^{(i)} + \mathbf{c}^{(j)}$ is received, then the decoder decides that the error $\mathbf{e}$ is the corresponding coset leader $\mathbf{a}^{(i)}$ and decodes $\mathbf{y}$ as the code word $\mathbf{x} = \mathbf{y} - \mathbf{e} = \mathbf{c}^{(j)}$; that is, $\mathbf{y}$ is decoded as the code word in the column of $\mathbf{y}$. The coset of $\mathbf{y}$ can be determined by evaluating the so-called syndrome of $\mathbf{y}$.

8.17.   Definition.   Let H be the parity-check matrix of a linear (n, k) code C. Then the vector $S(\mathbf{y}) = H\mathbf{y}^T$ of length $n - k$ is called the *syndrome* of $\mathbf{y}$.

8.18.   Theorem.   *For* $\mathbf{y}, \mathbf{z} \in \mathbb{F}_q^n$ *we have*:

(i) $S(\mathbf{y}) = \mathbf{0}$ *if and only if* $\mathbf{y} \in C$;
(ii) $S(\mathbf{y}) = S(\mathbf{z})$ *if and only if* $\mathbf{y} + C = \mathbf{z} + C$.

Proof.   (i) follows immediately from the definition of C in terms of H. For (ii) note that $S(\mathbf{y}) = S(\mathbf{z})$ if and only if $H\mathbf{y}^T = H\mathbf{z}^T$ if and only if $H(\mathbf{y} - \mathbf{z})^T = \mathbf{0}$ if and only if $\mathbf{y} - \mathbf{z} \in C$ if and only if $\mathbf{y} + C = \mathbf{z} + C$. $\square$

If $\mathbf{e} = \mathbf{y} - \mathbf{c}$, $\mathbf{c} \in C$, $\mathbf{y} \in \mathbb{F}_q^n$, then

$$S(\mathbf{y}) = S(\mathbf{c} + \mathbf{e}) = S(\mathbf{c}) + S(\mathbf{e}) = S(\mathbf{e}) \tag{8.2}$$

and $\mathbf{y}$ and $\mathbf{e}$ are in the same coset. The coset leader of that coset also has the same syndrome. We have the following decoding algorithm.

8.19.   Coset-Leader Algorithm.   Let $C \subseteq \mathbb{F}_q^n$ be a linear (n, k) code and let y be the received vector. To correct errors in y, calculate $S(y)$ and find the coset leader, say e, with syndrome equal to $S(y)$. Then decode y as $x = y - e$. Here x is the code word with minimum distance to y.

8.20.   Example. Let C be a binary linear $(4, 2)$ code with generator matrix G and parity-check matrix H:

$$G = \begin{pmatrix} 1 & 0 & 1 & 0 \\ 0 & 1 & 1 & 1 \end{pmatrix}, \qquad H = \begin{pmatrix} 1 & 1 & 1 & 0 \\ 0 & 1 & 0 & 1 \end{pmatrix}.$$

The corresponding array of cosets is:

message row	00	10	01	11	
code words	0000	1010	0111	1101	$\begin{pmatrix} 0 \\ 0 \end{pmatrix}$
	1000	0010	1111	0101	$\begin{pmatrix} 1 \\ 0 \end{pmatrix}$
other cosets	0100	1110	0011	1001	$\begin{pmatrix} 1 \\ 1 \end{pmatrix}$
	0001	1011	0110	1100	$\begin{pmatrix} 0 \\ 1 \end{pmatrix}$

$\underbrace{\qquad}_{\text{coset leaders}}$ $\qquad\qquad$ syndromes

If $y = 1110$ is received, we could look where in the array y occurs. But for large arrays this is very time consuming. Therefore we find $S(y)$ first—namely, $S(y) = Hy^T = \begin{pmatrix} 1 \\ 1 \end{pmatrix}$—and decide that the error is equal to the coset leader 0100 that also has syndrome $\begin{pmatrix} 1 \\ 1 \end{pmatrix}$. The original code word was most likely the word 1010 and the original message was 10.   □

In large linear codes it is practically impossible to find coset leaders with minimum weight; for example, a linear $(50, 20)$ code over $\mathbb{F}_2$ has some 10^9 cosets. Therefore it is necessary to construct special codes in order to overcome such difficulties. First we note the following.

8.21.   Theorem. *In a binary linear (n, k) code with parity-check matrix H the syndrome is the sum of those columns of H that correspond to positions where errors have occurred.*

Proof. Let $y \in \mathbb{F}_2^n$ be the received vector, $y = x + e$, $x \in C$; then from (8.2) we have $S(y) = He^T$. Let $i_1, i_2, \ldots$ be the error coordinates in e, say $e = 0 \cdots 01_{i_1} 0 \cdots 01_{i_2} 0 \cdots$, then $S(y) = h_{i_1} + h_{i_2} + \cdots$, where h_i denotes the ith column of H.   □

If all columns of H are different, then a single error in the ith position of the transmitted word yields $S(\mathbf{y}) = \mathbf{h}_i$, thus one error can be corrected. To simplify the process of error location, the following class of codes is useful.

8.22. Definition. A binary code C_m of length $n = 2^m - 1$, $m \geqslant 2$, with an $m \times (2^m - 1)$ parity-check matrix H is called a *binary Hamming code* if the columns of H are the binary representations of the integers $1, 2, \ldots, 2^m - 1$.

8.23. Lemma. *C_m is a 1-error-correcting code of dimension $2^m - m - 1$.*

Proof. By definition of the parity-check matrix H of C_m, the rank of H is m. Also, any two columns of H are linearly independent. Since H contains with any two of its columns also their sum, the minimum distance of C_m equals 3 by Lemma 8.14. Thus C_m is 1-error-correcting by Theorem 8.12.   □

8.24. Example. Let C_3 be the $(7, 4)$ Hamming code with parity-check matrix

$$H = \begin{pmatrix} 0 & 0 & 0 & 1 & 1 & 1 & 1 \\ 0 & 1 & 1 & 0 & 0 & 1 & 1 \\ 1 & 0 & 1 & 0 & 1 & 0 & 1 \end{pmatrix}.$$

If the syndrome of a received word $\mathbf{y}$ is, say, $S(\mathbf{y}) = (1 \quad 0 \quad 1)^T$, then we know that an error must have occurred in the fifth position, since 101 is the binary representation of 5.   □

Hamming codes can also be defined in the nonbinary case—that is, over arbitrary finite fields $\mathbb{F}_q$. Here the parity-check matrix H is an $m \times (q^m - 1)/(q - 1)$ matrix that has pairwise linearly independent columns. Such a matrix defines a linear $((q^m - 1)/(q - 1), (q^m - 1)/(q - 1) - m)$ code of minimum distance 3.

Next we describe some relationships between the length n of code words, the number k of information or message symbols, and the minimum distance d_C of a linear code over $\mathbb{F}_q$.

8.25. Theorem (Hamming Bound). *Let C be a t-error-correcting code over $\mathbb{F}_q$ of length n with M code words. Then*

$$M\left(1 + \binom{n}{1}(q-1) + \cdots + \binom{n}{t}(q-1)^t\right) \leqslant q^n.$$

Proof. There are $\binom{n}{m}(q-1)^m$ vectors with n coordinates in $\mathbb{F}_q$ of weight m. The balls of radius t centered at the code words are all pairwise disjoint and each of the M balls contains

$$1 + \binom{n}{1}(q-1) + \cdots + \binom{n}{t}(q-1)^t$$

vectors of all the q^n vectors in $\mathbb{F}_q^n$.   □

8.26. Theorem (Plotkin Bound). *For a linear (n, k) code C over $\mathbb{F}_q$ of minimum distance d_C we have*

$$d_C \leqslant \frac{nq^{k-1}(q-1)}{q^k - 1}.$$

Proof. Let $1 \leqslant i \leqslant n$ be such that C contains a code word with nonzero ith component. Let D be the subspace of C consisting of all code words with ith component zero. In C/D there are q elements which correspond to q choices for the ith component of a code word. Thus $|C|/|D| = |C/D|$ implies $|D| = q^{k-1}$. By counting along the components, the sum of the weights of the code words in C is then seen to be $\leqslant nq^{k-1}(q-1)$. The minimum distance d_C of the code is the minimum nonzero weight and therefore must satisfy the inequality given in the theorem since the total number of code words of nonzero weight is $q^k - 1$. $\square$

8.27. Theorem (Gilbert-Varshamov Bound). *There exists a linear (n, k) code over $\mathbb{F}_q$ with minimum distance $\geqslant d$ whenever*

$$q^{n-k} > \sum_{i=0}^{d-2} \binom{n-1}{i}(q-1)^i.$$

Proof. We prove this theorem by constructing an $(n-k) \times n$ parity-check matrix H for such a code. We choose the first column of H as any nonzero $(n-k)$-tuple over $\mathbb{F}_q$. The second column is any $(n-k)$-tuple over $\mathbb{F}_q$ that is not a scalar multiple of the first column. In general, suppose $j-1$ columns have been chosen so that any $d-1$ of them are linearly independent. There are at most

$$\sum_{i=0}^{d-2} \binom{j-1}{i}(q-1)^i$$

vectors obtained by linear combinations of $d-2$ or fewer of these $j-1$ columns. If the inequality of the theorem holds, then it will be possible to choose a jth column that is linearly independent of any $d-2$ of the first $j-1$ columns. The construction can be carried out in such a way that H has rank $n-k$. The resulting code has minimum distance $\geqslant d$ by Lemma 8.14. $\square$

We define the dual code of a given linear code C by means of the following concepts. Let $\mathbf{u} = (u_1, \ldots, u_n)$, $\mathbf{v} = (v_1, \ldots, v_n) \in \mathbb{F}_q^n$, then $\mathbf{u} \cdot \mathbf{v} = u_1 v_1 + \cdots + u_n v_n$ denotes the *dot product* of $\mathbf{u}$ and $\mathbf{v}$. If $\mathbf{u} \cdot \mathbf{v} = 0$, then $\mathbf{u}$ and $\mathbf{v}$ are called *orthogonal*.

8.28. Definition. Let C be a linear (n, k) code over $\mathbb{F}_q$. Then its *dual* (or *orthogonal*) *code* $C^\perp$ is defined as

$$C^\perp = \{\mathbf{u} \in \mathbb{F}_q^n : \mathbf{u} \cdot \mathbf{v} = 0 \quad \text{for all } \mathbf{v} \in C\}.$$

The code C is a k-dimensional subspace of $\mathbb{F}_q^n$, the dimension of $C^\perp$

is $n - k$. $C^\perp$ is a linear $(n, n - k)$ code. It is easy to show that $C^\perp$ has generator matrix H if C has parity-check matrix H and that $C^\perp$ has parity-check matrix G if C has generator matrix G.

Considerable information on a code is obtained from the weight enumeration. For instance, to determine decoding error probabilities or in certain decoding algorithms it is important to know the distribution of the weights of code words. There is a fundamental connection between the weight distribution of a linear code and of its dual code. This will be derived in the following theorem.

8.29. Definition. Let A_i denote the number of code words $\mathbf{c} \in C$ of weight i, $0 \leqslant i \leqslant n$. Then the polynomial

$$A(x, y) = \sum_{i=0}^{n} A_i x^i y^{n-i}$$

in the indeterminates x and y over the complex numbers is called the *weight enumerator* of C.

We shall need characters of finite fields, as discussed in Chapter 5.

8.30. Definition. Let χ be a nontrivial additive character of $\mathbb{F}_q$ and let $\mathbf{v} \cdot \mathbf{u}$ denote the dot product of $\mathbf{v}, \mathbf{u} \in \mathbb{F}_q^n$. We define for fixed $\mathbf{v} \in \mathbb{F}_q^n$ the mapping $\chi_{\mathbf{v}} : \mathbb{F}_q^n \to \mathbb{C}$ by

$$\chi_{\mathbf{v}}(\mathbf{u}) = \chi(\mathbf{v} \cdot \mathbf{u}) \quad \text{for } \mathbf{u} \in \mathbb{F}_q^n.$$

If V is a vector space over $\mathbb{C}$ and f a mapping from $\mathbb{F}_q^n$ into V, then we define $g_f : \mathbb{F}_q^n \to V$ by

$$g_f(\mathbf{u}) = \sum_{\mathbf{v} \in \mathbb{F}_q^n} \cdot \chi_{\mathbf{v}}(\mathbf{u}) f(\mathbf{v}) \quad \text{for } \mathbf{u} \in \mathbb{F}_q^n.$$

8.31. Lemma. *Let E be a subspace of $\mathbb{F}_q^n$, $E^\perp$ its orthogonal complement, $f : \mathbb{F}_q^n \to V$ a mapping from $\mathbb{F}_q^n$ into a vector space V over $\mathbb{C}$ and χ a nontrivial additive character of $\mathbb{F}_q$. Then*

$$\sum_{\mathbf{u} \in E} g_f(\mathbf{u}) = |E| \sum_{\mathbf{v} \in E^\perp} f(\mathbf{v}).$$

Proof.

$$\sum_{\mathbf{u} \in E} g_f(\mathbf{u}) = \sum_{\mathbf{u} \in E} \sum_{\mathbf{v} \in \mathbb{F}_q^n} \chi_{\mathbf{v}}(\mathbf{u}) f(\mathbf{v}) = \sum_{\mathbf{v} \in \mathbb{F}_q^n} \sum_{\mathbf{u} \in E} \chi(\mathbf{v} \cdot \mathbf{u}) f(\mathbf{v})$$

$$= |E| \sum_{\mathbf{v} \in E^\perp} f(\mathbf{v}) + \sum_{\mathbf{v} \notin E^\perp} \sum_{c \in \mathbb{F}_q} \sum_{\substack{\mathbf{u} \in E \\ \mathbf{v} \cdot \mathbf{u} = c}} \chi(c) f(\mathbf{v}).$$

For fixed $\mathbf{v} \notin E^\perp$, $\mathbf{u} \in E \mapsto \mathbf{v} \cdot \mathbf{u}$ is a nontrivial linear functional on E, thus

$$\sum_{\mathbf{u} \in E} g_f(\mathbf{u}) = |E| \sum_{\mathbf{v} \in E^\perp} f(\mathbf{v}) + \frac{|E|}{q} \sum_{\mathbf{v} \notin E^\perp} f(\mathbf{v}) \sum_{c \in \mathbb{F}_q} \chi(c) = |E| \sum_{\mathbf{v} \in E^\perp} f(\mathbf{v}),$$

by using (5.9). $\qquad\square$

We apply this lemma with V as the space of polynomials in two indeterminates x and y over $\mathbb{C}$ and the mapping f defined as $f(\mathbf{v}) = x^{w(\mathbf{v})}y^{n-w(\mathbf{v})}$, where $w(\mathbf{v})$ denotes the weight of $\mathbf{v} \in \mathbb{F}_q^n$.

8.32. Theorem (MacWilliams Identity). *Let C be a linear (n,k) code over $\mathbb{F}_q$ and $C^\perp$ its dual code. If $A(x,y)$ is the weight enumerator of C and $A^\perp(x,y)$ is the weight enumerator of $C^\perp$, then*

$$A^\perp(x,y) = q^{-k}A(y-x, y+(q-1)x).$$

Proof. Let $f: \mathbb{F}_q^n \to \mathbb{C}[x,y]$ be as given above, then the weight enumerator of $C^\perp$ is

$$A^\perp(x,y) = \sum_{\mathbf{v} \in C^\perp} f(\mathbf{v}).$$

Let g_f be as in Definition 8.30 and for $v \in \mathbb{F}_q$ define

$$|v| = \begin{cases} 1 & \text{if } v \neq 0, \\ 0 & \text{if } v = 0. \end{cases}$$

For $\mathbf{u} = (u_1, \dots, u_n) \in \mathbb{F}_q^n$ we have

$$g_f(\mathbf{u}) = \sum_{\mathbf{v} \in \mathbb{F}_q^n} \chi(\mathbf{v} \cdot \mathbf{u}) x^{w(\mathbf{v})} y^{n-w(\mathbf{v})}$$

$$= \sum_{v_1, \dots, v_n \in \mathbb{F}_q} \chi(u_1 v_1 + \cdots + u_n v_n) x^{|v_1| + \cdots + |v_n|} y^{(1-|v_1|) + \cdots + (1-|v_n|)}$$

$$= \sum_{v_1, \dots, v_n \in \mathbb{F}_q} \prod_{i=1}^{n} \left[\chi(u_i v_i) x^{|v_i|} y^{1-|v_i|} \right]$$

$$= \prod_{i=1}^{n} \sum_{v \in \mathbb{F}_q} \left[\chi(u_i v) x^{|v|} y^{1-|v|} \right].$$

For $u_i = 0$ we have $\chi(u_i v) = \chi(0) = 1$, hence the corresponding factor in the product is $(q-1)x + y$. For $u_i \neq 0$ the corresponding factor is

$$y + x \sum_{v \in \mathbb{F}_q^*} \chi(v) = y - x.$$

Therefore,

$$g_f(\mathbf{u}) = (y-x)^{w(\mathbf{u})} (y+(q-1)x)^{n-w(\mathbf{u})}.$$

Lemma 8.31 implies

$$|C|A^\perp(x,y) = |C| \sum_{\mathbf{v} \in C^\perp} f(\mathbf{v}) = \sum_{\mathbf{u} \in C} g_f(\mathbf{u}) = A(y-x, y+(q-1)x).$$

Finally, $|C| = q^k$ by hypothesis. $\square$

8.33. Corollary. *Let $x = z$ and $y = 1$ in the weight enumerators $A(x,y)$ and $A^\perp(x,y)$ and denote the resulting polynomials by $A(z)$ and*

$A^\perp(z)$, respectively. *Then the MacWilliams identity can be written in the form*

$$A^\perp(z) = q^{-k}(1+(q-1)z)^n A\left(\frac{1-z}{1+(q-1)z}\right).$$

8.34.   Example.   Let C_m be the binary Hamming code of length $n = 2^m - 1$ and dimension $n - m$ over $\mathbb{F}_2$. The dual code $C_m^\perp$ has as its generator matrix the parity-check matrix H of C_m, which consists of all nonzero column vectors of length m over $\mathbb{F}_2$. $C_m^\perp$ consists of the zero vector and $2^m - 1$ vectors of weight 2^{m-1}. Thus the weight enumerator of $C_m^\perp$ is

$$y^n + (2^m - 1)x^{2^{m-1}}y^{2^{m-1}-1}.$$

By Theorem 8.32 the weight enumerator for C_m is given by

$$A(x, y) = \frac{1}{n+1}\left[(y+x)^n + n(y-x)^{(n+1)/2}(y+x)^{(n-1)/2}\right].$$

Let $A(z) = A(z,1)$—that is, $A(z) = \sum_{i=0}^n A_i z^i$—then one can verify that $A(z)$ satisfies the differential equation

$$(1-z^2)\frac{dA(z)}{dz} + (1+nz)A(z) = (1+z)^n$$

with initial condition $A(0) = A_0 = 1$. This is equivalent to

$$iA_i = \binom{n}{i-1} - A_{i-1} - (n-i+2)A_{i-2} \quad \text{for } i = 2,3,\ldots,n$$

with initial conditions $A_0 = 1$, $A_1 = 0$.   □

2.   CYCLIC CODES

Cyclic codes are a special class of linear codes that can be implemented fairly simply and whose mathematical structure is reasonably well known.

8.35.   Definition.   A linear (n, k) code C over $\mathbb{F}_q$ is called *cyclic* if $(a_0, a_1, \ldots, a_{n-1}) \in C$ implies $(a_{n-1}, a_0, \ldots, a_{n-2}) \in C$.

From now on we impose the restriction $\gcd(n, q) = 1$ and let $(x^n - 1)$ be the ideal generated by $x^n - 1 \in \mathbb{F}_q[x]$. Then all elements of $\mathbb{F}_q[x]/(x^n - 1)$ can be represented by polynomials of degree less than n and clearly this residue class ring is isomorphic to $\mathbb{F}_q^n$ as a vector space over $\mathbb{F}_q$. An isomorphism is given by

$$(a_0, a_1, \ldots, a_{n-1}) \leftrightarrow a_0 + a_1 x + \cdots + a_{n-1}x^{n-1}.$$

Because of this isomorphism, we denote the elements of $\mathbb{F}_q[x]/(x^n - 1)$ either as polynomials of degree $< n$ modulo $x^n - 1$ or as vectors or words over $\mathbb{F}_q$. We introduce multiplication of polynomials modulo $x^n - 1$ in the usual way; that is, if $f \in \mathbb{F}_q[x]/(x^n - 1)$, $g_1, g_2 \in \mathbb{F}_q[x]$, then $g_1 g_2 = f$ means that $g_1 g_2 \equiv f \bmod(x^n - 1)$.

A cyclic (n, k) code C can be obtained by multiplying each message of k coordinates (identified with a polynomial of degree $< k$) by a fixed polynomial $g(x)$ of degree $n - k$ with $g(x)$ a divisor of $x^n - 1$. The polynomials $g(x), xg(x), \ldots, x^{k-1}g(x)$ correspond to code words of C. A generator matrix of C is given by

$$
G = \begin{pmatrix}
g_0 & g_1 & \cdots & g_{n-k} & 0 & 0 & \cdots & 0 \\
0 & g_0 & g_1 & \cdots & & g_{n-k} & 0 & \cdots & 0 \\
\vdots & \vdots & \vdots & & & & & & \vdots \\
0 & 0 & 0 & \cdots & 0 & g_0 & g_1 & \cdots & g_{n-k}
\end{pmatrix},
$$

where $g(x) = g_0 + g_1 x + \cdots + g_{n-k} x^{n-k}$. The rows of G are obviously linearly independent and rank $(G) = k$, the dimension of C. If

$$
h(x) = (x^n - 1)/g(x) = h_0 + h_1 x + \cdots + h_k x^k,
$$

then we see that the matrix

$$
H = \begin{pmatrix}
0 & 0 & \cdots & 0 & h_k & h_{k-1} & \cdots & & h_0 \\
0 & 0 & \cdots & 0 & h_k & h_{k-1} & & \cdots & h_0 & 0 \\
\vdots & \vdots & & & & & & & \vdots \\
h_k & h_{k-1} & \cdots & & h_0 & 0 & & \cdots & & 0
\end{pmatrix}
$$

is a parity-check matrix for C. The code with generator matrix H is the dual code of C, which is again cyclic.

Since we are using the terminologies of vectors $(a_0, a_1, \ldots, a_{n-1})$ and polynomials $a_0 + a_1 x + \cdots + a_{n-1} x^{n-1}$ over $\mathbb{F}_q$ synonymously, we can interpret C as a subset of the factor ring $\mathbb{F}_q[x]/(x^n - 1)$.

8.36. Theorem. *The linear code C is cyclic if and only if C is an ideal of $\mathbb{F}_q[x]/(x^n - 1)$.*

Proof. If C is an ideal and $(a_0, a_1, \ldots, a_{n-1}) \in C$, then also

$$
x(a_0 + a_1 x + \cdots + a_{n-1} x^{n-1}) = (a_{n-1}, a_0, \ldots, a_{n-2}) \in C.
$$

Conversely, if $(a_0, a_1, \ldots, a_{n-1}) \in C$ implies $(a_{n-1}, a_0, \ldots, a_{n-2}) \in C$, then for every $a(x) \in C$ we have $xa(x) \in C$, hence also $x^2 a(x) \in C$, $x^3 a(x) \in C$, and so on. Therefore also $b(x)a(x) \in C$ for any polynomial $b(x)$; that is, C is an ideal.   □

Every ideal of $\mathbb{F}_q[x]/(x^n - 1)$ is principal; in particular, every non-zero ideal C is generated by the monic polynomial of lowest degree in the ideal, say $g(x)$, where $g(x)$ divides $x^n - 1$.

8.37.   Definition.   Let $C = (g(x))$ be a cyclic code. Then $g(x)$ is called the *generator polynomial* of C and $h(x) = (x^n - 1)/g(x)$ is called the *parity-check polynomial* of C.

Let $x^n - 1 = f_1(x)f_2(x)\cdots f_m(x)$ be the decomposition of $x^n - 1$ into monic irreducible factors over $\mathbb{F}_q$. Since we assume $\gcd(n, q) = 1$, there are no multiple factors. If $f_i(x)$ is irreducible over $\mathbb{F}_q$, then $(f_i(x))$ is a maximal ideal and the cyclic code generated by $f_i(x)$ is called a *maximal cyclic code*. The code generated by $(x^n - 1)/f_i(x)$ is called an *irreducible cyclic code*. We can find all cyclic codes of length n over $\mathbb{F}_q$ by factoring $x^n - 1$ as above and taking any of the $2^m - 2$ nontrivial monic factors of $x^n - 1$ as a generator polynomial.

If $h(x)$ is the parity-check polynomial of a cyclic code $C \subseteq \mathbb{F}_q[x]/(x^n - 1)$ and $v(x) \in \mathbb{F}_q[x]/(x^n - 1)$, then $v(x) \in C$ if and only if $v(x)h(x) \equiv 0 \bmod(x^n - 1)$. A message polynomial $a(x) = a_0 + a_1 x + \cdots + a_{k-1}x^{k-1}$ is encoded by C into $w(x) = a(x)g(x)$, where $g(x)$ is the generator polynomial of C. If we divide the received polynomial $v(x)$ by $g(x)$, and if there is a nonzero remainder, we know that an error occurs. The canonical generator matrix of C can be obtained as follows. Let $\deg((g(x)) = n - k$. Then there are unique polynomials $a_j(x)$ and $r_j(x)$ with $\deg(r_j(x)) < n - k$ such that

$$x^j = a_j(x)g(x) + r_j(x).$$

Consequently, $x^j - r_j(x)$ is a code polynomial, and so is $g_j(x) = x^k(x^j - r_j(x))$ considered modulo $x^n - 1$. The polynomials $g_j(x)$, $j = n - k, \ldots, n - 1$, are linearly independent and form the canonical generator matrix

$$(I_k, -R),$$

where I_k is the $k \times k$ identity matrix and R is the $k \times (n - k)$ matrix whose ith row is the vector of coefficients of $r_{n-k-1+i}(x)$.

8.38.   Example.   Let $n = 7$, $q = 2$. Then

$$x^7 - 1 = (x + 1)(x^3 + x + 1)(x^3 + x^2 + 1).$$

Thus $g(x) = x^3 + x^2 + 1$ generates a cyclic $(7,4)$ code with parity-check polynomial $h(x) = x^4 + x^3 + x^2 + 1$. The corresponding canonical generator matrix and parity-check matrix is, respectively,

$$G = \begin{pmatrix} 1 & 0 & 0 & 0 & 1 & 0 & 1 \\ 0 & 1 & 0 & 0 & 1 & 1 & 1 \\ 0 & 0 & 1 & 0 & 1 & 1 & 0 \\ 0 & 0 & 0 & 1 & 0 & 1 & 1 \end{pmatrix},$$

$$H = \begin{pmatrix} 1 & 1 & 1 & 0 & 1 & 0 & 0 \\ 0 & 1 & 1 & 1 & 0 & 1 & 0 \\ 1 & 1 & 0 & 1 & 0 & 0 & 1 \end{pmatrix}. \qquad \square$$

We recall from Chapter 6 that if $f \in \mathbb{F}_q[x]$ is a polynomial of the form

$$f(x) = f_0 + f_1 x + \cdots + f_k x^k, \quad f_0 \neq 0, f_k = 1,$$

then the solutions of the linear recurrence relation

$$\sum_{j=0}^{k} f_j a_{i+j} = 0, \quad i = 0, 1, \ldots,$$

are periodic of period n. The set of the n-tuples of the first n terms of each possible solution, considered as polynomials modulo $x^n - 1$, is the ideal generated by $g(x)$ in $\mathbb{F}_q[x]/(x^n - 1)$, where $g(x)$ is the reciprocal polynomial of $(x^n - 1)/f(x)$ of degree $n - k$. Thus *linear recurrence relations can be used to generate code words of cyclic codes*, and this generation process can be implemented on feedback shift registers.

8.39. Example. Let $f(x) = x^3 + x + 1$, a factor of $x^7 - 1$ over $\mathbb{F}_2$. The associated linear recurrence relation is $a_{i+3} + a_{i+1} + a_i = 0$, which gives rise to a $(7,3)$ cyclic code, which encodes $1\,1\,1$, say, as $1\,1\,1\,0\,0\,1\,0$. The generator polynomial is the reciprocal polynomial of $(x^7 - 1)/f(x)$; that is, $g(x) = x^4 + x^3 + x^2 + 1$.   □

Cyclic codes can also be described by prescribing certain roots of all code polynomials in a suitable extension field of $\mathbb{F}_q$. The requirement that all code polynomials are multiples of $g(x)$, a generator polynomial, simply means that they are all 0 at the roots of $g(x)$. Let $\alpha_1, \ldots, \alpha_s$ be elements of a finite extension field of $\mathbb{F}_q$ and $p_i(x)$ be the minimal polynomial of α_i over $\mathbb{F}_q$ for $i = 1, 2, \ldots, s$. Let $n \in \mathbb{N}$ be such that $\alpha_i^n = 1$, $i = 1, 2, \ldots, s$, and define $g(x) = \mathrm{lcm}(p_1(x), \ldots, p_s(x))$. Thus $g(x)$ divides $x^n - 1$. If $C \subseteq \mathbb{F}_q^n$ is the cyclic code with generator polynomial $g(x)$, then we have $v(x) \in C$ if and only if $v(\alpha_i) = 0$, $i = 1, 2, \ldots, s$. As an example of the concurrence of the description of a cyclic code by a generator polynomial or by roots of code polynomials we prove the following result, which uses the concept of equivalence of codes in Exercise 8.10.

8.40. Theorem. *The binary cyclic code of length $n = 2^m - 1$ for which the generator polynomial is the minimal polynomial over $\mathbb{F}_2$ of a primitive element of $\mathbb{F}_{2^m}$ is equivalent to the binary $(n, n - m)$ Hamming code.*

Proof. Let α denote a primitive element of $\mathbb{F}_{2^m}$ and let

$$p(x) = (x - \alpha)(x - \alpha^2) \cdots (x - \alpha^{2^{m-1}})$$

be the minimal polynomial of α over $\mathbb{F}_2$. We now consider the cyclic code C generated by $p(x)$. We construct an $m \times (2^m - 1)$ matrix H for which the jth column is $(c_0, c_1, \ldots, c_{m-1})^T$ if

$$\alpha^{j-1} = \sum_{i=0}^{m-1} c_i \alpha^i, \quad j = 1, 2, \ldots, 2^m - 1,$$

where $c_i \in \mathbb{F}_2$. If $\mathbf{a} = (a_0, a_1, \ldots, a_{n-1})$ and $a(x) = a_0 + a_1 x + \cdots + a_{n-1} x^{n-1} \in \mathbb{F}_2[x]$, then the vector $H\mathbf{a}^T$ corresponds to the element $a(\alpha)$ expressed in the basis $\{1, \alpha, \ldots, \alpha^{m-1}\}$. Consequently, $H\mathbf{a}^T = \mathbf{0}$ holds exactly when $p(x)$ divides $a(x)$, so H is a parity-check matrix of C. Since the columns of H are a permutation of the binary representations of the numbers $1, 2, \ldots, 2^m - 1$, the proof is complete. $\square$

8.41.   Example.   The polynomial $x^4 + x + 1$ is primitive over $\mathbb{F}_2$ and thus has a primitive element α of $\mathbb{F}_{16}$ as a root. If we use vector notation for the 15 elements $\alpha^j \in \mathbb{F}_{16}^*, j = 0, 1, \ldots, 14$, expressed in the basis $\{1, \alpha, \alpha^2, \alpha^3\}$ and we form a 4×15 matrix with these vectors as columns, then we get the parity-check matrix of a code equivalent to the $(15, 11)$ Hamming code. A message $(a_0, a_1, \ldots, a_{10})$ is encoded into a code polynomial

$$w(x) = a(x)(x^4 + x + 1),$$

where $a(x) = a_0 + a_1 x + \cdots + a_{10} x^{10}$. Now suppose the received polynomial contains one error; that is, $w(x) + x^{e-1}$ is received when $w(x)$ is transmitted. Then the syndrome is $w(\alpha) + \alpha^{e-1} = \alpha^{e-1}$ and the decoder is led to the conclusion that there is an error in the eth position. $\square$

8.42.   Theorem.   *Let $C \subseteq \mathbb{F}_q[x]/(x^n - 1)$ by a cyclic code with generator polynomial g and let $\alpha_1, \ldots, \alpha_{n-k}$ be the roots of g. Then $f \in \mathbb{F}_q[x]/(x^n - 1)$ is a code polynomial if and only if the coefficient vector $(f_0, \ldots, f_{n-1})$ of f is in the null space of the matrix*

$$H = \begin{pmatrix} 1 & \alpha_1 & \alpha_1^2 & \cdots & \alpha_1^{n-1} \\ \vdots & \vdots & \vdots & & \vdots \\ 1 & \alpha_{n-k} & \alpha_{n-k}^2 & \cdots & \alpha_{n-k}^{n-1} \end{pmatrix}. \tag{8.3}$$

Proof.   Let $f(x) = f_0 + f_1 x + \cdots + f_{n-1} x^{n-1}$; then $f(\alpha_i) = f_0 + f_1 \alpha_i + \cdots + f_{n-1} \alpha_i^{n-1} = 0$ for $1 \leqslant i \leqslant n - k$, that is,

$$\left(1, \alpha_i, \ldots, \alpha_i^{n-1}\right)(f_0, f_1, \ldots, f_{n-1})^T = 0 \quad \text{for } 1 \leqslant i \leqslant n - k,$$

if and only if $H(f_0, f_1, \ldots, f_{n-1})^T = \mathbf{0}$. $\square$

We recall from Section 1 that for error correction we have to determine the syndrome of the received word $\mathbf{y}$. In the case of cyclic codes, the syndrome, which is a column vector of length $n - k$, can often be replaced by a simpler entity serving the same purpose. For instance, let α be a primitive nth root of unity in $\mathbb{F}_{q^m}$ and let the generator polynomial g be the minimal polynomial of α over $\mathbb{F}_q$. Since g divides $f \in \mathbb{F}_q[x]/(x^n - 1)$ if and only if $f(\alpha) = 0$, it suffices to replace the matrix H in (8.3) by

$$H = \begin{pmatrix} 1 & \alpha & \alpha^2 & \cdots & \alpha^{n-1} \end{pmatrix}.$$

Then the role of the syndrome is played by $S(\mathbf{y}) = H\mathbf{y}^T$, and $S(\mathbf{y}) = y(\alpha)$ since $\mathbf{y} = (y_0, y_1, \ldots, y_{n-1})$ can be regarded as a polynomial $y(x)$ with coefficients y_i. In the following we use the notation **w** for a transmitted word and **v** for a received word, and we write $w(x)$ and $v(x)$, respectively, for the corresponding polynomials. Suppose $e^{(j)}(x) = x^{j-1}$ with $1 \leq j \leq n$ is an error polynomial with a single error, and let $\mathbf{v} = \mathbf{w} + \mathbf{e}^{(j)}$ be the received word. Then

$$v(\alpha) = w(\alpha) + e^{(j)}(\alpha) = e^{(j)}(\alpha) = \alpha^{j-1}.$$

$e^{(j)}(\alpha)$ is called the *error-location number*. $S(\mathbf{v}) = \alpha^{j-1}$ indicates the error uniquely, since $e^{(i)}(\alpha) \neq e^{(j)}(\alpha)$ for $1 \leq i \leq n$ with $i \neq j$.

Before describing a general class of cyclic codes and their decoding, we consider a special example to motivate the theory.

8.43.   Example.   Let $\alpha \in \mathbb{F}_{16}$ be a root of $x^4 + x + 1 \in \mathbb{F}_2[x]$, then α and α^3 have the minimal polynomials $m^{(1)}(x) = x^4 + x + 1$ and $m^{(3)}(x) = x^4 + x^3 + x^2 + x + 1$ over $\mathbb{F}_2$, respectively. Both $m^{(1)}(x)$ and $m^{(3)}(x)$ are divisors of $x^{15} - 1$. Hence we can define a binary cyclic code C with generator polynomial $g = m^{(1)} m^{(3)}$. Since g divides $f \in \mathbb{F}_2[x]/(x^{15} - 1)$ if and only if $f(\alpha) = f(\alpha^3) = 0$, it suffices to replace the matrix H in (8.3) by

$$H = \begin{pmatrix} 1 & \alpha & \alpha^2 & \cdots & \alpha^{14} \\ 1 & \alpha^3 & \alpha^6 & \cdots & \alpha^{42} \end{pmatrix}.$$

We shall show (see Theorem 8.45 and Example 8.47) that the minimum distance of C is ≥ 5, therefore C can correct up to 2 errors. C is a cyclic $(15, 7)$ code. Let

$$S_1 = \sum_{i=0}^{14} v_i \alpha^i \quad \text{and} \quad S_3 = \sum_{i=0}^{14} v_i \alpha^{3i}$$

be the components of $S(\mathbf{v}) = H\mathbf{v}^T$ Then $\mathbf{v} \in C$ if and only if $S(\mathbf{v}) = H\mathbf{v}^T = \mathbf{0}$ if and only if $S_1 = S_3 = 0$. If we use binary notation to represent elements of $\mathbb{F}_{16}$, then H attains the form

$$H = \begin{pmatrix} 1 & 0 & 0 & 0 & 1 & 0 & 0 & 1 & 1 & 0 & 1 & 0 & 1 & 1 & 1 \\ 0 & 1 & 0 & 0 & 1 & 1 & 0 & 1 & 0 & 1 & 1 & 1 & 1 & 0 & 0 \\ 0 & 0 & 1 & 0 & 0 & 1 & 1 & 0 & 1 & 0 & 1 & 1 & 1 & 1 & 0 \\ 0 & 0 & 0 & 1 & 0 & 0 & 1 & 1 & 0 & 1 & 0 & 1 & 1 & 1 & 1 \\ 1 & 0 & 0 & 0 & 1 & 1 & 0 & 0 & 0 & 1 & 1 & 0 & 0 & 0 & 1 \\ 0 & 0 & 0 & 1 & 1 & 0 & 0 & 0 & 1 & 1 & 0 & 0 & 0 & 1 & 1 \\ 0 & 0 & 1 & 0 & 1 & 0 & 0 & 1 & 0 & 1 & 0 & 0 & 1 & 0 & 1 \\ 0 & 1 & 1 & 1 & 1 & 0 & 1 & 1 & 1 & 1 & 0 & 1 & 1 & 1 & 1 \end{pmatrix}.$$

The columns of H are calculated as follows: the first four entries of the first column are the coefficients in $1 = 1 \cdot \alpha^0 + 0 \cdot \alpha^1 + 0 \cdot \alpha^2 + 0 \cdot \alpha^3$, the first four entries of the second column are the coefficients in $\alpha = 0 \cdot \alpha^0 + 1 \cdot \alpha^1 + 0 \cdot \alpha^2 + 0 \cdot \alpha^3$, and so on; the last four entries of the first column are the coefficients in $1 = 1 \cdot \alpha^0 + 0 \cdot \alpha^1 + 0 \cdot \alpha^2 + 0 \cdot \alpha^3$, the last four entries of the second column are the coefficients in $\alpha^3 = 0 \cdot \alpha^0 + 0 \cdot \alpha^1 + 0 \cdot \alpha^2 + 1 \cdot \alpha^3$, and so on. We use $\alpha^4 + \alpha + 1 = 0$ in the calculations.

Suppose the received vector $\mathbf{v} = (v_0, \ldots, v_{14})$ has at most two errors; for example, $e(x) = x^{a_1} + x^{a_2}$ with $0 \leqslant a_1, a_2 \leqslant 14$, $a_1 \neq a_2$. Then we have

$$S_1 = \alpha^{a_1} + \alpha^{a_2}, \qquad S_3 = \alpha^{3a_1} + \alpha^{3a_2}.$$

Let $\eta_1 = \alpha^{a_1}$, $\eta_2 = \alpha^{a_2}$ be the error-location numbers, then

$$S_1 = \eta_1 + \eta_2, \qquad S_3 = \eta_1^3 + \eta_2^3,$$

therefore

$$S_3 = S_1^3 + S_1^2 \eta_1 + S_1 \eta_1^2,$$

hence

$$1 + S_1 \eta_1^{-1} + \left(S_1^2 + S_3 S_1^{-1}\right) \eta_1^{-2} = 0.$$

If two errors occurred, then η_1^{-1} and η_2^{-1} are roots of the polynomial

$$s(x) = 1 + S_1 x + \left(S_1^2 + S_3 S_1^{-1}\right) x^2. \tag{8.4}$$

If only one error occurred, then $S_1 = \eta_1$ and $S_3 = \eta_1^3$, hence $S_1^3 + S_3 = 0$; that is,

$$s(x) = 1 + S_1 x. \tag{8.5}$$

If no error occurred, then $S_1 = S_3 = 0$ and the correct code word $\mathbf{w}$ has been received.

To summarize, we first evaluate the syndrome $S(\mathbf{v}) = H\mathbf{v}^{\mathsf{T}}$ of the received vector $\mathbf{v}$, then determine $s(x)$ and find the errors via the roots of $s(x)$. The polynomial in (8.5) has a root in $\mathbb{F}_{16}$ whenever $S_1 \neq 0$. If $s(x)$ in (8.4) has no roots in $\mathbb{F}_{16}$, then we know that the error $e(x)$ has more than two error locations and therefore cannot be corrected by the given $(15,7)$ code.

More specifically, suppose

$$\mathbf{v} = 100111000000000$$

is the received word. Then $S(\mathbf{v}) = \begin{pmatrix} S_1 \\ S_3 \end{pmatrix}$ is given by

$$S_1 = 1 + \alpha^3 + \alpha^4 + \alpha^5 = \alpha^2 + \alpha^3, \qquad S_3 = 1 + \alpha^9 + \alpha^{12} + \alpha^{15} = 1 + \alpha^2.$$

For the polynomial $s(x)$ in (8.4) we obtain

$$s(x) = 1 + (\alpha^2 + \alpha^3)x + \left[1 + \alpha + \alpha^2 + \alpha^3 + (1 + \alpha^2)(\alpha^2 + \alpha^3)^{-1}\right]x^2$$

$$= 1 + (\alpha^2 + \alpha^3)x + (1 + \alpha + \alpha^3)x^2.$$

We determine the roots of $s(x)$ by trial and error and find α and α^7 as roots. Hence we have $\eta_1^{-1} = \alpha$, $\eta_2^{-1} = \alpha^7$, thus $\eta_1 = \alpha^{14}$, $\eta_2 = \alpha^8$. Therefore, we know that errors must have occurred in the positions corresponding to x^8 and x^{14}, that is, in the 9th and 15th position of v. The transmitted code word must have been

$$\mathbf{w} = 100111001000001.$$

The code word **w** is decoded by dividing the corresponding polynomial by the generator polynomial g. This gives $1 + x^3 + x^5 + x^6$ with remainder 0. Hence the original message was 1001011.   □

8.44. Definition. Let b be a nonnegative integer and let $\alpha \in \mathbb{F}_{q^m}$ be a primitive nth root of unity, where m is the multiplicative order of q modulo n. A *BCH code* over $\mathbb{F}_q$ of length n and *designed distance* d, $2 \leqslant d \leqslant n$, is a cyclic code defined by the roots

$$\alpha^b, \alpha^{b+1}, \ldots, \alpha^{b+d-2}$$

of the generator polynomial.

If $m^{(i)}(x)$ denotes the minimal polynomial of α^i over $\mathbb{F}_q$, then the generator polynomial $g(x)$ of a BCH code is of the form

$$g(x) = \operatorname{lcm}\big(m^{(b)}(x), m^{(b+1)}(x), \ldots, m^{(b+d-2)}(x)\big).$$

Some special cases of the general Definition 8.44 are also important. If $b = 1$, the corresponding BCH codes are called *narrow-sense* BCH codes. If $n = q^m - 1$, the BCH codes are called *primitive*. If $n = q - 1$, a BCH code of length n over $\mathbb{F}_q$ is called a *Reed-Solomon code*.

8.45. Theorem. *The minimum distance of a BCH code of designed distance d is at least d.*

Proof. The BCH code is contained in the null space of the matrix

$$H = \begin{pmatrix} 1 & \alpha^b & \alpha^{2b} & \cdots & \alpha^{(n-1)b} \\ 1 & \alpha^{b+1} & \alpha^{2(b+1)} & \cdots & \alpha^{(n-1)(b+1)} \\ \vdots & \vdots & \vdots & & \vdots \\ 1 & \alpha^{b+d-2} & \alpha^{2(b+d-2)} & \cdots & \alpha^{(n-1)(b+d-2)} \end{pmatrix}.$$

We show that any $d-1$ columns of this matrix are linearly independent. Take the determinant of any $d-1$ distinct columns of H, then we obtain

$$\begin{vmatrix} \alpha^{bi_1} & \alpha^{bi_2} & \cdots & \alpha^{bi_{d-1}} \\ \alpha^{(b+1)i_1} & \alpha^{(b+1)i_2} & \cdots & \alpha^{(b+1)i_{d-1}} \\ \vdots & \vdots & \vdots & \\ \alpha^{(b+d-2)i_1} & \alpha^{(b+d-2)i_2} & \cdots & \alpha^{(b+d-2)i_{d-1}} \end{vmatrix}$$

$$
= \alpha^{b(i_1 + i_2 + \cdots + i_{d-1})}
\begin{vmatrix}
1 & 1 & \cdots & 1 \\
\alpha^{i_1} & \alpha^{i_2} & \cdots & \alpha^{i_{d-1}} \\
\vdots & \vdots & & \vdots \\
\alpha^{i_1(d-2)} & \alpha^{i_2(d-2)} & \cdots & \alpha^{i_{d-1}(d-2)}
\end{vmatrix}
$$

$$
= \alpha^{b(i_1 + i_2 + \cdots + i_{d-1})} \prod_{1 \leqslant k < j \leqslant d-1} (\alpha^{i_j} - \alpha^{i_k}) \neq 0.
$$

Therefore the minimum distance of the code is at least d.   □

8.46. Example. Let $m^{(1)}(x) = x^4 + x + 1$ be the minimal polynomial over $\mathbb{F}_2$ of a primitive element $\alpha \in \mathbb{F}_{16}$. We represent the powers α^i, $0 \leqslant i \leqslant 14$, as linear combinations of $1, \alpha, \alpha^2, \alpha^3$ and thus obtain a parity-check matrix H of a code equivalent to the $(15, 11)$ Hamming code:

$$
H = \begin{pmatrix}
1 & 0 & 0 & 0 & 1 & 0 & 0 & 1 & 1 & 0 & 1 & 0 & 1 & 1 & 1 \\
0 & 1 & 0 & 0 & 1 & 1 & 0 & 1 & 0 & 1 & 1 & 1 & 1 & 0 & 0 \\
0 & 0 & 1 & 0 & 0 & 1 & 1 & 0 & 1 & 0 & 1 & 1 & 1 & 1 & 0 \\
0 & 0 & 0 & 1 & 0 & 0 & 1 & 1 & 0 & 1 & 0 & 1 & 1 & 1 & 1
\end{pmatrix}
$$

$$
= (1 \quad \alpha \quad \alpha^2 \quad \alpha^3 \quad \alpha^4 \quad \alpha^5 \quad \alpha^6 \quad \alpha^7 \quad \alpha^8 \quad \alpha^9 \quad \alpha^{10} \quad \alpha^{11} \quad \alpha^{12} \quad \alpha^{13} \quad \alpha^{14}).
$$

This code can also be regarded as a narrow-sense BCH code of designed distance $d = 3$ over $\mathbb{F}_2$ (note that α^2 is also a root of $m^{(1)}(x)$). Its minimum distance is also 3, and it can therefore correct one error. In order to decode a received vector $\mathbf{v} \in \mathbb{F}_2^{15}$, we have to find the syndrome $H\mathbf{v}^T$. For this cyclic $(15, 11)$ code the syndrome is given as $v(\alpha)$ in the basis $\{1, \alpha, \alpha^2, \alpha^3\}$. It is obtained by dividing $v(x)$ by $m^{(1)}(x)$, say $v(x) = a(x)m^{(1)}(x) + r(x)$ with $\deg(r(x)) < 4$, for then $v(\alpha) = r(\alpha)$; that is, the components of the syndrome are equal to the coefficients of $r(x)$.

For instance, let

$$
\mathbf{v} = 010110001011101,
$$

then $r(x) = 1 + x$, hence

$$
H\mathbf{v}^T = (1100)^T = 1 + \alpha.
$$

Next we have to find the error $\mathbf{e}$ with weight $w(\mathbf{e}) \leqslant 1$ and having the same syndrome. Thus we must determine the exponent j, $0 \leqslant j \leqslant 14$, such that $\alpha^j = H\mathbf{v}^T$. In our numerical example $j = 4$, thus in the received vector $\mathbf{v}$ the fifth position is in error and the transmitted word was

$$
\mathbf{w} = 010100001011101.
$$
                                                                                              □

8.47. Example. Let $q = 2$, $n = 15$, and $d = 4$. Then $x^4 + x + 1$ is irreducible over $\mathbb{F}_2$ and its roots are primitive elements of $\mathbb{F}_{16}$. If α is such a root, then α^2 is a root, and α^3 is then a root of $x^4 + x^3 + x^2 + x + 1$. Thus a narrow-sense BCH code with $d = 4$ is generated by

$$g(x) = (x^4 + x + 1)(x^4 + x^3 + x^2 + x + 1).$$

This is also a generator for a BCH code with $d = 5$, since α^4 is a root of $x^4 + x + 1$. The dimension of this code is $15 - \deg(g(x)) = 7$. This code was considered in greater detail in Example 8.43. $\square$

BCH codes are very powerful since for any positive integer d we can construct a BCH code of minimum distance $\geqslant d$. To find a BCH code for a larger minimum distance, we have to increase the length n and hence increase the number m—that is, the degree of $\mathbb{F}_{q^m}$ over $\mathbb{F}_q$. A BCH code of designed distance $d \geqslant 2t + 1$ will correct t or fewer errors, but at the same time, in order to achieve the desired minimum distance, we must use code words of great length.

We describe now a general *decoding algorithm for BCH codes*. Let us denote by $w(x)$, $v(x)$, and $e(x)$ the transmitted code polynomial, the received polynomial, and the error polynomial, respectively, so that $v(x) = w(x) + e(x)$. First we have to obtain the syndrome of $\mathbf{v}$,

$$S(\mathbf{v}) = H\mathbf{v}^{\mathrm{T}} = (S_b, S_{b+1}, \dots, S_{b+d-2})^{\mathrm{T}},$$

where

$$S_j = v(\alpha^j) = w(\alpha^j) + e(\alpha^j) = e(\alpha^j) \quad \text{for } b \leqslant j \leqslant b + d - 2.$$

If $r \leqslant t$ errors occur, then

$$e(x) = \sum_{i=1}^{r} c_i x^{a_i},$$

where $a_1, \dots, a_r$ are distinct elements of $\{0, 1, \dots, n - 1\}$. The elements $\eta_i = \alpha^{a_i} \in \mathbb{F}_{q^m}$ are called *error-location numbers*, the elements $c_i \in \mathbb{F}_q^*$ are called *error values*. Thus we obtain for the syndrome of $\mathbf{v}$,

$$S_j = e(\alpha^j) = \sum_{i=1}^{r} c_i \eta_i^j \quad \text{for } b \leqslant j \leqslant b + d - 2.$$

Because of the computational rules in $\mathbb{F}_{q^m}$ we have

$$S_j^q = \left(\sum_{i=1}^{r} c_i \eta_i^j \right)^q = \sum_{i=1}^{r} c_i^q \eta_i^{jq} = \sum_{i=1}^{r} c_i \eta_i^{jq} = S_{jq}. \tag{8.6}$$

The unknown quantities are the pairs (η_i, c_i), $i = 1, \dots, r$, the coordinates S_j of the syndrome $S(\mathbf{v})$ are known since they can be calculated from the received vector $\mathbf{v}$. In the binary case any error is completely characterized by the η_i alone, since in this case all c_i are 1.

In the next stage of the decoding algorithm we determine the coefficients σ_i defined by the polynomial identity

$$\prod_{i=1}^{r} (\eta_i - x) = \sum_{i=0}^{r} (-1)^i \sigma_{r-i} x^i$$

$$= \sigma_r - \sigma_{r-1} x + \cdots + (-1)^r \sigma_0 x^r.$$

Thus $\sigma_0 = 1$ and $\sigma_1, \ldots, \sigma_r$ are the elementary symmetric polynomials in $\eta_1, \ldots, \eta_r$. Substituting η_i for x gives

$$(-1)^r \sigma_r + (-1)^{r-1} \sigma_{r-1} \eta_i + \cdots + (-1) \sigma_1 \eta_i^{r-1} + \eta_i^r = 0 \quad \text{for } i = 1, \ldots, r.$$

Multiplying by $c_i \eta_i^j$ and summing these equations for $i = 1, \ldots, r$ yields

$$(-1)^r \sigma_r S_j + (-1)^{r-1} \sigma_{r-1} S_{j+1} + \cdots + (-1) \sigma_1 S_{j+r-1} + S_{j+r} = 0$$

$$\text{for } j = b, b+1, \ldots, b+r-1.$$

8.48. Lemma. *The system of equations*

$$\sum_{i=1}^{r} c_i \eta_i^j = S_j, \quad j = b, b+1, \ldots, b+r-1,$$

in the unknowns c_i is solvable if the η_i are distinct elements of $\mathbb{F}_{q^m}^*$.

Proof. The determinant of the system is

$$\begin{vmatrix} \eta_1^b & \eta_2^b & \cdots & \eta_r^b \\ \eta_1^{b+1} & \eta_2^{b+1} & \cdots & \eta_r^{b+1} \\ \vdots & \vdots & & \vdots \\ \eta_1^{b+r-1} & \eta_2^{b+r-1} & \cdots & \eta_r^{b+r-1} \end{vmatrix} = \eta_1^b \eta_2^b \cdots \eta_r^b \prod_{1 \le i < j \le r} (\eta_j - \eta_i) \ne 0. \quad \square$$

8.49. Lemma. *The system of equations*

$$(-1)^r \sigma_r S_j + (-1)^{r-1} \sigma_{r-1} S_{j+1} + \cdots + (-1) \sigma_1 S_{j+r-1} + S_{j+r} = 0,$$

$$j = b, b+1, \ldots, b+r-1,$$

in the unknowns $(-1)^i \sigma_i$, $i = 1, 2, \ldots, r$, is solvable uniquely if and only if r errors occur.

Proof. The matrix of the system can be decomposed as follows:

$$\begin{pmatrix} S_b & S_{b+1} & \cdots & S_{b+r-1} \\ S_{b+1} & S_{b+2} & \cdots & S_{b+r} \\ \vdots & \vdots & & \vdots \\ S_{b+r-1} & S_{b+r} & \cdots & S_{b+2r-2} \end{pmatrix} = VDV^{\mathrm{T}},$$

where

$$V = \begin{pmatrix} 1 & 1 & \cdots & 1 \\ \eta_1 & \eta_2 & \cdots & \eta_r \\ \vdots & \vdots & & \vdots \\ \eta_1^{r-1} & \eta_2^{r-1} & \cdots & \eta_r^{r-1} \end{pmatrix}$$

and

$$D = \begin{pmatrix} c_1\eta_1^b & 0 & \cdots & 0 \\ 0 & c_2\eta_2^b & \cdots & 0 \\ \vdots & \vdots & & \vdots \\ 0 & 0 & \cdots & c_r\eta_r^b \end{pmatrix}.$$

The matrix of the given system of equations is nonsingular if and only if V and D are nonsingular. V as a Vandermonde matrix is nonsingular if and only if the η_i, $i = 1,\ldots,r$, are distinct and D is nonsingular if and only if all the η_i and c_i are nonzero. Both conditions are satisfied if and only if r errors occur.   □

We introduce the *error-locator polynomial* that is closely related to the considerations above:

$$s(x) = \prod_{i=1}^{r}(1 - \eta_i x) = \sum_{i=0}^{r}(-1)^i \sigma_i x^i,$$

where the σ_i are as above. The roots of $s(x)$ are $\eta_1^{-1}, \eta_2^{-1}, \ldots, \eta_r^{-1}$. In order to find these roots, we can use a search method due to Chien. First we want to know if α^{n-1} is an error-location number—that is, if $\alpha = \alpha^{-(n-1)}$ is a root of $s(x)$. To test this we form

$$- \sigma_1\alpha + \sigma_2\alpha^2 + \cdots + (-1)^r \sigma_r\alpha^r.$$

If this is equal to -1, then α^{n-1} is an error-location number since then $s(\alpha) = 0$. More generally, α^{n-m} is tested for $m = 1,2,\ldots,n$ in the same way. In the binary case, the discovery of error locations is equivalent to correcting errors. We summarize the BCH decoding algorithm, writing now τ_i for $(-1)^i \sigma_i$.

8.50. BCH Decoding. Suppose at most t errors occur in transmitting a code word **w**, using a BCH code of designed distance $d \geq 2t + 1$.

Step 1. Determine the syndrome of the received word **v**,

$$S(\mathbf{v}) = (S_b, S_{b+1}, \ldots, S_{b+d-2})^\mathrm{T}.$$

Let

$$S_j = \sum_{i=1}^{r} c_i \eta_i^j, \quad b \leqslant j \leqslant b + d - 2.$$

Step 2. Determine the maximum number $r \leqslant t$ such that the system of equations

$$S_{j+r} + S_{j+r-1}\tau_1 + \cdots + S_j\tau_r = 0, \quad b \leqslant j \leqslant b + r - 1,$$

in the τ_i has a nonsingular coefficient matrix, thus obtaining the number r of errors that have occurred. Then set up the error-locator polynomial

$$s(x) = \prod_{i=1}^{r}(1 - \eta_i x) = \sum_{i=0}^{r} \tau_i x^i.$$

Find the coefficients τ_i from the S_j.

Step 3. Solve $s(x) = 0$ by substituting the powers of α into $s(x)$. Thus find the error-location numbers η_i (Chien search).

Step 4. Introduce the η_i in the first r equations of Step 1 to determine the error values c_i. Then find the transmitted word $\mathbf{w}$ from $w(x) = v(x) - e(x)$.

8.51. Remark. We note that the difficult step in this algorithm is Step 2. There are various methods to perform this step, one possibility is to use the Berlekamp-Massey algorithm of Chapter 6 to determine the unknown coefficients τ_i in the linear recurrence relation for the S_j. $\qquad \Box$

8.52. Example. Consider a BCH code with designed distance $d = 5$ that is able to correct any single or double error. In this case, let $b = 1$, $n = 15$, $q = 2$. If $m^{(i)}(x)$ denotes the minimal polynomial of α^i over $\mathbb{F}_2$, where the primitive element $\alpha \in \mathbb{F}_{16}$ is a root of $x^4 + x + 1$, then

$$m^{(1)}(x) = m^{(2)}(x) = m^{(4)}(x) = m^{(8)}(x) = 1 + x + x^4,$$

$$m^{(3)}(x) = m^{(6)}(x) = m^{(12)}(x) = m^{(9)}(x) = 1 + x + x^2 + x^3 + x^4.$$

Therefore a generator polynomial of the BCH code will be

$$g(x) = m^{(1)}(x)m^{(3)}(x) = 1 + x^4 + x^6 + x^7 + x^8.$$

The code is a $(15, 7)$ code, with parity-check polynomial

$$h(x) = (x^{15} - 1)/g(x) = 1 + x^4 + x^6 + x^7.$$

We take the vectors corresponding to

$$g(x), xg(x), x^2g(x), x^3g(x), x^4g(x), x^5g(x), x^6g(x)$$

as the basis of the $(15,7)$ BCH code and obtain the generator matrix

$$G = \begin{pmatrix} 1 & 0 & 0 & 0 & 1 & 0 & 1 & 1 & 1 & 0 & 0 & 0 & 0 & 0 & 0 \\ 0 & 1 & 0 & 0 & 0 & 1 & 0 & 1 & 1 & 1 & 0 & 0 & 0 & 0 & 0 \\ 0 & 0 & 1 & 0 & 0 & 0 & 1 & 0 & 1 & 1 & 1 & 0 & 0 & 0 & 0 \\ 0 & 0 & 0 & 1 & 0 & 0 & 0 & 1 & 0 & 1 & 1 & 1 & 0 & 0 & 0 \\ 0 & 0 & 0 & 0 & 1 & 0 & 0 & 0 & 1 & 0 & 1 & 1 & 1 & 0 & 0 \\ 0 & 0 & 0 & 0 & 0 & 1 & 0 & 0 & 0 & 1 & 0 & 1 & 1 & 1 & 0 \\ 0 & 0 & 0 & 0 & 0 & 0 & 1 & 0 & 0 & 0 & 1 & 0 & 1 & 1 & 1 \end{pmatrix}.$$

Suppose now that the received word $\mathbf{v}$ is

$$1\ 0\ 0\ 1\ 0\ 0\ 1\ 1\ 0\ 0\ 0\ 0\ 1\ 0\ 0,$$

or as a polynomial,

$$v(x) = 1 + x^3 + x^6 + x^7 + x^{12}.$$

We calculate the syndrome according to Step 1, using (8.6) to simplify the work:

$$S_1 = e(\alpha) = v(\alpha) = 1,$$

$$S_2 = e(\alpha^2) = v(\alpha^2) = 1,$$

$$S_3 = e(\alpha^3) = v(\alpha^3) = \alpha^4,$$

$$S_4 = e(\alpha^4) = v(\alpha^4) = 1.$$

The largest possible system of linear equations in the unknowns τ_i (Step 2) is then of the form

$$S_2\tau_1 + S_1\tau_2 = S_3,$$

$$S_3\tau_1 + S_2\tau_2 = S_4,$$

or

$$\tau_1 + \tau_2 = \alpha^4,$$

$$\alpha^4\tau_1 + \tau_2 = 1.$$

This system clearly has a nonsingular coefficient matrix. Therefore two errors must have occurred—that is, $r = 2$. We solve this system of equations and obtain $\tau_1 = 1$, $\tau_2 = \alpha$. Substituting these values into $s(x)$ and recalling $\tau_0 = 1$ gives

$$s(x) = 1 + x + \alpha x^2.$$

As roots in $\mathbb{F}_{16}$ we find $\eta_1^{-1} = \alpha^8$, $\eta_2^{-1} = \alpha^6$, hence $\eta_1 = \alpha^7$, $\eta_2 = \alpha^9$. Therefore, we know that errors must have occurred in positions 8 and 10 of the code word. We correct these errors in the received polynomial and obtain

$$w(x) = v(x) - e(x)$$
$$= (1 + x^3 + x^6 + x^7 + x^{12}) - (x^7 + x^9)$$
$$= 1 + x^3 + x^6 + x^9 + x^{12}.$$

The corresponding code word is

$$1\ 0\ 0\ 1\ 0\ 0\ 1\ 0\ 0\ 1\ 0\ 0\ 1\ 0\ 0.$$

The initial message can be recovered by dividing the corrected polynomial —that is, the transmitted code polynomial $w(x)$—by $g(x)$. This gives

$$w(x)/g(x) = 1 + x^3 + x^4,$$

which yields the corresponding message word $1\ 0\ 0\ 1\ 1\ 0\ 0$. $\square$

3. GOPPA CODES

We generalize the narrow-sense BCH codes introduced in Section 2 to obtain an important class of linear codes which still allow an efficient decoding algorithm and which are also useful for applications in cryptography (see Chapter 9, Section 4). These codes meet the Gilbert-Varshamov bound in Theorem 8.27 at least asymptotically. To motivate the definition of this class of codes, we first go back to narrow-sense BCH codes and present another characterization of their code words.

We recall that narrow-sense BCH codes correspond to the special case $b = 1$ of Definition 8.44. A narrow-sense BCH code over $\mathbb{F}_q$ of length n and designed distance d is thus the cyclic code defined by the roots α, $\alpha^2, \ldots, \alpha^{d-1}$ of the generator polynomial, where $\alpha \in \mathbb{F}_{q^m}$ is a primitive nth root of unity. We characterize the code words of this code by using an identity in the polynomial ring $\mathbb{F}_{q^m}[x]$.

8.53. Lemma. $(c_0, c_1, \ldots, c_{n-1}) \in \mathbb{F}_q^n$ *is a code word of the narrow-sense BCH code over* $\mathbb{F}_q$ *defined by the roots* $\alpha, \alpha^2, \ldots, \alpha^{d-1}$ *of the generator polynomial if and only if*

$$\sum_{i=0}^{n-1} c_i \alpha^{i(d-1)} \frac{x^{d-1} - \alpha^{-i(d-1)}}{x - \alpha^{-i}} = 0. \tag{8.7}$$

Proof. By definition, $(c_0, c_1, \ldots, c_{n-1})$ is a code word of the given code if and only if

$$\sum_{i=0}^{n-1} c_i \alpha^{ij} = 0 \quad \text{for} \quad 1 \leqslant j \leqslant d - 1.$$

On the other hand, we have

$$\sum_{i=0}^{n-1} c_i \alpha^{i(d-1)} \frac{x^{d-1} - \alpha^{-i(d-1)}}{x - \alpha^{-i}} = \sum_{i=0}^{n-1} c_i \alpha^{i(d-1)} \sum_{j=0}^{d-2} \alpha^{-i(d-2-j)} x^j$$

$$= \sum_{j=1}^{d-1} \left(\sum_{i=0}^{n-1} c_i \alpha^{ij} \right) x^{j-1},$$

and so $(c_0, c_1, \ldots, c_{n-1})$ is a code word if and only if the identity (8.7) holds. $\square$

This result provides the motivation for the following definition of Goppa codes over $\mathbb{F}_q$.

8.54. Definition. Let $g(x)$ be a polynomial of degree t, $1 \leqslant t < n$, over an extension $\mathbb{F}_{q^m}$ of $\mathbb{F}_q$, and let $L = \{\gamma_0, \gamma_1, \ldots, \gamma_{n-1}\}$ be a set of n distinct elements of $\mathbb{F}_{q^m}$ such that $g(\gamma_i) \neq 0$ for $0 \leqslant i \leqslant n-1$. The *Goppa code* $\Gamma(L, g)$ over $\mathbb{F}_q$ with *Goppa polynomial* $g(x)$ is the set of all $(c_0, c_1, \ldots, c_{n-1}) \in \mathbb{F}_q^n$ such that the identity

$$\sum_{i=0}^{n-1} c_i g(\gamma_i)^{-1} \frac{g(x) - g(\gamma_i)}{x - \gamma_i} = 0 \tag{8.8}$$

holds in the polynomial ring $\mathbb{F}_{q^m}[x]$. If $g(x)$ is irreducible over $\mathbb{F}_{q^m}$, then $\Gamma(L, g)$ is called an *irreducible* Goppa code.

8.55. Example. If $g(x) = x^{d-1}$ and $L = \{\alpha^{-i} : i = 0, 1, \ldots, n-1\}$, where $\alpha \in \mathbb{F}_{q^m}$ is a primitive nth root of unity, then $\Gamma(L, g)$ is a narrow-sense BCH code over $\mathbb{F}_q$ of length n and designed distance d.   □

It is clear that $\Gamma(L, g)$ is a linear code, since the condition (8.8) defines a subspace of the vector space $\mathbb{F}_q^n$. We want to find a matrix such that the intersection of its null space with $\mathbb{F}_q^n$ is equal to $\Gamma(L, g)$. If

$$g(x) = \sum_{j=0}^{t} g_j x^j,$$

then

$$\frac{g(x) - g(\gamma)}{x - \gamma} = \sum_{j=0}^{t} g_j \frac{x^j - \gamma^j}{x - \gamma} = \sum_{s=0}^{t-1} \left(\sum_{j=s+1}^{t} g_j \gamma^{j-1-s} \right) x^s.$$

Putting $h_i = g(\gamma_i)^{-1}$ for $0 \leqslant i \leqslant n-1$, it follows that $(c_0, c_1, \ldots, c_{n-1}) \in \mathbb{F}_q^n$ satisfies (8.8) if and only if

$$\sum_{i=0}^{n-1} \left(h_i \sum_{j=s+1}^{t} g_j \gamma_i^{j-1-s} \right) c_i = 0 \quad \text{for} \quad 0 \leqslant s \leqslant t-1.$$

Therefore $\Gamma(L, g)$ is the intersection of $\mathbb{F}_q^n$ with the null space of the matrix

$$\begin{pmatrix} h_0 g_t & \cdots & h_{n-1} g_t \\ h_0(g_{t-1} + g_t \gamma_0) & \cdots & h_{n-1}(g_{t-1} + g_t \gamma_{n-1}) \\ \vdots & & \vdots \\ h_0 \sum_{j=1}^{t} g_j \gamma_0^{j-1} & \cdots & h_{n-1} \sum_{j=1}^{t} g_j \gamma_{n-1}^{j-1} \end{pmatrix}.$$

Since $g_t \neq 0$, we can use row operations to transform this into the matrix

$$H = \begin{pmatrix} g(\gamma_0)^{-1} & \cdots & g(\gamma_{n-1})^{-1} \\ g(\gamma_0)^{-1} \gamma_0 & \cdots & g(\gamma_{n-1})^{-1} \gamma_{n-1} \\ \vdots & & \vdots \\ g(\gamma_0)^{-1} \gamma_0^{t-1} & \cdots & g(\gamma_{n-1})^{-1} \gamma_{n-1}^{t-1} \end{pmatrix}, \tag{8.9}$$

for which the intersection of its null space with $\mathbb{F}_q^n$ is again $\Gamma(L, g)$. The entries of H are elements of $\mathbb{F}_{q^m}$. Each element of $\mathbb{F}_{q^m}$ has a unique representation in a fixed basis of $\mathbb{F}_{q^m}$ over $\mathbb{F}_q$. A matrix H' with entries in $\mathbb{F}_q$ having $\Gamma(L, g)$ as its null space can thus be obtained by replacing each entry of H by the column vector over $\mathbb{F}_q$ of length m that we get from the coefficients in that representation.

8.56. Theorem. *The dimension of the Goppa code $\Gamma(L, g)$ is at least $n - mt$ and its minimum distance is at least $t + 1$.*

Proof. The matrix H' described above is an $mt \times n$ matrix with entries in $\mathbb{F}_q$. Since $\Gamma(L, g)$ is the null space of H', the dimension of $\Gamma(L, g)$ is at least $n - mt$. For the second part consider the determinant of any t distinct columns of the matrix H in (8.9). After taking out obvious constant factors, such a determinant reduces to a Vandermonde determinant which is nonzero in view of the condition that the elements $\gamma_0, \gamma_1, \ldots, \gamma_{n-1}$ are distinct. Therefore any t columns of H are linearly independent, and so the minimum distance of $\Gamma(L, g)$ is at least $t + 1$. $\square$

In most applications one works with binary Goppa codes—that is, Goppa codes over $\mathbb{F}_2$. In this case the following improvement on the lower bound of the minimum distance can be obtained.

8.57. Theorem. *For a binary Goppa code whose Goppa polynomial has no multiple roots, the minimum distance is at least $2t + 1$.*

Proof. If $(c_0, c_1, \ldots, c_{n-1}) \in \mathbb{F}_2^n$ is a code word of weight $w > 0$ in the binary Goppa code $\Gamma(L, g)$, then $c_{i_1} = c_{i_2} = \cdots = c_{i_w} = 1$ with $0 \leqslant i_1 < i_2 < \cdots < i_w \leqslant n - 1$ and all other $c_i = 0$. If $L = \{\gamma_0, \gamma_1, \ldots, \gamma_{n-1}\} \subseteq \mathbb{F}_{2^m}$, define

$$f(x) = \prod_{j=1}^{w} (x - \gamma_{i_j}) \in \mathbb{F}_{2^m}[x].$$

From (8.8) we obtain

$$0 = f(x) \sum_{i=0}^{n-1} c_i g(\gamma_i)^{-1} \frac{g(x) - g(\gamma_i)}{x - \gamma_i} = f(x) \sum_{j=1}^{w} g(\gamma_{i_j})^{-1} \frac{g(x) - g(\gamma_{i_j})}{x - \gamma_{i_j}}$$

$$= \sum_{j=1}^{w} g(\gamma_{i_j})^{-1} (g(x) - g(\gamma_{i_j})) \prod_{\substack{h=1 \\ h \neq j}}^{w} (x - \gamma_{i_h}).$$

Considering the last polynomial modulo $g(x)$, we get

$$0 \equiv - \sum_{j=1}^{w} \prod_{\substack{h=1 \\ h \neq j}}^{w} (x - \gamma_{i_h}) \equiv -f'(x) \bmod g(x),$$

and so $g(x)$ divides the derivative $f'(x)$. Since we are working in characteristic 2, $f'(x)$ contains only even powers and is thus the square of a polynomial in $\mathbb{F}_{2^m}[x]$. Now $g(x)$ has no multiple roots by hypothesis, hence it follows that $g(x)^2$ divides $f'(x)$. Consequently,

$$w - 1 \geqslant \deg(f'(x)) \geqslant 2t,$$

and so any nonzero code word has weight at least $2t + 1$. $\square$

8.58. Example. We describe the binary irreducible Goppa code $\Gamma(L,g)$ with Goppa polynomial $g(x) = x^2 + x + 1$ and $L = \mathbb{F}_8 = \{0, 1, \alpha, \ldots, \alpha^6\}$, where α is a primitive element of $\mathbb{F}_8$ satisfying $\alpha^3 + \alpha + 1 = 0$. From Theorems 8.56 and 8.57 we get the following information on the parameters of this code: length $n = 8$, dimension $k \geqslant n - mt = 2$, and minimum distance $d \geqslant 2t + 1 = 5$. Furthermore, $\Gamma(L,g)$ is the intersection of $\mathbb{F}_2^n$ with the null space of the matrix

$$
H = \begin{pmatrix} g(0)^{-1} & g(1)^{-1} & \cdots & g(\alpha^6)^{-1} \\ g(0)^{-1}0 & g(1)^{-1}1 & \cdots & g(\alpha^6)^{-1}\alpha^6 \end{pmatrix}
$$

$$
= \begin{pmatrix} 1 & 1 & \alpha^2 & \alpha^4 & \alpha^2 & \alpha & \alpha & \alpha^4 \\ 0 & 1 & \alpha^3 & \alpha^6 & \alpha^5 & \alpha^5 & \alpha^6 & \alpha^3 \end{pmatrix}
$$

obtained from (8.9). Using the basis $\{1, \alpha, \alpha^2\}$ of $\mathbb{F}_8$ over $\mathbb{F}_2$, we get the corresponding binary matrix

$$
H' = \begin{pmatrix}
1 & 1 & 0 & 0 & 0 & 0 & 0 & 0 \\
0 & 0 & 0 & 1 & 0 & 1 & 1 & 1 \\
0 & 0 & 1 & 1 & 1 & 0 & 0 & 1 \\
0 & 1 & 1 & 1 & 1 & 1 & 1 & 1 \\
0 & 0 & 1 & 0 & 1 & 1 & 0 & 1 \\
0 & 0 & 0 & 1 & 1 & 1 & 1 & 0
\end{pmatrix}
$$

having $\Gamma(L,g)$ as its null space. Since H' has rank 6, we have $k = 2$, and H' is a parity-check matrix of $\Gamma(L,g)$. The linear $(8,2)$ code $\Gamma(L,g)$ consists of the following four code words:

$$
\begin{aligned}
& 0\,0\,0\,0\,0\,0\,0\,0, \quad 0\,0\,1\,1\,1\,1\,1\,1, \\
& 1\,1\,0\,0\,1\,0\,1\,1, \quad 1\,1\,1\,1\,0\,1\,0\,0.
\end{aligned}
$$

Thus it has minimum distance $d = 5$. A generator matrix of this code is

$$
G = \begin{pmatrix} 1 & 1 & 0 & 0 & 1 & 0 & 1 & 1 \\ 0 & 0 & 1 & 1 & 1 & 1 & 1 & 1 \end{pmatrix}. \qquad \square
$$

We discuss now a *decoding algorithm for Goppa codes*. We note that if this algorithm is applied in the special case of a narrow-sense BCH code, then it yields an algorithm that is different from the BCH decoding algorithm described in Section 2. Let $\Gamma(L,g)$ be a Goppa code over $\mathbb{F}_q$ with Goppa polynomial $g(x)$ of degree $t \geqslant 2$. We suppose for simplicity that $L \subseteq \mathbb{F}_{q^m}^*$ — that is, $\gamma_i \neq 0$ for $0 \leqslant i \leqslant n - 1$. By Example 8.55, this condition is in particular satisfied for narrow-sense BCH codes. It follows from Theorems 8.12 and 8.56 that $\Gamma(L,g)$ can correct up to $\lfloor t/2 \rfloor$ errors. To correct errors, we take the received word $\mathbf{v}$ and the matrix H in (8.9) and calculate the syndrome

$$
S(\mathbf{v}) = H\mathbf{v}^{\mathrm{T}} = (S_0, S_1, \ldots, S_{t-1})^{\mathrm{T}}. \tag{8.10}
$$

If $S(\mathbf{v}) = \mathbf{0}$, then $\mathbf{v}$ is a code word and no error correction is needed. If $S(\mathbf{v}) \neq \mathbf{0}$, we assume that r errors have occurred, where $1 \leqslant r \leqslant \lfloor t/2 \rfloor$. Let the distinct elements $a_1, \ldots, a_r$ of $\{0, 1, \ldots, n-1\}$ denote the error locations and let $c_1, \ldots, c_r \in \mathbb{F}_q^*$ be the corresponding *error values*. We define the *error-location numbers* $\eta_i = \gamma_{a_i} \in \mathbb{F}_{q^m}$ for $1 \leqslant i \leqslant r$.

Decoding means determining the pairs (η_i, c_i), $1 \leqslant i \leqslant r$, given the components S_j, $0 \leqslant j \leqslant t-1$, of the syndrome. From (8.10) we get

$$S_j = \sum_{i=1}^{r} c_i g(\eta_i)^{-1} \eta_i^j \quad \text{for} \quad 0 \leqslant j \leqslant t-1.$$

With these S_j we set up the *syndrome polynomial*

$$f(x) = \sum_{j=0}^{t-1} S_j x^j.$$

Furthermore, we need the *error-locator polynomial*

$$s(x) = \prod_{i=1}^{r} (1 - \eta_i x)$$

and the *error-evaluator polynomial*

$$u(x) = \sum_{i=1}^{r} c_i g(\eta_i)^{-1} \prod_{\substack{h=1 \\ h \neq i}}^{r} (1 - \eta_h x).$$

As usual, an empty product is identified with the constant 1. We note that $u(x)$ and $s(x)$ are relatively prime since

$$u(\eta_i^{-1}) = c_i g(\eta_i)^{-1} \prod_{\substack{h=1 \\ h \neq i}}^{r} (1 - \eta_h \eta_i^{-1}) \neq 0 \quad \text{for} \quad 1 \leqslant i \leqslant r. \qquad (8.11)$$

8.59. Lemma. *The congruence*

$$u(x) \equiv s(x) f(x) \bmod x^t$$

holds in the ring of polynomials over $\mathbb{F}_{q^m}$.

Proof. Since $s(0) = 1$, $s(x)$ has a multiplicative inverse in the ring $\mathbb{F}_{q^m}[[x]]$ of formal power series over $\mathbb{F}_{q^m}$ by Theorem 6.37. Then

$$\frac{u(x)}{s(x)} = \sum_{i=1}^{r} \frac{c_i g(\eta_i)^{-1}}{1 - \eta_i x} = \sum_{i=1}^{r} c_i g(\eta_i)^{-1} \sum_{j=0}^{\infty} \eta_i^j x^j$$

$$= \sum_{j=0}^{\infty} \left(\sum_{i=1}^{r} c_i g(\eta_i)^{-1} \eta_i^j \right) x^j = f(x) + x^t B(x)$$

for some $B(x) \in \mathbb{F}_{q^m}[[x]]$, and so

$$u(x) = s(x) f(x) + x^t B_1(x)$$

for some $B_1(x) \in \mathbb{F}_{q^m}[[x]]$. A comparison of terms of sufficiently large degrees shows that $B_1(x)$ is actually a polynomial, and this yields the desired congruence.   □

The congruence in Lemma 8.59 can be solved by using the Euclidean algorithm (see p. 22) with the polynomials $r_{-1}(x) = x^t$ and $r_0(x) = f(x)$. This algorithm yields

$$r_{h-1}(x) = q_{h+1}(x)r_h(x) + r_{h+1}(x), \quad \deg(r_{h+1}(x)) < \deg(r_h(x)),$$
$$\text{for } h = 0, 1, \ldots, s-1,$$
$$r_{s-1}(x) = q_{s+1}(x)r_s(x).$$

We define recursively the polynomials

$$z_{-1}(x) = 0, \quad z_0(x) = 1,$$
$$z_h(x) = z_{h-2}(x) - q_h(x)z_{h-1}(x) \quad \text{for} \quad h = 1, 2, \ldots, s.$$

The following properties are shown by straightforward induction:

$$r_h(x) \equiv z_h(x)f(x) \bmod x^t \quad \text{for} \quad h = -1, 0, \ldots, s, \tag{8.12}$$

$$\deg(z_h(x)) = t - \deg(r_{h-1}(x)) \quad \text{for} \quad h = 0, 1, \ldots, s. \tag{8.13}$$

The polynomials $s(x)$ and $u(x)$ are now determined by the following result.

8.60. Lemma. *The error-locator polynomial $s(x)$ and the error-evaluator polynomial $u(x)$ are given by*

$$s(x) = z_b(0)^{-1}z_b(x),$$
$$u(x) = z_b(0)^{-1}r_b(x),$$

where b is the least index such that $\deg(r_b(x)) < t/2$.

Proof. We have $\deg(s(x)) = r$ and $\deg(u(x)) \leqslant r - 1$. If $d(x) = \gcd(x^t, f(x))$, then $d(x)$ divides $u(x)$ by Lemma 8.59 and so $\deg(r_s(x)) = \deg(d(x)) \leqslant \deg(u(x))$. It follows that there exists an index h, $0 \leqslant h \leqslant s$, such that

$$\deg(r_h(x)) \leqslant \deg(u(x)), \deg(r_{h-1}(x)) \geqslant \deg(u(x)) + 1.$$

From (8.13) we obtain

$$\deg(z_h(x)) = t - \deg(r_{h-1}(x)) \leqslant t - \deg(u(x)) - 1.$$

Lemma 8.59 and (8.12) yield

$$u(x) \equiv s(x)f(x) \bmod x^t, \quad r_h(x) \equiv z_h(x)f(x) \bmod x^t,$$

hence

$$u(x)z_h(x) \equiv r_h(x)s(x) \bmod x^t.$$

The polynomial on the left-hand side has degree $\leqslant t - 1$, and the one on the right-hand side has degree $\leqslant \deg(u(x)) + r \leqslant 2r - 1 \leqslant 2\lfloor t/2 \rfloor - 1 \leqslant t - 1$. Thus we have in fact the identity

$$u(x)z_h(x) = r_h(x)s(x). \tag{8.14}$$

Consequently, $u(x)$ divides $r_h(x)s(x)$, and since $u(x)$ and $s(x)$ are relatively prime, $u(x)$ divides $r_h(x)$. But $0 \leqslant \deg(r_h(x)) \leqslant \deg(u(x))$, hence $u(x) = \beta r_h(x)$ for some $\beta \in \mathbb{F}_{q^m}^*$. It follows then from (8.14) that $s(x) = \beta z_h(x)$, and from $s(0) = 1$ we get $\beta = z_h(0)^{-1}$. It remains to show that $h = b$. Since $\deg(r_h(x)) = \deg(u(x)) < t/2$, it is clear that $h \geqslant b$. If we had $h > b$, then

$$\deg(r_{h-1}(x)) \leqslant \deg(r_b(x)) < t/2,$$

and so by (8.13),

$$\lfloor t/2 \rfloor \geqslant \deg(s(x)) = \deg(z_h(x)) = t - \deg(r_{h-1}(x)) > t/2,$$

a contradiction.   □

We may summarize this decoding algorithm for Goppa codes in the following way.

8.61. Decoding of Goppa Codes.   Suppose at most $\lfloor t/2 \rfloor$ errors occur in transmitting a code word $\mathbf{w}$, using a Goppa code $\Gamma(L, g)$ over $\mathbb{F}_q$ with Goppa polynomial of degree $t \geqslant 2$ and $L \subseteq \mathbb{F}_{q^m}^*$.

Step 1.   Determine the syndrome

$$S(\mathbf{v}) = (S_0, S_1, \ldots, S_{t-1})^T$$

of the received word $\mathbf{v}$ by (8.10) and set up the syndrome polynomial

$$f(x) = \sum_{j=0}^{t-1} S_j x^j.$$

If $f(x) = 0$, no errors have occurred, so $\mathbf{w} = \mathbf{v}$.
If $f(x) \neq 0$, proceed to Step 2.

Step 2.   Carry out the Euclidean algorithm with $r_{-1}(x) = x^t$ and $r_0(x) = f(x)$ and stop as soon as $\deg(r_b(x)) < t/2$. Put

$$s(x) = z_b(0)^{-1} z_b(x), \; u(x) = z_b(0)^{-1} r_b(x).$$

Step 3.   Determine the error-location numbers η_i as the multiplicative inverses of the roots of $s(x)$.

Step 4.   Determine the error values c_i from (8.11)—that is,

$$c_i = u(\eta_i^{-1}) g(\eta_i) \prod_{\substack{h=1 \\ h \neq i}}^{r} (1 - \eta_h \eta_i^{-1})^{-1}.$$

Subtract c_i from the component of $\mathbf{v}$ indicated by the error-location number η_i to obtain the transmitted word $\mathbf{w}$.

8.62. Example.   We solve the decoding problem in Example 8.52 by the decoding algorithm for Goppa codes. According to Example 8.55, the narrow-sense BCH code in Example 8.52 is equal to the binary Goppa code $\Gamma(L, g)$ with $g(x) = x^4$ and $L = \{\gamma_0, \gamma_1, \ldots, \gamma_{14}\}$, where $\gamma_i = \alpha^{-i}$ for $0 \leqslant i \leqslant 14$ and $\alpha \in \mathbb{F}_{16}$

is a root of $x^4 + x + 1$. Let the received word $\mathbf{v}$ be

$$1\ 0\ 0\ 1\ 0\ 0\ 1\ 1\ 0\ 0\ 0\ 0\ 1\ 0\ 0.$$

Using the 4×15 matrix H obtained from (8.9), we get the syndrome

$$S(\mathbf{v}) = H\mathbf{v}^T = (S_0, S_1, S_2, S_3)^T$$

with

$$S_0 = 1, \quad S_1 = \alpha^4, \quad S_2 = 1, \quad S_3 = 1.$$

This leads to the syndrome polynomial

$$f(x) = x^3 + x^2 + \alpha^4 x + 1.$$

Now carry out the Euclidean algorithm with $r_{-1}(x) = x^4$ and $r_0(x) = f(x)$ and stop as soon as $\deg(r_b(x)) < 2$. This yields

$$x^4 = (x+1)(x^3 + x^2 + \alpha^4 x + 1) + (\alpha x^2 + \alpha x + 1),$$

$$x^3 + x^2 + \alpha^4 x + 1 = \alpha^{14} x(\alpha x^2 + \alpha x + 1) + (\alpha^9 x + 1).$$

Thus $b = 2$ and $s(x) = z_2(0)^{-1} z_2(x)$. Since $q_1(x) = x + 1$ and $q_2(x) = \alpha^{14} x$, we calculate recursively

$$z_{-1}(x) = 0, \quad z_0(x) = 1,$$
$$z_1(x) = z_{-1}(x) - q_1(x)z_0(x) = x + 1,$$
$$z_2(x) = z_0(x) - q_2(x)z_1(x) = \alpha^{14} x^2 + \alpha^{14} x + 1.$$

Therefore

$$s(x) = \alpha^{14} x^2 + \alpha^{14} x + 1.$$

The roots of $s(x)$ are α^7 and α^9, hence $\eta_1 = \alpha^{-7} = \gamma_7$ and $\eta_2 = \alpha^{-9} = \gamma_9$. It follows that the errors have occurred in positions 8 and 10 of the transmitted code word. By observing that for a binary code the corresponding error values can only be $c_1 = c_2 = 1$, or by a direct calculation of c_1 and c_2 from the formula in Step 4 of 8.61 with $u(x) = z_2(0)^{-1} r_2(x) = \alpha^9 x + 1$, we find the transmitted code word

$$1\ 0\ 0\ 1\ 0\ 0\ 1\ 0\ 0\ 1\ 0\ 0\ 1\ 0\ 0$$

in accordance with the result in Example 8.52.   □

EXERCISES

8.1. Determine all code words, the minimum distance, and a parity-check matrix of the binary linear $(5, 3)$ code that is defined by the generator matrix

$$G = \begin{pmatrix} 0 & 1 & 0 & 0 & 1 \\ 0 & 0 & 1 & 0 & 1 \\ 1 & 0 & 0 & 1 & 1 \end{pmatrix}.$$

8.2.   Prove: a linear code can detect s or fewer errors if and only if its minimum distance is $\geq s+1$.

8.3.   Prove that the Hamming distance is a metric on $\mathbb{F}_q^n$.

8.4.   Let H be a parity-check matrix of a linear code. Prove that the code has minimum distance d if and only if any $d-1$ columns of H are linearly independent and there exist d linearly dependent columns.

8.5.   If a linear (n, k) code has minimum distance d, prove that $n - k + 1 \geq d$ (Singleton bound).

8.6.   Let G_1 and G_2 be generator matrices for a linear (n_1, k) code and (n_2, k) code with minimum distance d_1 and d_2, respectively. Show that the linear codes with generator matrices

$$\begin{pmatrix} G_1 & 0 \\ 0 & G_2 \end{pmatrix} \quad \text{and} \quad (G_1, \quad G_2)$$

are $(n_1 + n_2, 2k)$ codes and $(n_1 + n_2, k)$ codes, respectively, with minimum distances $\min(d_1, d_2)$ and $d \geq d_1 + d_2$, respectively.

8.7.   Prove: given k and d, then for a binary linear (n, k) code to have minimum distance $d = d_0$ we must have

$$n \geq d_0 + d_1 + \cdots + d_{k-1},$$

where $d_{i+1} = \lfloor (d_i + 1)/2 \rfloor$ for $i = 0, 1, \ldots, k - 2$. Here $\lfloor x \rfloor$ denotes the largest integer $\leq x$.

8.8.   A code $C \subseteq \mathbb{F}_q^n$ is called *perfect* if for some integer t the balls $B_t(\mathbf{c})$ of radius t centered at code words $\mathbf{c}$ are pairwise disjoint and "fill" the space $\mathbb{F}_q^n$ — that is,

$$\bigcup_{\mathbf{c} \in C} B_t(\mathbf{c}) = \mathbb{F}_q^n.$$

Prove that in the binary case all Hamming codes and all repetition codes of odd length are perfect codes.

8.9.   Using the definition of Exercise 8.8, prove that all Hamming codes over $\mathbb{F}_q$ are perfect.

8.10.   Two linear (n, k) codes C_1 and C_2 over $\mathbb{F}_q$ are called *equivalent* if the code words of C_1 can be obtained from the code words of C_2 by applying a fixed permutation to the coordinate places of all words in C_2. Let G be a generator matrix for a linear code C. Show that any permutation of the rows of G or any permutation of the columns of G gives a generator matrix of a linear code which is equivalent to C.

8.11.   Use the definition of equivalent codes in Exercise 8.10 to show that the binary linear codes with generator matrices

$$G_1 = \begin{pmatrix} 1 & 1 & 1 & 0 \\ 0 & 1 & 1 & 0 \\ 0 & 0 & 1 & 1 \end{pmatrix} \quad \text{and} \quad G_2 = \begin{pmatrix} 1 & 0 & 1 & 1 \\ 0 & 1 & 1 & 1 \\ 1 & 0 & 0 & 1 \end{pmatrix},$$

respectively, are equivalent.

8.12.   Let C be a linear (n, k) code. Prove that the dimension of $C^{\perp}$ is $n - k$.

8.13.   Prove that $(C^{\perp})^{\perp} = C$ for any linear code C.

8.14.   Prove $(C_1 + C_2)^{\perp} = C_1^{\perp} \cap C_2^{\perp}$ for any linear codes C_1, C_2 over $\mathbb{F}_q$ of the same length.

8.15.   If C is the binary $(n, 1)$ repetition code, prove that $C^{\perp}$ is the $(n, n - 1)$ parity-check code.

8.16.   Determine a generator matrix and all code words of the $(7, 3)$ code which is dual to the binary Hamming code C_3.

8.17.   Determine the dual code $C^{\perp}$ to the code given in Exercise 8.1. Find the table of cosets of $\mathbb{F}_2^5$ modulo $C^{\perp}$, determine the coset leaders and syndromes. If $y = 01001$ is a received word, which message was probably sent?

8.18.   Apply Theorem 8.32 to the binary linear code $C = \{000, 011, 101, 110\}$; that is, find its dual code, determine the weight enumerators, and verify the MacWilliams identity.

8.19.   Let C be a binary linear (n, k) code with weight enumerator

$$A(x, y) = \sum_{i=0}^{n} A_i x^i y^{n-i}$$

and let

$$A^{\perp}(x, y) = \sum_{i=0}^{n} A_i^{\perp} x^i y^{n-i}$$

be the weight enumerator of the dual code $C^{\perp}$. Show the following identity for $r = 0, 1, \ldots$:

$$\sum_{i=0}^{n} i^r A_i = \sum_{i=0}^{n} (-1)^i A_i^{\perp} \sum_{t=0}^{r} t! S(r, t) 2^{k-t} \binom{n-i}{n-t},$$

where

$$S(r, t) = \frac{1}{t!} \sum_{j=0}^{t} (-1)^{t-j} \binom{t}{j} j^r$$

is a Stirling number of the second kind and the binomial coefficient $\binom{m}{h}$ is defined to be 0 whenever $h > m$ or $h < 0$. Write down the identity for $r = 0$, 1, and 2.

8.20.   Let $n = (q^m - 1)/(q - 1)$ and β a primitive nth root of unity in $\mathbb{F}_{q^m}$, $m \geqslant 2$. Prove that the null space of the matrix $H = (1 \quad \beta \quad \beta^2 \quad \cdots \quad \beta^{n-1})$ is a code over $\mathbb{F}_q$ with minimum distance at least 3 if and only if $\gcd(m, q - 1) = 1$.

8.21.   Let α be a primitive element of $\mathbb{F}_9$ with minimal polynomial $x^2 - x - 1$ over $\mathbb{F}_3$. Find a generator polynomial for a BCH code of length 8 and dimension 4 over $\mathbb{F}_3$. Determine the minimum distance of this code.

8.22. Find a generator polynomial for a BCH code of dimension 12 and designed distance $d = 5$ over $\mathbb{F}_2$.

8.23. Determine the dimension of a 5-error-correcting BCH code over $\mathbb{F}_3$ of length 80.

8.24. Find the generator polynomial for a 3-error-correcting binary BCH code of length 15 by using the primitive element α of $\mathbb{F}_{16}$ with $\alpha^4 = \alpha^3 + 1$.

8.25. Determine a generator polynomial g for a $(31, 31 - \deg(g))$ binary BCH code with designed distance $d = 9$.

8.26. Let m and t be any two positive integers. Show that there exists a binary BCH code of length $2^m - 1$ which corrects all combinations of t or fewer errors using not more than mt control symbols.

8.27. Describe a Reed-Solomon $(15, 13)$ code over $\mathbb{F}_{16}$ by determining its generator polynomial and the number of errors it will correct.

8.28. Prove that the minimum distance of a Reed-Solomon code with generator polynomial

$$g(x) = \prod_{i=1}^{d-1} (x - \alpha^i)$$

is equal to d.

8.29. Determine if the dual of an arbitrary BCH code is a BCH code. Is the dual of an arbitrary Reed-Solomon code a Reed-Solomon code?

8.30. Find the error locations in Example 8.43, given that the syndrome of a received vector is $(10010110)^T$. Find a generator matrix for this code.

8.31. Let a binary 2-error-correcting BCH code of length 31 be defined by the root α of $x^5 + x^2 + 1$ in $\mathbb{F}_{32}$. Suppose the received word has the syndrome $(1110011101)^T$. Find the error polynomial.

8.32. Let α be a primitive element of $\mathbb{F}_{16}$ with $\alpha^4 = \alpha + 1$, and let $g(x) = x^{10} + x^8 + x^5 + x^4 + x^2 + x + 1$ be the generator polynomial of a binary $(15, 5)$ BCH code. Suppose the word $\mathbf{v} = 000101100100011$ is received. Determine the corrected code word and the message word.

8.33. A code C is called *reversible* if $(a_0, a_1, \ldots, a_{n-1}) \in C$ implies $(a_{n-1}, \ldots, a_1, a_0) \in C$. (a) Prove that a cyclic code $C = (g(x))$ is reversible if and only if with each root of $g(x)$ also the reciprocal value of that root is a root of $g(x)$. (b) Prove that any cyclic code over $\mathbb{F}_q$ of length n is reversible if -1 is a power of q modulo n.

8.34. Given a cyclic (n, k) code, a linear $(n - m, k - m)$ code is obtained by omitting the last m rows and columns in the generator matrix of the cyclic code described prior to Theorem 8.36. Show that the resulting code is in general not cyclic, but that it has at least the same minimum distance as the original code. (*Note*: Such an $(n - m, k - m)$ code is called a *shortened cyclic code*.)

8.35. Let $\Gamma(L, g)$ be a Goppa code over $\mathbb{F}_q$ with $L = \{\gamma_0, \gamma_1, \ldots, \gamma_{n-1}\} \subseteq \mathbb{F}_{q^m}$. Prove that $(c_0, c_1, \ldots, c_{n-1}) \in \mathbb{F}_q^n$ is a code word of $\Gamma(L, g)$ if and only if the congruence

$$\sum_{i=0}^{n-1} \frac{c_i}{x - \gamma_i} \equiv 0 \quad \mod g(x)$$

holds, where $1/(x - \gamma_i)$ is interpreted as the multiplicative inverse of $x - \gamma_i$ in the residue class ring $\mathbb{F}_{q^m}[x]/(g)$.

8.36. Let $\Gamma(L, g)$ be a Goppa code over $\mathbb{F}_q$ whose Goppa polynomial of degree t has t distinct roots $\beta_1, \ldots, \beta_t$ in a suitable extension of $\mathbb{F}_{q^m}$, and let $L = \{\gamma_0, \gamma_1, \ldots, \gamma_{n-1}\} \subseteq \mathbb{F}_{q^m}$. Prove that $\Gamma(L, g)$ is the intersection of $\mathbb{F}_q^n$ with the null space of the $t \times n$ matrix whose entry in the jth row and ith column is $(\beta_j - \gamma_{i-1})^{-1}$ for $1 \leqslant j \leqslant t$, $1 \leqslant i \leqslant n$.

8.37. Prove that the minimum distance of a binary irreducible Goppa code with Goppa polynomial of degree t is at least $2t + 1$.

8.38. Determine the dimension of the binary Goppa code $\Gamma(L, g)$ with $L = \mathbb{F}_{16}^*$, $g(x) = x^2 + x + \alpha^3$, and α a primitive element of $\mathbb{F}_{16}$.

8.39. Determine the dimension of the binary Goppa code $\Gamma(L, g)$ with $L = \mathbb{F}_{16}$ and $g(x) = x^3 + x + 1$. Find also a generator matrix for this code.

8.40. Determine the transmitted code word in Exercise 8.32 by the algorithm in Section 3.

8.41. Determine the transmitted code word in Example 8.43 by the algorithm in Section 3.

8.42. Determine the transmitted code word in Example 8.46 by the algorithm in Section 3.

8.43. Let $r_{-1}(x)$ and $r_0(x)$ be two nonzero polynomials over a field F with $\deg(r_{-1}(x)) \geqslant \deg(r_0(x))$. The Euclidean algorithm yields

$$r_{h-1}(x) = q_{h+1}(x) r_h(x) + r_{h+1}(x), \quad \deg(r_{h+1}(x)) < \deg(r_h(x)),$$
$$\text{for } h = 0, 1, \ldots, s - 1,$$

$$r_{s-1}(x) = q_{s+1}(x) r_s(x).$$

Define recursively the polynomials

$$z_{-1}(x) = 0, \quad z_0(x) = 1,$$
$$z_h(x) = z_{h-2}(x) - q_h(x) z_{h-1}(x) \quad \text{for } h = 1, 2, \ldots, s.$$

Prove the following properties:

(a) $r_h(x) \equiv z_h(x) r_0(x) \mod r_{-1}(x)$ for $h = -1, 0, \ldots, s$;
(b) $z_h(x) r_{h-1}(x) - z_{h-1}(x) r_h(x) = (-1)^h r_{-1}(x)$ for $h = 0, 1, \ldots, s$;
(c) $\deg(z_h(x)) = \deg(r_{-1}(x)) - \deg(r_{h-1}(x))$ for $h = 0, 1, \ldots, s$.

8.44. An *alternant code* A over $\mathbb{F}_q$ is defined as follows. Let $h_1, \ldots, h_n$ be arbitrary elements of $\mathbb{F}_{q^m}^*$ and let $\alpha_1, \ldots, \alpha_n$ be distinct elements of $\mathbb{F}_{q^m}$. Fix an integer t with $1 \leqslant t < n$. Then A consists of all vectors in $\mathbb{F}_q^n$ that are in the null space of the $t \times n$ matrix

$$\begin{pmatrix} h_1 & h_2 & \cdots & h_n \\ h_1\alpha_1 & h_2\alpha_2 & \cdots & h_n\alpha_n \\ h_1\alpha_1^2 & h_2\alpha_2^2 & \cdots & h_n\alpha_n^2 \\ \vdots & \vdots & & \vdots \\ h_1\alpha_1^{t-1} & h_2\alpha_2^{t-1} & \cdots & h_n\alpha_n^{t-1} \end{pmatrix}.$$

Show that any Goppa code is an alternant code. Prove that the dimension of A is at least $n - mt$ and that its minimum distance is at least $t + 1$.

Chapter 9

Cryptology

In this chapter we consider some aspects of cryptology that have received considerable attention over the last few years. Cryptology is concerned with the designing and the breaking of systems for the communication of secret information. Such systems are called cryptosystems or cipher systems or ciphers. The designing aspect is called cryptography, the breaking is referred to as cryptanalysis. The rapid development of computers, the electronic transmission of information, and the advent of electronic transfer of funds all contributed to the evolution of cryptology from a government monopoly that deals with military and diplomatic communications to a major concern of business. The concepts have changed from conventional (private-key) cryptosystems to public-key cryptosystems that provide privacy and authenticity in communication via transfer of messages. Cryptology as a science is in its infancy since it is still searching for appropriate criteria for security and measures of complexity of cryptosystems.

Conventional cryptosystems date back to the ancient Spartans and Romans. One elementary cipher, the Caesar cipher, was used by Julius Caesar and consists of a single key $K = 3$ such that a message M is transformed into $M + 3$ modulo 26, where the integers $0, 1, \ldots, 25$ represent the letters $A, B, \ldots,$ Z of the alphabet. An obvious generalization of this cipher leads to the substitution ciphers often named after de Vigenère, a French cryptographer of the 16th century. Mechanical cipher devices based on such cryptosystems started to appear in the 19th century and were widely used in both World Wars.

Significant advances in cryptanalysis, for instance the breaking of the German ENIGMA cipher in World War II, have led to the necessity of developing more sophisticated cryptosystems, some of which will be described in this chapter.

In Section 1 a general background on cryptology is given and the distinction between conventional and public-key cryptosystems is discussed. The most secure cryptosystem is the one-time pad in which a random string of bits is added modulo 2 to a binary message. Since this requires very long keys, one has come up with the notion of a stream cipher in which a shorter key generates long strings of bits. This concept is studied in more detail in Section 2.

Some very recent developments in cryptography are based on the use of discrete exponentiation in finite fields. A scrutiny of these cipher systems from the viewpoint of the cryptanalyst leads to a study of the inverse function—that is, the index or discrete logarithm in finite fields. In particular, it becomes necessary to analyze the computational complexity of the discrete logarithm. Various applications of discrete exponentiation and discrete logarithms to cryptology and several algorithms for the calculation of discrete logarithms are presented in Section 3. Two more cryptosystems, one based on Goppa codes and one on polynomial interpolation in finite fields, are discussed in Section 4.

1. BACKGROUND

Cryptosystems are designed to transform plaintext messages into ciphertexts. The particular transformations applied at any given time are controlled by the key of the cryptosystem used at that time. In conventional cryptosystems this key is supposed to be known to both the legitimate sender and the legitimate receiver, but not to the attacker (or cryptanalyst) who wants to break the cryptosystem.

The general structure of a *cryptosystem* can be described as follows. The main ingredients are an *enciphering scheme E* (for encryption), a *deciphering scheme D* (for decryption), a *key K*, the *plaintext message* (or simply *plaintext* or *message*) M, and the *ciphertext C*. Given a plaintext message M and a key K, the enciphering scheme produces the ciphertext $C = E_K(M)$ which is transmitted. The deciphering scheme recovers M by $D_K(C) = M$. One basic requirement is that E_K be injective—that is, E_K should transform distinct messages into distinct ciphertexts. In this notation the parameter K remains fixed for a considerable number of messages. If only one key K is involved, the system is called a *conventional* (or *single-key*) cryptosystem.

An attacker is assumed to have full knowledge of the general form of the enciphering and deciphering schemes, has access to a number of plaintext–ciphertext pairs produced by the cryptosystem, and has additional inform-

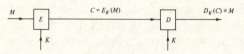

FIGURE 9.1 A cryptosystem.

ation such as language statistics (letter frequencies and so on) and an idea about the general context of the communication. The attacker does not know the key K and has the task to produce the best estimate M' of M. Breaking a system means determining the key K.

As most current data are stored, transmitted, and processed in binary form, cryptosystems over the binary alphabet $\mathbb{F}_2 = \{0, 1\}$ are of particular importance, but other alphabets such as $\mathbb{F}_q$ are also possible. Thus both plaintext and ciphertext are often given in the form of a string of 0's and 1's (or *bits*). If the plaintext string is broken into blocks of fixed length and then enciphered on a block-by-block basis, the corresponding scheme is called a *block cipher*. In this chapter—as in this whole book—we are mainly interested in material directly connected with finite fields, and accordingly we will be concentrating on certain types of cryptosystems. However, we mention one of the commercially widely used block ciphers, the DES (Data Encryption Standard), which is the official system adopted by the National Bureau of Standards of the United States and used by most U.S. Federal Departments. It is a cryptosystem with 64-bit data blocks and a 64-bit key; 56 bits of the key are true key bits, the remaining 8 bits are used for error detection.

The main disadvantage of conventional cryptosystems is that they require the advance establishment of a secret (or private) key between every pair of correspondents. This makes proper management of the keys a crucial problem for the security of the system. Key management is increasingly difficult if a large number of correspondents are involved in a communication system, because then it will be even harder to ensure key secrecy. In 1976 Diffie and Hellman suggested how to overcome some of these problems by introducing *public-key cryptosystems*. Public-key cryptosystems ensure that subscribers who have never met or communicated before could have instant secure communication. In general terms, each subscriber places an enciphering procedure E into a public directory to be used by other subscribers while keeping secret his corresponding deciphering procedure D. These procedures, applied to message M or ciphertext C, must have the following properties:

(i) If $C = E(M)$, then $M = D(C)$; hence

$$D(E(M)) = M \text{ for each } M.$$

(ii) E and D must be fast and easy to apply.

(iii) E can be made public without revealing D—that is, deriving D from E must be computationally infeasible.

For instance, if A wants to send a message M to B, he looks up B's public enciphering method E_B and transmits $C = E_B(M)$ to B in the open. Only B can decipher C, since only B knows the secret deciphering method D_B to apply to C.

Privacy or security of messages is not the only problem area in cryptology. It is also important that the correspondents or subscribers can be authenticated. For example, A has to be able to convince B that it is really from A the message came. The log-on procedure on computers is also an obvious example of *authentication*. The problem area of authentication or of *digital signatures* is increasingly important as computer networks, electronic mail, and similar communication systems grow. Digital signature features can be attained by public-key cryptosystems if we add a fourth property:

(iv) D can be applied to every M, and if $S = D(M)$, then $M = E(S)$; hence

$$E(D(M)) = M \text{ for each } M.$$

With this property, subscriber A can sign his message to B by first forming his message-dependent signature $S = D_A(M)$ and then computing $C = E_B(S)$. Only B can recover S by applying the secret deciphering method D_B to C. Then B computes $E_A(S) = E_A(D_A(M)) = M$, by using A's public enciphering method E_A. Now B can be satisfied that M came from A since no other person would have used A's secret deciphering method D_A to compute $S = D_A(M)$.

Public-key cryptosystems can be implemented by using trapdoor one-way functions. A function f is said to be *one-way* if f is easy to compute and invertible, but it is computationally infeasible to compute the inverse function f^{-1} from a complete description of f. A function f is *trapdoor one-way* if f^{-1} is easy to compute once certain private trapdoor information is known, but without this information f would be one-way. An example of a trapdoor one-way function is contained in the RSA cryptosystem (see Section 3); it is based on exponentiation and the difficulty of factorization of integers. Another trapdoor one-way function is based on the difficulty of the general decoding problem for linear error-correcting codes (see the Goppa-code cryptosystem in Section 4).

A major problem area in cryptology is to find appropriate criteria for the complexity of a cryptosystem that will replace the present unsatisfactory method of "certifying" a cryptosystem as secure through heuristics or concentrated man/computer years of efforts rather than rigorous proof. Computational complexity theory seems to offer a suitable framework for doing that, since there one can classify problems as "hard". A problem is said to belong to the class P (for polynomial time) if there exists a deterministic algorithm that will solve every instance of the problem in a running time bounded by some polynomial in the number of bits needed for the binary representation of the problem parameters. Problems that can be solved in polynomial time by a nondeterministic algorithm—that is, by an algorithm in which random choices are allowed in each step—make up the class NP (for nondeterministic polynomial time). Clearly, P is a subclass of NP. It is a fundamental open question of complexity theory whether $P = NP$. Particularly interesting problems in the class NP from the viewpoint of complexity theory are the *NP-complete* problems, which have the property that if any one problem of the NP-complete class is found to be in P, then all of NP belongs to P. Examples of NP-complete problems are the graph coloring problem, the traveling salesman problem, and the knapsack packing problem.

The security of public-key cryptosystems is based on the computational infeasibility of performing certain tasks—such as factoring integers, decoding linear codes, or finding discrete logarithms in finite fields—with the best algorithms and the best hardware publicly available. Of course, there may be secret advances in software or hardware we do not know about.

2. STREAM CIPHERS

The simplest and most secure of all cryptosystems is the *one-time pad*. Suppose the message is given as a string of bits—that is, of elements of $\mathbb{F}_2$. Then a long random string of bits is formed; this is the key which is known to sender and receiver. The sender adds this key to the message, using addition in $\mathbb{F}_2$. At the receiving end the key is again added in $\mathbb{F}_2$ to the enciphered message to recover the original message. The key string must be at least as long as the message string and is used only once. This is a perfect, unbreakable cipher since all the different key strings and all possible messages are equally likely. The major disadvantage of this cryptosystem is that it requires as much key as there is data to be sent. So it is restricted to sending only important messages.

In a *stream cipher* one uses a much smaller key as the seed to produce longer key strings—or even infinite key sequences—which are then added to the message string. One possibility is to use feedback shift registers where certain initial values suffice to produce infinite linear recurring sequences in $\mathbb{F}_2$ (compare with Chapter 6, Section 1). Before we consider a specific cryptosystem, we list some general properties a stream cipher should have:

(i)   The number of possible keys must be large enough so that an exhaustive search for the key is not feasible.
(ii) The infinite sequences must have a guaranteed minimum length for their periods which exceeds the length of the message strings.
(iii) The ciphertext must appear to be random.

There are a number of properties a random sequence of bits should satisfy. We refer to Chapter 7, Section 4, for more details.

On first glance it would seem that certain homogeneous linear recurring sequences σ in $\mathbb{F}_2$ are good candidates for key sequences. We know from Theorem 6.33 that if the characteristic polynomial of σ is primitive over $\mathbb{F}_2$ of degree k and σ is not the zero sequence, then σ has least period $2^k - 1$, which can be made arbitrarily large as k varies. There are many such primitive polynomials available, namely $\phi(2^k-1)/k$. These maximal period sequences σ (see Definition 6.32) of least period 2^k-1 satisfy the basic randomness requirements imposed on sequences, as we have shown in Chapter 7, Section 4. Nevertheless, linear recurring sequences are not suitable for constructing secure cryptosystems, since it follows from the discussion on p. 231 that if we know that such a sequence has a characteristic polynomial of degree $\leqslant k$, then any $2k$ consecutive terms determine a characteristic polynomial and thus the entire sequence. In spite of the proven insecurity of this cryptosystem, it is quite popular, perhaps because the large periods 2^k-1 create an illusion of strength.

Because of this weakness of linear recurring sequences, we have to consider pseudorandom generators of higher complexity. One possibility is to increase complexity by appropriately combining linear recurring sequences. We shall only describe one such approach, namely the construction of multiplexed sequences which may be used as building blocks in a cryptosystem in the category of stream ciphers. Here a multiplexer, which is a many-input-one-output system, is used to produce a multiplexed sequence from two given linear recurring sequences. The construction can be carried out over any finite prime field $\mathbb{F}_p$.

9.1. Definition. A *multiplexed sequence* $u_0, u_1, \ldots$ in $\mathbb{F}_p$ is constructed as follows:

(i) Let $s_0, s_1, \cdots$ be a kth-order and $t_0, t_1, \cdots$ an mth-order maximal period sequence in $\mathbb{F}_p$.

(ii) Choose an integer h in the range $1 \leqslant h \leqslant k$ such that $p^h \leqslant m$ if $h < k$ and $p^h - 1 \leqslant m$ if $h = k$.

(iii) Choose integers $j_1, \cdots, j_h$ with $0 \leqslant j_1 < j_2 < \cdots < j_h \leqslant k - 1$.
For $n = 0, 1, \cdots$ consider the h-tuple $(s_{n+j_1}, \cdots, s_{n+j_h})$ of elements of $\mathbb{F}_p$ and interpret it as the digital representation in the base p of an integer $b_n \in I_h$, where

$$I_h = \{0, 1, \cdots, p^h - 1\} \qquad \text{if } h < k,$$

$$I_h = \{1, 2, \cdots, p^h - 1\} \qquad \text{if } h = k.$$

(iv) Choose an injective mapping ψ from I_h into $\{0, 1, \cdots, m - 1\}$.

(v) With these choices of $h, j_1, \cdots, j_h$, and ψ we set

$$u_n = t_{n + \psi(b_n)} \qquad \text{for } n = 0, 1, \cdots.$$

Some comments on this definition are in order. We note that, if $h < k$, then all elements of $\mathbb{F}_p^h$ appear among the h-tuples $(s_{n+j_1}, \cdots, s_{n+j_h})$ as n varies from 0 to $p^k - 2$, and so b_n runs exactly through the values in I_h. If $h = k$, then necessarily $j_i = i - 1$ for $1 \leqslant i \leqslant k$, and the k-tuples $(s_n, \cdots, s_{n+k-1})$ are just the state vectors of the sequence $s_0, s_1, \cdots$; the fact that b_n runs exactly through the values in I_k follows therefore from Theorem 7.43. We note also that the existence of an injection ψ in (iv) is guaranteed by the condition on m in (ii). The definition in (v) says that we obtain the multiplexed sequence by scrambling the terms of the sequence $t_0, t_1, \cdots$ in a way that is controlled by the sequence $s_0, s_1, \cdots$.

9.2. Example. Let $p = 2$, and let $s_0, s_1, \cdots$ and $t_0, t_1, \cdots$ be the maximal period sequences in $\mathbb{F}_2$ with

$$s_{n+3} = s_{n+1} + s_n \qquad \text{for } n = 0, 1, \cdots,$$

$$t_{n+4} = t_{n+3} + t_n \qquad \text{for } n = 0, 1, \cdots,$$

and initial state vectors $(1, 0, 0)$ and $(1, 0, 0, 0)$, respectively. The first sequence has least period 7 and the terms in the period are

$$1 \quad 0 \quad 0 \quad 1 \quad 0 \quad 1 \quad 1.$$

The second sequence has least period 15 and the terms in the period are

$$1 \quad 0 \quad 0 \quad 0 \quad 1 \quad 1 \quad 1 \quad 1 \quad 0 \quad 1 \quad 0 \quad 1 \quad 1 \quad 0 \quad 0.$$

Now choose $h = 2$, $j_1 = 0$, $j_2 = 1$, and define the injective mapping ψ from $\{0, 1, 2, 3\}$ into itself by

$$\psi(0) = 1, \quad \psi(1) = 2, \quad \psi(2) = 3, \quad \psi(3) = 0.$$

The sequence $b_0, b_1, \cdots$ of integers in Definition 9.1(iii) has least period 7 and the terms in the period are

$$2 \quad 0 \quad 1 \quad 2 \quad 1 \quad 3 \quad 3.$$

Consequently, the first few terms of the multiplexed sequence $u_0, u_1, \cdots$ are

$$0 \quad 0 \quad 1 \quad 1 \quad 1 \quad 1 \quad 0 \quad 1 \quad 1 \quad 0 \quad 0 \quad 1 \quad 0 \quad 0 \quad 0 \cdots.$$

The diagrammatic representation of the two feedback shift registers and the multiplexer is given in Fig. 9.2. The delay elements of the first feedback shift register are labeled by A_0, A_1, A_2 and those of the second feedback shift register are labeled by B_0, B_1, B_2, B_3. $\square$

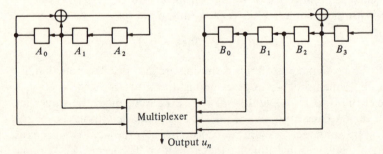

FIGURE 9.2   The switching circuit for Example 9.2.

9.3. Theorem. *The multiplexed sequence* $u_0, u_1, \cdots$ *is periodic and its least period divides* $\mathrm{lcm}(p^k - 1, p^m - 1)$.

Proof. Put $r = \mathrm{lcm}(p^k - 1, p^m - 1)$. Since r is a multiple of the period $p^k - 1$ of $s_0, s_1, \cdots$, we have

$$(s_{n+j_1}, \cdots, s_{n+j_h}) = (s_{n+r+j_1}, \cdots, s_{n+r+j_h})$$

and so $b_n = b_{n+r}$ for all $n \geqslant 0$. Since r is a multiple of the period $p^m - 1$ of $t_0, t_1, \cdots$, we obtain

$$u_{n+r} = t_{n+r+\psi(b_{n+r})} = t_{n+r+\psi(b_n)} = t_{n+\psi(b_n)} = u_n$$

for all $n \geqslant 0$. The rest follows from Lemmas 6.4 and 6.6. □

The following property of certain decimations of multiplexed sequences can be applied to obtain further information on the least period. We use again the notation for decimations introduced in Chapter 7, Section 4— that is, if σ is a sequence with terms $s_0, s_1, \cdots$, then the decimated sequence $\sigma_d^{(i)}$ is obtained by taking every dth term of σ, starting from s_i.

9.4. Lemma. *If* v *denotes the multiplexed sequence* $u_0, u_1, \cdots$ *and* τ *denotes the sequence* $t_0, t_1, \cdots$ *in Definition 9.1(i), then*

$$v_d^{(i)} = \tau_d^{(j(i))} \qquad for\ i = 0, 1, \cdots,$$

where $d = p^k - 1$ *and* $j(i) = i + \psi(b_i)$.

Proof. The terms of $v_d^{(i)}$ are the elements u_{nd+i}, $n = 0, 1, \cdots$. Since $d = p^k - 1$ is the least period of the sequence $s_0, s_1, \cdots$ in Definition 9.1(i), we have $b_{nd+i} = b_i$ by the construction in Definition 9.1(iii), and so

$$u_{nd+i} = t_{nd+i+\psi(b_{nd+i})} = t_{nd+i+\psi(b_i)}$$

for all $n \geqslant 0$, which is the desired result. □

9.5. Lemma. *For any integers* $a \geqslant 2$, $k \geqslant 1$, *and* $m \geqslant 1$ *we have* $\gcd(a^k - 1, a^m - 1) = a^{\gcd(k,m)} - 1$.

Proof. If $b = \gcd(k, m)$, then it is clear that $a^b - 1$ divides $c = \gcd(a^k - 1, a^m - 1)$. Now write $k = dm + e$ with integers $d \geqslant 0$ and $0 \leqslant e < m$. Then

$$a^k - 1 = (a^{dm} - 1)a^e + (a^e - 1),$$

and so c divides $a^e - 1$. Continuing this process in analogy with the Euclidean algorithm for k and m, we find that c divides $a^b - 1$, hence $c = a^b - 1$. □

9.6. Theorem. *If* $\gcd(k, m) = 1$, *then the least period of the multiplexed sequence* $u_0, u_1, \cdots$ *is a multiple of* $(p^m - 1)/(p - 1)$.

Proof. We apply Lemma 9.4 with $i = 0$. Then $v_d^{(0)} = \tau_d^{(j)}$ with $d = p^k - 1$ and $j = j(0)$. Put $K = \mathbb{F}_p$ and $F = \mathbb{F}_{p^m}$. Since τ is an mth-order maximal period sequence in K, it follows from Theorem 6.24 that there exists a primitive element α of F and a $\theta \in F^*$ such that

$$t_n = \mathrm{Tr}_{F/K}(\theta \alpha^n) \qquad for\ n = 0, 1, \cdots.$$

The terms w_n of $\tau_d^{(j)}$ are thus given by

$$w_n = t_{nd+j} = \text{Tr}_{F/K}(\gamma\beta^n) \qquad \text{for } n = 0, 1, \cdots, \tag{9.1}$$

where $\beta = \alpha^d \in F^*$ and $\gamma = \theta\alpha^j \in F^*$. By Theorem 1.15(ii), the order of β in the multiplicative group F^* is

$$\frac{p^m - 1}{\gcd(d, p^m - 1)} = \frac{p^m - 1}{\gcd(p^k - 1, p^m - 1)} = \frac{p^m - 1}{p - 1},$$

where we used Lemma 9.5 and the hypothesis $\gcd(k, m) = 1$ in the last step. Let $f(x)$ be the minimal polynomial of β over K. Then it follows from (9.1) and the calculation in the proof of Theorem 6.24 that $f(x)$ is a characteristic polynomial of the linear recurring sequence $\tau_d^{(j)}$. We claim that $\tau_d^{(j)}$ is not the zero sequence. We have

$$\frac{p^m - 1}{p - 1} \geqslant p^{m-1}. \tag{9.2}$$

Furthermore, the elements $\gamma\beta^n$, $0 \leqslant n < (p^m - 1)/(p - 1)$, are $(p^m - 1)/(p - 1)$ distinct elements of F^*. Since there are just $p^{m-1} - 1$ elements $\delta \in F^*$ with $\text{Tr}_{F/K}(\delta) = 0$ by Theorem 2.23(iii), it follows from (9.1) and (9.2) that $w_n = \text{Tr}_{F/K}(\gamma\beta^n) \neq 0$ for some n. Thus, indeed, $\tau_d^{(j)}$ is not the zero sequence, and since $f(x)$ is irreducible over K by Theorem 3.33(i), the least period of $\tau_d^{(j)}$ is equal to $\text{ord}(f(x)) = (p^m - 1)/(p - 1)$ according to Theorems 6.28 and 3.3. If r is the least period of $u_0, u_1, \ldots$, then r is a period of the decimated sequence $v_d^{(0)} = \tau_d^{(j)}$, and so Lemma 6.4 shows that r is divisible by $(p^m - 1)/(p - 1)$. $\square$

It can be proved that if $p = 2$, $\gcd(k, m) = 1$, and $m > 1$, then the least period of the multiplexed sequence $u_0, u_1, \cdots$ is equal to $(2^k - 1)(2^m - 1)$. For instance, the least period of the multiplexed sequence in Example 9.2 is equal to $(2^3 - 1)(2^4 - 1) = 105$.

As to the application of multiplexed sequences in stream ciphers, it appears that such sequences may be quite complex, but further research will be needed in order to establish that their complexity is sufficiently high.

3. DISCRETE LOGARITHMS

Let b be a primitive element of $\mathbb{F}_q$ and let a be a nonzero element of $\mathbb{F}_q$. Then the *index* of a with respect to the base b is the uniquely determined integer r, $0 \leqslant r < q - 1$, for which $a = b^r$. We use the notation $r = \text{ind}_b(a)$, or simply $r = \text{ind}(a)$ if b is kept fixed. The index of a is also called the *discrete logarithm* of a. The discrete exponential function $\exp_b(r) = \exp(r) = b^r$ and the discrete logarithm form a pair of inverse functions of each other; compare also with Chapter 10, Section 1. Their use for cryptography depends on the apparent one-way nature of the discrete exponential function: it is easy to compute, but appears hard to invert.

The discrete exponential function $\exp(r) = b^r$ in $\mathbb{F}_q$ can be calculated for $1 \leqslant r < q - 1$ by an analog of the repeated squaring technique discussed after Theorem 4.13, which is often called the *square and multiply* technique in the present context. In detail, we first compute the elements $b, b^2, b^4, \cdots, b^{2^e}$ by repeated squaring, where 2^e is the largest power of 2 that is $\leqslant r$. Then b^r is obtained by multiplying together an appropriate combination of these elements. For instance, to get b^{27} one would multiply together the elements b, b^2, b^8, and b^{16}. A simple analysis shows that the calculation of b^r requires at most $2\lfloor \log_2 q \rfloor$ multiplications in $\mathbb{F}_q$, where $\log_2$ denotes the real logarithm to the base 2.

Until recently, the inverse problem of computing discrete logarithms in $\mathbb{F}_q$ was believed to be much harder, since for one of the best algorithms available then the required number of arithmetic operations in $\mathbb{F}_q$ was of the order of magnitude $q^{1/2}$. If q is sufficiently large, say $q > 2^{100}$, exponentiation in $\mathbb{F}_q$ might justly have been regarded as a one-way function. However, great progress has recently been achieved in the computation of discrete logarithms, which makes it necessary to construct cryptosystems based on discrete exponentiation in a careful manner in order to protect them against attacks by these recent algorithms. We now describe some cryptographic applications of discrete exponentiation and then present some discrete logarithm algorithms.

9.7. Example. The following is a cryptosystem for message transmission in $\mathbb{F}_q$. Let M, K, and C denote the plaintext message, the key, and the ciphertext, respectively, where $M, C \in \mathbb{F}_q^*$, K is an integer with $1 \leqslant K \leqslant q - 2$ and $\gcd(K, q - 1) = 1$, and q is a large prime power. The last condition on K makes it possible to solve the congruence

$$KD \equiv 1 \bmod (q - 1) \tag{9.3}$$

for the integer D. We encipher by computing

$$C = M^K \tag{9.4}$$

and decipher by

$$C^D = M. \tag{9.5}$$

Both operations are easily performed. To find the key, however, is as hard as finding discrete logarithms since (9.4) is equivalent to

$$K \operatorname{ind}(M) \equiv \operatorname{ind}(C) \bmod (q - 1). \tag{9.6}$$

Even if we know a plaintext–ciphertext pair M and C, computing K can be expected to be difficult for large q. From (9.6) we see that M must be a primitive element of $\mathbb{F}_q$ so that M and C determine K uniquely. We also observe that there is a wide choice for the key K since for $q > 2$ there are $\phi(q - 1)$ integers K that satisfy $1 \leqslant K \leqslant q - 2$ and $\gcd(K, q - 1) = 1$. Primes of the form $q = 2p + 1$, p also a prime, are the most promising values of q to use in order to get a secure system. For primes q we may view M and C as integers with $1 \leqslant M, C \leqslant q - 1$, and then (9.4) and (9.5) are replaced by the congruences

$$C \equiv M^K \bmod q, \qquad C^D \equiv M \bmod q. \qquad \square$$

The cryptosystem in Example 9.7 can be made into a new system by replacing congruences modulo a large prime q by congruences modulo a product n of two large primes p and q. Such a cryptosystem was proposed by Rivest, Shamir, and Adleman and is now known as the *RSA cryptosystem*. Instead of using (9.3), we now find D from

$$KD \equiv 1 \bmod \phi(n) \tag{9.7}$$

in this generalized system, where we assume $\gcd(K, \phi(n)) = 1$. The security of the RSA cryptosystem is based on keeping the factors of n secret and depends on the difficulty of factoring large integers into primes. Normally n would be a product of two primes with approximately 100 decimal digits each. At present, the factorization of arbitrary integers is computationally feasible only if they have at most about 70 decimal digits. The RSA cryptosystem is an example of a public-key cryptosystem with public keys K and n that do not compromise the secret deciphering key D. Only by knowing the factors of n it is possible to solve (9.7) for D.

Of special interest in Example 9.7 is the finite field $\mathbb{F}_{2^m}$, where $2^m - 1$ is a large Mersenne prime, because with this choice all the keys K with $1 \leqslant K \leqslant 2^m - 2$ can be used. The field with 2^{127} elements attracted particular attention. It is generated by the primitive trinomial $x^{127} + x + 1$ over $\mathbb{F}_2$ and is used to implement a cryptosystem with discrete exponentiation. This particular system has recently been shown to be totally insecure; compare also with the discussion following Example 9.13.

9.8. Example. An application of discrete exponentiation to computer systems is the following. In such systems users' passwords are stored in specially protected files so that only authorized users have access to them. This can be achieved by utilizing discrete exponentiation as a candidate for a one-way function f by creating a public file of pairs $(i, f(p_i))$, where i denotes the user's log-on name and p_i is the user's password. $\qquad \square$

9.9. Example. Discrete exponentiation can be used to create a well-known key-exchange system, the *Diffie-Hellman scheme*. Suppose users A and B wish to communicate by using a standard high-speed cryptosystem such as DES, but they do not have a common key. They choose random integers h and k, respectively, where $2 \leqslant h, k \leqslant q - 2$. Let b be a primitive element of $\mathbb{F}_q$. Then A sends b^h to B, while B transmits b^k to A. Both take b^{hk} as their common key, which can be computed by A as $(b^k)^h$ and by B as $(b^h)^k$. It is an unsolved problem to generate b^{hk} from knowledge of b^h and b^k only, without computing either h or k. The public-key cryptosystems that are known today have the disadvantage that they are rather slow. Therefore their main use is for the distribution of keys for conventional cryptosystems. □

9.10. Example. Consider the following conventional system for message transmission. User A wishes to send a message m—regarded as a nonzero element of the publicly known field $\mathbb{F}_q$—to user B. Then A chooses a random integer h, where $1 \leqslant h \leqslant q - 1$ and $\gcd(h, q - 1) = 1$, and transmits $x = m^h$ to B. User B chooses a random integer k, where $1 \leqslant k \leqslant q - 1$ and $\gcd(k, q - 1) = 1$, and sends $y = x^k = m^{hk}$ to A. Now A forms $z = y^{h'}$, where $hh' \equiv 1 \bmod (q - 1)$, and sends z to B. Then B only has to compute $z^{k'}$ to recover m, where $kk' \equiv 1 \bmod (q - 1)$, since

$$z = y^{h'} = m^{hkh'} = m^k \text{ and } z^{k'} = m^{kk'} = m.$$

This three-pass procedure between A and B is also known as the *no-key algorithm*, where users A and B keep their own respective key pairs (h, h') and (k, k') secret. □

9.11. Example. Consider the following public-key cryptosystem for message transmission. Let b be a primitive element of $\mathbb{F}_q$, where q and b are known publicly. Let A's public key be the element $b^h \in \mathbb{F}_q$, where h is kept secret by A. If B wants to send a message $m \in \mathbb{F}_q^*$ to A, then B selects a random integer k, $1 \leqslant k \leqslant q - 2$, and transmits the pair (b^k, mb^{hk}) to A. Since A knows h, he can compute $b^{hk} = (b^k)^h$ and so recover m. This cryptosystem could be broken by computing h or k with an efficient discrete logarithm algorithm. □

9.12. Example. The following is a digital signature scheme using discrete exponentiation. If user A wishes to attach a digital signature to a message m with $1 \leqslant m \leqslant p - 1$, he publishes a prime p, a primitive element b of $\mathbb{F}_p$ identified with an integer, and an integer c, $1 \leqslant c \leqslant p - 1$, obtained from a secret random integer h such that $c \equiv b^h \bmod p$. To sign m, A provides a pair (r, s) of integers with $1 \leqslant r \leqslant p - 1$, $0 \leqslant s \leqslant p - 1$, such that

$$b^m \equiv c^r r^s \bmod p.$$

The integer r is generated from a random integer k with $\gcd(k, p - 1) = 1$ by computing $r \equiv b^k \bmod p$. Then s has to satisfy

$$b^m \equiv b^{hr} b^{ks} \equiv b^{hr + ks} \bmod p,$$

which is equivalent to

$$m \equiv hr + ks \bmod (p - 1).$$

The unique solution s of this congruence is easily obtained by A since he knows h, r, and k. If an attacker could compute h from c by using a discrete logarithm algorithm, this digital signature scheme would be insecure. □

Before we describe several discrete logarithm algorithms for $\mathbb{F}_q$, we make a few general observations. As above, we repeatedly use the fact that arithmetic in the exponents is done modulo $q - 1$ since $b^{q-1} = b^0 = 1$ for any primitive element b of $\mathbb{F}_q$. In the case of a prime field $\mathbb{F}_p$ it is often convenient to identify elements of $\mathbb{F}_p$ with integers, so that identities in $\mathbb{F}_p$ are also written as congruences modulo p.

Next we observe that it is not difficult to find the discrete logarithms of arbitrary elements of $\mathbb{F}_q^*$ under the assumption that discrete logarithms are "easy" to calculate (or known) for a relatively small portion of all the elements of $\mathbb{F}_q^*$. For suppose it is easy to compute $\mathrm{ind}_b(a) = \mathrm{ind}(a)$ for a set E of $\varepsilon(q - 1)$ special elements $a \in \mathbb{F}_q^*$, where $0 < \varepsilon < 1$. If a given $a_0 \in \mathbb{F}_q^*$ is not in E, take a uniform random sample t_1 from $\{0, 1, \cdots, q - 2\}$ and define $a_1 = a_0 b^{t_1}$. If $a_1 \in E$, so that $\mathrm{ind}(a_1)$ is easy to compute, then

$$\mathrm{ind}(a_0) \equiv \mathrm{ind}(a_1) - t_1 \bmod (q - 1).$$

Otherwise, take independent and uniform random samples $t_2, t_3, \cdots$ from $\{0, 1, \cdots, q - 2\}$ until an $a_i = a_0 b^{t_i}$ in E is found. Then $\mathrm{ind}(a_0)$ can be calculated by subtraction. Note that the probability that all the elements $a_0, a_1, \cdots, a_k$ are outside of E is $(1 - \varepsilon)^{k+1}$, which rapidly becomes small.

As an illustration consider the case of a prime field $\mathbb{F}_p$. If the discrete logarithms in $\mathbb{F}_p$ of the first n primes $2 = p_1 < \cdots < p_n < p$ are known and if an integer a satisfies

$$a \equiv \prod_{j=1}^{n} p_j^{e_j} \bmod p,$$

then

$$\mathrm{ind}(a) \equiv \sum_{j=1}^{n} e_j \, \mathrm{ind}(p_j) \bmod (p - 1).$$

Integers which factor completely into small primes are called "smooth". In the case described here it is easy to compute $\mathrm{ind}(a)$ if a is smooth and the values $\mathrm{ind}(p_j)$ are known. The set of smooth integers is then an example of a set E from above. The density of smooth integers is crucial in the analysis of several discrete logarithm algorithms.

We first present the *Silver-Pohlig-Hellman algorithm* for computing discrete logarithms in $\mathbb{F}_q$. The main point to be made here is to show that if $q - 1$ factors into small primes, then the discrete logarithms can be calculated rather efficiently, and so cryptosystems based on discrete exponentiation in $\mathbb{F}_q$ are insecure for such q. Let

$$q - 1 = \prod_{i=1}^{k} p_i^{e_i}$$

be the prime factor decomposition of $q - 1$, where $p_1 < p_2 < \cdots < p_k$ are the distinct prime factors. We wish to find the value $r = \mathrm{ind}_b(a)$ such that $a = b^r$, where b is a primitive element of $\mathbb{F}_q$. The value of r will be determined modulo

$p_i^{e_i}$ for $i = 1, 2, \cdots, k$ and the results will then be combined by the Chinese Remainder Theorem for integers (see Exercise 1.13) to obtain r modulo $q - 1$, which completely determines r since $0 \leqslant r < q - 1$. Suppose

$$r \equiv \sum_{j=0}^{e_i-1} s_j p_i^j \bmod p_i^{e_i} \qquad \text{with } 0 \leqslant s_j \leqslant p_i - 1. \qquad (9.8)$$

In order to determine s_0 we form

$$a^{(q-1)/p_i} = b^{(q-1)r/p_i} = c_i^r = c_i^{s_0},$$

where $c_i = b^{(q-1)/p_i}$ is a primitive p_ith root of unity in $\mathbb{F}_q$. Therefore there are only p_i possible values for $a^{(q-1)/p_i}$ corresponding to $s_0 = 0, 1, \cdots, p_i - 1$. The resulting value uniquely determines s_0. The next digit s_1 in (9.8) is obtained by letting

$$d = ab^{-s_0} = b^{r_1}, \qquad \text{where } r_1 = \sum_{j=1}^{e_i-1} s_j p_i^j.$$

Then

$$d^{(q-1)/p_i^2} = b^{(q-1)r_1/p_i^2} = c_i^{r_1/p_i} = c_i^{s_1}$$

uniquely determines s_1. This method is continued to find all the s_j in (9.8). It can be shown that this algorithm has a running time of order at most $p_k^{1/2}(\log q)^2$, where p_k is the largest prime factor of $q - 1$. Therefore the algorithm is most efficient if $q - 1$ only has small prime factors.

9.13. Example. Let $q = 17$, then $b = 3$ is a primitive element of $\mathbb{F}_{17}$. We wish to find $r = \mathrm{ind}_3(a)$ for $a = -2$ by the Silver-Pohlig-Hellman algorithm. Since $q - 1 = 2^4$, we only have to work with the prime factor $p_1 = 2$. We calculate $c_1 = b^{(q-1)/2} = -1$. Write

$$r = s_0 + s_1 \cdot 2 + s_2 \cdot 2^2 + s_3 \cdot 2^3 \qquad \text{with } s_j = 0 \text{ or } 1.$$

Now

$$a^{(q-1)/2} = (-2)^8 = 1 = c_1^{s_0},$$

and so $s_0 = 0$. Then $d = ab^0 = -2$ and

$$d^{(q-1)/4} = (-2)^4 = -1 = c_1^{s_1},$$

and so $s_1 = 1$. Then $e = ab^{-2} = -4$ and

$$e^{(q-1)/8} = (-4)^2 = -1 = c_1^{s_2},$$

and so $s_2 = 1$. Now $f = ab^{-6} = 1$, hence $a = b^6$, and so $\mathrm{ind}_3(-2) = 6$. $\qquad \square$

The fact that the Silver-Pohlig-Hellman algorithm is less efficient if $q - 1$ has a large prime factor has led to the idea that fields $\mathbb{F}_{2^n}$ be employed, where $2^n - 1$ is a Mersenne prime. Such fields are also easy to implement. At the time of this writing 29 Mersenne primes are known, the largest one being

$2^{132049} - 1$, but the case of $2^{127} - 1$ is of particular interest since the field with 2^{127} elements has been used in practical hardware implementations. Unfortunately, the resulting cryptosystem is completely unsafe since the discrete logarithm algorithms given below can be carried out rapidly. If one uses $\mathbb{F}_{2^n}$ for a cryptosystem based on discrete exponentiation, an attacker would need access to a modern supercomputer in order to break a system based on such finite fields for $n \geqslant 400$. Within the next ten years it is recommended to choose $n \geqslant 800$ or even $n \geqslant 2000$ if one wishes to take into account developments in large special-purpose machines or improvements of algorithms. Another disadvantage of fields $\mathbb{F}_{2^n}$ for cryptographic applications is that there are few fields of this type, in the sense that there is only one field of order 2^n, but there are many prime fields $\mathbb{F}_p$ of comparable order since there are many primes p with $2^{n-1} < p < 2^n$. This also lessens the security of a cryptosystem. For large primes p it appears that fields of the form $\mathbb{F}_{p^n}$ do not offer increased security over fields $\mathbb{F}_p$. If we take a system whose main objective is key exchange and which is based on discrete exponentiation in $\mathbb{F}_p$, such as the Diffie-Hellman scheme in Example 9.9, and compare it with a public-key cryptosystem like RSA, then the former seems preferable since one can use keys that are about half as long for the same level of security.

We now discuss another discrete logarithm algorithm for $\mathbb{F}_q$, the *index-calculus algorithm*, one variant of which is due to Blake, Fuji-Hara, Mullin, and Vanstone. This algorithm works best for $q = 2^n$, but it can also be carried out for $q = p^n$ with p prime and $n \geqslant 2$. Let $\mathbb{F}_q$ with $q = p^n$ be defined by the irreducible polynomial $f(x)$ over $\mathbb{F}_p$ of degree n. Since $\mathbb{F}_q$ is isomorphic to the residue class ring $\mathbb{F}_p[x]/(f)$, all elements of $\mathbb{F}_q$ can be uniquely represented as polynomials over $\mathbb{F}_p$ of degree $< n$, with the arithmetic being polynomial arithmetic modulo $f(x)$; compare with Chapter 1, Section 3. This identification will be used throughout the rest of this section. Suppose $b(x)$ is a primitive element of $\mathbb{F}_q$. The algorithm to find the discrete logarithm $\mathrm{ind}(a(x))$ to the base $b(x)$ of an arbitrary nonzero element $a(x)$ of $\mathbb{F}_q$ consists of two stages.

In the initial stage we compute the discrete logarithms to the base $b(x)$ of all elements of a chosen subset V of $\mathbb{F}_q$. The set V usually consists of all the monic irreducible polynomials over $\mathbb{F}_p$ of degree $\leqslant m$, where the integer $m < n$ is determined according to certain probability computations described later. We suppose that $\mathrm{ind}(d)$ is known for all $d \in \mathbb{F}_p^*$. This is trivially satisfied for $p = 2$ since the only possibility is $d = 1$ and then $\mathrm{ind}(d) = \mathrm{ind}(1) = 0$. For $p > 2$ we use the observation that $b = b(x)^{(q-1)/(p-1)}$ is a primitive element of $\mathbb{F}_p$ and

$$\mathrm{ind}_{b(x)}(d) = \frac{q-1}{p-1} \mathrm{ind}_b(d) \quad \text{for all} \quad d \in \mathbb{F}_p^*.$$

For small values of p, $\mathrm{ind}_b(d)$ can be obtained by direct calculation. For large p we may use, for instance, the Silver-Pohlig-Hellman algorithm to compute these discrete logarithms.

9.14 Index-Calculus Algorithm: Initial Stage. Choose a random integer, t, $1 \leqslant t \leqslant q - 2$, and form the polynomial $c(x) \in \mathbb{F}_p[x]$ determined by

$$c(x) \equiv b(x)^t \bmod f(x), \quad \deg(c(x)) < n.$$

Then factor $c(x)$ into irreducible polynomials over $\mathbb{F}_p$, using techniques of Chapter 4 if necessary. If all the monic irreducible factors are elements of V, so that

$$c(x) = d \prod_{v \in V} v(x)^{e_v(c)}$$

with $d \in \mathbb{F}_p^*$ is the canonical factorization in $\mathbb{F}_p[x]$, then

$$t \equiv \operatorname{ind}(d) + \sum_{v \in V} e_v(c) \operatorname{ind}(v(x)) \bmod (q - 1).$$

As soon as we obtain more than $|V|$ independent congruences of this type, we expect that the corresponding system in the unknowns $\operatorname{ind}(v(x))$, $v \in V$, will determine these discrete logarithms uniquely modulo $q - 1$ for all $v \in V$.

The initial stage depends on the possibility to factor $c(x)$ in the way stated above and to obtain sufficiently many independent congruences. This stage is independent of $a(x)$ and can be used for other computations in $\mathbb{F}_q$. The second stage of the algorithm is based on the principles described in the discussion following Example 9.12. In the earlier illustration the set E of elements with easily computable discrete logarithms was formed by the smooth integers. The role of the smooth integers is now played by those polynomials over $\mathbb{F}_p$ all of whose irreducible factors are elements of V—that is, all of whose irreducible factors have degree $\leqslant m$. Note that the discrete logarithms of the elements of V are known from the precomputation in the initial stage. These discrete logarithms serve thus as a data base for the second stage of the algorithm, and as mentioned earlier this data base should also include the discrete logarithms of all elements of $\mathbb{F}_p^*$.

9.15. Index-Calculus Algorithm: Second Stage. To compute $\operatorname{ind}(a(x))$ to the base $b(x)$, choose a random integer t, $0 \leqslant t \leqslant q - 2$, and form the polynomial $a_1(x) \in \mathbb{F}_p[x]$ determined by

$$a_1(x) \equiv a(x)b(x)^t \bmod f(x), \quad \deg(a_1(x)) < n.$$

Then factor $a_1(x)$ into irreducible polynomials over $\mathbb{F}_p$, using techniques of Chapter 4 if necessary. If all the monic irreducible factors are elements of V, so that

$$a_1(x) = d \prod_{v \in V} v(x)^{e_v(a_1)}$$

with $d \in \mathbb{F}_p^*$ is the canonical factorization in $\mathbb{F}_p[x]$, then $\operatorname{ind}(a(x))$ is determined by

$$\operatorname{ind}(a(x)) \equiv \operatorname{ind}(d) + \sum_{v \in V} e_v(a_1) \operatorname{ind}(v(x)) - t \bmod (q - 1).$$

If $a_1(x)$ does not have the desired type of factorization, choose other values of t until this type of factorization is obtained.

For the analysis of both stages of the algorithm it is important to study the probability $P(n, m)$ that a nonzero polynomial over $\mathbb{F}_p$ of degree $< n$ has all its irreducible factors in $\mathbb{F}_p[x]$ of degree $\leqslant m$. We have

$$P(n, m) = \frac{1}{p^n - 1} \sum_{k=0}^{n-1} N(k, m),$$

where $N(k, m)$ is the number of polynomials over $\mathbb{F}_p$ of degree k that have all their irreducible factors in $\mathbb{F}_p[x]$ of degree $\leqslant m$. A recurrence relation for evaluating $N(k, m)$ is given in Exercise 9.14. For $p = 2$ this leads after lengthy calculations to the formula

$$P(n, m) = A(n, m) \left(\frac{m}{n} \right)^{B(n, m) n / m},$$

where $A(n, m)$ and $B(n, m)$ tend to 1 for $n \to \infty$ and $n^{1/100} \leqslant m \leqslant n^{99/100}$. Thus we will need roughly

$$P(n, m)^{-1} \approx \left(\frac{n}{m} \right)^{n/m} \tag{9.9}$$

choices of integers t before the second stage of the algorithm can find the discrete logarithm of $a(x)$. It is clear that m cannot be chosen too small, for otherwise the running time of the second stage would be exorbitantly long. On the other hand, if m is chosen too large, then the initial stage will require a very long running time. Thus one has to find a middle ground between these two extremes. For instance, in the important special case $p = 2$ and $n = 127$ the choice $m = 17$ is recommended; then 16510 discrete logarithms have to be precomputed in the initial stage since there are that many irreducible polynomials over $\mathbb{F}_2$ of degree $\leqslant 17$.

9.16. Example. To illustrate how the initial stage is carried out, we consider $\mathbb{F}_{64}$ defined by $f(x) = x^6 + x + 1 \in \mathbb{F}_2[x]$. Since $f(x)$ is a primitive polynomial over $\mathbb{F}_2$, we can take $b(x) = x$ as a primitive element of $\mathbb{F}_{64}$. Suppose the maximum degree m of irreducible polynomials in the set V is 2. So we have to find the discrete logarithms of $x, x + 1$, and $x^2 + x + 1$ to the base x. Clearly $\mathrm{ind}(x) = 1$. Now we choose integers t with $1 \leqslant t \leqslant 62$. A good choice is $t = 6$, since then

$$c(x) \equiv x^6 \equiv x + 1 \bmod f(x),$$

hence $\mathrm{ind}(x + 1) = 6$. Another good choice is $t = 32$, since

$$x^{64} \equiv x \equiv x^6 + 1 \equiv (x^3 + 1)^2 \bmod f(x)$$

implies

$$c(x) \equiv x^{32} \equiv x^3 + 1 \equiv (x + 1)(x^2 + x + 1) \bmod f(x).$$

This yields

$$32 \equiv \text{ind}(x+1) + \text{ind}(x^2 + x + 1) \equiv 6 + \text{ind}(x^2 + x + 1) \bmod 63,$$

hence $\text{ind}(x^2 + x + 1) = 26$.   □

9.17. Example. To demonstrate a simple case for the second stage, let $\mathbb{F}_{64}$ again be defined by $f(x) = x^6 + x + 1 \in \mathbb{F}_2[x]$ and let $b(x) = x$. Suppose $m = 2$, so that the discrete logarithms of the elements of V are known from Example 9.16. These values constitute our data base. We wish to find the discrete logarithm of $a(x) = x^4 + x^3 + x^2 + x + 1$ to the base x. We form

$$a_1(x) \equiv a(x)x^t \bmod f(x)$$

with a suitable t. The choice $t = 2$ yields

$$a_1(x) \equiv a(x)x^2 \equiv x^5 + x^4 + x^3 + x^2 + x + 1 \equiv (x^2 + x + 1)^2(x + 1) \bmod f(x),$$

hence all the irreducible factors are in V. Therefore

$$\text{ind}(a(x)) \equiv 2\,\text{ind}(x^2 + x + 1) + \text{ind}(x + 1) - 2$$

$$\equiv 2 \cdot 26 + 6 - 2 \equiv 56 \bmod 63,$$

and so $\text{ind}(x^4 + x^3 + x^2 + x + 1) = 56$.   □

The second stage of the index-calculus algorithm can be speeded up by using the Euclidean algorithm. Consider again the nonzero polynomial $a_1(x) \in \mathbb{F}_p[x]$ determined by

$$a_1(x) \equiv a(x)b(x)^t \bmod f(x), \qquad \deg(a_1(x)) < n. \tag{9.10}$$

The method in 9.15 is successful only if $a_1(x)$ has all its irreducible factors in $\mathbb{F}_p[x]$ of degree $\leqslant m$. The main idea is to replace this now by the following condition: there exist nonzero polynomials $w_1(x)$ and $w_2(x)$ over $\mathbb{F}_p$ with

$$w_2(x)a_1(x) \equiv w_1(x) \bmod f(x) \tag{9.11}$$

and $\deg(w_i(x)) \leqslant n/2$ for $i = 1, 2$ such that each $w_i(x)$ has all its irreducible factors in $\mathbb{F}_p[x]$ of degree $\leqslant m$. If such polynomials $w_i(x)$ can be found, then their canonical factorization in $\mathbb{F}_p[x]$ is of the form

$$w_i(x) = d_i \prod_{v \in V} v(x)^{e_v(w_i)} \quad \text{for} \quad i = 1, 2 \tag{9.12}$$

with $d_i \in \mathbb{F}_p^*$. It follows then from (9.10) and (9.11) that the discrete logarithm of $a(x)$ is determined by

$$\text{ind}(a(x)) \equiv \text{ind}(d_1 d_2^{-1}) + \sum_{v \in V} (e_v(w_1) - e_v(w_2))\,\text{ind}(v(x)) - t \bmod (q - 1). \tag{9.13}$$

Polynomials $w_1(x)$ and $w_2(x)$ satisfying (9.11) and the degree restriction above can be calculated by an application of the Euclidean algorithm that is similar to the procedure for decoding Goppa codes (see Chapter 8, Section 3). In detail, we use the Euclidean algorithm with the polynomials $r_{-1}(x) = f(x)$ and $r_0(x) = a_1(x)$. This yields

$$r_{h-1}(x) = q_{h+1}(x)r_h(x) + r_{h+1}(x), \quad \deg(r_{h+1}(x)) < \deg(r_h(x)),$$

$$\text{for} \quad h = 0, 1, \ldots, s-1,$$

$$r_{s-1}(x) = q_{s+1}(x)r_s(x).$$

Since $0 \leqslant \deg(a_1(x)) < n = \deg(f(x))$ and $f(x)$ is irreducible over $\mathbb{F}_p$, we have $\gcd(f(x), a_1(x)) = 1$ and so $\deg(r_s(x)) = 0$. Consequently, there exists a least index j, $0 \leqslant j \leqslant s$, such that $\deg(r_j(x)) \leqslant n/2$. Now calculate recursively the polynomials

$$z_{-1}(x) = 0, \quad z_0(x) = 1,$$

$$z_h(x) = z_{h-2}(x) - q_h(x)z_{h-1}(x) \quad \text{for} \quad h = 1, 2, \ldots, j.$$

By the generalizations of (8.12) and (8.13) shown in Exercises 8.43(a) and 8.43(c) we have

$$z_j(x)a_1(x) \equiv r_j(x) \bmod f(x)$$

and

$$\deg(z_j(x)) = \deg(f(x)) - \deg(r_{j-1}(x)) = n - \deg(r_{j-1}(x)).$$

From the minimality of j we get $\deg(r_{j-1}(x)) > n/2$, and so $\deg(z_j(x)) < n/2$. It follows that $w_1(x) = r_j(x)$ and $w_2(x) = z_j(x)$ are polynomials satisfying (9.11) and $\deg(w_1(x)) \leqslant n/2$, $\deg(w_2(x)) < n/2$. Moreover, $w_1(x)$ and $w_2(x)$ can be calculated very quickly.

The condition about the irreducible factors of $w_1(x)$ and $w_2(x)$ cannot be guaranteed by the algorithm above. However, we may heuristically estimate the probability that both $w_1(x)$ and $w_2(x)$ have the desired type of factorization in (9.12). Let $P(n, m)$ again denote the probability that a nonzero polynomial over $\mathbb{F}_p$ of degree $< n$ has all its irreducible factors in $\mathbb{F}_p[x]$ of degree $\leqslant m$. If we make the reasonable assumption that $w_1(x)$ and $w_2(x)$ behave like independently chosen random polynomials of degree $< \lfloor n/2 \rfloor + 1$, then the probability that $w_1(x)$ and $w_2(x)$ have all their irreducible factors in $\mathbb{F}_p[x]$ of degree $\leqslant m$ will be approximately $P(\lfloor n/2 \rfloor + 1, m)^2$. Using the approximation (9.9) in the case $p = 2$, we obtain that we will now need roughly

$$P(\lfloor n/2 \rfloor + 1, m)^{-2} \approx \left(\frac{n}{2m} \right)^{n/m}$$

choices of integers t in the second stage of the algorithm. This is a saving by a factor of approximately $2^{-n/m}$ over the corresponding expression in (9.9). Thus we can expect that this version of the second stage of the index-calculus algorithm will be significantly faster than the earlier one. We summarize this method as follows, assuming again that we already have a data base containing the discrete logarithms of all elements of $\mathbb{F}_p^*$ and of all elements of the set V consisting of the monic irreducible polynomials over $\mathbb{F}_p$ of degree $\leqslant m$.

9.18. Index-Calculus Algorithm: Improved Version of Second Stage. To compute $\mathrm{ind}(a(x))$ to the base $b(x)$, choose a random integer t, $0 \leqslant t \leqslant q - 2$, and form the polynomial $a_1(x) \in \mathbb{F}_p[x]$ determined by

$$a_1(x) \equiv a(x)b(x)^t \bmod f(x), \quad \deg(a_1(x)) < n.$$

Then carry out the Euclidean algorithm with $r_{-1}(x) = f(x)$ and $r_0(x) = a_1(x)$ and stop as soon as $\deg(r_j(x)) \leqslant n/2$. Put $w_1(x) = r_j(x)$ and $w_2(x) = z_j(x)$ and factor these polynomials into irreducible polynomials over $\mathbb{F}_p$, using techniques of Chapter 4 if necessary. If all the monic irreducible factors are elements of V, so that $w_1(x)$ and $w_2(x)$ are of the form (9.12), then $\mathrm{ind}(a(x))$ is determined by (9.13). If this condition on the monic irreducible factors of $w_1(x)$ and $w_2(x)$ is not satisfied, choose other values of t until this condition holds.

In the case $p = 2$ further improvements on the construction of the data base and on the speeding up of the second stage of the index-calculus algorithm were recently achieved by Coppersmith.

9.19. Example. We give a simple illustration of the algorithm in 9.18. Consider the finite field $\mathbb{F}_{32}$, so that $p = 2$ and $n = 5$. Let $\mathbb{F}_{32}$ be defined by the primitive polynomial $f(x) = x^5 + x^2 + 1$ over $\mathbb{F}_2$. Then we can take $b(x) = x$ as a primitive element of $\mathbb{F}_{32}$. We wish to find the discrete logarithm of $a(x) = x^3 + x + 1$ to the base x. Suppose the data base consists of the discrete logarithms of all irreducible polynomials over $\mathbb{F}_2$ of degree $\leqslant 2$:

$$\mathrm{ind}(x) = 1, \quad \mathrm{ind}(x + 1) = 18, \quad \mathrm{ind}(x^2 + x + 1) = 11.$$

Choose the integer $t = 0$, so that $a_1(x) = a(x)$, and carry out the Euclidean algorithm with $r_{-1}(x) = x^5 + x^2 + 1$ and $r_0(x) = x^3 + x + 1$. This yields

$$x^5 + x^2 + 1 = (x^2 + 1)(x^3 + x + 1) + x.$$

Since $r_1(x) = x$ satisfies $\deg(r_1(x)) \leqslant n/2$, we can already stop. Thus $w_1(x) = r_1(x) = x$ and $w_2(x) = z_1(x) = x^2 + 1 = (x + 1)^2$. It follows then from (9.13) that

$$\mathrm{ind}(x^3 + x + 1) \equiv \mathrm{ind}(x) - 2\,\mathrm{ind}(x + 1) \equiv 1 - 2 \cdot 18 \equiv 27 \bmod 31,$$

and so $\mathrm{ind}(x^3 + x + 1) = 27$. □

These discrete logarithm algorithms have diminished the security of cryptosystems based on discrete exponentiation. It is therefore of interest to design cryptosystems that use similar principles, but employ more complex operations than discrete exponentiation. This is carried out in the recent proposal of *FSR cryptosystems* by Niederreiter, where FSR stands for "feedback shift register". In these cryptosystems, discrete exponentiation is replaced by the operation of decimation for linear recurring sequences in finite fields (compare with Chapter 7, Section 4). The ciphertexts are strings of consecutive terms of linear recurring sequences that are obtained by decimation from message-dependent linear recurring sequences. The cryptanalysis amounts to inferring the value of the integer k from the knowledge of the polynomials $f(x) = \prod_{j=1}^{s}(x - \alpha_j)$ and $f_k(x) = \prod_{j=1}^{s}(x - \alpha_j^k)$ over $\mathbb{F}_q$, where both factorizations are in the splitting fields of the polynomials over $\mathbb{F}_q$. This problem is more difficult than determining discrete logarithms.

We now describe a public-key cryptosystem in which discrete logarithms are used for encryption. This cryptosystem due to Chor and Rivest is of the *knapsack type*—that is, it is based on the difficulty of recovering the summands from the value of their sum. The following auxiliary result is crucial.

9.20. Lemma. *Let p be a prime and $n \geq 2$ an integer. Then there exist integers $a_0, a_1, \ldots, a_{p-1}$ with $1 \leq a_i \leq p^n - 2$ for $0 \leq i \leq p - 1$ such that for any two distinct vectors $(h_0, h_1, \ldots, h_{p-1})$ and $(k_0, k_1, \ldots, k_{p-1})$ with nonnegative integral coordinates satisfying*

$$\sum_{i=0}^{p-1} h_i < n \quad and \quad \sum_{i=0}^{p-1} k_i < n \tag{9.14}$$

we have

$$\sum_{i=0}^{p-1} h_i a_i \not\equiv \sum_{i=0}^{p-1} k_i a_i \bmod (p^n - 1).$$

Proof. Consider the finite field $\mathbb{F}_q$ with $q = p^n$ and identify it as before with the residue class ring $\mathbb{F}_p[x]/(f)$, where $f(x)$ is an irreducible polynomial over $\mathbb{F}_p$ of degree n. Relative to a fixed primitive element of $\mathbb{F}_q$ set

$$a_i = \mathrm{ind}\,(x - i) \quad \text{for} \quad i = 0, 1, \ldots, p - 1.$$

Each a_i satisfies $1 \leq a_i \leq q - 2$. Now suppose $(h_0, h_1, \ldots, h_{p-1})$ and $(k_0, k_1, \ldots, k_{p-1})$ are two vectors with nonnegative integral coordinates satisfying (9.14) and

$$\sum_{i=0}^{p-1} h_i a_i \equiv \sum_{i=0}^{p-1} k_i a_i \bmod (q - 1).$$

Then

$$\mathrm{ind}\left(\prod_{i=0}^{p-1} (x - i)^{h_i}\right) \equiv \mathrm{ind}\left(\prod_{i=0}^{p-1} (x - i)^{k_i}\right) \bmod (q - 1),$$

and so

$$\prod_{i=0}^{p-1} (x - i)^{h_i} \equiv \prod_{i=0}^{p-1} (x - i)^{k_i} \bmod f(x).$$

Now (9.14) shows that on each side we have a polynomial over $\mathbb{F}_p$ of degree $< n$, hence

$$\prod_{i=0}^{p-1} (x - i)^{h_i} = \prod_{i=0}^{p-1} (x - i)^{k_i}.$$

Unique factorization in $\mathbb{F}_p[x]$ implies $h_i = k_i$ for $0 \leqslant i \leqslant p - 1$, and the desired result is established. $\square$

The cryptosystem is implemented as follows. Take a publicly known finite field $\mathbb{F}_q$ with $q = p^n$, p prime, $n \geqslant 2$, in which discrete logarithms can be efficiently computed. Choose a random irreducible polynomial $f(x)$ over $\mathbb{F}_p$ of degree n and a random primitive element $b(x)$ of $\mathbb{F}_q$. Relative to the base $b(x)$ compute

$$a_i = \text{ind}\,(x - i) \quad \text{for} \quad i = 0, 1, \ldots, p - 1$$

as in the proof of Lemma 9.20. Scramble the a_i by selecting a random permutation ψ of $\{0, 1, \ldots, p - 1\}$ and putting

$$c_i = a_{\psi(i)} \quad \text{for} \quad i = 0, 1, \ldots, p - 1.$$

Then publish $c_0, c_1, \ldots, c_{p-1}$ as the public key and keep $f(x), b(x)$, and ψ secret.

With this cryptosystem we can encipher binary messages $M = m_0 m_1 \ldots m_{p-1}$ of length p for which the number N of 1's is less than n. In detail, let $m_{i_1} = \ldots = m_{i_N} = 1$ and all other $m_i = 0$. Then encipher M as the integer $E(M)$ with $0 \leqslant E(M) \leqslant p^n - 2$ and

$$E(M) \equiv c_{i_1} + \cdots + c_{i_N} \quad \text{mod}\,(p^n - 1).$$

It follows from Lemma 9.20 that distinct messages are enciphered as distinct ciphertexts.

To decipher the ciphertext $s = E(M)$, we calculate the uniquely determined polynomial $g(x) \in \mathbb{F}_p[x]$ with $\deg(g(x)) < n$ and

$$g(x) \equiv b(x)^s \bmod f(x).$$

From the definition of the c_i and a_i we obtain

$$g(x) \equiv b(x)^{c_{i_1} + \cdots + c_{i_N}} \equiv (x - \psi(i_1))\ldots(x - \psi(i_N)) \bmod f(x).$$

Since $N < n = \deg(f(x))$, we have

$$g(x) = (x - \psi(i_1)) \cdots (x - \psi(i_N)).$$

Therefore $\psi(i_1), \ldots, \psi(i_N)$ can be determined as the roots of $g(x)$ in $\mathbb{F}_p$, which are obtained either by successive substitution of elements of $\mathbb{F}_p$ or by one of the root-finding algorithms in Chapter 4, Section 3. By applying the inverse permutation of ψ, we recover the positions $i_1, \ldots, i_N$ where the original message M has the bit 1.

9.21. Example. We illustrate the procedure with an example involving small parameters. Let $p = 5$ and $n = 3$, so that we are working in the finite field $\mathbb{F}_{125}$. Choose $f(x) = x^3 + x^2 + 2$, which is a primitive polynomial over $\mathbb{F}_5$. Therefore $b(x) = x$ is a primitive element of $\mathbb{F}_{125}$. The part of Table A in Chapter 10 pertaining to $\mathbb{F}_{125} = GF(5^3)$ refers precisely to this situation. Thus we can read off the values of the a_i from this table. This yields

$$a_0 = 1, \quad a_1 = 84, \quad a_2 = 80, \quad a_3 = 99, \quad a_4 = 29.$$

Let the permutation ψ of $\{0, 1, 2, 3, 4\}$ be given by

$$\psi(0) = 2, \quad \psi(1) = 4, \quad \psi(2) = 1, \quad \psi(3) = 0, \quad \psi(4) = 3,$$

so that

$$c_0 = 80, \quad c_1 = 29, \quad c_2 = 84, \quad c_3 = 1, \quad c_4 = 99.$$

If we wish to encipher the binary message $M = 10100$, then

$$E(M) \equiv c_0 + c_2 \equiv 80 + 84 \bmod 124,$$

and so $E(M) = 40$. To decipher $s = 40$, we calculate

$$g(x) \equiv x^{40} \equiv x^2 + 2x + 2 \bmod (x^3 + x^2 + 2)$$

from Table A, hence

$$g(x) = x^2 + 2x + 2 = (x - 1)(x - 2).$$

Therefore $N = 2$, $\psi(i_1) = 1$, and $\psi(i_2) = 2$, yielding $i_1 = 2$ and $i_2 = 0$, and we recover the original message M.   □

4. FURTHER CRYPTOSYSTEMS

We first describe a public-key cryptosystem that is based on binary irreducible Goppa codes (see Chapter 8, Section 3). This cryptosystem falls into the category of block ciphers. We recall from Theorems 8.56 and 8.57 that for any irreducible polynomial $g(x)$ over $\mathbb{F}_{2^m}$ of degree t and any integer n, where $2 \leqslant t < n \leqslant 2^m$, there exists a binary irreducible Goppa code $\Gamma(L, g)$ of length n and dimension $k \geqslant n - mt$ that is capable of correcting any pattern of t or fewer errors. All one has to do is to choose L as a subset of $\mathbb{F}_{2^m}$ of cardinality n. We note also that Goppa codes allow a fast decoding algorithm discussed in Chapter 8, Section 3.

    The *Goppa-code cryptosystem* is set up as follows. We choose integers t and n with $2 \leqslant t < n \leqslant 2^m$ and then randomly select a monic irreducible polynomial $g(x)$ over $\mathbb{F}_{2^m}$ of degree t. This is easy to do since the probability that a monic polynomial over $\mathbb{F}_{2^m}$ of degree t is irreducible is

$$N_{2^m}(t)2^{-mt} = \frac{1}{2^{mt}t}\sum_{d|t}\mu\left(\frac{t}{d}\right)2^{md} \approx \frac{1}{t}$$

according to Theorem 3.25. The irreducibility of a randomly chosen polynomial may be tested by applying one of the factorization algorithms in Chapter 4 to it. We consider now a t-error-correcting binary irreducible Goppa code $\Gamma(L, g)$ of length n and dimension $k \geqslant n - mt$. This code has a binary $(n - k) \times n$ parity-check matrix H. From H we can derive a binary $k \times n$ generator matrix G of $\Gamma(L, g)$; compare with Chapter 8, Sections 1 and 3. This generator matrix is scrambled by selecting at random a binary invertible $k \times k$ matrix S and an $n \times n$ permutation matrix P—that is, a matrix obtained from the identity matrix by exchanging rows—and forming the new generator matrix

$$G' = SGP.$$

This $k \times n$ matrix G' generates a binary linear code with the same length, dimension, and minimum distance as the Goppa code $\Gamma(L, g)$. The matrices G, S, and P are kept secret, whereas G' is made public. Let $\mathbf{z}$ denote a random binary vector of length n and weight $\leqslant t$ chosen by the sender. Then the cryptosystem is implemented as follows.

9.22. Goppa-Code Cryptosystem.
Enciphering: The plaintext data, given as k-bit blocks $\mathbf{x}$, are enciphered as vectors $\mathbf{y} = \mathbf{x}G' + \mathbf{z}$.
Deciphering: On receipt of $\mathbf{y}$ we compute $\mathbf{y}' = \mathbf{y}^{P^{-1}}$, which will be at a Hamming distance at most t from the code word $\mathbf{x}SG$ of the Goppa code $\Gamma(L, g)$. Decoding $\mathbf{y}'$ gives the corresponding message $\mathbf{x}' = \mathbf{x}S$, and the plaintext is recovered by computing $\mathbf{x} = \mathbf{x}'S^{-1}$.

9.23. Example. We present an example with very small parameters to demonstrate the procedure. However, the cryptosystem in this example does of course not offer any security. We choose $m = 3, n = 8, t = 2$, and use the Goppa code of dimension $k = 2$ in Example 8.58, with G being the generator matrix given there. Furthermore , let

$$S = \begin{pmatrix} 1 & 1 \\ 0 & 1 \end{pmatrix}, \quad P = P^{-1} = \begin{bmatrix} 1 & 0 & 0 & 0 & 0 & 0 & 0 & 0 \\ 0 & 0 & 1 & 0 & 0 & 0 & 0 & 0 \\ 0 & 1 & 0 & 0 & 0 & 0 & 0 & 0 \\ 0 & 0 & 0 & 1 & 0 & 0 & 0 & 0 \\ 0 & 0 & 0 & 0 & 1 & 0 & 0 & 0 \\ 0 & 0 & 0 & 0 & 0 & 0 & 1 & 0 \\ 0 & 0 & 0 & 0 & 0 & 1 & 0 & 0 \\ 0 & 0 & 0 & 0 & 0 & 0 & 0 & 1 \end{bmatrix}.$$

Then the public key is

$$G' = SGP = \begin{pmatrix} 1 & 0 & 1 & 0 & 1 & 1 & 0 & 1 \\ 0 & 1 & 0 & 1 & 1 & 1 & 1 & 1 \end{pmatrix}.$$

Let $\mathbf{z} = 1\ 0\ 0\ 0\ 0\ 0\ 0\ 0$. Then the plaintext $\mathbf{x} = 0\ 1$, say, is enciphered as $\mathbf{y} = 1\ 1\ 0\ 1\ 1\ 1\ 1\ 1$. On receipt of $\mathbf{y}$ the authorized receiver computes $\mathbf{y}' = \mathbf{y}P^{-1} = 1\ 0\ 1\ 1\ 1\ 1\ 1\ 1$ and decodes this to the code word $0\ 0\ 1\ 1\ 1\ 1\ 1\ 1$ with corresponding message $\mathbf{x}' = 0\ 1$. The plaintext is recovered as $\mathbf{x} = \mathbf{x}'S^{-1} = 0\ 1$. In this example decoding is performed merely as nearest neighbor decoding—that is, by comparing the received word $\mathbf{y}'$ with the nearest code word of Example 8.58 relative to the Hamming distance. For large parameters this approach will be infeasible, but then the efficient decoding algorithm for Goppa codes can do the job for us. □

In order to break this cryptosystem, an attacker would have to determine G from G' or he might try to recover $\mathbf{x}$ from $\mathbf{y}$ without knowing G. To find G seems to be a hopeless task if n and t are large enough, since there are so many possibilities for G, S, and P. If the attacker wants to find $\mathbf{x}$ from $\mathbf{y}$, this amounts to decoding a random looking linear (n, k) code in the presence of up to t errors. Such an attack is expected to be infeasible for large enough code parameters, since the general decoding problem for linear codes has been shown to be NP-complete. For example, if $m = 10$, $n = 2^{10} = 1024$, and $t = 50$, then there will be about 10^{149} possible Goppa polynomials and a gigantic number of choices for S and P. The dimension of the code will be at least 524 and will make brute-force attacks based on comparisons of $\mathbf{y}$ with code words or based on coset leaders impossible. This cryptosystem is not suitable for use in authentication, but its fast communication rate makes it attractive for data communication.

We now discuss another scheme for the communication of secret information. Suppose the secret is some data D—for instance, the key for a cryptosystem or a bank safe combination. Then D is divided into n pieces $D_1, \ldots, D_n$ in such a way that for some $k < n$:

(i) knowledge of any k or more pieces D_i makes computing D easy;
(ii) knowledge of any $k - 1$ or fewer pieces D_i makes determining D impossible because of insufficient information.

Such a scheme is called a (k, n) *threshold scheme*. It can be helpful in a variety of situations. For instance, if $n = 2k - 1$, the original data D can be recovered even when $\lfloor n/2 \rfloor = k - 1$ of the n pieces D_i are destroyed or lost, but an opponent cannot reconstruct D even when security breaches expose $k - 1$ of the remaining k pieces. Other advantages are that a hierarchical scheme is possible where the number of pieces D_i given to each user is proportional to the user's importance. Threshold schemes are well suited to situations where a group of mutually suspicious individuals with conflicting interests must cooperate. By choosing the parameters n and k properly, any sufficiently large majority can be given the authority to take some action, while any sufficiently large minority can be given the power to block it.

We describe a (k, n) threshold scheme that was originated by Shamir and by Blakley for $\mathbb{F}_p$ and $\mathbb{F}_{2^m}$, respectively. First we identify D with an element of a suitable finite field $\mathbb{F}_q$. The pieces D_i are derived—in a way to be specified—from a random polynomial

$$f(x) = a_{k-1}x^{k-1} + \cdots + a_1 x + a_0 \in \mathbb{F}_q[x]$$

of degree $k - 1$ whose constant term a_0 is D. Here q is a prime power larger than n and the number of possibilities for D. If one knows the polynomial $f(x)$, then it is easy to compute D by $D = f(0)$. The pieces D_i are obtained by evaluating $f(x)$ at n distinct elements $c_1, \ldots, c_n \in \mathbb{F}_q$—that is, $D_i = f(c_i)$ for $i = 1, 2, \ldots, n$. These c_i could be elements of $\mathbb{F}_q$ which do not have to be secret; they could be user identifiers. Since any k pairs (c_i, D_i) uniquely determine a polynomial of degree $\leqslant k - 1$, the polynomial $f(x)$ and therefore the secret data D can be reconstructed from k pieces, but not from fewer pieces. If the k pieces are denoted by $D_{i_1}, \ldots, D_{i_k}$, then $f(x)$ can be computed by the Lagrange interpolation formula in Theorem 1.71, which yields

$$f(x) = \sum_{s=1}^{k} D_{i_s} \prod_{\substack{t=1 \\ t \neq s}}^{k} (c_{i_s} - c_{i_t})^{-1} (x - c_{i_t}).$$

9.24. Example. Let $q = 8$, $n = 3$, and $k = 2$. Suppose we know that

$$D_1 = f(1) = \alpha + \alpha^2,$$

$$D_2 = f(\alpha) = \alpha,$$

where $\alpha \in \mathbb{F}_8$ is a root of $x^3 + x + 1$. Then $f(x)$ can be reconstructed as follows:

$$f(x) = (\alpha + \alpha^2)(1 - \alpha)^{-1}(x - \alpha) + \alpha(\alpha - 1)^{-1}(x - 1)$$

$$= \alpha(x + \alpha) + (1 + \alpha + \alpha^2)(x + 1)$$

$$= (1 + \alpha^2)x + 1 + \alpha.$$

Therefore the secret data is $D = f(0) = 1 + \alpha$. $\qquad\qquad\square$

EXERCISES

9.1. Let $s_0, s_1, \ldots$ and $t_0, t_1, \ldots$ be the impulse response sequences with characteristic polynomial $x^4 + x + 1$ and $x^5 + x^2 + 1$ over $\mathbb{F}_2$, respectively. Let $h = 2$, $j_1 = 0$, $j_2 = 1$, and $\psi: \{0, 1, 2, 3\} \to \{0, 1, 2, 3, 4\}$ be defined by $\psi(i) = i + 1$. Find the first 16 terms of the resulting multiplexed sequence.

9.2. If $x^{15} + x^8 + x^7 + x^2 + 1$ and $x^{16} + x^{15} + x^4 + x + 1$ are characteristic polynomials of two maximal period sequences in $\mathbb{F}_2$, respectively, what can you say about the least period of a multiplexed sequence based on these two sequences?

9.3. Suppose we know that a feedback shift register with 5 delay elements has been used to construct a binary sequence $s_0, s_1, \ldots$ and that the first 10 values of s_n are 0, 1, 0, 1, 0, 1, 1, 1, 0, 1. Find a characteristic polynomial of the sequence and determine the first 20 terms of the sequence.

9.4. In the notation of Definition 9.1, let $s_0, s_1, \ldots$ be a kth-order maximal period sequence in $\mathbb{F}_2$ with $k > 1$ and let $1 \leqslant h < k$. For $n = 0, 1, \ldots$ let $P_i(n)$, $i = 0, 1, \ldots, 2^h - 1$, be the 2^h Boolean monomials formed by taking all 2^h possible products of the terms $s_{n+j_1}, \ldots, s_{n+j_h}$. Let τ be a maximal period sequence in $\mathbb{F}_2$ with terms $t_0, t_1, \ldots$ and minimal polynomial $g(x) \in \mathbb{F}_2[x]$, and let $u_0, u_1, \ldots$ be the resulting multiplexed sequence. Prove that

$$u_n = \sum_{i=0}^{2^h-1} P_i(n)t_{n+N(i)} \quad \text{for} \quad n = 0, 1, \ldots,$$

where the integers $N(0), N(1), \ldots, N(2^h - 1)$ are completely determined by $\psi(0)$, $\psi(1), \ldots, \psi(2^h - 1)$. Moreover, prove that the 2^h shifted sequences $\tau^{(N(i))}$, $i = 0, 1, \ldots, 2^h - 1$, are linearly independent in the vector space $S(g(x))$ over $\mathbb{F}_2$ defined on p. 215.

9.5.   In the proof of Theorem 9.6 it is shown that if τ is an mth-order maximal period sequence in the finite prime field $\mathbb{F}_p$, then the decimated sequence $\tau_d^{(j)}$ with $d = p^k - 1$ and $\gcd(k, m) = 1$ has least period $(p^m - 1)/(p - 1)$. Prove more generally that if τ is an mth-order linear recurring sequence in an arbitrary finite field $\mathbb{F}_q$ with least period r and irreducible minimal polynomial, then for $j \geqslant 0$ and $d \geqslant 1$ we have:
(a) the decimated sequence $\tau_d^{(j)}$ is either the zero sequence or it has least period $r/\gcd(d, r)$;
(b) if $\gcd(d, r) \leqslant rq^{1-m}$, then $\tau_d^{(j)}$ has least period $r/\gcd(d, r)$.

9.6.   How many possible enciphering keys K are there in the cryptosystem in Example 9.7 if $q = 1987$?

9.7.   Let $p = 47, q = 59$, and $K = 157$ be parameters of an RSA cryptosystem. Find the deciphering parameter D.

9.8.   Let $q = p = 13$ and $b = 2$. Demonstrate by a numerical example how each of the schemes in Examples 9.9, 9.10, 9.11, and 9.12 works.

9.9.   Show that in the public-key cryptosystem in Example 9.11 it is not advisable to use the same value of k for enciphering more than one message block.

9.10.   Use the Silver-Pohlig-Hellman algorithm for $q = 73$ and $b = 5$ to find the discrete logarithm of $a = 7$.

9.11.   In Example 9.16 let $m = 3$. Find the discrete logarithms to the base x for all polynomials in V.

9.12.   Refer to Example 9.17 and find the discrete logarithm of $a(x) = x^4 + x^3 + x^2 + 1$ to the base x.

9.13.   Suppose $\mathbb{F}_{32}$ is defined by $f(x) = x^5 + x^2 + 1$ over $\mathbb{F}_2$ and a data base for the second stage of the index-calculus algorithm is given as in Example 9.19, so that $m = 2$. Compute the discrete logarithm of $a(x) = x^4 + x + 1$ to the base x. Repeat the calculation under the assumption that $m = 1$.

9.14.   For an arbitrary finite field $\mathbb{F}_q$ let $N(k, m)$ be the number of polynomials over $\mathbb{F}_q$ of degree k all of whose irreducible factors in $\mathbb{F}_q[x]$ are of degree $\leqslant m$. Define $N(k, 0) = q - 1$ if $k = 0$, $N(k, 0) = 0$ if $k \neq 0$, and $N(k, m) = 0$ if $k < 0$ and $m \geqslant 0$. Let $N_q(n)$ be the number of monic irreducible polynomials over $\mathbb{F}_q$ of degree n (see Theorem 3.25). For $k, m \geqslant 1$ prove the recurrence

$$N(k, m) = \sum_{n=1}^{m} \sum_{i \geqslant 1} N(k - in, n - 1)\binom{i + N_q(n) - 1}{i}.$$

9.15.   Let $r_{-1}(x)$ and $r_0(x)$ be two nonzero polynomials over a field F with
        $\deg(r_{-1}(x)) \geqslant \deg(r_0(x))$ and $d(x) = \gcd(r_{-1}(x), r_0(x))$, and let k be an
        integer with $\deg(d(x)) \leqslant k < \deg(r_{-1}(x))$. In the notation of Exercise
        8.43, prove that there exists a unique index j, $0 \leqslant j \leqslant s$, such that
        $\deg(r_j(x)) \leqslant k$ and $\deg(z_j(x)) \leqslant \deg(r_{-1}(x)) - k - 1$.

9.16.   In several cryptosystems over $\mathbb{F}_q$ involving discrete exponentiation it
        is necessary to generate enciphering keys e and deciphering keys d
        with e, $d \in \mathbb{Z}$ and $ed \equiv 1 \bmod (q-1)$. Show that such keys can be
        generated in the following way. Let $ab \equiv 1 \bmod (q-1)$, where a has
        the largest possible multiplicative order N modulo $q-1$, and let r be
        randomly chosen from $\{0, 1, \ldots, N-1\}$. Then $e \equiv a^r \bmod (q-1)$ and
        $d \equiv b^r \bmod (q-1)$ are multiplicative inverses of each other modulo
        $q-1$.

9.17.   Prove that a polynomial $x^m + x^{m-1} + \cdots + x + 1$ is irreducible over $\mathbb{F}_2$
        and its roots α^{2^i}, $i = 0, 1, \ldots, m-1$, form a normal basis of $\mathbb{F}_{2^m}$ over $\mathbb{F}_2$ if
        and only if $m+1$ is prime and 2 is a primitive element of $\mathbb{F}_{m+1}$. (*Note*:
        Such all-one polynomials permit an attractive implementation of
        discrete exponentiation in a normal basis of $\mathbb{F}_{2^m}$ over $\mathbb{F}_2$.)

9.18.   Let b be a primitive element of the finite prime field $\mathbb{F}_p$, $p > 2$, and let
        $a \in \mathbb{F}_p^*$. Prove that $\text{ind}_b(a)$ is determined as an element of $\mathbb{F}_p$ by the
        formula

$$\text{ind}_b(a) = -1 + \sum_{j=1}^{p-2} (b^{-j} - 1)^{-1} a^j.$$

        (*Hint*: Use the Lagrange Interpolation Formula in Theorem 1.71.)

9.19.   Prove that the formula in Exercise 9.18 reduces to

$$\text{ind}_b(a) = \sum_{j=1}^{p-2} (1 - b^j)^{-1} a^j$$

        provided that $a \neq 1$. (*Note*: The formulas for discrete logarithms in
        Exercises 9.18 and 9.19 are of theoretical interest, but useless for
        computational purposes.)

9.20.   Let $f(x)$ be an irreducible polynomial over $\mathbb{F}_q$ of degree n and let $g(x)$ be
        an arbitrary polynomial over $\mathbb{F}_q$. Prove that the degree of any nonzero
        divisor of $f(g(x))$ in $\mathbb{F}_q[x]$ is a multiple of n. (*Note*: This property is
        useful in the initial stage of the index-calculus algorithm.)

9.21.   Let p, n, $f(x)$, and ψ be as in Example 9.21 and choose the primi-
        tive element $b(x) = 4x^2 + 3$ of $\mathbb{F}_{125}$. Encipher the binary message
        $M = 01010$ and then decipher it again.

9.22.   Prove that Lemma 9.20 holds also if p is a prime power.

9.23.   Show by a counterexample that Lemma 9.20 does not hold in general if
        (9.14) is replaced by the condition that $\sum_{i=0}^{p-1} h_i \leqslant n$ and $\sum_{i=0}^{p-1} k_i \leqslant n$.
        (*Hint*: Consider $n = 2$ and $p = 2$ or 3.)

9.24.   In Example 9.23 use as the matrix P the matrix obtained from the identity matrix by exchanging rows one and eight, let

$$S = \begin{pmatrix} 1 & 0 \\ 1 & 1 \end{pmatrix}$$

and $z = 1\ 1\ 0\ 0\ 0\ 0\ 0\ 0$. Encipher the plaintext $x = 0\ 1$ with the Goppa-code cryptosystem and decipher the result by using nearest-neighbor decoding and Example 8.58.

9.25. Explain why the Goppa-code cryptosystem cannot be used for digital signatures.

9.26. Let $k = 3$ and $n = 5$ be the parameters of a threshold scheme based on $\mathbb{F}_8$ as in Example 9.24. Suppose the following pairs (c_i, D_i) are known: $(1, 0)$, $(\alpha, 0)$, $(\alpha^2, 1 + \alpha)$. Reconstruct the polynomial $f(x)$ over $\mathbb{F}_8$ and thus find the secret data D.

9.27. A threshold scheme is given by the parameters $q = 17, n = 5$, and $k = 3$. Suppose $f(1) = 8$, $f(2) = 7$, and $f(3) = 10$ are three known pairs (c_i, D_i). Find the secret D.

9.28. Show how discrete exponentiation can be used in a threshold scheme. Also design a scheme in which $n \geqslant 2$ mutually suspicious users are all needed to encipher a common secret (for instance, a classified document), but each individual user should be able to gain access to the secret (read the document) and decipher individually.

9.29. A cryptosystem due to L. S. Hill is based on linear transformations of the residue class ring $R_n = \mathbb{Z}/(n)$. The plaintext is represented as a k-tuple in R_n^k, enciphering is a nonsingular linear transformation and deciphering is its inverse.

(a) Suppose $n = 29$, $k = 2$, and the linear transformation is $P \to AP \pmod{29}$, where $P \in R_{29}^2$ and

$$A = \begin{pmatrix} 1 & 3 \\ 2 & 4 \end{pmatrix}.$$

Encipher the message "CRYPTOGRAPHY IS FUN." under the assumption that the letters A to Z are denoted by 0 to 25, a period by 26, a comma by 27, and a blank space by 28.

(b) This cryptosystem can also be based on linear transformations of $\mathbb{F}_q$. Let $q = 27, k = 3$, choose a nonsingular 3×3 matrix with entries in $\mathbb{F}_{27}$ and encipher the message "CIPHER", where $A = 0$, $B = \alpha^0$, $C = \alpha^1, \ldots$ for a primitive element α of $\mathbb{F}_{27}$.

(c) Let A be an $m \times m$ matrix with integer entries, let $\mathbf{b}, \mathbf{x} \in \mathbb{Z}^m$, and let n be a positive integer. Prove that $A\mathbf{x} \equiv \mathbf{b} \bmod n$ has a unique solution $\mathbf{x}$ modulo n (where a congruence between vectors is interpreted coordinatewise) if and only if $\gcd(\det(A), n) = 1$. Find a similar condition for the existence of a unique solution of the equation $A\mathbf{x} = \mathbf{b}$ over $\mathbb{F}_q$.

Chapter 10

Tables

In this chapter we collect tables that facilitate the computation in finite fields and tables of irreducible and primitive polynomials. The description of these tables is given in Sections 1 and 2, respectively.

1. COMPUTATION IN FINITE FIELDS

Multiplication and division of nonzero elements of $\mathbb{F}_q$ can be performed using a notion analogous to logarithms. We speak of the *index* or *discrete logarithm*. If b is a primitive element of $\mathbb{F}_q$, then for any $a \in \mathbb{F}_q^*$ there exists a unique integer r with $0 \leqslant r < q - 1$ such that $a = b^r$. We write $r = \text{ind}_b(a)$, or simply $r = \text{ind}(a)$ if b is kept fixed. The index function satisfies the following basic rules:

$$\text{ind}(ac) \equiv \text{ind}(a) + \text{ind}(c) \bmod (q - 1),$$

$$\text{ind}(ac^{-1}) \equiv \text{ind}(a) - \text{ind}(c) \bmod (q - 1).$$

The inverse function of the index function, corresponding to taking antilogarithms, is denoted by $\exp_b$ or simply $\exp$, and we have:

$$\exp(r) = b^r, \quad \exp(\text{ind}(a)) = a, \quad \text{ind}(\exp(r)) = r.$$

Given a table of the ind and exp function, it is easy to carry out addition, subtraction, multiplication, and division in $\mathbb{F}_q$. Addition and subtraction are performed by using the vector space structure of $\mathbb{F}_q$ over its prime subfield $\mathbb{F}_p$, multiplication and division are performed by using the rules for the index function and the exp and ind table to convert from one notation to the other. Table A provides a complete list of the nonzero elements and their indices for the finite fields $\mathbb{F}_q$ with q composite and $q \leqslant 128$. In the exp column the parentheses and commas of the vector notation for the following element of $\mathbb{F}_q$ with $q = p^n$ have been dropped:

$$a = (a_1, \ldots, a_n) = a_1 b^{n-1} + a_2 b^{n-2} + \cdots + a_n, \quad 0 \leqslant a_i < p.$$

In Table A we use $GF(p^n)$ to denote the finite field with p^n elements.

10.1.   Example.   As an example for the use of Table A we calculate

$$[(b+1) + (2b+2)b](b+2)^{-1} + b$$

in the field $\mathbb{F}_9$. Working with the portion of the table pertaining to this field, we get

$$\mathrm{ind}((2b+2)b) \equiv \mathrm{ind}(2b+2) + \mathrm{ind}(b) \equiv 3 + 1 \equiv 4 \bmod 8,$$

$$(2b+2)b = \exp(4) = 2.$$

Thus, $(b+1) + (2b+2)b = b$ and

$$\mathrm{ind}\big([(b+1) + (2b+2)b](b+2)^{-1}\big) \equiv \mathrm{ind}(b) - \mathrm{ind}(b+2) \equiv 1 - 6 \equiv 3 \bmod 8,$$

$$[(b+1) + (2b+2)b](b+2)^{-1} = \exp(3) = 2b+2.$$

The final result is $(2b+2) + b = 2$.   □

Table B affords another possibility of doing arithmetic in finite fields. In the first two columns it provides a table of *Jacobi's logarithm* $L(n)$ for the fields $\mathbb{F}_{2^k}$ with $2 \leqslant k \leqslant 6$ (compare with Exercise 2.8). The symbol $n \rightarrow s$ means that $L(n) = s$ with respect to a fixed primitive element b. In characteristic 2 the value $L(0)$ is undefined. The elements b^n are multiplied in the obvious way and added according to the rule

$$b^m + b^n = b^{m+L(n-m)}$$

given in Exercise 2.8. The symbol "+" preceding the value of n indicates that b^n is a primitive element.

10.2.   Example.   We use Table B to calculate

$$(b^6 + b^{25} + b^{44})(1 + b^{35})^{-1} + b^{28}$$

in the field $\mathbb{F}_{64}$. We have $b^6 + b^{25} = b^{6+L(19)} = b^{40}$ and $b^{40} + b^{44} = b^{40+L(4)} = b^{72}$. Since $1 + b^{35} = b^{L(35)} = b^{31}$, we get

$$(b^6 + b^{25} + b^{44})(1 + b^{35})^{-1} = b^{72} b^{-31} = b^{41}.$$

Furthermore, since the argument of the function L and the exponent of b are considered modulo 63, we obtain $b^{41} + b^{28} = b^{41+L(-13)} = b^{41+L(50)} =$

$b^{101} = b^{38}$, which is the final result and happens to be a primitive element of $\mathbb{F}_{64}$. □

The remainder of Table B provides information about *minimal* and *characteristic polynomials* and about *dual bases*. We take the lines

$$+20 \rightarrow 26[100001] \quad 26 \quad 6 \quad 49 \quad 29 \quad 9 \quad 46:19$$
$$21 \rightarrow 42[101011] \qquad [11]$$

from the table for $\mathbb{F}_{64}$ over $\mathbb{F}_2$ as illustrations. The symbol $[a_1 \; a_2 \cdots a_m]$ indicates that $x^m + a_1 x^{m-1} + a_2 x^{m-2} + \cdots + a_m$ is the characteristic polynomial of the element with respect to the given field extension. Thus, $x^6 + x^5 + 1$ is the characteristic polynomial of b^{20} over $\mathbb{F}_2$ and $x^6 + x^5 + x^3 + x + 1$ is that of b^{21} over $\mathbb{F}_2$. If b^n is a defining element of the extension, then the set of integers between the characteristic polynomial and the colon describes the dual basis of the polynomial basis determined by b^n. If b^n is not a defining element, then the minimal polynomial of b^n with respect to the given extension is listed in the bracket notation explained above. For instance, b^{20} is a defining element of $\mathbb{F}_{64}$ over $\mathbb{F}_2$ and the dual basis of the polynomial basis $\{1, b^{20}, b^{40}, b^{60}, b^{80}, b^{100}\}$ is $\{b^{26}, b^6, b^{49}, b^{29}, b^9, b^{46}\}$. On the other hand, b^{21} is not a defining element of $\mathbb{F}_{64}$ over $\mathbb{F}_2$ and the minimal polynomial of b^{21} over $\mathbb{F}_2$ is $x^2 + x + 1$, so that $b^{21} \in \mathbb{F}_4$. If b^n is not only a defining element, but also determines a *normal basis* of the given extension, then the integer after the colon describes the element determining the *dual normal basis*. For instance, b^{20} determines the normal basis

$$\{b^{20}, (b^{20})^2, (b^{20})^4, (b^{20})^8, (b^{20})^{16}, (b^{20})^{32}\}$$

of $\mathbb{F}_{64}$ over $\mathbb{F}_2$, and its dual basis is given by

$$\{b^{19}, (b^{19})^2, (b^{19})^4, (b^{19})^8, (b^{19})^{16}, (b^{19})^{32}\}.$$

Elements in subfields except $\mathbb{F}_2$ are denoted in the table by capital letters whose meaning becomes clear upon inspection of the data for minimal polynomials. For example, in the table for $\mathbb{F}_{64}$ the letter X stands for $b^{21} \in \mathbb{F}_4$ and D stands for $b^{27} \in \mathbb{F}_8$.

TABLE A

$GF(2^2)$

exp	ind
01	0
10	1
11	2

$GF(2^3)$

exp	ind
001	0
010	1
100	2
101	3
111	4
011	5
110	6

$GF(2^4)$

exp	ind
0001	0
0010	1
0100	2
1000	3
1001	4
1011	5
1111	6
0111	7
1110	8
0101	9
1010	10
1101	11
0011	12
0110	13
1100	14

$GF(2^5)$

exp	ind
00001	0
00010	1
00100	2
01000	3
10000	4
01001	5
10010	6
01101	7
11010	8
11101	9
10011	10
01111	11
11110	12
10101	13
00011	14
00110	15
01100	16
11000	17
11001	18
11011	19
11111	20
10111	21
00111	22
01110	23
11100	24
10001	25
01011	26
10110	27
00101	28
01010	29
10100	30

$GF(2^6)$

exp	ind
000001	0
000010	1
000100	2
001000	3
010000	4
100000	5
100001	6
100011	7
100111	8
101111	9
111111	10
011111	11
111110	12
011101	13
111010	14
010101	15
101010	16
110101	17
001011	18
010110	19
101100	20
111001	21
010011	22
100110	23
101101	24
111011	25
010111	26
101110	27
111101	28
011011	29
110110	30
001101	31
011010	32
110100	33
001001	34
010010	35
100100	36
101001	37
110011	38
000111	39
001110	40
011100	41
111000	42
010001	43
100010	44
100101	45
101011	46
110111	47
001111	48
011110	49
111100	50
011001	51
110010	52
000101	53
001010	54
010100	55
101000	56
110001	57
000011	58
000110	59
001100	60
011000	61
110000	62

$GF(2^7)$

exp	ind
0000001	0
0000010	1
0000100	2
0001000	3
0010000	4
0100000	5
1000000	6
0000011	7
0000110	8
0001100	9
0011000	10
0110000	11
1100000	12
1000011	13
0000101	14
0001010	15
0010100	16
0101000	17
1010000	18
0100011	19
1000110	20
0001111	21
0011110	22
0111100	23
1111000	24
1110011	25
1100101	26
1001001	27
0010001	28
0100010	29
1000100	30
0001011	31
0010110	32
0101100	33
1011000	34
0110011	35
1100110	36
1001111	37
0011101	38
0111010	39
1110100	40
1101011	41
1010101	42
0101001	43
1010010	44
0100111	45
1001110	46
0011111	47
0111110	48
1111100	49
1111011	50
1110110	51
1101001	52
1010001	53

exp	ind	exp	ind	exp	ind	exp	ind
GF(2⁷)		GF(2⁷)		GF(3³)		GF(3⁴)	

exp	ind	exp	ind	exp	ind	exp	ind
0100001	54	1110110	101	112	8	1102	27
1000010	55	1101111	102	222	9	0021	28
0000111	56	1011101	103	121	10	0210	29
0001110	57	0111001	104	012	11	2100	30
0011100	58	1110010	105	120	12	2002	31
0111000	59	1100111	106	002	13	1022	32
1110000	60	1001101	107	020	14	2221	33
1100011	61	0011001	108	200	15	0212	34
1000101	62	0110010	109	201	16	2120	35
0001001	63	1100100	110	211	17	2202	36
0010010	64	1001011	111	011	18	0022	37
0100100	65	0010101	112	110	19	0220	38
1001000	66	0101010	113	202	20	2200	39
0010011	67	1010100	114	221	21	0002	40
0100110	68	0101011	115	111	22	0020	41
1001100	69	1010110	116	212	23	0200	42
0011011	70	0101111	117	021	24	2000	43
0110110	71	1011110	118	210	25	1002	44
1101100	72	0111111	119			2021	45
1011011	73	1111110	120			1212	46
0110101	74	1111111	121	GF(3⁴)		1121	47
1101010	75	1111101	122			0211	48
1010111	76	1111001	123	0001	0	2110	49
0101101	77	1110001	124	0010	1	2102	50
1011010	78	1100001	125	0100	2	2022	51
0110111	79	1000001	126	1000	3	1222	52
1101110	80			2001	4	1221	53
1011111	81	GF(3²)		1012	5	1211	54
0111101	82			2121	6	1111	55
1111010	83	01	0	2212	7	0111	56
		10	1	0122	8	1110	57
		21	2	1220	9	0101	58
1110111	84	22	3	1201	10	1010	59
1101101	85	02	4	1011	11	2101	60
1011001	86	20	5	2111	12	2012	61
0110001	87	12	6	2112	13	1122	62
1100010	88	11	7	2122	14	0221	63
1000111	89			2222	15	2210	64
0001101	90			0222	16	0102	65
0011010	91	GF(3³)		2220	17	1020	66
0110100	92			0202	18	2201	67
1101000	93	001	0	2020	19	0012	68
1010011	94	010	1	1202	20	0120	69
0100101	95	100	2	1021	21	1200	70
1001010	96	102	3	2211	22	1001	71
0010111	97	122	4	0112	23	2011	72
0101110	98	022	5	1120	24	1112	73
1011100	99	220	6	0201	25	0121	74
0111011	100	101	7	2010	26		

TABLE A (*Cont.*)

exp	ind	exp	ind	exp	ind	exp	ind
$GF(3^4)$		$GF(5^3)$		$GF(5^3)$		$GF(5^3)$	
1210	75	212	13	330	61	143	109
1101	76	421	14	004	62	333	110
0011	77	312	15	040	63	034	111
0110	78	324	16	400	64	340	112
1100	79	444	17	102	65	104	113
		042	18	423	66	443	114
$GF(5^2)$		420	19	332	67	032	115
		302	20	024	68	320	116
01	0	224	21	240	69	404	117
10	1	041	22	201	70	142	118
43	2	410	23	311	71	323	119
42	3	202	24	314	72	434	120
32	4	321	25	344	73	442	121
44	5	414	26	144	74	022	122
02	6	242	27	343	75	220	123
20	7	221	28	134	76		
31	8	011	29	243	77		
34	9	110	30	231	78	$GF(7^2)$	
14	10	003	31	111	79		
33	11	030	32	013	80	01	0
04	12	300	33	130	81	10	1
40	13	204	34	203	82	64	2
12	14	341	35	331	83	53	3
13	15	114	36	014	84	56	4
23	16	043	37	140	85	16	5
11	17	430	38	303	86	54	6
03	18	402	39	234	87	66	7
30	19	122	40	141	88	03	8
24	20	123	41	313	89	30	9
21	21	133	42	334	90		
41	22	233	43	044	91	45	10
22	23	131	44	440	92	12	11
		213	45	002	93	14	12
$GF(5^3)$		431	46	020	94	34	13
		412	47	200	95	15	14
001	0	222	48	301	96	44	15
010	1	021	49	214	97	02	16
100	2	210	50	441	98	20	17
403	3	401	51	012	99	51	18
132	4	112	52	120	100	36	19
223	5	023	53	103	101		
031	6	230	54	433	102	35	20
310	7	101	55	432	103	25	21
304	8	413	56	422	104	31	22
244	9	232	57	322	105	55	23
241	10	121	58	424	106	06	24
211	11	113	59	342	107	60	25
411	12	033	60	124	108	13	26

exp	ind	exp	ind	exp	ind	exp	ind
GF(7²)		*GF*(11²)		*GF*(11²)		*GF*(11²)	
24	27	07	12	30	49	06	84
21	28	70	13			60	85
61	29	46	14			52	86
		25	15			89	87
23	30	38	16	81	50	1*	88
11	31	51	17	4*	51	94	89
04	32	79	18	65	52		
40	33	26	19	*2	53	63	90
32	34			37	54	82	91
65	35	48	20	41	55	5*	92
63	36	45	21	85	56	59	93
43	37	15	22	8*	57	49	94
62	38	44	23	2*	58	55	95
33	39	05	24	88	59	09	96
		50	25			90	97
05	40	69	26	0*	60	23	98
50	41	32	27	*0	61	18	99
26	42	*1	28	17	62		
41	43	27	29	64	63	74	100
42	44			92	64	86	101
52	45	58	30	43	65	9*	102
46	46	39	31	*5	66	13	103
22	47	61	32	67	67	24	104
		62	33	12	68	28	105
		72	34	14	69	68	106
GF(11²)		66	35			22	107
		02	36	34	70	08	108
01	0	20	37	11	71	80	109
10	1	98	38	04	72		
*4	2	*3	39	40	73		
57	3			75	74	3*	110
29	4	47	40	96	75	71	111
78	5	35	41	83	76	56	112
16	6	21	42	6*	77	19	113
54	7	*8	43	42	78	84	114
*9	8	97	44	95	79	7*	115
*7	9	93	45			36	116
		53	46	73	80	31	117
87	10	99	47	76	81	91	118
**	11	03	48	*6	82	33	119
				77	83		

The symbol * denotes the element 10 in $\mathbb{F}_{11}$.

TABLE B

$\mathbb{F}_4$ over $\mathbb{F}_2$	$\mathbb{F}_8$ over $\mathbb{F}_2$
0 → * [01] [1]	0 → * [111] [1]
+ 1 → 2 [11]2 0:1	+ 1 → 5 [101]4 3 5:1
+ 2 → 1 [11]1 0:2	+ 2 → 3 [101]1 6 3:2
	+ 3 → 2 [011]0 6 3:-
	+ 4 → 6 [101]2 5 6:4
	+ 5 → 1 [011]0 3 5:-
	+ 6 → 4 [011]0 5 6:-

$\mathbb{F}_{16}$ over $\mathbb{F}_2$		over $\mathbb{F}_4$	
0 → * [0001] [1]		[01] [1]	
+ 1 → 4 [0011]14 2 1 0: -		[1X] 4 0: 1	
+ 2 → 8 [0011]13 4 2 0: -		[1Y] 8 0: 2	
3 → 14 [1111]14 10 1 2:11		[Y1] 2 5:13	
+ 4 → 1 [0011]11 8 4 0: -		[1X] 1 0: 4	
5 → 10 [0101] [11]		[0Y] [X]	
6 → 13 [1111]13 5 2 4: 7		[X1] 4 10:11	
+ 7 → 9 [1001] 9 2 10 1: 6		[XX] 8 10:12	
+ 8 → 2 [0011] 7 1 8 0: -		[1Y] 2 0: 8	
9 → 7 [1111] 7 5 8 1:13		[X1] 1 10:14	
10 → 5 [0101] [11]		[0X] [Y]	
+ 11 → 12 [1001]12 1 5 8: 3		[YY] 4 5: 6	
12 → 11 [1111]11 10 4 8:14		[Y1] 8 5: 7	
+ 13 → 6 [1001] 6 8 10 4: 9		[XX] 2 10: 3	
+ 14 → 3 [1001] 3 4 5 2:12		[YY] 1 5: 9	

$\mathbb{F}_{32}$ over $\mathbb{F}_2$

0 * [10011] [1]	
+ 1 → 19 [10111]16 3 6 5 17: 1	
+ 2 → 7 [10111] 1 6 12 10 3: 2	
+ 3 → 11 [01001] 0 28 25 6 3: -	
+ 4 → 14 [10111] 2· 12 24 20 6: 4	
+ 5 → 29 [01111] 4 28 5 14 9: -	
+ 6 → 22 [01001] 0 25 19 12 6: -	
+ 7 → 2 [00101]27 20 17 10 3: -	
+ 8 → 28 [10111] 4 24 17 9 12: 8	
+ 9 → 15 [01111] 1 7 9 19 10: -	
+ 10 → 27 [01111] 8 25 10 28 18: -	
+ 11 → 3 [11101]23 12 7 6 3:15	
+ 12 → 13 [01001] 0 19 7 24 12: -	
+ 13 → 12 [11101]30 17 28 24 12:29	
+ 14 → 4 [00101]23 9 3 20 6: -	
+ 15 → 9 [11011]26 20 5 19 10:11	

F_{32} over F_2

$+16 \rightarrow 25$ [10111] 8 17   3 18 24:16
$+17 \rightarrow 21$ [01001] 0 14 28   3 17: -
$+18 \rightarrow 30$ [01111] 2 14 18   7 20: -
$+19 \rightarrow \ 1$ [00101]29 10 24   5 17: -

$+20 \rightarrow 23$ [01111]16 19 20 25   5: -
$+21 \rightarrow 17$ [11101]27   6 19   3 17:23
$+22 \rightarrow \ 6$ [11101]15 24 14 12   6:30
$+23 \rightarrow 20$ [11011]13 10 18 25   5:21
$+24 \rightarrow 26$ [01001] 0   7 14 17 24: -
$+25 \rightarrow 16$ [00101]30   5 12 18 24: -
$+26 \rightarrow 24$ [11101]29   3 25 17 24:27
$+27 \rightarrow 10$ [11011]22   5   9 28 18:26
$+28 \rightarrow \ 8$ [00101]15 18   6   9 12: -
$+29 \rightarrow \ 5$ [11011]11 18 20 14   9:13

$+30 \rightarrow 18$ [11011]21   9 10   7 20:22

F_{64} over F_2	over F_4	over F_8
$0 \rightarrow$ * [010101] [1]	[111] [1]	[01] [1]
$+\ \ 1 \rightarrow \ 8$ [101101]44 43 58 54 53 45: -	[XXX]47 54 27: 8	[$1A$] 8 0: 1
$+\ \ 2 \rightarrow 16$ [101101]25 23 53 45 43 27: -	[YYY]31 45 54:16	[$1B$]16 0: 2
$3 \rightarrow 53$ [010111]60 47 46 43 3 0: -	[$0Y1$] 0 6 3: -	[FD]42 18:39
$+\ \ 4 \rightarrow 32$ [101101]50 46 43 27 23 54: -	[XXX]62 27 45:32	[$1C$] 32 0: 4
$+\ \ 5 \rightarrow 38$ [100001]38 33 28 23 18 43:52	[XYY]23 56 49: -	[CF] 4 27:59
$6 \rightarrow 43$ [010111]57 31 29 23 6 0: -	[$0X1$] 0 12 6: -	[DE]21 36:15
$7 \rightarrow 62$ [001001]42 35 28 0 56 49: -	[$00X$] 0 56 49: -	[$E1$] 2 9:25
$+\ \ 8 \rightarrow \ 1$ [101101]37 29 23 54 46 45: -	[YYY]61 54 27: 1	[$1A$] 1 0: 8
$9 \rightarrow 45$ [010001] [101]	[101] 36 27 45: 9	[$0B$] [A]
$+10 \rightarrow 13$ [100001]13 3 56 46 36 23:41	[YXX]46 49 35: -	[AD] 8 54:55
$+11 \rightarrow 51$ [110011]37 14 3 55 36 48:11	[$X1Y$]55 48 24:25	[AC]16 54:56
$12 \rightarrow 23$ [010111]51 62 58 46 12 0: -	[$0Y1$] 0 24 12: -	[EF]42 9:30
$+13 \rightarrow 10$ [100111]10 7 59 46 33 23:17	[XYX]43 49 35: -	[AE]32 54:58
$14 \rightarrow 61$ [001001]21 7 56 0 49 35: -	[$00Y$] 0 49 35: -	[$F1$] 4 18:50
$15 \rightarrow 44$ [110101]51 36 46 31 47 3:15	[$Y01$] 18 3 33:57	[CA]21 27: 6
$+16 \rightarrow \ 2$ [101101]11 58 46 45 29 27: -	[XXX]59 45 54: 2	[$1B$] 2 0:16
$+17 \rightarrow 41$ [100001]41 24 7 53 36 58:13	[XYY]53 14 28: -	[AD] 1 54:62
$18 \rightarrow 27$ [010001] [101]	[101] 9 54 27:18	[$0C$] [B]
$+19 \rightarrow 34$ [100111]34 49 62 43 24 53:20	[XYX]58 28 56: -	[BF] 8 45:46
$+20 \rightarrow 26$ [100001]26 6 49 29 9 46:19	[XYY]29 35 7: -	[BE]16 45:47
$21 \rightarrow 42$ [101011] [11]	[$XY1$] [X]	[11] 42 0:21
$+22 \rightarrow 39$ [110011]11 28 6 47 9 33:22	[$Y1X$]47 33 48:50	[BA]32 45:49
$+23 \rightarrow 12$ [000011]40 29 6 46 23 0: -	[$1XY$]31 24 12:58	[EB] 4 9:41
$24 \rightarrow 46$ [010111]39 61 53 29 24 0: -	[$0X1$] 0 48 24: -	[FD]21 18:60
$+25 \rightarrow 30$ [110011]44 49 24 62 36 6:25	[$Y1X$]62 6 3:11	[AC] 2 54: 7

TABLE B   (*Cont.*)

$\mathbb{F}_{64}$	over $\mathbb{F}_2$	over $\mathbb{F}_4$	over $\mathbb{F}_8$
+ 26 → 20	[100111]20 14 55 29 3 46:34	[YXY]23 35 7: -	[BF] 1 45:53
27 → 18	[000101] [011]	[011] 0 54 27: -	[OE] [D]
28 → 59	[001001]42 14 49 0 35 7: -	[00X] 0 35 7: -	[D1] 8 36:37
+ 29 → 48	[000011]34 53 24 58 29 0: -	[1XY]61 33 48:43	[DA]16 36:38
30 → 25	[110101]39 9 29 62 31 6:30	[X01]36 6 3:51	[AB]42 54:12
+ 31 → 35	[011011]25 29 61 0 24 56: -	[11X]46 28 56: -	[DD]32 36:40
+ 32 → 4	[101101]22 53 29 27 58 54: -	[YYY]55 27 45: 4	[1C] 4 0:32
33 → 58	[010111]30 55 23 53 33 0: -	[0X1] 0 3 33: -	[EF]21 9:51
+ 34 → 19	[100001]19 48 14 43 9 53:26	[YXX]43 28 56: -	[BE] 2 45:61
35 → 31	[001001]21 49 14 0 28 56: -	[00Y] 0 28 56: -	[D1] 1 36:44
36 → 54	[010001] [101]	[101] 18 45 54:36	[OA] [C]
+ 37 → 57	[110101]50 7 33 59 18 24:37	[Y1X]59 24 12:44	[CB] 8 27:28
+ 38 → 5	[100111] 5 35 61 23 48 43:40	[YXY]53 56 49: -	[CD]16 27:29
39 → 22	[110101]57 18 23 47 55 33:39	[X01] 9 33 48:60	[BC]42 45: 3
+ 40 → 52	[100001]52 12 35 58 18 29:38	[YXX]58 7 14: -	[CF]32 27:31
+ 41 → 17	[100111]17 56 31 53 12 58:10	[YXY]29 14 28: -	[AE] 4 54:23
42 → 21	[101011] [11]	[YX1] [Y]	[11] 21 0:42
+ 43 → 6	[000011]20 46 3 23 43 0: -	[1YX]47 12 6:29	[DA] 2 36:52
+ 44 → 15	[110011]22 56 12 31 18 3:44	[X1Y]31 3 33:37	[CB] 1 27:35
45 → 9	[000101] [011]	[011] 0 27 45: -	[OD] [F]
+ 46 → 24	[000011]17 58 12 29 46 0: -	[1YX] 62 48 24:53	[FC] 8 18:19
+ 47 → 49	[011011]44 46 62 0 12 28: -	[11Y] 23 14 28: -	[FF]16 18:20
48 → 29	[010111]15 59 43 58 48 0: -	[0Y1] 0 33 48: -	[DE]42 36:57
49 → 47	[001001]42 56 7 0 14 28: -	[00X] 0 14 28: -	[F1] 32 18:22
+ 50 → 60	[110011]25 35 48 61 9 12:50	[X1Y]61 12 6:22	[BA] 4 45:14
51 → 11	[110101]60 9 43 55 59 48:51	[Y01]36 48 24:30	[AB]21 54:33
+ 52 → 40	[100111]40 28 27 58 6 29: 5	[XYX]46 7 14: -	[CD] 2 27:43
+ 53 → 3	[000011]10 23 33 43 53 0: -	[1XY]55 6 3:46	[FC] 1 18:26
54 → 36	[000101] [011]	[011] 0 45 54: -	[OF] [E]
+ 55 → 56	[011011]22 23 31 0 6 14: -	[11X]43 7 14: -	[EE] 8 9:10
56 → 55	[001001]21 28 35 0 7 14: -	[00Y] 0 7 14: -	[E1]16 9:11
57 → 37	[110101]30 36 53 59 61 24:57	[X01]18 24 12:15	[CA]42 27:48
+ 58 → 33	[000011] 5 43 48 53 58 0: -	[1YX] 59 3 33:23	[EB]32 9:13
+ 59 → 28	[011011]11 43 47 0 3 7: -	[11Y] 53 35 7: -	[DD] 4 36: 5
60 → 50	[110101]15 18 58 61 62 12:60	[Y01] 9 12 6:39	[BC]21 45:24
+ 61 → 14	[011011]37 53 55 0 33 35: -	[11X]58 49 35: -	[EF] 2 18:34
+ 62 → 7	[011011]50 58 59 0 48 49: -	[11Y]29 56 49: -	[EE] 1 9:17

2. TABLES OF IRREDUCIBLE POLYNOMIALS

Table C lists all monic irreducible polynomials of degree n over prime fields $\mathbb{F}_p$ for small values of n and p. The extent of the table may be summarized as follows: $p = 2$ and $n \leqslant 11$, $p = 3$ and $n \leqslant 7$, $p = 5$ and $n \leqslant 5$, $p = 7$ and $n \leqslant 4$. The polynomial $a_0 x^n + a_1 x^{n-1} + \cdots + a_n$ is abbreviated in the form $a_0 a_1 \cdots a_n$ with $a_0 = 1$. The left-hand column, headed by the value of n, lists all monic irreducible polynomials f for the degree n and the modulus p concerned. The right-hand column, headed by e, contains the corresponding value of $\mathrm{ord}(f)$.

Table D lists one primitive polynomial over $\mathbb{F}_2$ for each degree $n \leqslant 100$. In this table only the degrees of the separate terms in the polynomial are given; thus $6\ 1\ 0$ stands for $x^6 + x + 1$.

Table E lists all primitive polynomials $x^2 + a_1 x + a_2$ of degree 2 over $\mathbb{F}_p$ for $11 \leqslant p \leqslant 31$. For smaller primes all quadratic primitive polynomials can be obtained from Table C by locating the polynomials f over $\mathbb{F}_p$ with $\mathrm{ord}(f) = p^2 - 1$.

Table F lists one primitive polynomial of degree n over $\mathbb{F}_p$ for all values of $n \geqslant 2$ and p with $p < 50$ and $p^n < 10^9$. The polynomial $x^n + a_1 x^{n-1} + a_2 x^{n-2} + \cdots + a_n$ is listed in the form $a_1 a_2 \cdots a_n$.

TABLE C

Irreducible Polynomials for the Modulus 2

	e		e		e		e
n = 1	e	10111001	127	1000011011	511	1111100011	511
		10111111	127	1000100001	511	1111101001	511
10	1	11000001	127	1000101101	511	1111111011	511
11	1	11001011	127	1000110011	511		
		11010011	127	1001001011	73	n = 10	e
n = 2	e	11010101	127	1001011001	511		
		11100101	127	1001011111	511	10000001001	1023
111	3	11101111	127	1001100101	73	10000001111	341
		11110001	127	1001101001	511	10000011011	1023
n = 3	e	11110111	127	1001101111	511	10000011101	341
		11111101	127	1001110111	511	10000100111	1023
1011	7			1001111101	511	10000101101	1023
1101	7	n = 8	e	1010000111	511	10000110101	93
				1010010101	511	10001000111	341
n = 4	e	100011011	51	1010011001	73	10001010011	341
		100011101	255	1010100011	511	10001100011	341
10011	15	100101011	255	1010100101	511	10001100101	1023
11001	15	100101101	255	1010101111	511	10001101111	1023
11111	5	100111001	17	1010110111	511	10010000001	1023
		100111111	85	1010111101	511	10010001011	1023
n = 5	e	101001101	255	1011001111	511	10010011001	341
		101011111	255	1011010001	511	10010101001	33
100101	31	101100011	255	1011011011	511	10010101111	341
101001	31	101100101	255	1011110101	511	10011000101	1023
101111	31	101101001	255	1011111001	511	10011001001	341
110111	31	101110001	255	1100000001	73	10011010111	1023
111011	31	101110111	85	1100010011	511	10011100111	1023
111101	31	101111011	85	1100010101	511	10011101101	341
		110000111	255	1100011111	511	10011110011	1023
n = 6	e	110001011	85	1100100011	511	10011111111	1023
		110001101	255	1100110001	511	10100001011	93
1000011	63	110011111	51	1100111011	511	10100001101	1023
1001001	9	110100011	85	1101001001	73	10100011001	1023
1010111	21	110101001	255	1101001111	511	10100011111	341
1011011	63	110110001	51	1101011011	511	10100100011	1023
1100001	63	110111101	85	1101100001	511	10100110001	1023
1100111	63	111000011	255	1101101011	511	10100111101	1023
1101101	63	111001111	255	1101101101	511	10101000011	1023
1110011	63	111010111	17	1101110011	511	10101010111	1023
1110101	21	111011101	85	1101111111	511	10101100001	93
		111100111	255	1110000101	511	10101100111	341
n = 7	e	111110011	51	1110001111	511	10101101011	1023
		111110101	255	1110100001	73	10110000101	1023
10000011	127	111111001	85	1110110101	511	10110001111	1023
10001001	127			1110111001	511	10110010111	1023
10001111	127	n = 9	e	1111000111	511	10110011011	341
10010001	127			1111001011	511	10110100001	1023
10011101	127	1000000011	73	1111001101	511	10110101011	341
10100111	127	1000010001	511	1111010101	511	10110111001	341
10101011	127	1000010111	73	1111011001	511	10111000001	341

Irreducible Polynomials for the Modulus 2							
10111000111	1023	11111011011	1023	100111100101	2047	101111101101	2047
10111100101	1023	11111101011	341	100111101111	89	110000001011	2047
10111110111	1023	11111110011	1023	100111110111	2047	110000001101	2047
10111111011	1023	11111111001	1023	101000000001	2047	110000011001	2047
11000010011	1023	11111111111	11	101000000111	2047	110000011111	2047
11000010101	1023			101000010011	2047	110000110001	89
11000100011	33	$n=11$	e	101000010101	2047	110001010111	2047
11000100101	1023			101000101001	2047	110001100001	2047
11000110001	341	100000000101	2047	101001001001	2047	110001101011	2047
11000110111	1023	100000010111	2047	101001100001	2047	110001110011	2047
11001000011	1023	100000101011	2047	101001101101	2047	110001110101	23
11001001111	1023	100000101101	2047	101001111001	2047	110010000101	2047
11001010001	341	100001000111	2047	101001111111	2047	110010001001	2047
11001011011	1023	100001100011	2047	101010000101	2047	110010010111	2047
11001111001	1023	100001100101	2047	101010010001	2047	110010011011	2047
11001111111	1023	100001110001	2047	101010011101	2047	110010011101	2047
11010000101	93	100001111011	2047	101010100111	2047	110010110011	2047
11010001001	1023	100010001101	2047	101010101011	2047	110010111111	2047
11010100111	93	100010010101	2047	101010110011	2047	110011000111	2047
11010101101	341	100010011111	2047	101010110101	2047	110011001101	2047
11010110101	1023	100010101001	2047	101011010101	2047	110011010011	2047
11010111111	341	100010110001	2047	101011011111	2047	110011010101	2047
11011000001	1023	100011000011	89	101011100011	23	110011100011	2047
11011001101	341	100011001111	2047	101011101001	2047	110011101001	2047
11011010011	1023	100011010001	2047	101011101111	2047	110011110111	2047
11011011111	1023	100011100001	2047	101011110001	2047	110100000011	2047
11011110111	341	100011100111	2047	101011111011	2047	110100001111	2047
11011111101	1023	100011101011	2047	101100000011	2047	110100011101	2047
11100001111	341	100011110101	2047	101100001001	2047	110100100111	2047
11100010001	341	100100001101	2047	101100010001	2047	110100101101	2047
11100010111	1023	100100010011	2047	101100110011	2047	110101000001	2047
11100011101	1023	100100100101	2047	101100111111	2047	110101000111	2047
11100100001	1023	100100101001	2047	101101000001	2047	110101010011	2047
11100101011	93	100100110111	89	101101001011	2047	110101011001	2047
11100110101	341	100100111011	2047	101101011001	2047	110101100011	2047
11100111001	1023	100100111101	2047	101101011111	2047	110101101111	2047
11101000111	1023	100101000101	2047	101101100101	2047	110101110001	2047
11101001101	1023	100101001001	2047	101101101111	2047	110110010011	2047
11101010101	1023	100101010001	2047	101101111101	2047	110110011111	2047
11101011001	1023	100101011011	2047	101110000111	2047	110110101001	2047
11101100011	1023	100101110011	2047	101110001011	2047	110110111011	2047
11101111011	341	100101110101	2047	101110010011	2047	110110111101	2047
11101111101	1023	100101111111	2047	101110010101	2047	110111001001	2047
11110000001	341	100110000011	2047	101110101111	2047	110111010111	2047
11110000111	341	100110001111	2047	101110110111	2047	110111011011	2047
11110001101	1023	100110101011	2047	101110111101	2047	110111100001	2047
11110010011	1023	100110101101	2047	101111001001	2047	110111100111	2047
11110101001	341	100110111001	2047	101111011011	2047	110111110101	2047
11110110001	1023	100111000111	2047	101111011101	2047	110111111111	89
11111000101	341	100111011001	2047	101111100111	2047	111000000101	2047

TABLE C (*Cont.*)

Irreducible Polynomials for the Modulus 2

111000011101	2047	111001111011	2047	111011111001	2047	111110010001	2047
111000100001	2047	111001111101	2047	111100001011	2047	111110010111	2047
111000100111	2047	111010000001	2047	111100011001	2047	111110011011	2047
111000101011	2047	111010010011	2047	111100110001	2047	111110100111	2047
111000110011	2047	111010011111	2047	111100110111	2047	111110101101	2047
111000111001	2047	111010100011	2047	111101011101	2047	111110110101	2047
111001000111	2047	111010111011	2047	111101101011	2047	111111001101	2047
111001001011	2047	111011001001	89	111101101101	2047	111111010011	2047
111001010101	2047	111011001111	2047	111101110101	2047	111111100101	2047
111001011111	2047	111011011101	2047	111101111001	89	111111101001	2047
111001110001	2047	111011110011	2047	111110000011	2047	111111111011	89

Irreducible Polynomials for the Modulus 3

n = 1	e								
		12101	40	120001	242	1011022	728	1111112	728
		12112	80	120011	242	1011122	728	1111222	728
10	1	12121	10	120022	121	1012001	182	1112011	91
11	2	12212	80	120202	121	1012012	728	1112201	182
12	1	*n = 5*	*e*	120212	121	1012021	364	1112222	728
				120221	242	1012112	728	1120102	728
n = 2	*e*	100021	242	121012	121	1020001	52	1120121	91
		100022	121	121111	242	1020101	52	1120222	728
101	4	100112	121	121112	121	1020112	728	1121012	728
112	8	100211	242	121222	121	1020122	728	1121102	728
122	8	101011	242	122002	121	1021021	364	1121122	104
		101012	121	122021	242	1021102	56	1121212	728
n = 3	*e*	101102	121	122101	242	1021112	728	1121221	364
		101122	121	122102	121	1021121	91	1122001	91
1021	26	101201	242	122201	22	1022011	364	1122002	104
1022	13	101221	242	122212	121	1022102	56	1122122	104
1102	13	102101	242			1022111	182	1122202	728
1112	13	102112	121	*n = 6*	*e*	1022122	728	1122221	364
1121	26	102122	11			1100002	728	1200002	728
1201	26	102202	121	1000012	728	1100012	56	1200022	56
1211	26	102211	242	1000022	728	1100111	364	1200121	364
1222	13	102221	22	1000111	364	1101002	728	1201001	364
				1000121	364	1101011	28	1201111	182
n = 4	*e*	110002	121	1000201	52	1101101	364	1201121	182
		110012	121	1001012	728	1101112	728	1201201	364
10012	80	110021	242	1001021	364	1101212	728	1201202	728
10022	80	110101	242	1001101	91	1102001	364	1202002	728
10102	16	110111	242	1001122	104	1102111	91	1202021	28
10111	40	110122	121	1001221	182	1102201	364	1202101	364
10121	40	111011	242	1002011	364	1102202	728	1202122	728
10202	16	111121	242	1002022	728	1110001	364	1202222	728
11002	80	111211	242	1002101	182	1110011	364	1210001	364
11021	20	111212	121	1002112	104	1110122	728	1210021	364
11101	40	112001	242	1002211	91	1110202	728	1210112	728
11111	5	112022	121	1010201	52	1110221	182	1210202	728
11122	80	112102	11	1010212	728	1111012	728	1210211	91
11222	80	112111	242	1010222	728	1111021	182	1211021	182
12002	80	112201	242	1011001	91	1111111	7	1211201	91
12011	20	112202	121	1011011	364			1211212	728

Irreducible Polynomials for the Modulus 3

	e		e		e		e		e
1212011	91	10022021	2186	10202012	1093	11021122	1093	11201222	1093
1212022	728	10022101	2186	10210001	2186	11021201	2186	11202002	1093
1212121	14	10022212	1093	10210121	2186	11021212	1093	11202121	2186
1212122	728	10100011	2186	10210202	1093	11022101	2186	11202211	2186
1212212	728	10100012	1093	10211101	2186	11022122	1093	11202212	1093
1220102	728	10100102	1093	10211111	2186	11022211	2186	11210002	1093
1220111	182	10100122	1093	10211122	1093	11022221	2186	11210011	2186
1220212	728	10100201	2186	10211221	2186	11100002	1093	11210021	2186
1221001	182	10100221	2186	10212011	2186	11100022	1093	11210101	2186
1221002	104	10101101	2186	10212022	1093	11100121	2186	11211001	2186
1221112	104	10101112	1093	10212101	2186	11100212	1093	11211022	1093
1221202	728	10101202	1093	10212112	1093	11101012	1093	11211122	1093
1221211	364	10101211	2186	10212212	1093	11101022	1093	11211212	1093
1222022	728	10102102	1093	10220002	1093	11101102	1093	11211221	2186
1222102	728	10102201	2186	10220101	2186	11101111	2186	11212012	1093
1222112	104	10110022	1093	10220222	1093	11101121	2186	11212112	1093
1222211	364	10110101	2186	10221122	1093	11102002	1093	11212212	1093
1222222	728	10110211	2186	10221202	1093	11102111	2186	11220001	2186
		10111001	2186	10221212	1093	11102222	1093	11220112	1093
n = 7	e	10111102	1093	10221221	2186	11110001	2186	11220211	2186
		10111121	2186	10222012	1093	11110012	1093	11221022	1093
10000102	1093	10111201	2186	10222021	2186	11110111	2186	11221102	1093
10000121	2186	10112002	1093	10222111	2186	11110112	1093	11221112	1093
10000201	2186	10112012	1093	10222202	1093	11110211	2186	11221121	2186
10000222	1093	10112021	2186	10222211	2186	11110222	1093	11222011	2186
10001011	2186	10112111	2186	11000101	2186	11111011	2186	11222102	1093
10001012	1093	10112122	1093	11000222	1093	11111021	2186	11222122	1093
10001102	1093	10120021	2186	11001022	1093	11111201	2186	11222201	2186
10001111	2186	10120112	1093	11001112	1093	11111222	1093	11222221	2186
10001201	2186	10120202	1093	11001211	2186	11112011	2186	12000121	2186
10001212	1093	10121002	1093	11002012	1093	11112221	2186	12000202	1093
10002112	1093	10121102	1093	11002022	1093	11120102	1093	12001021	2186
10002122	1093	10121201	2186	11002121	2186	11120111	2186	12001112	1093
10002211	2186	10121222	1093	11002202	1093	11120122	1093	12001211	2186
10002221	2186	10122001	2186	11010001	2186	11120212	1093	12002011	2186
10010122	1093	10122011	2186	11010022	1093	11120221	2186	12002021	2186
10010222	1093	10122022	1093	11010121	2186	11121001	2186	12002101	2186
10011002	1093	10122212	1093	11010221	2186	11121101	2186	12002222	1093
10011101	2186	10122221	2186	11011111	2186	11121202	1093	12010021	2186
10011211	2186	10200001	2186	11011202	1093	11122021	2186	12010022	1093
10012001	2186	10200002	1093	11012002	1093	11122112	1093	12010102	1093
10012022	1093	10200101	2186	11012102	1093	11122201	2186	12010121	2186
10012111	2186	10200112	1093	11012212	1093	11122222	1093	12010201	2186
10012202	1093	10200202	1093	11020021	2186	11200201	2186	12010211	2186
10020121	2186	10200211	2186	11020022	1093	11200202	1093	12011102	1093
10020221	2186	10201021	2186	11020102	1093	11201012	1093	12011111	2186
10021001	2186	10201022	1093	11020112	1093	11201021	2186	12011212	1093
10021112	1093	10201121	2186	11020201	2186	11201101	2186	12011221	2186
10021202	1093	10201222	1093	11020222	1093	11201111	2186	12012112	1093
10022002	1093	10202011	2186	11021111	2186	11201221	2186	12012122	1093

TABLE C (*Cont.*)

Irreducible Polynomials for the Modulus 3

12012202	1093	12101201	2186	12112211	2186	12201121	2186	12212122	1093
12012221	2186	12101212	1093	12120002	1093	12201122	1093	12212201	2186
12020002	1093	12101222	1093	12120011	2186	12201202	1093	12212221	2186
12020021	2186	12102001	2186	12120112	1093	12201212	1093	12220001	2186
12020122	1093	12102121	2186	12120121	2186	12202001	2186	12220012	1093
12020222	1093	12102212	1093	12120211	2186	12202111	2186	12220022	1093
12021101	2186	12110111	2186	12120212	1093	12202112	1093	12220202	1093
12021212	1093	12110122	1093	12121012	1093	12202222	1093	12221002	1093
12022001	2186	12110201	2186	12121022	1093	12210002	1093	12221021	2186
12022111	2186	12110212	1093	12121102	1093	12210112	1093	12221111	2186
12022201	2186	12110221	2186	12121121	2186	12210211	2186	12221122	1093
12100001	2186	12111002	1093	12122012	1093	12211021	2186	12221221	2186
12100021	2186	12111101	2186	12122122	1093	12211201	2186	12222011	2186
12100111	2186	12111202	1093	12200101	2186	12211211	2186	12222101	2186
12100222	1093	12112022	1093	12200102	1093	12211222	1093	12222211	2186
12101011	2186	12112102	1093	12201011	2186	12212012	1093		
12101021	2186	12112121	2186	12201022	1093	12212102	1093		

Irreducible Polynomials for the Modulus 5

$n=1$	e	1113	124	$n=4$	e	11013	624	12022	624	13102	208
		1114	31			11023	624	12033	624	13121	52
10	1	1131	62	10002	16	11024	104	12042	624	13124	312
11	2	1134	31	10003	16	11032	624	12102	208	13131	26
12	4	1141	62	10014	312	11041	52	12121	13	13133	624
13	4	1143	124	10024	312	11042	624	12123	624	13201	78
14	1	1201	62	10034	312	11101	78	12131	52	13203	624
		1203	124	10044	312	11113	624	12134	312	13232	624
$n=2$	e	1213	124	10102	48	11114	312	12201	39	13234	312
		1214	31	10111	78	11124	104	12203	624	13241	156
102	8	1222	124	10122	624	11133	208	12211	156	13302	624
103	8	1223	124	10123	624	11142	208	12222	624	13314	104
111	3	1242	124	10132	624	11202	624	12224	312	13322	624
112	24	1244	31	10133	624	11212	624	12302	624	13323	208
123	24	1302	124	10141	39	11213	208	12311	39	13334	312
124	12	1304	31	10203	48	11221	156	12312	208	13341	78
133	24	1311	62	10221	39	11222	208	12324	312	13342	208
134	12	1312	124	10223	208	11234	104	12332	624	13401	156
141	6	1322	124	10231	78	11244	312	12333	208	13413	208
142	24	1323	124	10233	208	11301	156	12344	104	13423	624
		1341	62	10303	48	11303	624	12401	156	13424	312
$n=3$	e	1343	124	10311	156	11321	39	12414	104	13432	208
		1403	124	10313	208	11342	624	12422	208	13444	104
1011	62	1404	31	10341	156	11344	312	12433	624	14004	312
1014	31	1411	62	10343	208	11402	208	12434	312	14011	52
1021	62	1412	124	10402	48	11411	13	12443	208	14012	624
1024	31	1431	62	10412	624	11414	312	13004	312	14022	624
1032	124	1434	31	10413	624	11441	52	13012	624	14033	624
1033	124	1442	124	10421	156	11443	624	13023	624	14034	104
1042	124	1444	31	10431	156	12004	312	13031	13	14043	624
1043	124			10442	624	12013	624	13032	624	14101	39
1101	62			10443	624	12014	104	13043	624	14112	208
1102	124			11004	312	12021	26	13044	104	14123	208

Irreducible Polynomials for the Modulus 5

14134	104	101033	284	103014	781	110123	3124	112034	781	114014	781
14143	624	101103	3124	103022	3124	110131	1562	112104	781	114024	781
14144	312	101104	781	103023	3124	110142	3124	112113	3124	114033	3124
14202	624	101141	1562	103101	1562	110144	781	112133	3124	114044	781
14214	312	101142	3124	103104	781	110202	284	112142	3124	114102	3124
14224	104	101203	3124	103111	1562	110213	3124	112143	284	114132	3124
14231	156	101204	781	103112	3124	110232	3124	112201	1562	114141	1562
14232	208	101212	3124	103143	3124	110243	3124	112212	3124	114201	1562
14242	624	101213	284	103144	71	110244	781	112214	781	114204	71
14243	208	101301	1562	103211	1562	110301	1562	112234	781	114233	3124
14301	156	101302	3124	103212	3124	110303	3124	112241	1562	114242	3124
14303	624	101312	284	103221	1562	110322	3124	112243	3124	114314	781
14312	624	101313	3124	103223	3124	110331	1562	112301	1562	114321	1562
14314	312	101401	1562	103232	3124	110333	3124	112311	1562	114322	3124
14331	78	101402	3124	103233	3124	110343	3124	112313	3124	114331	1562
14402	208	101443	3124	103313	3124	110403	3124	112314	781	114343	3124
14411	52	101444	781	103314	781	110411	1562	112323	3124	114401	1562
14413	624	102001	1562	103322	3124	110421	1562	112334	71	114403	3124
14441	26	102004	781	103324	781	110432	3124	112342	3124	114424	781
14444	312	102012	3124	103332	3124	110441	1562	112422	3124	114431	22
		102013	3124	103333	3124	110442	3124	112433	3124	114434	781
n = 5	e	102021	1562	103401	1562	110444	781	112441	1562	114442	3124
		102024	781	103404	781	111003	284	113002	3124	120003	3124
100041	1562	102112	3124	103413	3124	111013	3124	113004	781	120013	3124
100042	3124	102114	781	103414	781	111021	1562	113034	781	120042	3124
100043	3124	102121	1562	103441	142	111022	3124	113044	781	120104	71
100044	781	102122	3124	103442	3124	111024	781	113103	3124	120111	1562
100102	3124	102131	1562	104021	1562	111032	3124	113111	1562	120134	781
100114	781	102134	781	104024	781	111044	781	113134	781	120141	1562
100124	71	102202	3124	104031	142	111102	3124	113142	284	120143	3124
100132	3124	102203	3124	104034	71	111114	781	113143	3124	120201	1562
100143	3124	102211	1562	104101	1562	111123	3124	113211	1562	120212	3124
100201	1562	102213	3124	104103	3124	111124	44	113222	3124	120222	3124
100212	3124	102242	284	104111	142	111224	781	113224	71	120234	781
100222	284	102244	781	104114	781	111231	1562	113231	1562	120242	3124
100231	1562	102302	3124	104202	3124	111234	781	113241	1562	120243	3124
100244	781	102303	3124	104204	781	111301	1562	113243	284	120244	781
100304	781	102312	3124	104241	1562	111311	1562	113304	781	120321	1562
100313	3124	102314	781	104243	3124	111312	3124	113312	3124	120332	3124
100323	284	102341	1562	104301	1562	111324	71	113321	1562	120343	3124
100334	781	102343	284	104303	3124	111334	781	113323	3124	120344	781
100341	1562	102411	1562	104342	3124	111401	142	113324	781	120401	1562
100403	3124	102413	3124	104344	781	111404	781	113332	3124	120402	3124
100411	1562	102423	3124	104402	3124	111423	3124	113342	3124	120424	781
100421	142	102424	781	104404	781	111431	1562	113412	3124	120431	1562
100433	3124	102431	1562	104411	1562	111433	3124	113422	3124	120432	3124
100442	3124	102434	781	104414	71	111442	3124	113434	781	120441	1562
101022	3124	103002	3124	110004	781	112012	3124	114001	1562	121002	3124
101023	3124	103003	3124	110014	781	112023	3124	114011	1562	121012	3124
101032	284	103011	1562	110041	1562	112032	3124	114012	3124	121013	3124

TABLE C   (*Cont.*)

Irreducible Polynomials for the Modulus 5

121014	781	123034	781	130103	3124	132042	3124	134022	3124	141023	3124
121023	3124	123102	3124	130104	781	132102	3124	134023	3124	141024	781
121031	1562	123113	3124	130121	1562	132111	1562	134031	1562	141033	3124
121043	3124	123114	781	130133	3124	132122	3124	134042	3124	141041	1562
121102	3124	123133	3124	130134	781	132123	3124	134103	3124	141101	1562
121103	284	123141	1562	130144	781	132124	781	134111	1562	141104	71
121131	1562	123142	3124	130224	781	132131	1562	134113	3124	141122	3124
121144	781	123224	781	130233	3124	132141	1562	134122	284	141132	3124
121201	1562	123231	1562	130241	1562	132204	781	134132	3124	141134	781
121202	3124	123242	3124	130242	3124	132213	3124	134201	1562	141143	3124
121223	3124	123303	3124	130304	781	132232	3124	134212	3124	141204	781
121232	44	123311	1562	130313	3124	132241	142	134224	781	141213	3124
121233	3124	123331	1562	130323	3124	132244	781	134302	3124	141214	781
121244	781	123341	142	130331	1562	132311	1562	134303	284	141221	142
121304	781	123344	781	130341	1562	132321	1562	134324	781	141231	1562
121334	781	123402	3124	130342	3124	132332	3124	134333	3124	141313	44
121342	3124	123411	1562	130343	3124	132413	3124	134334	781	141321	1562
121413	3124	123412	3124	130401	142	132421	1562	134341	1562	141331	1562
121422	3124	123413	3124	130414	781	132422	284	134411	22	141334	781
121424	781	123421	1562	130431	1562	132433	3124	134422	3124	141403	3124
121432	3124	123433	284	130442	3124	132443	3124	134432	3124	141411	1562
121441	1562	123444	781	130444	781	132444	71	134433	3124	141422	3124
122003	3124	124001	142	131003	3124	133011	1562	140001	1562	142013	3124
122004	781	124011	1562	131011	1562	133024	781	140011	1562	142022	3124
122033	3124	124022	3124	131012	3124	133031	1562	140044	781	142031	1562
122043	3124	124023	3124	131013	3124	133032	3124	140102	3124	142033	3124
122112	3124	124024	781	131022	3124	133103	3124	140114	781	142123	3124
122123	284	124034	781	131034	781	133112	3124	140124	781	142132	3124
122124	781	124043	3124	131042	3124	133113	3124	140133	3124	142144	781
122132	3124	124114	11	131112	3124	133114	781	140141	1562	142204	781
122141	142	124123	3124	131121	1562	133124	781	140143	3124	142211	1562
122142	3124	124132	3124	131123	3124	133132	284	140144	781	142212	3124
122214	781	124133	3124	131133	3124	133141	1562	140202	3124	142214	781
122224	781	124202	284	131144	781	133202	3124	140204	781	142222	3124
122233	3124	124203	3124	131201	1562	133214	781	140223	3124	142231	142
122301	1562	124221	1562	131231	1562	133234	781	140232	3124	142243	3124
122312	3124	124231	1562	131243	3124	133241	1562	140234	781	142304	781
122333	3124	124232	3124	131303	3124	133244	71	140242	3124	142311	1562
122341	1562	124244	781	131304	781	133321	1562	140303	284	142313	3124
122344	71	124304	781	131322	3124	133334	781	140312	3124	142331	1562
122403	3124	124313	3124	131332	3124	133343	3124	140333	3124	142342	3124
122414	781	124321	1562	131333	44	133403	3124	140341	1562	142344	781
122421	1562	124402	3124	131341	1562	133411	1562	140342	3124	142401	1562
122422	3124	124412	3124	131402	284	133412	3124	140422	3124	142412	3124
122423	3124	124414	781	131403	3124	133432	3124	140434	781	142432	3124
122434	781	124423	284	131434	781	133443	3124	140441	1562	142442	284
122444	781	124433	3124	131441	1562	133444	781	140443	3124	142443	3124
123014	781	130002	3124	132001	1562	134004	71	141002	284	143001	1562
123021	1562	130012	3124	132002	3124	134014	781	141012	3124	143003	3124
123033	3124	130043	3124	132032	3124	134021	1562	141021	1562	143031	1562

Irreducible Polynomials for the Modulus 5

143041	1562	143224	781	143344	781	144013	3124	144131	1562	144301	142
143113	3124	143233	3124	143402	3124	144014	781	144134	11	144304	781
143123	3124	143243	3124	143414	781	144021	1562	144143	3124	144332	3124
143131	1562	143314	781	143431	1562	144032	3124	144211	1562	144343	3124
143201	1562	143321	142	143442	3124	144041	1562	144223	3124	144403	3124
143213	3124	143323	3124	143443	284	144102	3124	144224	781	144433	3124
143221	1562	143334	781	144004	781	144104	781	144234	781	144444	781
143222	3124	143342	284	144011	1562	144121	1562	144242	3124		

Irreducible Polynomials for the Modulus 7

n = 1	e	1021	38	1304	342	1552	342	10135	2400	10524	1200
		1026	19	1306	57	1556	57	10145	2400	10525	2400
10	1	1032	342	1311	38	1563	171	10151	200	10531	80
11	2	1035	171	1314	342	1564	342	10161	400	10533	2400
12	6	1041	114	1322	342	1565	171	10162	600	10536	800
13	3	1046	57	1325	171	1566	57	10203	96	10541	80
14	6	1052	342	1333	171	1604	342	10205	96	10543	2400
15	3	1055	171	1334	342	1606	57	10211	200	10546	800
16	1	1062	342	1335	171	1612	342	10214	1200	10554	1200
		1065	171	1336	19	1615	171	10224	600	10555	2400
n = 2	e	1101	114	1341	38	1621	114	10236	800	10565	2400
		1103	171	1343	171	1623	171	10246	800	10603	96
101	4	1112	342	1352	342	1632	342	10254	600	10606	32
102	12	1115	171	1354	342	1636	19	10261	200	10613	2400
104	12	1124	342	1362	342	1641	114	10264	1200	10621	400
113	48	1126	57	1366	57	1644	342	10305	96	10623	2400
114	24	1131	38	1401	114	1653	171	10306	32	10632	240
116	16	1135	171	1403	171	1654	342	10316	800	10635	2400
122	24	1143	171	1413	171	1655	171	10322	1200	10636	800
123	48	1146	57	1416	19	1656	57	10326	800	10642	240
125	48	1151	114	1422	342	1662	342	10333	2400	10645	2400
131	8	1152	342	1425	171	1664	342	10334	240	10646	800
135	48	1153	171	1431	38			10335	2400	10651	400
136	16	1154	342	1432	342	n = 4	e	10343	2400	10653	2400
141	8	1163	171	1433	171			10344	240	10663	2400
145	48	1165	171	1434	342	10011	400	10345	2400	11001	400
146	16	1201	38	1444	342	10012	1200	10352	1200	11003	480
152	24	1203	171	1446	19	10014	1200	10356	800	11013	2400
153	48	1214	342	1453	171	10023	480	10366	800	11026	800
155	48	1216	57	1455	171	10025	480	10405	96	11031	400
163	48	1223	171	1461	114	10026	160	10406	32	11042	75
164	24	1226	57	1465	171	10053	480	10412	1200	11054	300
166	16	1233	171	1504	342	10055	480	10414	600	11056	800
		1235	171	1506	19	10056	160	10422	600	11062	1200
n = 3	e	1242	342	1511	114	10061	400	10433	2400	11063	2400
		1245	171	1513	171	10062	1200	10443	2400	11101	400
1002	18	1251	114	1521	114	10064	1200	10452	600	11103	2400
1003	9	1255	171	1524	342	10103	96	10462	1200	11105	2400
1004	18	1261	114	1532	342	10106	32	10464	600	11111	5
1005	9	1262	342	1534	342	10111	400	10503	96	11112	1200
1011	114	1263	171	1542	342	10112	600	10505	96	11124	75
1016	57	1264	342	1545	171	10121	200	10515	2400	11134	240

TABLE C *(Cont.)*

Irreducible Polynomials for the Modulus 7

11136	800	11556	800	12266	800	12665	480	13432	1200	14125	2400
11141	400	11562	1200	12303	2400	13004	1200	13434	1200	14132	1200
11152	1200	11566	160	12304	240	13005	480	13436	800	14145	2400
11153	2400	11602	240	12311	200	13011	400	13441	20	14156	800
11161	100	11605	2400	12323	2400	13015	2400	13443	2400	14165	2400
11163	480	11614	600	12325	2400	13022	300	13445	2400	14204	1200
11166	800	11625	2400	12332	120	13023	2400	13455	2400	14205	2400
11201	200	11626	800	12345	480	13031	50	13456	800	14206	800
11204	600	11631	40	12346	800	13044	1200	13465	2400	14211	400
11213	2400	11643	2400	12351	100	13053	2400	13501	80	14214	15
11223	2400	11646	160	12354	600	13065	2400	13506	800	14222	75
11225	2400	11652	600	12356	800	13103	2400	13512	600	14232	240
11232	60	11653	2400	12361	200	13106	800	13513	2400	14233	2400
11233	2400	11654	300	12363	2400	13115	2400	13516	800	14244	1200
11236	800	11664	600	12365	2400	13126	800	13521	200	14251	400
11241	400	11665	2400	12402	1200	13135	2400	13522	300	14255	2400
11244	1200	11666	800	12403	2400	13142	1200	13525	2400	14263	2400
11245	2400	12002	1200	12406	800	13151	400	13533	480	14264	300
11252	240	12006	160	12412	15	13155	2400	13535	2400	14265	480
11254	300	12016	800	12414	1200	13161	400	13544	120	14302	1200
11266	160	12025	2400	12421	25	13166	160	13553	2400	14314	150
11321	200	12032	1200	12431	80	13204	1200	13556	800	14325	2400
11323	2400	12044	75	12435	2400	13205	2400	13562	600	14335	2400
11324	150	12051	100	12442	1200	13206	800	13611	40	14341	25
11331	400	12055	2400	12454	1200	13213	2400	13612	1200	14346	800
11332	300	12064	1200	12456	800	13214	300	13616	800	14353	2400
11334	600	12066	800	12462	300	13215	480	13623	480	14354	1200
11351	200	12101	200	12465	2400	13221	400	13624	600	14361	400
11355	2400	12102	600	12466	160	13225	2400	13626	800	14363	480
11356	160	12116	800	12521	50	13234	1200	13641	100	14402	600
11362	120	12123	2400	12522	600	13242	240	13642	600	14404	600
11364	1200	12126	800	12526	800	13243	2400	13644	1200	14415	2400
11365	2400	12134	60	12531	200	13252	150	13652	75	14425	2400
11405	2400	12135	2400	12532	1200	13261	400	13654	600	14426	800
11406	800	12136	800	12534	300	13264	30	13655	2400	14431	20
11412	1200	12141	400	12552	600	13302	1200	14004	1200	14433	2400
11415	480	12142	1200	12553	2400	13311	400	14005	480	14435	2400
11422	1200	12143	2400	12555	480	13313	480	14015	2400	14442	1200
11423	2400	12151	100	12561	400	13323	2400	14023	2400	14444	1200
11434	1200	12154	240	12563	2400	13324	1200	14034	1200	14446	800
11443	2400	12165	480	12564	120	13331	50	14041	25	14451	80
11455	2400	12203	2400	12601	400	13336	800	14052	300	14452	300
11463	2400	12205	2400	12612	150	13345	2400	14053	2400	14463	480
11504	1200	12213	480	12626	800	13355	2400	14061	400	14501	80
11511	50	12214	1200	12636	800	13364	75	14065	2400	14506	800
11523	2400	12224	1200	12643	2400	13402	600	14103	2400	14512	600
11533	2400	12226	800	12644	75	13404	600	14106	800	14523	2400
11542	75	12231	400	12652	1200	13413	480	14111	400	14526	800
11545	2400	12246	800	12655	2400	13421	80	14116	160	14534	120
11551	400	12253	2400	12664	1200	13422	300	14121	400	14543	480

Irreducible Polynomials for the Modulus 7

14545	2400	15121	100	15353	2400	15622	1200	16204	600	16453	2400
14551	200	15124	240	15355	2400	15625	2400	16216	160	16462	1200
14552	300	15131	400	15361	200	15633	2400	16222	240	16465	480
14555	2400	15132	1200	15402	1200	15634	150	16224	300	16504	1200
14562	600	15133	2400	15403	2400	15646	800	16231	400	16512	1200
14563	2400	15144	60	15406	800	15656	800	16234	1200	16516	160
14566	800	15145	2400	15412	300	15662	75	16235	2400	16521	400
14622	150	15146	800	15415	2400	16001	400	16242	60	16526	800
14624	600	15153	2400	15416	160	16003	480	16243	2400	16532	150
14625	2400	15156	800	15424	1200	16012	1200	16246	800	16535	2400
14631	100	15166	800	15426	800	16013	2400	16253	2400	16543	2400
14632	600	15203	2400	15432	1200	16024	300	16255	2400	16553	2400
14634	1200	15205	2400	15441	80	16026	800	16263	2400	16561	25
14653	480	15216	800	15445	2400	16032	150	16312	120	16602	240
14654	600	15223	2400	15451	50	16041	400	16314	1200	16605	2400
14656	800	15236	800	15462	30	16056	800	16315	2400	16614	600
14661	40	15241	400	15464	1200	16063	2400	16321	200	16615	2400
14662	1200	15254	1200	15511	400	16101	400	16325	2400	16616	800
14666	800	15256	800	15513	2400	16103	2400	16326	160	16622	600
15002	1200	15263	480	15514	120	16105	2400	16341	400	16623	2400
15006	160	15264	1200	15522	600	16111	100	16342	300	16624	300
15014	1200	15303	2400	15523	2400	16113	480	16344	600	16633	2400
15016	800	15304	240	15525	480	16116	800	16351	200	16636	160
15021	100	15311	200	15541	200	16122	1200	16353	2400	16641	40
15025	2400	15313	2400	15542	1200	16123	2400	16354	75	16655	2400
15034	150	15315	2400	15544	300	16131	400	16405	2400	16656	800
15042	1200	15321	100	15551	25	16144	240	16406	800	16664	600
15055	2400	15324	600	15552	600	16146	800	16413	2400		
15066	800	15326	800	15556	800	16154	150	16425	2400		
15101	200	15335	480	15601	400	16161	10	16433	2400		
15102	600	15336	800	15614	1200	16162	1200	16444	1200		
15115	480	15342	120	15615	480	16201	200	16452	1200		

TABLE D

1	0						51	6	3	1	0		
2	1	0					52	3	0				
3	1	0					53	6	2	1	0		
4	1	0					54	6	5	4	3	2	0
5	2	0					55	6	2	1	0		
6	1	0					56	7	4	2	0		
7	1	0					57	5	3	2	0		
8	4	3	2	0			58	6	5	1	0		
9	4	0					59	6	5	4	3	1	0
10	3	0					60	1	0				
11	2	0					61	5	2	1	0		
12	6	4	1	0			62	6	5	3	0		
13	4	3	1	0			63	1	0				
14	5	3	1	0			64	4	3	1	0		
15	1	0					65	4	3	1	0		
16	5	3	2	0			66	8	6	5	3	2	0
17	3	0					67	5	2	1	0		
18	5	2	1	0			68	7	5	1	0		
19	5	2	1	0			69	6	5	2	0		
20	3	0					70	5	3	1	0		
21	2	0					71	5	3	1	0		
22	1	0					72	6	4	3	2	1	0
23	5	0					73	4	3	2	0		
24	4	3	1	0			74	7	4	3	0		
25	3	0					75	6	3	1	0		
26	6	2	1	0			76	5	4	2	0		
27	5	2	1	0			77	6	5	2	0		
28	3	0					78	7	2	1	0		
29	2	0					79	4	3	2	0		
30	6	4	1	0			80	7	5	3	2	1	0
31	3	0					81	4	0				
32	7	5	3	2	1	0	82	8	7	6	4	1	0
33	6	4	1	0			83	7	4	2	0		
34	7	6	5	2	1	0	84	8	7	5	3	1	0
35	2	0					85	8	2	1	0		
36	6	5	4	2	1	0	86	6	5	2	0		
37	5	4	3	2	1	0	87	7	5	1	0		
38	6	5	1	0			88	8	5	4	3	1	0
39	4	0					89	6	5	3	0		
40	5	4	3	0			90	5	3	2	0		
41	3	0					91	7	6	5	3	2	0
42	5	4	3	2	1	0	92	6	5	2	0		
43	6	4	3	0			93	2	0				
44	6	5	2	0			94	6	5	1	0		
45	4	3	1	0			95	6	5	4	2	1	0
46	8	5	3	2	1	0	96	7	6	4	3	2	0
47	5	0					97	6	0				
48	7	5	4	2	1	0	98	7	4	3	2	1	0
49	6	5	4	0			99	7	5	4	0		
50	4	3	2	0			100	8	7	2	0		

TABLE E

	$p=11$	$n=2$	$q=121$	$120=2^3\cdot3\cdot5$		$\phi(120)/2=16$	
a_1	a_2	a_1	a_2	a_1	a_2	a_1	a_2
4	2	2	6	1	7	1	8
5	2	3	6	4	7	3	8
6	2	8	6	7	7	8	8
7	2	9	6	10	7	10	8

	$p=13$	$n=2$	$q=169$	$168=2^3\cdot3\cdot7$		$\phi(168)/2=24$	
a_1	a_2	a_1	a_2	a_1	a_2	a_1	a_2
1	2	2	6	2	7	4	11
4	2	3	6	3	7	5	11
6	2	4	6	6	7	6	11
7	2	9	6	7	7	7	11
9	2	10	6	10	7	8	11
12	2	11	6	11	7	9	11

	$p=17$	$n=2$	$q=289$	$288=2^5\cdot3^2$		$\phi(288)/2=48$	
a_1	a_2	a_1	a_2	a_1	a_2	a_1	a_2
1	3	2	6	1	10	2	12
6	3	6	6	3	10	3	12
7	3	8	6	4	10	5	12
10	3	9	6	13	10	12	12
11	3	11	6	14	10	14	12
16	3	15	6	16	10	15	12
3	5	1	7	2	11	4	14
5	5	4	7	7	11	6	14
8	5	5	7	8	11	7	14
9	5	12	7	9	11	10	14
12	5	13	7	10	11	11	14
14	5	16	7	15	11	13	14

	$p=19$	$n=2$	$q=361$	$360=2^3\cdot3^2\cdot5$		$\phi(360)/2=48$	
a_1	a_2	a_1	a_2	a_1	a_2	a_1	a_2
1	2	10	3	3	13	11	14
4	2	11	3	4	13	12	14
7	2	12	3	6	13	13	14
8	2	18	3	9	13	18	14
11	2	2	10	10	13	4	15
12	2	4	10	13	13	5	15
15	2	6	10	15	13	6	15
18	2	9	10	16	13	9	15
1	3	10	10	1	14	10	15
7	3	13	10	6	14	13	15
8	3	15	10	7	14	14	15
9	3	17	10	8	14	15	15

	$p=23$	$n=2$	$q=529$	$528=2^4\cdot3\cdot11$		$\phi(528)/2=80$			
a_1	a_2	a_1	a_2	a_1	a_2	a_1	a_2	a_1	a_2
2	5	2	10	1	14	3	17	4	20
4	5	3	10	3	14	4	17	7	20
5	5	6	10	5	14	6	17	8	20
8	5	10	10	10	14	11	17	10	20
15	5	13	10	13	14	12	17	13	20
18	5	17	10	18	14	17	17	15	20

TABLE E   (*Cont.*)

			$p = 23$	$n = 2$	$q = 529$	$528 = 2^4 \cdot 3 \cdot 11$		$\phi(528)/2 = 80$	

a_1	a_2	a_1	a_2	a_1	a_2	a_1	a_2	a_1	a_2
19	5	20	10	20	14	19	17	16	20
21	5	21	10	22	14	20	17	19	20
1	7	3	11	5	15	1	19	5	21
2	7	7	11	9	15	2	19	6	21
4	7	8	11	10	15	7	19	7	21
9	7	9	11	11	15	11	19	9	21
14	7	14	11	12	15	12	19	14	21
19	7	15	11	13	15	16	19	16	21
21	7	16	11	14	15	21	19	17	21
22	7	20	11	18	15	22	19	18	21

			$p = 29$	$n = 2$	$q = 841$	$840 = 2^3 \cdot 3 \cdot 5 \cdot 7$		$\phi(840)/2 = 96$			

a_1	a_2	a_1	a_2	a_1	a_2	a_1	a_2	a_1	a_2	a_1	a_2
5	2	1	8	6	11	7	15	2	19	5	26
7	2	7	8	9	11	9	15	4	19	6	26
11	2	10	8	10	11	11	15	7	19	8	26
14	2	14	8	11	11	12	15	8	19	12	26
15	2	15	8	18	11	17	15	21	19	17	26
18	2	19	8	19	11	18	15	22	19	21	26
22	2	22	8	20	11	20	15	25	19	23	26
24	2	28	8	23	11	22	15	27	19	24	26
1	3	3	10	1	14	4	18	3	21	2	27
2	3	5	10	3	14	8	18	4	21	3	27
9	3	9	10	8	14	13	18	6	21	6	27
14	3	10	10	13	14	14	18	12	21	13	27
15	3	19	10	16	14	15	18	17	21	16	27
20	3	20	10	21	14	16	18	23	21	23	27
27	3	24	10	26	14	21	18	25	21	26	27
28	3	26	10	28	14	25	18	26	21	27	27

			$p = 31$	$n = 2$	$q = 961$	$960 = 2^6 \cdot 3 \cdot 5$		$\phi(960)/2 = 128$							

a_1	a_2	a_1	a_2	a_1	a_2	a_1	a_2	a_1	a_2	a_1	a_2	a_1	a_2	a_1	a_2
2	3	2	11	1	12	1	13	1	17	2	21	1	22	1	24
5	3	3	11	3	12	4	13	2	17	5	21	4	22	3	24
6	3	4	11	4	12	6	13	3	17	7	21	5	22	4	24
7	3	5	11	10	12	8	13	6	17	8	21	7	22	5	24
8	3	6	11	11	12	9	13	7	17	11	21	9	22	7	24
10	3	9	11	12	12	10	13	8	17	12	21	10	22	8	24
14	3	11	11	14	12	12	13	9	17	13	21	14	22	12	24
15	3	15	11	15	12	13	13	11	17	15	21	15	22	13	24
16	3	16	11	16	12	18	13	20	17	16	21	16	22	18	24
17	3	20	11	17	12	19	13	22	17	18	21	17	22	19	24
21	3	22	11	19	12	21	13	23	17	19	21	21	22	23	24
23	3	25	11	20	12	22	13	24	17	20	21	22	22	24	24
24	3	26	11	21	12	23	13	25	17	23	21	24	22	26	24
25	3	27	11	27	12	25	13	28	17	24	21	26	22	27	24
26	3	28	11	28	12	27	13	29	17	26	21	27	22	28	24
29	3	29	11	30	12	30	13	30	17	29	21	30	22	30	24

TABLE F

p^n	$a_1a_2a_3\cdots a_n$	p^n	$a_1a_2a_3\cdots a_n$	p^n	$a_1a_2\cdots a_{n-1}a_n$
2^2	11	5^2	12	19^2	1 2
2^3	101	5^3	102	19^3	10 16
2^4	1001	5^4	1013	19^4	100 2
2^5	01001	5^5	00102	19^5	0001 16
2^6	100001	5^6	100002	19^6	00001 3
2^7	0000011	5^7	1000002	19^7	010000 9
2^8	11000011	5^8	00101003		
2^9	000100001	5^9	011000003	23^2	1 7
2^{10}	0010000001	5^{10}	1010000003	23^3	10 16
2^{11}	01000000001	5^{11}	10000000002	23^4	001 11
2^{12}	110000010001	5^{12}	000010010003	23^5	1000 18
2^{13}	1100100000001			23^6	10000 7
2^{14}	11000000000101				
2^{15}	100000000000001	7^2	13	29^2	1 3
2^{16}	1010000000010001	7^3	112	29^3	01 18
2^{17}	00100000000000001	7^4	1103	29^4	100 2
2^{18}	000000100000000001	7^5	10004	29^5	0100 26
2^{19}	1100100000000000001	7^6	110003	29^6	00001 3
2^{20}	00100000000000000001	7^7	0100004		
2^{21}	010000000000000000001	7^8	10000003		
2^{22}	1000000000000000000001	7^9	100001002	31^2	1 12
2^{23}	00001000000000000000001	7^{10}	1100000003	31^3	01 28
2^{24}	110000100000000000000001			31^4	100 13
2^{25}	0010000000000000000000001			31^5	0100 20
2^{26}	11000100000000000000000001	11^2	17	31^6	10000 12
2^{27}	110010000000000000000000001	11^3	105		
2^{28}	0010000000000000000000000001	11^4	0012		
2^{29}	01000000000000000000000000001	11^5	01109	37^2	1 5
		11^6	100017	37^3	10 24
		11^7	1000005	37^4	001 2
		11^8	00010012	37^5	0001 32
3^2	12				
3^3	201				
3^4	1002	13^2	1 2	41^2	1 12
3^5	10101	13^3	10 7	41^3	01 35
3^6	100002	13^4	101 2	41^4	001 17
3^7	1010001	13^5	0101 11	41^5	1000 35
3^8	00100002	13^6	10100 6		
3^9	010100001	13^7	001000 6	43^2	1 3
3^{10}	1010000002	13^8	0110000 2	43^3	01 40
3^{11}	10000010001			43^4	001 20
3^{12}	100010000002			43^5	1000 40
3^{13}	1000001000001	17^2	1 3		
3^{14}	10000000000002	17^3	01 14		
3^{15}	100000000010001	17^4	100 5	47^2	1 13
3^{16}	0000001000000002	17^5	1000 14	47^3	10 42
3^{17}	10000000100000001	17^6	10000 3	47^4	100 5
3^{18}	100000000000100002	17^7	000100 14	47^5	0001 42

BIBLIOGRAPHY

Note. We list only textbooks suggested for further reading and some basic research articles. A more detailed bibliography can be found in:

Lidl, R., and Niederreiter, H.: *Finite Fields,* Encyclopedia of Math. and Its Appl., vol. 20, Addison-Wesley, Reading, Mass., 1983; now published by Cambridge University Press.

Chapter 1

Books on Abstract Algebra:

Birkhoff, G., and MacLane, S.: *A Survey of Modern Algebra,* 4th ed., Macmillan, New York, 1977.

Fraleigh, J. B.: *A First Course in Abstract Algebra,* Addison-Wesley, Reading, Mass., 1982.

Herstein, I. N.: *Topics in Algebra,* 2nd ed., Xerox College Publ., Lexington, Mass., 1975.

Lang, S.: *Algebra,* Addison-Wesley, Reading, Mass., 1971.

Rédei, L.: *Algebra,* Pergamon Press, London, 1967.

van der Waerden, B. L.: *Algebra,* vol. 1, 7th ed., Springer-Verlag, Berlin, 1966.

Books on Applied Algebra:

Birkhoff, G., and Bartee, T. C.: *Modern Applied Algebra,* McGraw-Hill, New York, 1970.

Dornhoff, L. L., and Hohn, F. E.: *Applied Modern Algebra,* Macmillan, New York, 1978.

Lidl, R., and Pilz, G.: *Applied Abstract Algebra,* Springer-Verlag, New York, 1984.

Chapter 2

Dickson, L. E.: *Linear Groups with an Exposition of the Galois Field Theory*, Teubner, Leipzig, 1901; Dover, New York, 1958.

Herstein, I. N.: *Noncommutative Rings*, Carus Math. Monographs, no. 15, Math. Assoc. of America, Washington, D.C., 1968.

Hoffman, K., and Kunze, R.: *Linear Algebra*, 2nd ed., Prentice-Hall, Englewood Cliffs, N.J., 1971.

Jacobson, N.: *Lectures in Abstract Algebra*, vol. 3: *Theory of Fields and Galois Theory*, Springer-Verlag, New York, 1980; originally published by Van Nostrand, New York, 1964.

Chapter 3

Albert, A. A.: *Fundamental Concepts of Higher Algebra*, Univ. of Chicago Press, Chicago, 1956.

Berlekamp, E.R.: *Algebraic Coding Theory*, McGraw-Hill, New York, 1968.

MacWilliams, F. J., and Sloane, N. J. A.: *The Theory of Error-Correcting Codes*, North-Holland, Amsterdam, 1977.

Ore, O.: On a special class of polynomials, *Trans. Amer. Math. Soc.* **35**, 559–584 (1933); Errata, *ibid.* **36**, 275 (1934).

Ore, O.: Contributions to the theory of finite fields, *Trans. Amer. Math. Soc.* **36**, 243–274 (1934).

Chapter 4

Berlekamp, E. R.: *Algebraic Coding Theory*, McGraw-Hill, New York, 1968.

Berlekamp, E. R.: Factoring polynomials over large finite fields, *Math. Comp.* **24**, 713–735 (1970).

Cantor, D. G., and Zassenhaus, H.: A new algorithm for factoring polynomials over finite fields, *Math. Comp.* **36**, 587–592 (1981).

Knuth, D.E.: *The Art of Computer Programming*, vol. 2: *Seminumerical Algorithms*, 2nd ed., Addison-Wesley, Reading, Mass., 1981.

McEliece, R. J.: Factorization of polynomials over finite fields, *Math. Comp.* **23**, 861–867 (1969).

Rabin, M. O.: Probabilistic algorithms in finite fields, *SIAM J. Computing* **9**, 273–280 (1980).

Zassenhaus, H.: On Hensel factorization I, *J. Number Theory* **1**, 291–311 (1969).

Chapter 5

Hasse, H.: *Vorlesungen über Zahlentheorie*, 2nd ed., Springer-Verlag, Berlin, 1964.

Ireland, K., and Rosen, M.: *A Classical Introduction to Modern Number Theory*, Springer-Verlag, New York, 1982.

Chapter 6

Berlekamp, E. R.: *Algebraic Coding Theory*, McGraw-Hill, New York, 1968.
Fillmore, J. P., and Marx, M.L.: Linear recursive sequences, *SIAM Rev.* **10**, 342–353 (1968).
Golomb, S. W.: *Shift Register Sequences*, Aegean Park Press, Laguna Hills, Cal., 1982.
Massey, J. L.: Shift-register synthesis and BCH decoding, *IEEE Trans. Information Theory* **15**, 122–127 (1969).
Niederreiter, H.: On the cycle structure of linear recurring sequences, *Math. Scand.* **38**, 53–77 (1976).
Zierler, N.: Linear recurring sequences, *J. Soc. Indust. Appl. Math.* **7**, 31–48 (1959).

Chapter 7

Finite Geometries:
Albert, A.A., and Sandler, R.: *An Introduction to Finite Projective Planes*, Holt, Rinehart and Winston, New York, 1968.
Dembowski, P.: *Finite Geometries*, 2nd ed., Springer-Verlag, Berlin, 1977.
Hirschfeld, J. W. P.: *Projective Geometries over Finite Fields*, Clarendon Press, Oxford, 1979.
Hughes, D. R., and Piper, F. C.: *Projective Planes*, Springer-Verlag, New York, 1973.

Combinatorics:
Beth, T., Jungnickel, D., and Lenz, H.: *Design Theory.*, Bibliographisches Institut, Mannheim, 1985.
Brualdi, R. A.: *Introductory Combinatorics*, North-Holland, Amsterdam, 1977.
Dénes, J., and Keedwell, A. D.: *Latin Squares and Their Applications*, Academic Press, New York, 1974.
Hall, M., Jr.: *Combinatorial Theory*, Blaisdell, Waltham, Mass., 1967.
Raghavarao, D.: *Constructions and Combinatorial Problems in Design of Experiments*, Wiley, New York, 1971.
Ryser, H. J.: *Combinatorial Mathematics*, Carus Math. Monographs, no. 14, Math. Assoc. of America, New York, 1963.
Storer, T.: *Cyclotomy and Difference Sets*, Markham, Chicago, 1967.

Linear Modular Systems:
Arbib, M. A., Falb, P. L., and Kalman, R. E.: *Topics in Mathematical System Theory*, McGraw-Hill, New York, 1968.
Dornhoff, L. L., and Hohn, F. E.: *Applied Modern Algebra*, Macmillan, New York, 1978.
Zadeh, L. A., and Polak, E.: *System Theory*, McGraw-Hill, New York, 1969.

Pseudorandom Sequences:
Golomb, S. W.: *Shift Register Sequences*, Aegean Park Press, Laguna Hills, Cal., 1982.
Knuth, D. E.: *The Art of Computer Programming*, vol. 2: *Seminumerical Algorithms*, 2nd ed., Addison-Wesley, Reading, Mass., 1981.
Niederreiter, H.: The performance of k-step pseudorandom number generators

under the uniformity test, *SIAM J. Sci. Statist. Computing* **5**, 798–810 (1984).

Niederreiter, H.: Distribution properties of feedback shift register sequences, *Problems of Control and Information Theory*, to appear.

Tausworthe, R. C.: Random numbers generated by linear recurrence modulo two, *Math. Comp.* **19**, 201–209 (1965).

Zierler, N.: Linear recurring sequences, *J. Soc. Indust. Appl. Math.* **7**, 31–48 (1959).

Chapter 8

Berlekamp, E.R.: *Algebraic Coding Theory*, McGraw-Hill, New York, 1968.

Blake, I. F., and Mullin, R. C.: *The Mathematical Theory of Coding*, Academic Press, New York, 1975.

MacWilliams, F. J., and Sloane, N. J. A.: *The Theory of Error-Correcting Codes*, North-Holland, Amsterdam, 1977.

McEliece, R. J.: *The Theory of Information and Coding*, Encyclopedia of Math. and Its Appl., vol. 3, Addison-Wesley, Reading, Mass., 1977; now published by Cambridge University Press.

Peterson, W. W., and Weldon, E. J., Jr.: *Error-Correcting Codes*, 2nd ed., M.I.T. Press, Cambridge, Mass., 1972.

Pless, V.: *Introduction to the Theory of Error-Correcting Codes*, Wiley, New York, 1982.

van Lint, J. H.: *Introduction to Coding Theory*, Springer-Verlag, New York, 1982.

Chapter 9

Books:

Beker, H., and Piper, F.: *Cipher Systems. The Protection of Communications*, Northwood Books, London, 1982.

Denning, D. E. R.: *Cryptography and Data Security*, Addison-Wesley, Reading, Mass., 1983.

Kahn, D.: *The Codebreakers*, Weidenfeld & Nicholson, London, 1967.

Konheim, A. G.: *Cryptography. A Primer*, Wiley, New York, 1981.

Meyer, C. H., and Matyas, S. M.: *Cryptography. A New Dimension in Computer Data Security*, Wiley, New York, 1982.

Articles:

Blake, I. F., Fuji-Hara, R., Mullin, R. C., and Vanstone, S.A.: Computing logarithms in finite fields of characteristic two, *SIAM J. Algebraic Discrete Methods* **5**, 276–285 (1984).

Chor, B., and Rivest, R. L.: A knapsack type public key cryptosystem based on arithmetic in finite fields, *Proc. CRYPTO '84*, to appear.

Coppersmith, D.: Fast evaluation of logarithms in fields of characteristic two, *IEEE Trans. Information Theory* **30**, 587–594 (1984).

Diffie, W., and Hellman, M. E.: New directions in cryptography, *IEEE Trans. Information Theory* **22**, 644–654 (1976).

ElGamal, T.: A public key cryptosystem and a signature scheme based on discrete logarithms, *IEEE Trans. Information Theory*, to appear.

Jennings, S. M.: Multiplexed sequences: Some properties of the minimum polynomial, *Cryptography* (T. Beth, ed.). Lecture Notes in Computer Science, vol. 149, pp. 189–206, Springer-Verlag, Berlin, 1983.

Lempel, A.: Cryptology in transition, *ACM Computing Surveys* **11**, 285–303 (1979).

McEliece, R. J.: A public-key cryptosystem based on algebraic coding theory, DSN Progress Report 42–44, Jet Propulsion Lab., Pasadena, Cal., 1978.

Niederreiter, H.: A public-key cryptosystem based on shift register sequences, *Proc. EUROCRYPT '85*, to appear

Odlyzko, A. M.: Discrete logarithms in finite fields and their cryptographic significance, *Proc. EUROCRYPT '84*, to appear.

Pohlig, S. C., and Hellman, M. E.: An improved algorithm for computing logarithms over $GF(p)$ and its cryptographic significance, *IEEE Trans. Information Theory* **24**, 106–110 (1978).

Rivest, R. L., Shamir, A., and Adleman, L.: A method for obtaining digital signatures and public-key cryptosystems, *Comm. ACM* **21**, 120–126 (1978).

Chapter 10

Alanen, J. D., and Knuth, D. E.: Tables of finite fields, *Sankhyā Ser. A* **26**, 305–328 (1964).

Church, R.: Tables of irreducible polynomials for the first four prime moduli, *Ann. of Math.* (2) **36**, 198–209 (1935).

Conway, J. H.: A tabulation of some information concerning finite fields, *Computers in Mathematical Research* (R. F. Churchhouse and J.-C. Herz, eds.), pp.37–50, North-Holland, Amsterdam, 1968.

Marsh, R. W.: *Table of Irreducible Polynomials over GF(2) through Degree 19*, Office of Techn. Serv., U.S. Dept. of Commerce, Washington, D.C., 1957.

Stahnke, W.: Primitive binary polynomials, *Math. Comp.* **27**, 977–980 (1973).

Watson, E. J.: Primitive polynomials (mod 2), *Math. Comp.* **16**, 368–369 (1962).

List of Symbols

Note. Symbols that appear only in a restricted context are not listed. Wherever appropriate, a page reference is given.

$\mathbb{N}$	the set of natural numbers (= positive integers)
$\mathbb{Z}$	the set of integers
$\mathbb{Q}$	the set of rational numbers
$\mathbb{R}$	the set of real numbers
$\mathbb{C}$	the set of complex numbers
$S_1 \times \cdots \times S_n$	the set of all n-tuples $(s_1, \ldots, s_n)$ with $s_i \in S_i$ for $1 \leqslant i \leqslant n$
S^n	the set of all n-tuples $(s_1, \ldots, s_n)$ with $s_i \in S$ for $1 \leqslant i \leqslant n$
$\lvert S \rvert$	the cardinality (= number of elements) of the finite set S
$[s]$	the equivalence class of s, 4

$\bar{z}$	the complex conjugate of z
$\lvert z \rvert$	the absolute value of z
$\log z$	the natural logarithm of z
$e(t)$	$e^{2\pi it}$ for $t \in \mathbb{R}$
$\lfloor t \rfloor$	the greatest integer $\leqslant t \in \mathbb{R}$
$\max(k_1, \ldots, k_n)$	the maximum of $k_1, \ldots, k_n$
$\min(k_1, \ldots, k_n)$	the minimum of $k_1, \ldots, k_n$
$\gcd(k_1, \ldots, k_n)$	the greatest common divisor of $k_1, \ldots, k_n$
$\mathrm{lcm}(k_1, \ldots, k_n)$	the least common multiple of $k_1, \ldots, k_n$
$\dbinom{k}{i}$	binomial coefficient

$a \equiv b \bmod n$	a congruent to b modulo n, 4
$\phi(n)$	Euler's function of n, 7
$\mu(n)$	Moebius function of n, 83
$\left(\dfrac{c}{p}\right)$	Legendre symbol, 167
A^{T}	the transpose of the matrix A
$\det(A)$	the determinant of the matrix A
$\mathrm{Tr}(A)$	the trace of the matrix A
$\mathrm{rank}(A)$	the rank of the matrix A
$D_n^{(r)}$	Hankel determinant, 229
$\dim(V)$	the dimension of the vector space V
$|G|$	the order of the finite group G, 5
$\langle a \rangle$	the cyclic group generated by a, 4, 6
aH	the left coset of the group element a modulo the subgroup H, 6
G/H	the factor group of the group G modulo the normal subgroup H, 9
$N(S)$	the normalizer of the nonempty subset S of a group, 10
$\ker f$	the kernel of the homomorphism f, 9, 14
(a)	the principal ideal generated by a, 13
$[a], a+j$	the residue class of the ring element a modulo the ideal J, 13
$a \equiv b \bmod J$	congruence of ring elements a, b modulo the ideal J, 13
R/J	the residue class ring of the ring R modulo the ideal J, 13
$\mathbb{Z}_n$	the group of integers modulo n, 5
$\mathbb{Z}/(n)$	the ring of integers modulo n, 14
$GL(k, \mathbb{F}_q)$	the general linear group of nonsingular $k \times k$ matrices over $\mathbb{F}_q$, 191
$R[x]$	the polynomial ring over the ring R, 19
$R[x_1, \ldots, x_n]$	the ring of polynomials over the ring R in n indeterminates, 28
$\deg(f)$	the degree of the polynomial f, 20, 29
$D(f)$	the discriminant of the polynomial f, 35
$\mathrm{ord}(f)$	the order of the polynomial f, 75
f'	the derivative of the polynomial f, 27
f^*	the reciprocal polynomial of f, 79
$R(f, g)$	the resultant of the polynomials f and g, 36
$\gcd(f_1, \ldots, f_n)$	the greatest common divisor of the polynomials $f_1, \ldots, f_n$, 22
$\mathrm{lcm}(f_1, \ldots, f_n)$	the least common multiple of the polynomials $f_1, \ldots, f_n$, 23
$f_1(x) \vee \cdots \vee f_h(x)$	224

$L_1(x) \otimes L_2(x)$ symbolic multiplication of linearized polynomials $L_1(x)$ and $L_2(x)$, 105

$Q_n(x)$ the nth cyclotomic polynomial, 60

$\sigma_k(x_1, \ldots, x_n)$ the kth elementary symmetric polynomial in n indeterminates, 29

$K(M)$ the extension of K obtained by adjoining M, 30

$[L:K]$ the degree of the field L over K, 32

$K^{(n)}$ the nth cyclotomic field over K, 59

$E^{(n)}$ the set of nth roots of unity over K, 59

$\mathbb{F}_q, GF(q)$ the finite field of order q, 45

$\mathbb{F}_q^*$ the multiplicative group of nonzero elements of $\mathbb{F}_q$, 46

$\mathrm{Tr}_{F/K}(\alpha)$ the trace of $\alpha \in F$ over K, 50

$\mathrm{Tr}_F(\alpha)$ the absolute trace of $\alpha \in F$, 50

$\mathrm{N}_{F/K}(\alpha)$ the norm of $\alpha \in F$ over K, 53

$\Delta_{F/K}(\alpha_1, \ldots, \alpha_m)$ the discriminant of $\alpha_1, \ldots, \alpha_m \in F$ over K, 57

$\mathrm{ind}_b(a)$ the index (or discrete logarithm) of a with respect to the base b, 346

$\exp_b(r)$ the discrete exponential function to the base b, 346

$N_q(d)$ the number of monic irreducible polynomials in $\mathbb{F}_q[x]$ of degree d, 82

$I(q, n; x)$ the product of all monic irreducible polynomials in $\mathbb{F}_q[x]$ of degree n, 85

$\Phi_q(f)$ the number of polynomials in $\mathbb{F}_q[x]$ whose degree is less than $\deg(f)$ and which are relatively prime to $f \in \mathbb{F}_q[x]$, 113

$\mathbb{F}_q[[x]]$ the ring of formal power series over $\mathbb{F}_q$, 204

$S(f(x))$ the set of all homogeneous linear recurring sequences in $\mathbb{F}_q$ with characteristic polynomial $f(x)$, 215

$\sigma_d^{(h)}$ sequence obtained by decimation of the sequence σ, 285

$\sigma^{(h)}$ sequence obtained by shifting the sequence σ, 287

$\hat{G}$ the set of characters of the finite abelian group G, 163

$\bar{\chi}$ the conjugate of the character χ, 163

χ_0 the trivial additive character of $\mathbb{F}_q$, 166

χ_1 the canonical additive character of $\mathbb{F}_q$, 166

ψ_0 the trivial multiplicative character of $\mathbb{F}_q$, 167

η the quadractic character of $\mathbb{F}_q$ (q odd), 167

$G(\psi, \chi)$ Gaussian sum, 168

$AG(2, K)$ the affine plane over the field K, 254

$PG(2, K)$ the projective plane over the field K, 254

$AG(m, \mathbb{F}_q)$ affine geometry over $\mathbb{F}_q$, 262

$PG(m, \mathbb{F}_q)$ projective geometry over $\mathbb{F}_q$, 260

$d(\mathbf{x}, \mathbf{y})$	the Hamming distance between $\mathbf{x}$ and $\mathbf{y}$, 303
$w(\mathbf{x})$	the Hamming weight of $\mathbf{x}$, 303
d_C	the minimum distance of the linear code C, 304
$C^\perp$	the dual code of G, 308
$S(\mathbf{y})$	the syndrome of $\mathbf{y}$, 305
$\Gamma(L, g)$	Goppa code, 326
$\square$	end of proof, end of example, end of remark

Index

adder, 186, 187, 273
affine geometry, 262, 263
affine multiple, 103
affine plane, 253–255
affine polynomial, 103, 105, 126, 128
   see also q-polynomial(s)
affine subspace, 105
algebraic structure, 2
algebraic system, 1, 2
alternant code, 336, 337
annihilating polynomial, 56
annihilator, 165, 181
Artin lemma, 55
q-associate, 106–108
   canonical factorization of, 108
   conventional, 106
   linearized, 106, 126
authentication, 341
automorphism, 8, 49, 50, 70
   inner, 8

balanced incomplete block design, 263–265,
   269, 295, 296
basis, 50, 54–59, 71, 114, 115
   complementary, 54
   dual, *see* dual basis
   normal, *see* normal basis
   polynomial, 55
   self-dual, 54, 71
BCH code, 299, 318–326, 335
   narrow-sense, 318, 325, 326
   primitive, 318
Berlekamp–Massey algorithm, 231–235, 323

Berlekamp's algorithm, 130–134, 140
BIBD, *see* balanced incomplete block design
binary complementation, 221
binary operation, 2
   associative, 2
   closure property of, 2
binomial, 115–118, 127, 160
binomial theorem, 37
bits, 281, 340
block cipher, 340
block design, *see* balanced incomplete block
   design

Caesar cipher, 338
canonical factorization, 24
   of q-associate, 108
   see also factorization
Cayley–Hamilton theorem, 56
Cayley table, 5
center
   of division ring, 66
   of group, 10
character, 163–168
   additive, 166
   annihilating, 165
   canonical additive, 166
   conjugate, 163
   lifting of, 173, 182
   multiplicative, 167
   nontrivial, 163
   orthogonality relations, 165, 167, 168
   product, 163
   quadratic, 167

trivial, 163
trivial additive, 166
trivial multiplicative, 167
character group, 163, 182
characteristic, 16
characteristic matrix, 272
characteristic polynomial
    of element, 50, 70, 91–93, 369, 374–376
    for linear operator, 56
    of matrix, *see* matrix
    reciprocal, 207
    of sequence, *see* linear recurring
        sequenc(s)
characterizing matrices, 272
character sum, 162, 180, 181, 236–240, 250
Chien search, 322, 323
Chinese remainder theorem, 38, 40
cipher, 338
    block, 340
    Caesar, 338
    stream, 342
    substitution, 338
    see also cryptosystem
cipher system, 338
    see also cryptosystem
ciphertext, 339
class equation, 10, 66
code, 300, 301
    alternant, 336, 337
    BCH, *see* BCH code
    binary, 302
    cyclic, *see* cyclic code
    dimension of, 302
    dual, *see* dual code
    equivalent, 333
    Goppa, *see* Goppa code
    Hamming, *see* Hamming code
    length of, 302
    linear, 302–311, 333, 334
    minimum distance of, *see* minimum
        distance
    orthogonal, *see* dual code
    parity-check, 302, 334
    perfect, 333
    Reed–Soloman, 318, 335
    repetition, 302, 333, 334
    reversible, 335
    systematic, 302
code polynomial, 313–315
code vector, 302
code word, 300–302
coding scheme, 300, 301
coefficient, 19, 28, 202
    leading, 20
collinear points, 255
companion matrix, 63, 64, 93, 195, 279
complete quadrangle, 257
component
    forced, 276

free, 276
congruence, 4, 6, 13
    left, 6
conic, 258
    degenerate, 258
    nondegenerate, 258
    tangent of, 258
conjugacy class, 10
conjugate, 49, 50
    of set, 8
constant adder, 186, 187
constant multiplier, 186, 187, 273
constant term, 20
control symbol, 301
conventional cryptosystem, 338–340, 349
correlation coefficient, 282–285
correlation test, 282
coset, 6, 7
    left, 6
    right, 6
coset leader, 305
coset-leader algorithm, 305, 306
cryptanalysis, 338
cryptography, 338
cryptology, 338
cryptosystem, 338–340
    conventional, 338–340, 349
    DES, 340
    FSR, 357, 358
    Goppa-code, 360–362
    Hill, 366
    knapsack-type, 358–360
    public-key, 340, 341
    RSA, 348
    single-key, 339
cycle, 277
    length of, 277
    pure, 277
cycle sum, 278–281
cycle term, 278
cyclic code, 311–325
    irreducible, 313
    maximal, 313
    shortened, 335
cyclic group, *see* group
cyclic vector, 56
cyclotomic field, 59, 61, 62, 72
cyclotomic polynomial, 60–62, 64, 66, 72,
        73, 84–86, 96, 97, 124, 128, 138, 139

Davenport–Hasse theorem, 173
de Bruijn sequence, 246
decimated sequence, 285–287, 297, 298, 345,
        364
decimation, 285–287, 297, 298, 345, 358, 364
deciphering scheme, 339
decoding algorithm
    for BCH code, 320–325, 328, 331, 332
    for Goppa code, 328–332

for linear code, 304–306
decoding scheme, 300
degree
    of algebraic element, 31
    of extension, 32
    formal, 36
    of polynomial, 20, 29
delay element, 186, 187, 273
derivative, 27, 40, 41, 70
Desarguesian plane, 256–259
Desargues's theorem, 255–257, 260
DES cryptosystem, 340
design, 262
design of experiments, 269
diagonalization algorithm, 145–147
difference equation, see linear recurrence
        relation
difference set, 265–267, 296
Diffie–Hellman scheme, 348
digital method, 288
digital signature, 341, 349
discrete exponential function, 346–349, 367
discrete logarithm, 346, 347, 358, 359, 365,
        367
discrete logarithm algorithm, 349–357
discriminant
    of elements, 57, 58, 71, 72
    of polynomial, 35–37, 122
distribution test, 282, 283
distributive laws, 11, 12
divison algorithm, 20
divison ring, 12, 65–69
divisor, 17
dot product, 308
dual basis, 54, 71, 369, 374–376
dual code, 308–311, 334, 335

element
    algebraic, 31
    associate, 17
    binary, 15
    conjugate, 8
    defining, 30
    identity, 2
    inverse, 2
    multiple of, 3
    order of, 6, 7
    power of, 3, 69, 181
    prime, 17
    primitive, see primitive element
    unity, 2
    zero, 11
enciphering scheme, 339
endomorphism, 8
epimorphism, 8
equivalence class, 4
equivalence relation, 4
error-correcting code, 303
    see also code

error-evaluator polynomial, 329, 330
error-location number, 316, 320, 329
error-locator polynomial, 322, 329, 330
error value, 320, 329
error vector, 303
error word, 303
Euclidean algorithm, 22, 38, 330, 336, 355,
        356
Euler's function, 7, 37
exponential sum, 162–184, 236–240, 250
exponent of polynomial, see order
extension (field), 30–35
    algebraic, 31
    degree of, 32
    finite, 32
    simple, 30, 33, 34

factor group, 9
factorization
    of integers, 78
    of polynomials, 23, 24, 29, 39, 97, 98, 108,
        116–118, 120, 129–150
    symbolic, 108, 109
factor ring, 13
Fano plane, 253, 254, 263
feedback shift register, 186–188, 193, 314
Fermat's little theorem, 37
Fibonacci sequence, 246
field, 12
    cyclotomic, see cyclotomic field
    finite, see finite field
    prime, 30
    see also extension (field), splitting field
finite affine geometry, 262, 263
finite euclidean geometry, 262
finite field, 15, 45
    automorphism, 49, 50, 70
    characterization of, 43–47
    computation in, 367–369
    definition, 14
    existence and uniqueness, 45
    multiplicative group of, 46, 47, 69
finite-state system, 271, 272
k-flat(s), 259–262, 295
    cycle of, 261
    at infinity, 262
    parallel, 262
flip-flop, see delay element
formal power series, 202
    ring of, 204
Fourier coefficient, 171
Fourier expansion, 171
FSR cryptosystem, 357, 358
fundamental theorem on symmetric
        polynomials, 29

Galois field, 15, 45
    see also finite field

Gaussian sum, 168–180, 182, 183, 238,
    242–244, 250
general linear group, 191, 192
general response formula, 275, 276
generating function, 202, 207–209, 219
generator, 4
generator matrix, 303, 312, 333
    canonical, 302, 303, 313
generator polynomial, 313
Gilbert–Varshamov bound, 308, 325
Goppa code, 326–332, 336, 337, 360, 361
    irreducible, 326, 336, 360
Goppa-code cryptosystem, 360–362
Goppa polynomial, 326
group, 2
    abelian, 2
    commutative, 2
    cyclic, 3, 7, 163
    finite, 5
    general linear, 191, 192
    infinite, 5
    of integers modulo n, 5
    order of, 5, 7
group code, 302

Hadamard matrix, 269–271, 296
    normalized, 270, 296
Hamming bound, 307
Hamming code, 307, 311, 314, 315, 333
    binary, 307, 311, 314, 315, 333
Hamming distance, 303
Hamming weight, 303
Hankel determinant, 229–231, 249
Hill cryptosystem, 366
homomorphism, 8, 14
homomorphism theorem
    for groups, 10
    for rings, 14, 15
hyperplane, 259, 262, 266

ideal, 13
    maximal, 17
    prime, 17
    principal, 13
identity element, 2
impulse response sequence, 193–195, 199,
    215
incidence matrix, 263, 264
incidence relation, 252–254, 262
indeterminate, 19
index-calculus algorithm, 352–357
index function, 346, 367
    see also discrete logarithm
index of subgroup, 7
index table, 63, 368, 370–373
initial state vector, 188
initial value, 186
input alphabet, 271, 272

input space, 272
input symbol, 272
integral domain, 12
interpolation, 28, 41, 363
inverse element, 2
irreducible polynomial, 23–25, 28, 31, 47,
    48, 75, 76, 82–91, 97, 98, 115, 118–128,
    160, 183, 377–387
isomorphism, 8, 14

Jacobi's logarithm, 69, 368, 374–376

kernel of homomorphism
    group, 9
    ring, 14
key, 339, 340
key-exchange system, 348
knapsack-type cryptosystem, 358–360
Kronecker's method, 39

Lagrange interpolation formula, 28, 41, 363
Latin square(s), 267–269, 296
    mutually orthogonal, 267–269, 296
    normalized, 296
    orthogonal, 267–269, 296
law of quadratic reciprocity, 179, 183
Legendre symbol, 167, 179
line(s)
    at infinity, 255
    parallel, 255
linearized polynomial, 98–114, 126–128
    see also q-associate, q-polynomial(s)
linear modular system, 272–281, 296, 297
    characteristic matrix, 272
    characterizing matrices, 272
    order, 272
linear recurrence relation, 186, 314
    characteristic polynomial, 195–201,
        207–211, 214, 215, 226–228
    homogeneous, 186
    inhomogeneous, 186
    order, 186
linear recurring sequence(s), 186
    addition, 215, 218–221
    binary complementation, 221
    characteristic polynomial, 195–201,
        207–211, 214, 215, 226–228
    characterization, 228–231
    decimation, see decimation
    distribution properties, 235–245, 250
    families of, 215–228
    homogeneous, 186
    inhomogeneous, 186
    least period, 189–195, 199, 200, 212, 213,
        220–223, 227, 247, 248
    minimal polynomial, 211–215, 218–223,
        230–235, 247–249
    multiplication, 224–228

order, 186
  reciprocal characteristic polynomial, 207
  scalar multiplication, 215
LMS, *see* linear modular system

MacWilliams identity, 310, 311
magic square, 296
matrix
  associated with sequence, 191–195, 199
  characteristic polynomial of, 93, 160, 278,
    279
  elementary block of, 279, 280
  Hadamard, *see* Hadamard matrix
  minimal polynomial of, 278–280
  rational canonical form of, 279
matrix of polynomials, 143–147
  diagonalization, 145–147
  equivalence, 144
  nonsingular, 144
  normalized, 146
  unimodular, 144
maximal ideal, 17
maximal period sequence, 201, 240, 241,
    246, 282–288, 297, 298, 343
Mersenne prime, 348, 351, 352
message symbol, 300, 301
minimal polynomial
  of element, 31, 86, 87, 91–97, 369,
    374–376
  for linear operator, 56
  of matrix, 278–280
  of sequence, *see* linear recurring
    sequence(s)
minimum distance, 304, 308, 318, 319, 327,
    328, 333–337
q-modulus, 109, 110, 114
Moebius function, 83, 124
Moebius inversion formula, 83, 84
multiplexed sequence, 343–346, 363, 364
multiplexer, 343

nearest neighbor decoding, 303
Newton's formula, 29, 30
next-state function, 272
no-key algorithm, 349
non-Desarguesian plane, 256, 257
norm, 53, 54, 70, 71
  transitivity of, 54
normal basis, 55–59, 71, 115, 127, 365, 369,
    374–376
  self-dual, 71, 127
normal basis theorem, 56, 57, 59, 115
normalization method, 288
normalizer, 10
NP-complete problem, 342, 362

one-time pad, 342
one-way function, 341

operation, *see* Binary operation
order
  of character, 170, 182
  of element, 6, 7
  of group, 5, 7
  of linear modular system, 272
  of linear recurrence relation, 186
  of linear recurring sequence, 186
  multiplicative mod n, 76, 87
  of polynomial, 75–82, 122, 123, 199–201,
    212, 377–387
  of projective plane, 253–257
  of state, 277, 278, 281
orthogonal code, *see* dual code
orthogonality relations, 165, 167, 168
orthogonal vectors, 308
output alphabet, 272
output function, 272
output space, 272
output symbol, 272

Pappus theorem, 255–257
parity-check code, 302, 334
parity-check equation, 301
parity-check matrix, 302
parity-check polynomial, 313
partition, 4
pencil, 257, 258, 262
period of polynomial, *see* order
period of sequence, 189–191
  least, 189 (*see also* linear recurring
    sequence(s))
permutation matrix, 360, 361
plaintext, 339
plaintext message, 339
Plotkin bound, 308
polynomial(s), 19, 28
  affine, *see* affine polynomial
  affine multiple of, 103
  canonical factorization of, 24
  characteristic, *see* characteristic
    polynomial
  constant, 20
  cyclotomic, *see* cyclotomic polynomial
  defining, 31
  degree of, 20, 29
  derivative of, 27, 40, 41, 70
  discriminant of, *see* discriminant
  division of, 20, 21
  elementary symmetric, 29
  exponent of, *see* order
  factorization of, *see* factorization
  formal degree of, 36
  greatest common divisor of, 22, 38, 39,
    109
  homogeneous, 29
  irreducible, *see* irreducible polynomial
  least common multiple of, 22, 23, 39

linearized, *see* linearized polynomial
matrix of, *see* matrix of polynomials
minimal, *see* minimal polynomial
monic, 20
order of, *see* order
pairwise relatively prime, 22
period of, *see* order
primitive, *see* primitive polynomial
product of, 19
reciprocal, 79, 123
reciprocal characteristic, 207
reducible, 23
f-reducing, 131–137
relatively prime, 22
resultant of, *see* resultant
root of, *see* root(s)
self-reciprocal, 123, 128
splitting of, 34, 35
sum of, 19
symmetric, 29
zero, 19
q-polynomial(s), 99, 101–103, 106–114, 126
affine, 103, 105, 126, 128
greatest common symbolic divisor of, 109
minimal, 111–113, 126
symbolically irreducible, 108, 110
symbolic division of, 106, 107
symbolic multiplication of, 105, 106
see also q-associate, linearized polynomial
polynomial basis, 55
polynomial ring, 19, 28, 204
preperiod of sequence, 189, 245, 247
prime element, 17
prime field, 30
prime ideal, 17
primitive element, 47, 49, 59, 63, 80, 96, 97, 182, 368
primitive polynomial, 80–82, 87, 96–98, 121, 123, 377, 388–391
q-primitive root, 110–114
principal ideal, 13
principal ideal domain, 17
principle of substitution, 27
probabilistic root-finding algorithm, 151
projective correspondence, 258
projective geometry, *see* projective space
projective plane, 252–259, 262–265, 295
Desarguesian, 256–259
finite, 252–259, 262–265, 295
non-Desarguesian, 256, 257
order of, 253–257
projective space, 259–263, 266, 295
finite, 260–263, 266, 295
pseudorandom sequence of bits, 282
public-key cryptosystem, 340, 341

quadratic reciprocity, *see* law of quadratic reciprocity

quotient group, 9

random sequence
of bits, 281, 282, 342
of real numbers, 288
rational canonical form of matrix, 279
reducible polynomial, 23
f-reducing polynomial, 131–137
reduction mod f, 25
Reed–Solomon code, 318, 335
reflexivity of relation, 4
repeated squaring technique, 148, 149, 151, 347
repetition code, 302, 333, 334
residue class, 13
residue class ring, 13, 25
resultant, 36, 37, 42, 127, 140
ring, 11–17
of algebraic integers, 179
characteristic of, 16
commutative, 11
of formal power series, 204
with identity, 11
of polynomials, *see* polynomial ring
root(s), 27, 28, 34–36, 150–159, 161
of affine polynomial, 103–105, 107
of irreducible polynomial, 48
of linearized polynomial, 99, 101–103
multiple, 27, 35, 40
multiplicity of, 27, 41
q-primitive, 110–114
simple, 27
root adjunction, 33–35
root-finding algorithm, 101–105, 150–159
probabilistic, 151
root of unity, 59–62, 72
primitive, 60–62, 72
RSA cryptosystem, 348
Run, 297

secant, 258
sequence
decimated, *see* decimated sequence
impulse response, *see* impulse response sequence
least period of, 189
maximal period, *see* maximal period sequence
multiplexed, 343–346, 363, 364
periodic, 189, 247
period of, 189–191
preperiod of, 189, 245, 247
pseudorandom, 282
random, *see* random sequence
shifted, 287, 288, 297, 298
ultimately periodic, 189
of uniform pseudorandom numbers, 288–294

of uniform random numbers, 288
zero, 215
see also linear recurring sequence(s)
serial test, 282, 294
Shannon's theorem, 299
shifted sequence, 287, 288, 297, 298
Silver–Pohlig–Hellman algorithm, 350–352
single-key cryptosystem, 339
Singleton bound, 333
skew field, *see* division ring
smooth integer, 350
m-space, *see* projective space
splitting field, 35, 48, 134
existence and uniqueness, 35
square and multiply technique, 347
state, 272, 273
order of, 277, 278, 281
state graph, 277, 278, 296
path in, 277
state set, 272
state space, 272
state vector, 188, 191, 193–195, 214, 215
initial, 188
modified, 192
Steiner triple system, 263
Stickelberger's theorem, 177–179
stream cipher, 342
subfield, 30
criterion for, 45, 46
maximal, 67, 68
prime, 30
proper, 30
subgroup, 6
generated by element, 6
generated by subset, 6
index of, 7
nontrivial, 6
normal, 9
trivial, 6
subring, 13
substitution cipher, 338
symmetric polynomial, 29

elementary, 29
symmetry of relation, 4
syndrome, 305, 306, 315
syndrome polynomial, 329

tactical configuration, 262, 263
symmetric, 262
tangent, 258
Tausworthe method, 288
term of polynomial, 29
test for randomness, 282, 288, 289
theorem of Pappus, 255–257
threshold scheme, 362, 363
trace, 50–53, 70, 71
absolute, 50
transitivity of, 52, 53
transitivity of relation, 4
trapdoor one-way function, 341
trinomial, 118–122, 127, 128
irreducible, 118, 119, 121, 122, 127, 128
primitive, 121

uniformity test, 289, 290
uniform pseudorandom numbers, 288–294
uniform random numbers, 288
unique factorization, 23, 24, 29
unit, 17
unity element, 2

Waring's formula, 30
Wedderburn's theorem, 65–69, 256
weight, 303
weight enumerator, 309–311, 334
Wilson's theorem, 37

Zassenhaus algorithm, 142, 143
zero divisor, 12
zero element, 11
zero of polynomial, 27, 42
see also root(s)
zero polynomial, 19
zero sequence, 215